PURE AND APPLIED MATHEMATICS

Plemelj—*Problems in the Sense of Riemann and Klein*
Ribenboim—*Algebraic Numbers*
Ribénboim—*Rings and Modules*
Richtmyer and Morton—*Difference Methods for Initial-Value Problems*, 2nd Edition
Rudin—*Fourier Analysis on Groups*
Sansone—*Orthogonal Functions*
Siegel—*Topics in Complex Function Theory*
      Volume 1—Elliptic Functions and Uniformization Theory
      Volume 2—Automorphic Functions and Abelian Integrals
      Volume 3—Abelian Functions and Modular Functions of Several Variables
Stoker—*Differential Geometry*
Stoker—*Nonlinear Vibrations in Mechanical and Electrical Systems*
Stoker—*Water Waves*
Tricomi—*Integral Equations*
Wasow—*Asymptotic Expansions for Ordinary Differential Equations*
Yosida—*Lectures on Differential and Integral Equations*
Zemanian—*Generalized Integral Transformations*

PERTURBATION METHODS

PERTURBATION METHODS

ALI HASAN NAYFEH

Professor of Engineering Science and Mechanics
Virginia Polytechnic Institute and State University
Blacksburg, Virginia 24061

A WILEY-INTERSCIENCE PUBLICATION

JOHN WILEY & SONS, New York · London · Sydney · Toronto

Library of Congress Cataloging in Publication Data:

Nayfeh, Ali Hasan, 1933–
  Perturbation methods.

  (Series in pure and applied mathematics)
  "A Wiley-Interscience publication."
  1. Perturbation (Mathematics).   2. Asymptotic expansions.
3. Differential equations—Numerical solutions. I. Title.

QA221.N38   629.1'01'515   72-8068
ISBN 0-471-63059-4

Printed in the United States of America

10 9 8 7 6

PREFACE

Many of the problems faced today by physicists, engineers, and applied mathematicians involve difficulties, such as nonlinear governing equations, variable coefficients, and nonlinear boundary conditions at complex known or unknown boundaries, which preclude their solutions exactly. To solve these problems we are forced to resort to a form of approximation, a numerical solution, or a combination of both. Foremost among the approximation techniques is the systematic method of perturbations (asymptotic expansions) in terms of a small or a large parameter or coordinate. This book is concerned only with these perturbation techniques.

According to these perturbation techniques, the solution of the full problem is represented by the first few terms of a perturbation expansion, usually the first two terms. Although these perturbation expansions may be divergent, they can be more useful for a qualitative as well as a quantitative representation of the solution than expansions that are uniformly and absolutely convergent.

It is the rule rather than the exception that the straightforward (pedestrian) expansions in powers of a parameter have limited regions of validity and break down in certain regions called regions of nonuniformity. To render these expansions uniformly valid, investigators working in different branches of physics, engineering, and applied mathematics have developed a number of techniques. Some of these techniques are radically different, while others are different interpretations of the same basic idea.

The purpose of this book is to present in a unified way an account of some of these techniques, pointing out their similarities, differences, and advantages, as well as their limitations. The different techniques are described using examples which start with model simple ordinary equations that can be solved exactly and progress toward complex partial differential equations. The examples are drawn from different branches of physics and engineering. For each example a short description of the physical problem is first presented.

The different techniques are described as formal procedures without any attempt at justifying them rigorously. In fact, there are no rigorous mathematical justifications available yet for the expansions obtained for some of the complex examples treated in this book.

At the end of each chapter, a number of exercises have been included, which progress in complexity and provide further references.

The reader need not understand the physical bases of the examples used to describe the techniques, but it is assumed that he has a knowledge of basic calculus as well as the elementary properties of ordinary and partial differential equations.

Chapter 1 presents the notations, definitions, and manipulations of asymptotic expansions. The sources of nonuniformity in perturbation expansions are classified and discussed in Chapter 2. Chapter 3 deals with the method of strained coordinates where uniformity is achieved by expanding the dependent as well as the independent variables in terms of new independent parameters. Chapter 4 describes the methods of matched and composite asymptotic expansions; the first method expresses the solution in terms of several expansions valid in different regions but related by matching procedures, while the second method expresses the solution in terms of a single expansion valid everywhere. In Chapter 5 the idea of fast and slow variables is used in conjunction with the variation of parameters method to study the slow variations of the amplitudes and phases of weakly nonlinear waves and oscillations. The methods of Chapter 3, 4, and 5 are generalized in Chapter 6 into one of three versions of the method of multiple scales. Chapter 7 treats available methods for obtaining asymptotic solutions of linear ordinary and partial differential equations.

My first technical debt is to Dr. W. S. Saric and to my brothers Dr. Adnan Nayfeh and Mr. Munir Nayfeh for their comments and encouragement throughout the writing of this book. I am indebted to several colleagues for helpful comments and criticism, including in particular Drs. D. T. Mook, D. P. Telionis, A. A. Kamel, and B. H. Stephan and Messers O. R. Asfar and M. S. Tsai. This book would not have been written without the patience and encouragement of my wife, and the insistence of my parents Hasan and Khadrah, in spite of their illiteracy, that I acquire a higher education. Therefore I dedicate this book to my parents and wife.

ALI HASAN NAYFEH

Blacksburg, Virginia
May 1972

CONTENTS

1. Introduction   **1**

   1.1.   Parameter Perturbations, 1
       1.1.1.   An Algebraic Equation, 2
       1.1.2.   The van der Pol Oscillator, 3
   1.2.   Coordinate Perturbations, 4
       1.2.1.   The Bessel Equation of Zeroth Order, 5
       1.2.2.   A Simple Example, 6
   1.3.   Order Symbols and Gauge Functions, 7
   1.4.   Asymptotic Expansions and Sequences, 9
       1.4.1.   Asymptotic Series, 9
       1.4.2.   Asymptotic Expansions, 12
       1.4.3.   Uniqueness of Asymptotic Expansions, 14
   1.5.   Convergent versus Asymptotic Series, 15
   1.6.   Nonuniform Expansions, 16
   1.7.   Elementary Operations on Asymptotic Expansions, 18
   Exercises, 19

2. Straightforward Expansions and Sources of Nonuniformity   **23**

   2.1.   Infinite Domains, 24
       2.1.1.   The Duffing Equation, 24
       2.1.2.   A Model for Weak Nonlinear Instability, 25
       2.1.3.   Supersonic Flow Past a Thin Airfoil, 26
       2.1.4.   Small Reynolds Number Flow Past a Sphere, 28
   2.2.   A Small Parameter Multiplying the Highest Derivative, 31
       2.2.1.   A Second-Order Example, 31
       2.2.2.   High Reynolds Number Flow Past a Body, 33
       2.2.3.   Relaxation Oscillations, 34
       2.2.4.   Unsymmetrical Bending of Prestressed Annular Plates, 35

2.3. Type Change of a Partial Differential Equation, 37
    2.3.1. A Simple Example, 38
    2.3.2. Long Waves on Liquids Flowing down Incline Planes, 38
2.4. The Presence of Singularities, 42
    2.4.1. Shift in Singularity, 42
    2.4.2. The Earth-Moon-Spaceship Problem, 43
    2.4.3. Thermoelastic Surface Waves, 45
    2.4.4. Turning Point Problems, 48
2.5. The Role of Coordinate Systems, 49
Exercises, 52

3. The Method of Strained Coordinates   **56**

3.1. The Method of Strained Parameters, 58
    3.1.1. The Lindstedt-Poincaré Method, 58
    3.1.2. Transition Curves for the Mathieu Equation, 60
    3.1.3. Characteristic Exponents for the Mathieu Equation (Whittaker's Method), 62
    3.1.4. The Stability of the Triangular Points in the Elliptic Restricted Problem of Three Bodies, 64
    3.1.5. Characteristic Exponents for the Triangular Points in the Elliptic Restricted Problem of Three Bodies, 66
    3.1.6. A Simple Linear Eigenvalue Problem, 68
    3.1.7. A Quasi-Linear Eigenvalue Problem, 71
    3.1.8. The Quasi-Linear Klein-Gordon Equation, 76
3.2. Lighthill's Technique, 77
    3.2.1. A First-Order Differential Equation, 79
    3.2.2. The One-Dimensional Earth-Moon-Spaceship Problem, 82
    3.2.3. A Solid Cylinder Expanding Uniformly in Still Air, 83
    3.2.4. Supersonic Flow Past a Thin Airfoil, 86
    3.2.5. Expansions by Using Exact Characteristics—Nonlinear Elastic Waves, 89
3.3. Temple's Technique, 94
3.4. Renormalization Technique, 95
    3.4.1. The Duffing Equation, 95
    3.4.2. A Model for Weak Nonlinear Instability, 96
    3.4.3. Supersonic Flow Past a Thin Airfoil, 97
    3.4.4. Shift in Singularity, 98
3.5. Limitations of the Method of Strained Coordinates, 98
    3.5.1. A Model for Weak Nonlinear Instability, 99

       3.5.2. A Small Parameter Multiplying the Highest Derivative, 100

       3.5.3. The Earth-Moon-Spaceship Problem, 102

   Exercises, 103

4. The Methods of Matched and Composite Asymptotic Expansions   **110**

   4.1. The Method of Matched Asymptotic Expansions, 111

       4.1.1. Introduction—Prandtl's Technique, 111

       4.1.2. Higher Approximations and Refined Matching Procedures, 114

       4.1.3. A Second-Order Equation with Variable Coefficients, 122

       4.1.4. Reynolds' Equation for a Slider Bearing, 125

       4.1.5. Unsymmetrical Bending of Prestressed Annular Plates, 128

       4.1.6. Thermoelastic Surface Waves, 133

       4.1.7. The Earth-Moon-Spaceship Problem, 137

       4.1.8. Small Reynolds Number Flow Past a Sphere, 139

   4.2. The Method of Composite Expansions, 144

       4.2.1. A Second-Order Equation with Constant Coefficients, 145

       4.2.2. A Second-Order Equation with Variable Coefficients, 148

       4.2.3. An Initial Value Problem for the Heat Equation, 150

       4.2.4. Limitations of the Method of Composite Expansions, 153

   Exercises, 154

5. Variation of Parameters and Methods of Averaging   **159**

   5.1. Variation of Parameters, 159

       5.1.1. Time-Dependent Solutions of the Schrödinger Equation, 160

       5.1.2. A Nonlinear Stability Example, 162

   5.2. The Method of Averaging, 164

       5.2.1. Van der Pol's Technique, 164

       5.2.2. The Krylov–Bogoliubov Technique, 165

       5.2.3. The Generalized Method of Averaging, 168

   5.3. Struble's Technique, 171

   5.4. The Krylov–Bogoliubov–Mitropolski Technique, 174

       5.4.1. The Duffing Equation, 175

       5.4.2. The van der Pol Oscillator, 176

       5.4.3. The Klein–Gordon Equation, 178

5.5.   The Method of Averaging by Using Canonical Variables, 179
    5.5.1.   The Duffing Equation, 182
    5.5.2.   The Mathieu Equation, 183
    5.5.3.   A Swinging Spring, 185
5.6.   Von Zeipel's Procedure, 189
    5.6.1.   The Duffing Equation, 192
    5.6.2.   The Mathieu Equation, 194
5.7.   Averaging by Using the Lie Series and Transforms, 200
    5.7.1.   The Lie Series and Transforms, 201
    5.7.2.   Generalized Algorithms, 202
    5.7.3.   Simplified General Algorithms, 206
    5.7.4.   A Procedure Outline, 208
    5.7.5.   Algorithms for Canonical Systems, 212
5.8.   Averaging by Using Lagrangians, 216
    5.8.1.   A Model for Dispersive Waves, 217
    5.8.2.   A Model for Wave–Wave Interaction, 219
    5.8.3.   The Nonlinear Klein–Gordon Equation, 221
Exercises, 223

6.   The Method of Multiple Scales   **228**

6.1.   Description of the Method, 228
    6.1.1.   Many-Variable Version (The Derivative–Expansion Procedure), 236
    6.1.2.   The Two-Variable Expansion Procedure, 240
    6.1.3.   Generalized Method—Nonlinear Scales, 241
6.2.   Applications of the Derivative-Expansion Method, 243
    6.2.1.   The Duffing Equation, 243
    6.2.2.   The van der Pol Oscillator, 245
    6.2.3.   Forced Oscillations of the van der Pol Equation, 248
    6.2.4.   Parametric Resonances—The Mathieu Equation, 253
    6.2.5.   The van der Pol Oscillator with Delayed Amplitude Limiting, 257
    6.2.6.   The Stability of the Triangular Points in the Elliptic Restricted Problem of Three Bodies, 259
    6.2.7.   A Swinging Spring, 262
    6.2.8.   A Model for Weak Nonlinear Instability, 264
    6.2.9.   A Model for Wave–Wave Interaction, 266
    6.2.10. Limitations of the Derivative-Expansion Method, 269
6.3.   The Two-Variable Expansion Procedure, 270
    6.3.1.   The Duffing Equation, 270
    6.3.2.   The van der Pol Oscillator, 272

6.3.3. The Stability of the Triangular Points in the Elliptic Restricted Problem of Three Bodies, 275

6.3.4. Limitations of This Technique, 275

6.4. Generalized Method, 276

6.4.1. A Second-Order Equation with Variable Coefficients, 276

6.4.2. A General Second-Order Equation with Variable Coefficients, 280

6.4.3. A Linear Oscillator with a Slowly Varying Restoring Force, 282

6.4.4. An Example with a Turning Point, 284

6.4.5. The Duffing Equation with Slowly Varying Coefficients, 286

6.4.6. Reentry Dynamics, 291

6.4.7. The Earth-Moon-Spaceship Problem, 295

6.4.8. A Model for Dispersive Waves, 298

6.4.9. The Nonlinear Klein–Gordon Equation, 301

6.4.10. Advantages and Limitations of the Generalized Method, 302

Exercises, 303

7. Asymptotic Solutions of Linear Equations   308

7.1. Second-Order Differential Equations, 309

7.1.1. Expansions Near an Irregular Singularity, 309

7.1.2. An Expansion of the Zeroth-Order Bessel Function for Large Argument, 312

7.1.3. Liouville's Problem, 314

7.1.4. Higher Approximations for Equations Containing a Large Parameter, 315

7.1.5. A Small Parameter Multiplying the Highest Derivative, 317

7.1.6. Homogeneous Problems with Slowly Varying Coefficients, 318

7.1.7. Reentry Missile Dynamics, 320

7.1.8. Inhomogeneous Problems with Slowly Varying Coefficients, 321

7.1.9. Successive Liouville–Green (WKB) Approximations, 324

7.2. Systems of First-Order Ordinary Equations, 325

7.2.1. *Expansions Near an Irregular Singular Point,* 326

7.2.2. *Asymptotic Partitioning of Systems of Equations,* 327

7.2.3. *Subnormal Solutions,* 331

7.2.4. *Systems Containing a Parameter,* 332

7.2.5. *Homogeneous Systems with Slowly Varying Co-efficients,* 333

7.3. Turning Point Problems, 335

7.3.1. *The Method of Matched Asymptotic Expansions,* 336

7.3.2. *The Langer Transformation,* 339

7.3.3. *Problems with Two Turning Points,* 342

7.3.4. *Higher-Order Turning Point Problems,* 345

7.3.5. *Higher Approximations,* 346

7.3.6. *An Inhomogeneous Problem with a Simple Turning Point—First Approximation,* 352

7.3.7. *An Inhomogeneous Problem with a Simple Turning Point—Higher Approximations,* 354

7.3.8. *An Inhomogeneous Problem with a Second-Order Turning Point,* 356

7.3.9. *Turning Point Problems about Singularities,* 358

7.3.10. *Turning Point Problems of Higher Order,* 360

7.4. Wave Equations, 360

7.4.1. *The Born or Neumann Expansion and The Feynman Diagrams,* 361

7.4.2. *Renormalization Techniques,* 367

7.4.3. *Rytov's Method,* 373

7.4.4. *A Geometrical Optics Approximation,* 374

7.4.5. *A Uniform Expansion at a Caustic,* 377

7.4.6. *The Method of Smoothing,* 380

Exercises, 382

References and Author Index   387

Subject Index   417

PERTURBATION METHODS

CHAPTER 1

Introduction

Most of the physical problems facing engineers, physicists, and applied mathematicians today exhibit certain essential features which preclude exact analytical solutions. Some of these features are nonlinearities, variable coefficients, complex boundary shapes, and nonlinear boundary conditions at known or, in some cases, unknown boundaries. Even if the exact solution of a problem can be found explicitly, it may be useless for mathematical and physical interpretation or numerical evaluation. Examples of such problems are Bessel functions of large argument and large-order and doubly periodic functions. Thus, in order to obtain information about solutions of equations, we are forced to resort to approximations, numerical solutions, or combinations of both. Foremost among the approximation methods are perturbation (asymptotic) methods which are the subject of this book. According to these techniques, the solution is represented by the first few terms of an asymptotic expansion, usually not more than two terms. The expansions may be carried out in terms of a parameter (small or large) which appears naturally in the equations, or which may be artificially introduced for convenience. Such expansions are called *parameter perturbations*. Alternatively, the expansions may be carried out in terms of a coordinate (either small or large); these are called *coordinate perturbations*. Examples of parameter and coordinate expansions and their essential characteristics are presented in Sections 1.1 and 1.2. To formalize the concepts of limits and error estimates, definitions of order symbols and other notations are introduced in Section 1.3. Section 1.4 contains definitions of an asymptotic expansion, an asymptotic sequence, and a power series, while Section 1.5 presents a comparison of convergent and asymptotic series. Uniform and nonuniform asymptotic expansions are then defined in Section 1.6. A short summary of operations with asymptotic expansions is given in Section 1.7.

1.1. Parameter Perturbations

Many physical problems involving the function $u(x, \epsilon)$ can be represented mathematically by the differential equation $L(u, x, \epsilon) = 0$ and the boundary

1

condition $B(u, \epsilon) = 0$, where x is a scalar or vector independent variable and ϵ is a parameter. In general, this problem cannot be solved exactly. However, if there exists an $\epsilon = \epsilon_0$ (ϵ can be scaled so that $\epsilon_0 = 0$) for which the above problem can be solved exactly or more readily, one seeks to find the solution for small ϵ in, say, powers of ϵ; that is

$$u(x; \epsilon) = u_0(x) + \epsilon u_1(x) + \epsilon^2 u_2(x) + \cdots \qquad (1.1.1)$$

where u_n is independent of ϵ and $u_0(x)$ is the solution of the problem for $\epsilon = 0$. One then substitutes this expansion into $L(u, x, \epsilon) = 0$ and $B(u, \epsilon) = 0$, expands for small ϵ, and collects coefficients of each power of ϵ. Since these equations must hold for all values of ϵ, each coefficient of ϵ must vanish independently because sequences of ϵ are linearly independent. These usually are simpler equations governing u_n, which can be solved successively. This is demonstrated in the next two examples.

1.1.1. AN ALGEBRAIC EQUATION

Let us consider first the solution of the algebraic equation

$$u = 1 + \epsilon u^3 \qquad (1.1.2)$$

for small ϵ. If $\epsilon = 0$, $u = 1$. For ϵ small, but different from zero, we let

$$u = 1 + \epsilon u_1 + \epsilon^2 u_2 + \epsilon^3 u_3 + \cdots \qquad (1.1.3)$$

and (1.1.2) becomes

$$\epsilon u_1 + \epsilon^2 u_2 + \epsilon^3 u_3 + \cdots = \epsilon(1 + \epsilon u_1 + \epsilon^2 u_2 + \epsilon^3 u_3 + \cdots)^3 \qquad (1.1.4)$$

Expanding for small ϵ, we rewrite (1.1.4) as

$$\epsilon u_1 + \epsilon^2 u_2 + \epsilon^3 u_3 + \cdots = \epsilon[1 + 3\epsilon u_1 + 3\epsilon^2(u_2 + u_1{}^2) + \cdots] \qquad (1.1.5)$$

Collecting coefficients of like powers of ϵ, we have

$$\epsilon(u_1 - 1) + \epsilon^2(u_2 - 3u_1) + \epsilon^3(u_3 - 3u_2 - 3u_1{}^2) + \cdots = 0 \qquad (1.1.6)$$

Since this equation is an identity in ϵ, each coefficient of ϵ vanishes independently. Thus

$$u_1 - 1 = 0 \qquad (1.1.7)$$

$$u_2 - 3u_1 = 0 \qquad (1.1.8)$$

$$u_3 - 3u_2 - 3u_1{}^2 = 0 \qquad (1.1.9)$$

The solution of (1.1.7) is

$$u_1 = 1 \qquad (1.1.10)$$

Then the solution of (1.1.8) is

$$u_2 = 3u_1 = 3 \tag{1.1.11}$$

and the solution of (1.1.9) is

$$u_3 = 3u_2 + 3u_1{}^2 = 12 \tag{1.1.12}$$

Therefore (1.1.3) becomes

$$u = 1 + \epsilon + 3\epsilon^2 + 12\epsilon^3 + \cdots \tag{1.1.13}$$

where the ellipsis dots stand for all terms with powers of ϵ^n for which $n \geq 4$. Thus (1.1.13) is an approximation to the solution of (1.1.2), which is equal to 1 when $\epsilon \equiv 0$.

1.1.2.   THE VAN DER POL OSCILLATOR

As a second example, we consider van der Pol's (1922) equation

$$\frac{d^2u}{dt^2} + u = \epsilon(1 - u^2)\frac{du}{dt} \tag{1.1.14}$$

for small ϵ. If $\epsilon = 0$ this equation reduces to

$$\frac{d^2u}{dt^2} + u = 0 \tag{1.1.15}$$

with the general solution

$$u = a \cos (t + \varphi) \tag{1.1.16}$$

where a and φ are constants. To determine an improved approximation to the solution of (1.1.14), we seek a perturbation expansion of the form

$$u(t; \epsilon) = u_0(t) + \epsilon u_1(t) + \epsilon^2 u_2(t) + \cdots \tag{1.1.17}$$

where the ellipsis dots stand for terms proportional to powers of ϵ greater than 2. Substituting this expansion into (1.1.14), we have

$$\frac{d^2u_0}{dt^2} + u_0 + \epsilon\left(\frac{d^2u_1}{dt^2} + u_1\right) + \epsilon^2\left(\frac{d^2u_2}{dt^2} + u_2\right) + \cdots$$

$$= \epsilon[1 - (u_0 + \epsilon u_1 + \epsilon^2 u_2 + \cdots)^2]\left[\frac{du_0}{dt} + \epsilon\frac{du_1}{dt} + \epsilon^2\frac{du_2}{dt} + \cdots\right] \tag{1.1.18}$$

Expanding for small ϵ, we obtain

$$\frac{d^2u_0}{dt^2} + u_0 + \epsilon\left(\frac{d^2u_1}{dt^2} + u_1\right) + \epsilon^2\left(\frac{d^2u_2}{dt^2} + u_2\right) + \cdots$$

$$= \epsilon(1 - u_0{}^2)\frac{du_0}{dt} + \epsilon^2\left[(1 - u_0{}^2)\frac{du_1}{dt} - 2u_0u_1\frac{du_0}{dt}\right] + \cdots \tag{1.1.19}$$

Since u_n is independent of ϵ and (1.1.19) is valid for all small values of ϵ, the coefficients of like powers of ϵ must be the same on both sides of this equation. Equating the coefficients of like powers of ϵ on both sides of (1.1.19), we have

Coefficient of ϵ^0

$$\frac{d^2u_0}{dt^2} + u_0 = 0 \qquad (1.1.20)$$

Coefficient of ϵ

$$\frac{d^2u_1}{dt^2} + u_1 = (1 - u_0{}^2)\frac{du_0}{dt} \qquad (1.1.21)$$

Coefficient of ϵ^2

$$\frac{d^2u_2}{dt^2} + u_2 = (1 - u_0{}^2)\frac{du_1}{dt} - 2u_0u_1\frac{du_0}{dt} \qquad (1.1.22)$$

Note that (1.1.20) is the same as (1.1.15) and its general solution is given by (1.1.16); that is

$$u_0 = a\cos(t + \varphi) \qquad (1.1.23)$$

Substituting for u_0 into (1.1.21) gives

$$\frac{d^2u_1}{dt^2} + u_1 = -[1 - a^2\cos^2(t + \varphi)]a\sin(t + \varphi)$$

Using the trigonometric identity

$$\cos^2(t + \varphi)\sin(t + \varphi) = \frac{\sin(t + \varphi) + \sin 3(t + \varphi)}{4}$$

we can rewrite this equation as

$$\frac{d^2u_1}{dt^2} + u_1 = \frac{a^3 - 4a}{4}\sin(t + \varphi) + \tfrac{1}{4}a^3\sin 3(t + \varphi) \qquad (1.1.24)$$

Its particular solution is

$$u_1 = -\frac{a^3 - 4a}{8}t\cos(t + \varphi) - \tfrac{1}{32}a^3\sin 3(t + \varphi) \qquad (1.1.25)$$

With u_0 and u_1 known the right-hand side of (1.1.22) is known, and one can solve it for u_2 in a similar fashion. The usefulness of such an expansion is the subject of this book.

1.2. Coordinate Perturbations

If the physical problem is represented mathematically by a differential equation $L(u, x) = 0$ subject to the boundary conditions $B(u) = 0$, where x

is a scalar, and if $u(x)$ takes a known form u_0 as $x \to x_0$ (x_0 is scaled to 0 or ∞), one attempts to determine the deviation of u from u_0 for x near x_0 in terms of powers of x if $x_0 = 0$, or x^{-1} if $x_0 = \infty$. This technique is demonstrated by the next two examples.

1.2.1. THE BESSEL EQUATION OF ZEROTH ORDER
We consider the solutions of

$$x \frac{d^2 y}{dx^2} + \frac{dy}{dx} + xy = 0 \qquad (1.2.1)$$

This equation has a regular singular point at $x = 0$, which suggests that a power series solution for y can be obtained using the method of Frobenius (e.g., Ince, 1926, Section 16.1). Thus we let

$$y = \sum_{m=0}^{\infty} a_m x^{\mu+m} \qquad (1.2.2)$$

where the number μ and the coefficients a_m must be determined so that (1.2.2) is a solution of (1.2.1).

Substituting (1.2.2) into (1.2.1) gives

$$\sum_{m=0}^{\infty} (\mu + m)(\mu + m - 1)a_m x^{\mu+m-1}$$

$$+ \sum_{m=0}^{\infty} (\mu + m)a_m x^{\mu+m-1} + \sum_{m=0}^{\infty} a_m x^{\mu+m+1} = 0$$

or

$$\sum_{m=0}^{\infty} (\mu + m)^2 a_m x^{\mu+m-1} + \sum_{m=0}^{\infty} a_m x^{\mu+m+1} = 0 \qquad (1.2.3)$$

which can be written as

$$\mu^2 a_0 x^{\mu-1} + (\mu + 1)^2 a_1 x^{\mu} + \sum_{m=2}^{\infty} (\mu + m)^2 a_m x^{\mu+m-1} + \sum_{m=0}^{\infty} a_m x^{\mu+m+1} = 0$$

Replacing m by $m + 2$ in the first summation of this equation, we can rewrite it as

$$\mu^2 a_0 x^{\mu-1} + (\mu + 1)^2 a_1 x^{\mu} + \sum_{m=0}^{\infty} [(\mu + m + 2)^2 a_{m+2} + a_m] x^{\mu+m+1} = 0 \quad (1.2.4)$$

Since (1.2.4) is an identity in x, the coefficient of each power of x must vanish independently; that is

$$\mu^2 a_0 = 0 \qquad (1.2.5)$$

$$(\mu + 1)^2 a_1 = 0 \qquad (1.2.6)$$

$$(\mu + m + 2)^2 a_{m+2} + a_m = 0, \qquad m = 0, 1, 2, \ldots \qquad (1.2.7)$$

The first equation demands that $\mu = 0$ if $a_0 \neq 0$; then (1.2.6) gives $a_1 = 0$ and (1.2.7) gives

$$a_{m+2} = -\frac{a_m}{(\mu + m + 2)^2}, \qquad m = 0, 1, 2, \ldots \qquad (1.2.8)$$

Therefore

$$a_{2m+1} = 0, \qquad m = 1, 2, 3, \ldots,$$

$$a_2 = -\frac{a_0}{2^2}, \qquad a_4 = \frac{a_0}{2^2 \cdot 4^2}, \qquad a_6 = -\frac{a_0}{2^2 \cdot 4^2 \cdot 6^2}$$

$$a_{2n} = (-1)^n \frac{a_0}{2^2 \cdot 4^2 \cdot 6^2 \cdots (2n)^2} \qquad (1.2.9)$$

The solution thus obtained if $a_0 = 1$ is a Bessel function of zeroth order, and it is often denoted by J_0. Thus

$$J_0 = 1 - \frac{x^2}{2^2} + \frac{x^4}{2^2 \cdot 4^2} - \frac{x^6}{2^2 \cdot 4^2 \cdot 6^2} + \cdots$$

$$+ (-1)^n \frac{x^{2n}}{2^2 \cdot 4^2 \cdot 6^2 \cdots (2n)^2} + \cdots \qquad (1.2.10)$$

Since the ratio of the nth term to the $(n - 1)$th term is $-x^2/(2n)^2$ and tends to zero as $n \to \infty$ irrespective of the value and sign of x, the series (1.2.10) for J_0 converges uniformly and absolutely for all values of x.

An expansion valid for large values of x is obtained in Section 7.1.2 and compared with the above expansion in Section 1.5.

1.2.2. A SIMPLE EXAMPLE

As a second example, we consider the solution of

$$\frac{dy}{dx} + y = \frac{1}{x} \qquad (1.2.11)$$

for large x. For large x we seek a solution in the form

$$y = \sum_{m=1}^{\infty} a_m x^{-m} \qquad (1.2.12)$$

Substituting this expansion into (1.2.11) yields

$$\sum_{m=1}^{\infty} -m a_m x^{-m-1} + \sum_{m=2}^{\infty} a_m x^{-m} + (a_1 - 1)x^{-1} = 0 \qquad (1.2.13)$$

Replacing m by $m + 1$ in the second summation series, we can rewrite this equation as

$$(a_1 - 1)x^{-1} + \sum_{m=1}^{\infty} (a_{m+1} - ma_m)x^{-m-1} = 0 \tag{1.2.14}$$

Since this equation is an identity in x, the coefficient of each x^{-m} must vanish independently; that is

$$a_1 = 1, \qquad a_{m+1} = ma_m \quad \text{for} \quad m \geq 1 \tag{1.2.15}$$

Hence

$$a_2 = 1, \qquad a_3 = 2!, \qquad a_4 = 3!, \qquad a_n = (n-1)!$$

and (1.2.12) becomes

$$y = \frac{1}{x} + \frac{1!}{x^2} + \frac{2!}{x^3} + \frac{3!}{x^4} + \cdots + \frac{(n-1)!}{x^n} + \cdots \tag{1.2.16}$$

Since the ratio of the nth to the $(n-1)$th term is $(n-1)x^{-1}$ and it tends to infinity as $n \to \infty$ irrespective of the value of x, the series (1.2.16) diverges for all values of x. In spite of its divergence, this series is shown in Section 1.4 to be useful for numerical calculations, and it is called an *asymptotic series*.

1.3.   Order Symbols and Gauge Functions

Suppose we are interested in a function of the single real parameter ϵ, denoted by $f(\epsilon)$. In carrying out our approximations, we are interested in the limit of $f(\epsilon)$ as ϵ tends to zero, denoted by $\epsilon \to 0$. This limit might depend on whether ϵ tends to zero from below, denoted by $\epsilon \uparrow 0$, or from above, denoted by $\epsilon \downarrow 0$. If the limit of $f(\epsilon)$ exists (i.e., it does not have an essential singularity at $\epsilon = 0$ such as $\sin \epsilon^{-1}$), then there are three possibilities

$$\left.\begin{array}{l} f(\epsilon) \to 0 \\ f(\epsilon) \to A \\ f(\epsilon) \to \infty \end{array}\right\} \quad \text{as} \quad \epsilon \to 0, 0 < A < \infty \tag{1.3.1}$$

In the first and last cases, the rate at which $f(\epsilon) \to 0$ and $f(\epsilon) \to \infty$ is expressed by comparing $f(\epsilon)$ with known functions called *gauge functions*. The simplest and most useful of these are

$$\ldots, \epsilon^{-n}, \ldots, \epsilon^{-2}, \epsilon^{-1}, 1, \epsilon, \epsilon^2, \ldots, \epsilon^n, \ldots$$

In some cases these must be supplemented by

$$\log \epsilon^{-1}, \log(\log \epsilon^{-1}), e^{\epsilon^{-1}}, e^{-\epsilon^{-1}}, \text{ and so on}$$

Other gauge functions are

$$\sin \epsilon, \cos \epsilon, \tan \epsilon, \sinh \epsilon, \cosh \epsilon, \tanh \epsilon, \text{ and so on}$$

The behavior of a function $f(\epsilon)$ is compared with a gauge function $g(\epsilon)$ as $\epsilon \to 0$, employing either of the Landau symbols: O or o.

The Symbol O

We write

$$f(\epsilon) = O[g(\epsilon)] \quad \text{as} \quad \epsilon \to 0 \qquad (1.3.2)$$

if there exists a positive number A independent of ϵ and an $\epsilon_0 > 0$ such that

$$|f(\epsilon)| \leq A \, |g(\epsilon)| \quad \text{for all} \quad |\epsilon| \leq \epsilon_0 \qquad (1.3.3)$$

This condition can be replaced by

$$\lim_{\epsilon \to 0} \left| \frac{f(\epsilon)}{g(\epsilon)} \right| < \infty \qquad (1.3.4)$$

For example, as $\epsilon \to 0$

$$\sin \epsilon = O(\epsilon), \qquad \sin \epsilon^2 = O(\epsilon^2)$$
$$\sin 7\epsilon = O(\epsilon), \qquad \sin 2\epsilon - 2\epsilon = O(\epsilon^3)$$
$$\cos \epsilon = O(1), \qquad 1 - \cos \epsilon = O(\epsilon^2)$$
$$J_0(\epsilon) = O(1), \qquad J_0(\epsilon) - 1 = O(\epsilon^2)$$
$$\sinh \epsilon = O(\epsilon), \qquad \cosh \epsilon = O(1)$$
$$\tanh \epsilon = O(\epsilon), \qquad \tan \epsilon = O(\epsilon)$$
$$\coth \epsilon = O(\epsilon^{-1}), \quad \cot \epsilon = O(\epsilon^{-1})$$

If f is a function of another variable x in addition to ϵ, and $g(x, \epsilon)$ is a gauge function, we also write

$$f(x, \epsilon) = O[g(x, \epsilon)] \quad \text{as} \quad \epsilon \to 0 \qquad (1.3.5)$$

if there exists a positive number A independent of ϵ and an $\epsilon_0 > 0$ such that

$$|f(x, \epsilon)| \leq A \, |g(x, \epsilon)| \quad \text{for all} \quad |\epsilon| \leq \epsilon_0 \qquad (1.3.6)$$

If A and ϵ_0 are independent of x, the relationship (1.3.5) is said to hold *uniformly*. For example

$$\sin (x + \epsilon) = O(1) = O[\sin (x)] \quad \text{uniformly as} \quad \epsilon \to 0$$

while

$$e^{-\epsilon t} - 1 = O(\epsilon) \quad \text{nonuniformly as} \quad \epsilon \to 0$$
$$\sqrt{x + \epsilon} - \sqrt{x} = O(\epsilon) \quad \text{nonuniformly as} \quad \epsilon \to 0$$

The Symbol o

We write

$$f(\epsilon) = o[g(\epsilon)] \quad \text{as} \quad \epsilon \to 0 \tag{1.3.7}$$

if for every positive number δ, independent of ϵ, there exists an ϵ_0 such that

$$|f(\epsilon)| \leq \delta\,|g(\epsilon)| \quad \text{for} \quad |\epsilon| \leq \epsilon_0 \tag{1.3.8}$$

This condition can be replaced by

$$\lim_{\epsilon \to 0} \left| \frac{f(\epsilon)}{g(\epsilon)} \right| = 0 \tag{1.3.9}$$

Thus as $\epsilon \to 0$

$$\sin \epsilon = o(1), \qquad \sin \epsilon^2 = o(\epsilon)$$

$$\cos \epsilon = o(\epsilon^{-1/2}), \qquad J_0(\epsilon) = o(\epsilon^{-1})$$

$$\coth \epsilon = o(\epsilon^{-3/2}), \qquad \cot \epsilon = o[\epsilon^{-(n+1)/n}] \quad \text{for positive } n$$

$$1 - \cos 3\epsilon = o(\epsilon), \qquad \exp(-\epsilon^{-1}) = o(\epsilon^n) \quad \text{for all } n$$

If $f = f(x, \epsilon)$ and $g = g(x, \epsilon)$, then (1.3.7) is said to hold uniformly if δ and ϵ_0 are independent of x. For example

$$\sin(x + \epsilon) = o(\epsilon^{-1/3}) \quad \text{uniformly as} \quad \epsilon \to 0$$

while

$$e^{-\epsilon t} - 1 = o(\epsilon^{1/2}) \quad \text{nonuniformly as} \quad \epsilon \to 0$$

$$\sqrt{x + \epsilon} - \sqrt{x} = o(\epsilon^{3/4}) \quad \text{nonuniformly as} \quad \epsilon \to 0$$

1.4. Asymptotic Expansions and Sequences

1.4.1. ASYMPTOTIC SERIES

We found in Section 1.2.2 that a particular solution of

$$\frac{dy}{dx} + y = \frac{1}{x} \tag{1.4.1}$$

is

$$y = \frac{1}{x} + \frac{1!}{x^2} + \frac{2!}{x^3} + \frac{3!}{x^4} + \cdots + \frac{(n-1)!}{x^n} + \cdots \tag{1.4.2}$$

which diverges for all values of x. To investigate whether this series is of any value for computing a particular solution of our equation, we determine the remainder if we truncate the series after n terms. To do this we note that a particular integral of our differential equation is given by

$$y = e^{-x} \int_{-\infty}^{x} x^{-1} e^{x} \, dx \tag{1.4.3}$$

which converges for negative x. Integrating (1.4.3) by parts, we find that

$$y = \frac{1}{x} + e^{-x} \int_{-\infty}^{x} x^{-2} e^{x} \, dx = \frac{1}{x} + \frac{1}{x^2} + 2e^{-x} \int_{-\infty}^{x} x^{-3} e^{x} \, dx$$

$$= \frac{1}{x} + \frac{1}{x^2} + \frac{2}{x^3} + 3! \, e^{-x} \int_{-\infty}^{x} x^{-4} e^{x} \, dx$$

$$= \frac{1}{x} + \frac{1!}{x^2} + \frac{2!}{x^3} + \frac{3!}{x^4} + \cdots + \frac{(n-1)!}{x^n}$$

$$+ n! \, e^{-x} \int_{-\infty}^{x} x^{-n-1} e^{x} \, dx \tag{1.4.4}$$

Therefore if we truncate the series after n terms, the remainder is

$$R_n = n! \, e^{-x} \int_{-\infty}^{x} x^{-n-1} e^{x} \, dx \tag{1.4.5}$$

which is a function of n and x. For the series to converge, $\lim_{n \to \infty} R_n$ must be zero. This is not true in our example; in fact, $R_n \to \infty$ as $n \to \infty$ so that the series diverges for all x in agreement with what we found in Section 1.2.2 using the ratio test. Therefore, if the series (1.4.2) is to be useful, n must be fixed. For negative x

$$|R_n| \le n! \, |x^{-n-1}| \, e^{-x} \int_{-\infty}^{x} e^{x} \, dx = \frac{n!}{|x^{n+1}|} \tag{1.4.6}$$

Thus the error committed in truncating the series after n terms is numerically less than the first neglected term, namely, the $(n+1)$th term. Moreover, as $|x| \to \infty$ with n fixed, $R_n \to 0$. Therefore, although the series (1.4.2) diverges, for a fixed n the first n terms in the series can represent y with an error which can be made arbitrarily small by taking $|x|$ sufficiently large. Such a series is called an *asymptotic series of the Poincaré type* (Poincaré, 1892) and is denoted by

$$y \sim \sum_{n=1}^{\infty} \frac{(n-1)!}{x^n} \quad \text{as} \quad |x| \to \infty \tag{1.4.7}$$

In general, given a series $\sum_{m=0}^{\infty} (a_m / x^m)$, where a_m is independent of x, we say that the series is an *asymptotic series* and write

$$y \sim \sum_{m=0}^{\infty} \frac{a_m}{x^m} \quad \text{as} \quad |x| \to \infty \tag{1.4.8}$$

if and only if

$$y = \sum_{m=0}^{n} \frac{a_m}{x^m} + o(|x|^{-n}) \quad \text{as} \quad |x| \to \infty \tag{1.4.9}$$

The condition (1.4.9) can be rewritten as

$$y = \sum_{m=0}^{n-1} \frac{a_m}{x^m} + O(|x|^{-n}) \quad \text{as} \quad |x| \to \infty \tag{1.4.10}$$

As another example of an asymptotic series, we consider, after Euler (1754), the evaluation of the integral

$$f(\omega) = \omega \int_0^{\infty} \frac{e^{-x}}{\omega + x} \, dx \tag{1.4.11}$$

for large positive ω. Since

$$\frac{\omega}{\omega + x} = \sum_{m=0}^{\infty} \frac{(-1)^m x^m}{\omega^m} \quad \text{if} \quad x < \omega \tag{1.4.12}$$

and

$$\int_0^{\infty} x^m e^{-x} \, dx = m! \tag{1.4.13}$$

$$f(\omega) = \sum_{m=0}^{\infty} \frac{(-1)^m m!}{\omega^m} \tag{1.4.14}$$

Since the ratio of the mth to the $(m-1)$th term, $-m\omega^{-1}$, tends to infinity as $m \to \infty$, the series (1.4.14) diverges for all values of ω.

To investigate whether (1.4.14) is an asymptotic series, we estimate the remainder if the series is truncated after the nth term. To do this we note that

$$\frac{\omega}{\omega + x} = \sum_{m=0}^{n-1} \frac{(-1)^m x^m}{\omega^m} + \frac{(-1)^n x^n}{\omega^{n-1}(\omega + x)} \tag{1.4.15}$$

Hence

$$f(\omega) = \sum_{m=0}^{n-1} \frac{(-1)^m m!}{\omega^m} + R_n \tag{1.4.16}$$

where

$$R_n = \frac{(-1)^n}{\omega^{n-1}} \int_0^{\infty} \frac{x^n e^{-x}}{\omega + x} \, dx < \frac{(-1)^n}{\omega^n} \int_0^{\infty} x^n e^{-x} \, dx = \frac{(-1)^n n!}{\omega^n} \tag{1.4.17}$$

Hence the error committed by truncating the series after the first n terms is numerically less than the first neglected term, and

$$f(\omega) = \sum_{m=0}^{n-1} \frac{(-1)^m m!}{\omega^m} + O(\omega^{-n}) \qquad (1.4.18)$$

Therefore the series (1.4.14) is an asymptotic series, and we write

$$f(\omega) \sim \sum_{m=0}^{\infty} \frac{(-1)^m m!}{\omega^m} \qquad (1.4.19)$$

1.4.2. ASYMPTOTIC EXPANSIONS

One does not need to use a power series to represent a function. Instead, one can use a general sequence of functions $\delta_n(\epsilon)$ as long as

$$\delta_n(\epsilon) = o[\delta_{n-1}(\epsilon)] \quad \text{as} \quad \epsilon \to 0 \qquad (1.4.20)$$

Such a sequence is called an *asymptotic sequence*. Examples of such asymptotic sequences are

$$\epsilon^n, \ \epsilon^{n/3}, \ (\log \epsilon)^{-n}, \ (\sin \epsilon)^n, \ (\cot \epsilon)^{-n} \qquad (1.4.21)$$

In terms of asymptotic sequences, we can define asymptotic expansions. Thus, given $\sum_{m=0}^{\infty} a_m \delta_m(\epsilon)$ where a_m is independent of ϵ and $\delta_m(\epsilon)$ is an asymptotic sequence, we say that this expansion is an *asymptotic expansion* and write

$$y \sim \sum_{m=0}^{\infty} a_m \delta_m(\epsilon) \quad \text{as} \quad \epsilon \to 0 \qquad (1.4.22)$$

if and only if

$$y = \sum_{m=0}^{n-1} a_m \delta_m(\epsilon) + O[\delta_n(\epsilon)] \quad \text{as} \quad \epsilon \to 0 \qquad (1.4.23)$$

Clearly, an asymptotic series is a special case of an asymptotic expansion.

As an example of an asymptotic expansion that is not an asymptotic power series, we return to the integral (1.4.11). Following van der Corput (1962), we represent $f(\omega)$ in terms of the factorial asymptotic sequence $[(\omega + 1)(\omega + 2) \cdots (\omega + n)]^{-1}$ as $\omega \to \infty$. To do this we note that

$$\frac{1}{\omega + x} = \frac{1}{\omega} - \frac{x}{\omega(\omega + x)}$$

$$= \frac{1}{\omega} - \frac{x}{\omega(\omega + 1)} + \frac{x(x - 1)}{\omega(\omega + 1)(\omega + x)}$$

$$= \frac{1}{\omega} - \frac{x}{\omega(\omega + 1)} + \frac{x(x - 1)}{\omega(\omega + 1)(\omega + 2)} - \frac{x(x - 1)(x - 2)}{\omega(\omega + 1)(\omega + 2)(\omega + x)}$$

$$(1.4.24)$$

In general

$$\frac{\omega}{\omega + x} = \sum_{m=0}^{n} \frac{(-1)^m x(x-1)\cdots(x+1-m)}{(\omega+1)(\omega+2)\cdots(\omega+m)}$$
$$+ \frac{(-1)^{n+1}x(x-1)\cdots(x-n)}{(\omega+1)(\omega+2)\cdots(\omega+n)(\omega+x)} \quad (1.4.25)$$

This equation can be proved by induction as follows. If (1.4.25) is valid for n, we show that it is valid for $(n+1)$. To do this we note that

$$\frac{\omega}{\omega + x} = \sum_{m=0}^{n} \frac{(-1)^m x(x-1)\cdots(x+1-m)}{(\omega+1)(\omega+2)\cdots(\omega+m)}$$
$$+ \frac{(-1)^{n+1}x(x-1)\cdots(x-n)}{(\omega+1)(\omega+2)\cdots(\omega+n+1)}$$
$$- \frac{(-1)^{n+1}x(x-1)\cdots(x-n)}{(\omega+1)(\omega+2)\cdots(\omega+n+1)}$$
$$+ \frac{(-1)^{n+1}x(x-1)\cdots(x-n)}{(\omega+1)(\omega+2)\cdots(\omega+n)(\omega+x)}$$

By combining the last two terms and extending the summation to $n+1$, we can rewrite this expression as

$$\frac{\omega}{\omega + x} = \sum_{m=0}^{n+1} \frac{(-1)^m x(x-1)\cdots(x+1-m)}{(\omega+1)(\omega+2)\cdots(\omega+m)}$$
$$+ (-1)^{n+2} \frac{x(x-1)\cdots(x-n-1)}{(\omega+1)(\omega+2)\cdots(\omega+n+1)(\omega+x)} \quad (1.4.26)$$

Thus if (1.4.25) is true for n, (1.4.26) shows that it is true for $n+1$. Since (1.4.25) is true for $n = 0, 1$, and 2 according to (1.4.24), it is true for $n = 3, 4, 5, \ldots$. Therefore it is true for all n.

Multiplying (1.4.25) by $\exp(-x)$ and integrating from $x = 0$ to $x = \infty$, we have

$$f(\omega) = \sum_{m=0}^{n} a_m \delta_m(\omega) + R_n(\omega) \quad (1.4.27)$$

where

$$a_m = \int_0^\infty x(x-1)\cdots(x-m+1)e^{-x}\, dx \quad (1.4.28)$$

$$\delta_m(\omega) = (-1)^m[(\omega+1)(\omega+2)\cdots(\omega+m)]^{-1} \quad (1.4.29)$$

$$R_n = -\delta_n(\omega)\int_0^\infty \frac{x(x-1)\cdots(x-n)}{\omega+n+x-n}e^{-x}\, dx \quad (1.4.30)$$

Since ω is a positive large number

$$|R_n| < |\delta_n(\omega)| \left| \int_0^\infty x(x-1) \cdots (x-n+1)e^{-x}\,dx \right|$$

$$= |a_n|\,|\delta_n(\omega)| \qquad (1.4.31)$$

Thus the error committed by keeping the first n terms is numerically less than the nth term, hence

$$f(\omega) = \sum_{m=0}^{n-1} a_m \delta_m(\omega) + O[\delta_n(\omega)] \qquad (1.4.32)$$

Since $\delta_m(\omega)$ is an asymptotic sequence as $\omega \to \infty$

$$f(\omega) \sim \sum_{m=0}^{\infty} a_m \delta_m(\omega) \quad \text{as} \quad \omega \to \infty \qquad (1.4.33)$$

1.4.3. UNIQUENESS OF ASYMPTOTIC EXPANSIONS

We have shown in the previous two sections that

$$f(\omega) \sim \sum_{m=0}^{\infty} \frac{(-1)^m m!}{\omega^m} \quad \text{as} \quad \omega \to \infty \qquad (1.4.34)$$

and

$$f(\omega) \sim \sum_{m=0}^{\infty} \frac{(-1)^m \displaystyle\int_0^\infty x(x-1)\cdots(x+1-m)e^{-x}\,dx}{(\omega+1)(\omega+2)\cdots(\omega+m)} \quad \text{as} \quad \omega \to \infty \qquad (1.4.35)$$

Thus the asymptotic representation of $f(\omega)$ as $\omega \to \infty$ is not unique. In fact, $f(\omega)$ can be represented by an infinite number of asymptotic expansions because there exists an infinite number of asymptotic sequences that can be used in the representation. However, given an asymptotic sequence $\delta_m(\omega)$, the representation of $f(\omega)$ in terms of this sequence is unique. In this case

$$f(\omega) \sim \sum_{m=0}^{\infty} a_m \delta_m(\omega) \quad \text{as} \quad \omega \to \infty \qquad (1.4.36)$$

where the a_m are uniquely given by

$$a_0 = \lim_{\omega \to \infty} \frac{f(\omega)}{\delta_0(\omega)}, \qquad a_1 = \lim_{\omega \to \infty} \frac{f(\omega) - a_0 \delta_0(\omega)}{\delta_1(\omega)}$$

$$a_n = \lim_{\omega \to \infty} \frac{f(\omega) - \displaystyle\sum_{m=0}^{n-1} a_m \delta_m(\omega)}{\delta_n(\omega)} \qquad (1.4.37)$$

1.5. Convergent versus Asymptotic Series

We found in Section 1.2.1 that one of the solutions of Bessel's equation

$$x \frac{d^2y}{dx^2} + \frac{dy}{dx} + xy = 0 \qquad (1.5.1)$$

is given by the series

$$J_0(x) = 1 - \frac{x^2}{2^2} + \frac{x^4}{2^2 \cdot 4^2} - \frac{x^6}{2^2 \cdot 4^2 \cdot 6^2} + \cdots + (-1)^n \frac{x^{2n}}{2^2 \cdot 4^2 \cdots (2n)^2} + \cdots \qquad (1.5.2)$$

which is uniformly and absolutely convergent for all values of x.

Another representation of J_0 can be obtained if we note that the change of variable

$$y = x^{-1/2} y_1 \qquad (1.5.3)$$

transforms (1.5.1) into

$$\frac{d^2y_1}{dx^2} + \left(1 + \frac{1}{4x^2}\right) y_1 = 0 \qquad (1.5.4)$$

As $x \to \infty$, this equation tends to

$$\frac{d^2y_1}{dx^2} + y_1 = 0 \qquad (1.5.5)$$

with the solutions

$$y_1 = e^{\pm ix} \qquad (1.5.6)$$

This suggests the transformation

$$y_1 = e^{ix} y_2 \qquad (1.5.7)$$

which gives

$$\frac{d^2y_2}{dx^2} + 2i \frac{dy_2}{dx} + \frac{1}{4x^2} y_2 = 0 \qquad (1.5.8)$$

This equation can be satisfied formally by

$$y_2 = 1 - \frac{1}{8x} i - \frac{1 \cdot 3^2}{8^2 \cdot 2! \cdot x^2} + \frac{1 \cdot 3^2 \cdot 5^2}{8^3 \cdot 3! \cdot x^3} i + \frac{1 \cdot 3^2 \cdot 5^2 \cdot 7^2}{8^4 \cdot 4! \cdot x^4} + \cdots \qquad (1.5.9)$$

By combining this series with that obtained by changing i into $-i$, we obtain the following two independent solutions

$$y^{(1)} \sim x^{-1/2}(u \cos x + v \sin x)$$
$$y^{(2)} \sim x^{-1/2}(u \sin x - v \cos x) \qquad (1.5.10)$$

where

$$u = 1 - \frac{1 \cdot 3^2}{8^2 \cdot 2! \cdot x^2} + \frac{1 \cdot 3^2 \cdot 5^2 \cdot 7^2}{8^4 \cdot 4! \cdot x^4} + \cdots$$

$$v = \frac{1}{8x} - \frac{1 \cdot 3^2 \cdot 5^2}{8^3 \cdot 3! \cdot x^3} + \cdots$$

(1.5.11)

To determine the connection between $J_0(x)$ and these two independent solutions, we use the integral representation

$$\pi J_0(x) = \int_0^\pi \cos\left(x \cos \theta\right) d\theta$$

(1.5.12)

and obtain (see Section 7.1.2)

$$J_0(x) \sim \sqrt{\frac{2}{\pi x}} \left[u \cos\left(x - \tfrac{1}{4}\pi\right) + v \sin\left(x - \tfrac{1}{4}\pi\right)\right]$$

(1.5.13)

The ratio test shows that y_2, u, and v, and hence the right-hand side of (1.5.13), are divergent for all values of x. However, for large x the leading terms in u and v decrease rapidly with increasing rank so that (1.5.13) is an asymptotic expansion for large x.

For small x the first few terms of (1.5.2) give fairly accurate results. In fact, the first 9 terms give a value for $J_0(2)$ correct to 11 significant figures. However, as x increases, the number of terms needed to yield the same accuracy increases rapidly. At $x = 4$, eight terms are needed to give an accuracy of three significant figures, whereas the first term of the asymptotic expansion (1.5.13) yields the same accuracy. As x increases further, an accurate result is obtained with far less labor by using the asymptotic divergent series (1.5.13).

1.6. Nonuniform Expansions

In parameter perturbations the quantities to be expanded can be functions of one or more variables besides the perturbation parameter. If we develop the asymptotic expansion of a function $f(x; \epsilon)$, where x is a scalar or vector variable independent of ϵ, in terms of the asymptotic sequence $\delta_m(\epsilon)$, we have

$$f(x; \epsilon) \sim \sum_{m=0}^\infty a_m(x)\delta_m(\epsilon) \quad \text{as} \quad \epsilon \to 0$$

(1.6.1)

where the coefficients a_m are functions of x only. This expansion is said to be *uniformly valid* if

$$f(x; \epsilon) = \sum_{m=0}^{N-1} a_m(x)\delta_m(\epsilon) + R_N(x; \epsilon)$$

(1.6.2a)

$$R_N(x; \epsilon) = O[\delta_N(\epsilon)] \qquad \text{uniformly for all } x \text{ of interest}$$

(1.6.2b)

Otherwise the expansion is said to be *nonuniformly valid* (it is often called a singular perturbation expansion). For the uniformity conditions (1.6.2) to hold, $a_m(x)\delta_m(\epsilon)$ must be small compared to the preceding term $a_{m-1}(x)\delta_{m-1}(\epsilon)$ for each m. Since $\delta_m(\epsilon) = o[\delta_{m-1}(\epsilon)]$ as $\epsilon \to 0$, we require that $a_m(x)$ *be no more singular than* $a_{m-1}(x)$ for all values of x of interest if the expansion is to be uniform. In other words, each term must be a small correction to the preceding term irrespective of the value of x.

A uniformly valid expansion is

$$\sin (x + \epsilon)$$

$$= \sin x \cos \epsilon + \cos x \sin \epsilon$$

$$= \sin x \left(1 - \frac{\epsilon^2}{2!} + \frac{\epsilon^4}{4!} - \frac{\epsilon^6}{6!} + \cdots\right) + \cos x \left(\epsilon - \frac{\epsilon^3}{3!} + \frac{\epsilon^5}{5!} - \frac{\epsilon^7}{7!} + \cdots\right)$$

$$= \sin x + \epsilon \cos x - \frac{\epsilon^2}{2!} \sin x - \frac{\epsilon^3}{3!} \cos x + \frac{\epsilon^4}{4!} \sin x + \frac{\epsilon^5}{5!} \cos x$$

$$- \frac{\epsilon^6}{6!} \sin x - \frac{\epsilon^7}{7!} \cos x + \cdots \quad \text{as} \quad \epsilon \to 0 \tag{1.6.3}$$

Note that the coefficients of all powers of ϵ are bounded for all values of x, hence $a_m(x)$ is no more singular than $a_{m-1}(x)$, and the expansion is uniformly valid as a consequence.

For a nonuniformly valid expansion, let us expand $f(x; \epsilon) = \sqrt{x + \epsilon}$ for small ϵ; that is

$$f(x; \epsilon) = \sqrt{x + \epsilon} = \sqrt{x}\left(1 + \frac{\epsilon}{x}\right)^{1/2} = \sqrt{x}\left(1 + \frac{\epsilon}{2x} - \frac{\epsilon^2}{8x^2} + \frac{\epsilon^3}{16x^3} + \cdots\right)$$

$$\tag{1.6.4}$$

Each term of this expansion except the first is singular at $x = 0$ and is more singular than the preceding one, hence the expansion is not uniformly valid. It breaks down near $x = 0$. An estimate of the size of the region of nonuniformity can sometimes be obtained by assuming two successive terms to be of the same order; that is

$$\frac{\epsilon}{2x} = O(1) \quad \text{gives} \quad x = O(\epsilon) \tag{1.6.5}$$

We can see this if we look at $[1 + (\epsilon/x)]^{1/2}$ whose Taylor series expansion converges only if $|\epsilon/x|$ is less than unity.

For a second nonuniformly valid expansion, we consider the expansion of $\exp(-\epsilon t)$ for small ϵ. It possesses the following uniformly convergent series for all t

$$e^{-\epsilon t} = \sum_{n=0}^{\infty} (-1)^n \frac{(\epsilon t)^n}{n!} \tag{1.6.6}$$

It is clear that $\exp{(-\epsilon t)}$ can be approximately represented by a finite number of terms only if the combination ϵt is small. Since ϵ is small, this means that $t = O(1)$. When t is as large as $O(\epsilon^{-1})$, ϵt is not small and the truncated series breaks down. For example, if $t = 2\epsilon^{-1}$, the first two terms give -1 for $\exp{(-2)}$. If the above series is truncated after a finite number of terms, it is found that the truncated series is satisfactory up to a certain value of t after which $\exp{(-\epsilon t)}$ and the truncated series differ from each other by a quantity which exceeds the prescribed limit of accuracy. Adding more terms to the truncated series increases the value of t to a new value t' for which this truncated series is satisfactory. However, for $t > t'$ the difference between $\exp{(-\epsilon t)}$ and the new truncated series again exceeds the prescribed accuracy. Thus all terms of the series are needed to give a satisfactory expansion for $\exp{(-\epsilon t)}$ for all t.

It is the rule rather than the exception that asymptotic expansions in terms of a parameter are not uniformly valid and break down in regions called *regions of nonuniformity*, which are sometimes referred to as *boundary layers*. Friedrichs (1955) discussed in a survey article the occurrence of these nonuniformities in different branches of mathematical physics. Most perturbation techniques were developed to render these nonuniform expansions uniformly valid. The sources of nonuniformity are discussed in Chapter 2 and the techniques of rendering nonuniform expansions uniform are taken up in the remaining chapters.

1.7. Elementary Operations on Asymptotic Expansions

To determine approximate solutions to differential and integral-differential equations, we assume expansions, substitute them into equations, and perform on them elementary operations such as addition, subtraction, exponentiation, integration, differentiation, and multiplication. These operations are carried out without justification although some of these expansions are divergent. Conditions under which these operations are justified are discussed by van der Corput (1956), Erdélyi (1956), and de Bruijn (1958).

Addition and subtraction are justified in general. For example, if

$$f(x; \epsilon) \sim \sum a_n(x)\varphi_n(\epsilon) \quad \text{as} \quad \epsilon \to 0$$
$$g(x; \epsilon) \sim \sum b_n(x)\varphi_n(\epsilon) \quad \text{as} \quad \epsilon \to 0 \tag{1.7.1}$$

where $\varphi_n(\epsilon)$ is an asymptotic sequence, then (Erdélyi, 1956)

$$\alpha f(x; \epsilon) + \beta g(x; \epsilon) \sim \sum [\alpha a_n(x) + \beta b_n(x)]\varphi_n(\epsilon) \tag{1.7.2}$$

Moreover, if $f(x; \epsilon)$ and $a_n(x)$ are integrable functions of x, then

$$\int_\alpha^x f(x; \epsilon)\, dx \sim \sum \varphi_n(\epsilon) \int_\alpha^x a_n(x)\, dx \quad \text{as} \quad \epsilon \to 0 \tag{1.7.3}$$

If $f(x; \epsilon)$ and $\varphi_n(\epsilon)$ are integrable functions of ϵ, then

$$\int_0^\epsilon f(x; \epsilon)\, d\epsilon \sim \sum a_n(x) \int_0^\epsilon \varphi_n(\epsilon)\, d\epsilon \quad \text{as} \quad \epsilon \to 0 \qquad (1.7.4)$$

Multiplication is not justified in general because in the formal product of $\sum a_n(x)\varphi_n(\epsilon)$ and $\sum b_n(x)\varphi_n(\epsilon)$ all products $\varphi_n(\epsilon)\varphi_m(\epsilon)$ occur, and it is generally not possible to arrange them so as to obtain an asymptotic sequence. In other words, multiplication is justified if the result is an asymptotic expansion. This is the case for all asymptotic sequences φ_n such that $\varphi_n\varphi_m$ either form an asymptotic sequence or possess asymptotic expansions. An important class of such sequences is the collection of powers of ϵ. Thus if

$$f(x; \epsilon) \sim \sum a_n(x)\epsilon^n \quad \text{as} \quad \epsilon \to 0$$
$$g(x; \epsilon) \sim \sum b_n(x)\epsilon^n \quad \text{as} \quad \epsilon \to 0 \qquad (1.7.5)$$

then

$$f(x; \epsilon)g(x; \epsilon) \sim \sum c_n(x)\epsilon^n \quad \text{as} \quad \epsilon \to 0 \qquad (1.7.6)$$

where

$$c_n(x) = \sum_{m=0}^n a_m(x)b_{n-m}(x) \qquad (1.7.7)$$

Exponentiation is not justified in general. When it is not justified, it leads to nonuniformities. For example

$$\sqrt{x + \epsilon} = \sqrt{x}\left(1 + \frac{1}{2}\frac{\epsilon}{x} - \frac{1}{8}\frac{\epsilon^2}{x^2} + \cdots\right) \quad \text{as} \quad \epsilon \to 0 \qquad (1.7.8)$$

is not justified when $\epsilon/x = O(1)$ because its right-hand side is nonuniform in the region $x = O(\epsilon)$. Similarly

$$\frac{1}{1 + \epsilon x} = 1 - \epsilon x + \epsilon^2 x^2 - \epsilon^3 x^3 + \cdots \quad \text{as} \quad \epsilon \to 0 \qquad (1.7.9)$$

is not justified when $\epsilon x = O(1)$ because its right-hand side is nonuniform for large x.

It is not justified in general to differentiate asymptotic expansions with respect to a variable such as x or with respect to the perturbation parameter ϵ. As in exponentiation, when it is not justified, differentiation leads to nonuniformities.

Exercises

1.1. Determine the order of the following expressions as $\epsilon \to 0$:

$$\sqrt{\epsilon(1 - \epsilon)}, \quad 4\pi^2\epsilon, \quad 1000\epsilon^{1/2}, \quad \ln(1 + \epsilon), \quad \frac{1 - \cos\epsilon}{1 + \cos\epsilon}, \quad \frac{\epsilon^{3/2}}{1 + \sin\epsilon}, \quad \frac{\epsilon^{1/2}}{1 - \cos\epsilon}, \quad \text{sech}^{-1}\epsilon,$$

$$e^{\tan\epsilon}, \quad \ln\left[1 + \frac{\ln(1 + 2\epsilon)}{\epsilon(1 - 2\epsilon)}\right], \quad \ln\left[1 + \frac{\ln\dfrac{1 + 2\epsilon}{\epsilon}}{1 - 2\epsilon}\right], \quad e^{-\cosh(1/\epsilon)}, \quad \int_0^\epsilon e^{-s^2}\, ds$$

1.2. Arrange the following in descending order for small ϵ:

$$\epsilon^2, \ \epsilon^{1/2}, \ \ln(\ln \epsilon^{-1}), \ 1, \ \epsilon^{1/2}\ln \epsilon^{-1}, \ \epsilon \ln \epsilon^{-1}, \ e^{-1/\epsilon}, \ \ln \epsilon^{-1}, \ \epsilon^{3/2}, \ \epsilon, \ \epsilon^2 \ln \epsilon^{-1}$$

1.3. Expand each of the following expressions for small ϵ and keep three terms:

(a) $\sqrt{1 - \frac{1}{2}\epsilon^2 t - \frac{1}{8}\epsilon^4 t}$

(b) $(1 + \epsilon \cos f)^{-1}$

(c) $(1 + \epsilon\omega_1 + \epsilon^2\omega_2)^{-2}$

(d) $\sin(s + \epsilon\omega_1 s + \epsilon^2\omega_2 s)$

(e) $\sin^{-1}\left(\dfrac{\epsilon}{\sqrt{1 + \epsilon}}\right)$

(f) $\ln \dfrac{1 + 2\epsilon - \epsilon^2}{\sqrt[3]{1 + 2\epsilon}}$

1.4. Let $\mu = \mu_0 + e\mu_1 + e^2\mu_2$ in $h = (3/2)[1 - \sqrt{1 - 3\mu(1 - \mu)}]$, expand for small e, and keep three terms.

1.5. Find a second-order expansion for the solution of

$$x = 1 + \epsilon x^2, \qquad \epsilon \ll 1$$

and compare it with the exact solution for $\epsilon = 0.1$ and $\epsilon = 0.001$.

1.6. Show that the asymptotic expansion for large x of

$$F(x) = \int_x^\infty \frac{\cos t}{t}\, dt$$

is

$$F(x) = \left(-\frac{1}{x} + \frac{2!}{x^3} - \frac{4!}{x^5} + \cdots\right)\sin x + \left(\frac{1}{x^2} - \frac{3!}{x^4} + \frac{5!}{x^6} - \cdots\right)\cos x$$

Does the series converge? Find an upper bound for the remainder and show that the remainder approaches zero faster than the last term in the expansion as $x \to \infty$.

1.7. If $x = s - (\epsilon/3s^2) - (3\epsilon^2/10s^4)$, find a two-term expansion for s when $x = 0$.

1.8. If $x = s + \epsilon(2 - (2/3)s^{1/2}) + (42/5)\epsilon^2 s^{1/2}$, show that the solution of $dx/ds = 0$ is

$$s = \frac{1}{9}\epsilon^2 - \frac{14}{5}\epsilon^3 + O(\epsilon^4)$$

and then find x corresponding to this value.

1.9. If

$$y_0(s) + \epsilon y_1(s) + \epsilon^2 y_2(s) + \cdots = A$$

and

$$1 = s + \epsilon x_1(s) + \epsilon^2 x_2(s) + \cdots$$

show that

$$s = 1 - \epsilon x_1(1) - \epsilon^2[x_2(1) - x_1'(1)x_1(1)] + \cdots$$

and then find $y_0(1)$, $y_1(1)$, and $y_2(1)$.

1.10. Consider the equation

$$y' + y = \epsilon y^2, \qquad y(0) = 1$$

(a) Determine a three-term expansion for small ϵ.

(b) Show that the exact solution is

$$y = e^{-x}[1 + \epsilon(e^{-x} - 1)]^{-1}$$

(c) Expand this exact solution for small ϵ and compare the result with (a).

(d) Is this expansion valid for all x?

1.11. Determine a coordinate expansion for

$$y'' - \left(\frac{2}{x^2} + \frac{1}{x^3}\right)y = 0$$

of the form

$$y = \sum_{n=0}^{\infty} a_n x^{-n+\sigma}$$

1.12. Determine second-order (three-term) expansions for

(a) $\ddot{u} + u = \epsilon u^2, \qquad \epsilon \ll 1$

(b) $\ddot{u} + u = -\epsilon \dot{u}$

with $u(0) = a$ and $\dot{u}(0) = 0$. Are these expansions uniformly valid?

1.13. Find a first-order (two-term) expansion valid for small ϵ for the solution of

$$s\frac{dx}{ds} = x + \epsilon y$$

$$s\frac{dy}{ds} = -(2 + x)y$$

$$y(1) = e^{-1}, \qquad x(1) = 1$$

1.14. Use the asymptotic expansion (1.5.13) to show that the large zeros ξ of $J_0(x)$ are the solutions of

$$\cot(\xi - \tfrac{1}{4}\pi) = -\frac{1}{8\xi} + \frac{33}{512\xi^3} + \cdots$$

and then show that

$$\xi = \tfrac{1}{4}\pi(4n + 3) + \frac{1}{2\pi(4n + 3)} + \cdots \text{ with } n \text{ an integer.}$$

1.15. Show that

$$u\ddot{u} + \dot{u} + tu = t^2$$

is satisfied by the expansions (Levinson, 1969)

$$u = -\frac{t^3}{6} + c_1 t + c_2 - 9t \ln t + O\left(\frac{\ln t}{t}\right)$$

$$u = t + (b_1 \sin t + b_2 \cos t)t^{-1/2} + O(t^{-1})$$

as $t \to \infty$, where c_i and b_i are constants.

1.16. Show that the equation of Bellman (1955)

$$(\ddot{u})^2 = \dot{u} + u$$

is satisfied by the expansions (Levinson, 1969)

$$u = ae^{-t} - a^2 e^{-2t} + O(e^{-3t})$$

$$u = \tfrac{1}{144}t^4 + \tfrac{1}{30}t^3 \ln t + c_1 t^3 + c_2 t^{-2} + O(t^2 \ln^2 t)$$

as $t \to \infty$, where a and c_i are constants.

CHAPTER 2

Straightforward Expansions and Sources of Nonuniformity

As pointed out in Section 1.6, it is the rule rather than the exception that expansions of the Poincaré type (straightforward expansions), such as

$$f(x, \epsilon) \sim \sum_{m=0}^{\infty} \delta_m(\epsilon) f_m(x)$$

where $\delta_m(\epsilon)$ is an asymptotic sequence in terms of the parameter ϵ, are nonuniformly valid and break down in regions called regions of nonuniformity. Some of the sources of nonuniformities are: infinite domain, small parameter multiplying the highest derivative, type change of a partial differential equation, and presence of singularities.

In the infinite domain case, the nonuniformity manifests itself in the presence of so-called secular terms such as $x^n \cos x$ and $x^n \sin x$, which make $f_m(x)/f_{m-1}(x)$ unbounded as x approaches infinity. In the case of the small parameter multiplying the highest derivative, the perturbation expansion cannot satisfy all the boundary and initial conditions, and the expansion thus is not valid in boundary and initial layers. Since the boundary and initial conditions required to form a well-posed problem depend on the type of the partial differential equation under consideration, nonuniformities might arise if the type of the perturbation equations is different from the type of the original equation. In the fourth class, singularities that are not part of the exact solution appear at some point in the expansion, generally becoming more pronounced in succeeding terms.

To illustrate how nonuniform expansions arise and how they are recognized, we give several examples for each source of nonuniformity. These examples also serve to describe the mechanics of carrying out parameter perturbations. Moreover, most of these examples are taken up again in the remaining chapters where they are rendered uniformly valid. To conclude this chapter, we discuss the role of coordinates (dependent as well as independent) in making the expansions uniform or nonuniform and the role of perturbation methods in choosing the coordinate systems that render the expansions uniformly valid.

2.1. Infinite Domains

2.1.1. THE DUFFING EQUATION

Consider the oscillations of a mass connected to a nonlinear spring described by Duffing's equation

$$\ddot{u} + u + \epsilon u^3 = 0, \qquad u(0) = a, \qquad \dot{u}(0) = 0 \qquad (2.1.1)$$

where ϵ is a small positive number. This problem admits the integral

$$\dot{u}^2 + u^2 + \epsilon \frac{u^4}{2} = \left(1 + \frac{\epsilon a^2}{2}\right) a^2 \qquad (2.1.2)$$

Equation (2.1.2) shows that u is bounded for all times when ϵ is positive.

Let us seek an approximate solution in the form of a Poincaré-type asymptotic expansion

$$u = \sum_{m=0}^{\infty} \epsilon^m u_m(t) \qquad (2.1.3)$$

Substituting into (2.1.1), expanding, and equating coefficients of equal powers of ϵ lead to the following problems for u_0 and u_1

$$\ddot{u}_0 + u_0 = 0, \qquad u_0(0) = a, \qquad \dot{u}_0(0) = 0 \qquad (2.1.4)$$

$$\ddot{u}_1 + u_1 = -u_0^3, \qquad u_1(0) = 0, \qquad \dot{u}_1(0) = 0 \qquad (2.1.5)$$

The solution for u_0 that satisfies the initial conditions is

$$u_0 = a \cos t \qquad (2.1.6)$$

Substituting for u_0 in (2.1.5) and using the trigonometric identity $\cos 3t = 4 \cos^3 t - 3 \cos t$ lead to

$$\ddot{u}_1 + u_1 = -a^3 \frac{\cos 3t + 3 \cos t}{4} \qquad (2.1.7)$$

The solution of (2.1.7) that satisfies the initial conditions in (2.1.5) is

$$u_1 = -\frac{3a^3}{8} t \sin t + \frac{a^3}{32} (\cos 3t - \cos t) \qquad (2.1.8)$$

Thus

$$u = a \cos t + \epsilon a^3 [-\tfrac{3}{8} t \sin t + \tfrac{1}{32}(\cos 3t - \cos t)] + O(\epsilon^2) \quad (2.1.9)$$

The two-term expansion above cannot approximate the solution as $t \to \infty$ because the term $t \sin t$ makes $u_1/u_0 \to \infty$ as $t \to \infty$; $t \sin t$ is called a *secular* term. It tends to infinity as $t \to \infty$, whereas u should be bounded for all t as discussed above. The variable t does not need to be infinite for (2.1.9) to break down; if $t = O(\epsilon^{-1})$, the second term is of the same order as the first, contrary to our assumption that ϵu_1 is a small correction to u_0 when we derived (2.1.9). When more terms in the series are calculated, secular terms of the form $t^n(\cos t, \sin t)$ are obtained. Although the resulting series is convergent, it is slowly convergent, and one cannot represent the solution for all t using a finite number of terms of this series.

The appearance of secular terms is characteristic of nonlinear oscillation problems, hence one should not expect to obtain a uniformly valid straightforward expansion in these cases.

2.1.2. A MODEL FOR WEAK NONLINEAR INSTABILITY

As a model for the weak nonlinear instability of a standing wave, we consider the problem

$$u_{tt} - u_{xx} - u = u^3 \tag{2.1.10}$$

$$u(x, 0) = \epsilon \cos kx, \qquad u_t(x, 0) = 0 \tag{2.1.11}$$

The initial conditions suggest an expansion of the form

$$u = \epsilon u_1 + \epsilon^2 u_2 + \epsilon^3 u_3 + \cdots \tag{2.1.12}$$

Substituting this expansion into (2.1.10) and (2.1.11) and equating coefficients of like powers of ϵ, we obtain

Order ϵ

$$u_{1tt} - u_{1xx} - u_1 = 0$$
$$u_1(x, 0) = \cos kx, \qquad u_{1t}(x, 0) = 0 \tag{2.1.13}$$

Order ϵ^2

$$u_{2tt} - u_{2xx} - u_2 = 0$$
$$u_2(x, 0) = u_{2t}(x, 0) = 0 \tag{2.1.14}$$

Order ϵ^3

$$u_{3tt} - u_{3xx} - u_3 = u_1{}^3$$
$$u_3(x, 0) = u_{3t}(x, 0) = 0 \tag{2.1.15}$$

The solution of the first-order problem (2.1.13) is

$$u_1 = \cos \sigma_1 t \cos kx, \qquad \sigma_1{}^2 = k^2 - 1 \tag{2.1.16}$$

Thus the wave is stable or unstable depending on whether k is greater or less than unity. The special case $k = 1$ separates stable from unstable waves.

The solution of the second-order problem (2.1.14) is $u_2 = 0$. Substituting for u_1 from (2.1.16) into (2.1.15) and solving the resulting problem, we have

$$u_3 = \frac{3}{128\sigma_1{}^2} [12\sigma_1 t \sin \sigma_1 t + \cos \sigma_1 t - \cos 3\sigma_1 t] \cos kx$$

$$+ \frac{1}{128k^2} [3(\cos \sigma_1 t - \cos \mu t) + k^2(\cos 3\sigma_1 t - \cos \mu t)] \cos 3kx \quad (2.1.17)$$

where $\mu^2 = 9k^2 - 1$. Therefore

$$u = \epsilon \cos \sigma_1 t \cos kx$$

$$+ \epsilon^3 \left[\frac{9}{32\sigma_1} t \sin \sigma_1 t \cos kx + \text{terms bounded as } t \to \infty \right] \quad (2.1.18)$$

Here again the straightforward expansion is invalid when $t = O(\epsilon^{-2})$ or larger because of the presence of the secular term $t \sin \sigma_1 t$.

2.1.3. SUPERSONIC FLOW PAST A THIN AIRFOIL

As a third example showing how an infinite domain is responsible for a nonuniformity in an expansion of the Poincaré type, let us consider a uniform inviscid supersonic flow past the thin symmetric airfoil shown in Figure 2-1.

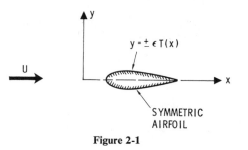

Figure 2-1

If we assume that the velocity vector $\mathbf{q} = U \, \mathbf{grad} \, (x + \phi)$, the equation governing the potential function ϕ for steady two-dimensional, irrotational, isentropic motion is

$$\phi_{yy} - B^2 \phi_{xx} = M^2 \left[\frac{\gamma - 1}{2} (2\phi_x + \phi_x{}^2 + \phi_y{}^2)(\phi_{xx} + \phi_{yy}) \right.$$

$$\left. + (2\phi_x + \phi_x{}^2)\phi_{xx} + 2(1 + \phi_x)\phi_y \phi_{xy} + \phi_y{}^2 \phi_{yy} \right] \quad (2.1.19)$$

where $B^2 = M^2 - 1$ and M is the free-stream Mach number. The normal velocity vanishes on the surface; that is, the flow must be tangent to the surface. Hence

$$\frac{\phi_y}{1 + \phi_x} = \epsilon T'(x) \quad \text{at} \quad y = \epsilon T(x), \qquad 0 \le x \le l \qquad (2.1.20)$$

where l is the airfoil chord. The boundary condition upstream is

$$\phi(x, y) = 0 \qquad (2.1.21)$$

Van Dyke (1952) obtained a second-order solution to (2.1.19) through (2.1.21) by iteration for small but finite ϵ. Let us obtain an expansion of the Poincaré type using ϵ as the perturbation parameter. Let

$$\phi = \epsilon \phi_1 + \epsilon^2 \phi_2 + \cdots \qquad (2.1.22)$$

Since ϵ is small, we can greatly simplify the problem by transferring the boundary condition (2.1.20) from $y = \epsilon T(x)$ to $y = 0$ by using the following Taylor series expansion

$$\phi(x, \epsilon T) = \phi(x, 0) + \epsilon T \phi_y(x, 0) + \tfrac{1}{2}\epsilon^2 T^2 \phi_{yy}(x, 0) + \cdots$$

Thus we rewrite (2.1.20) as

$$\frac{\phi_y(x, 0) + \epsilon T \phi_{yy}(x, 0) + \cdots}{1 + \phi_x(x, 0) + \cdots} = \epsilon T'(x), \qquad 0 \le x \le l \qquad (2.1.23)$$

Substituting (2.1.22) into (2.1.19), (2.1.21), and (2.1.23), expanding for small ϵ, and equating coefficients of equal powers of ϵ, we have

Order ϵ

$$\phi_{1yy} - B^2 \phi_{1xx} = 0 \qquad (2.1.24)$$

$$\phi_{1y}(x, 0) = T'(x), \qquad 0 \le x \le l \qquad (2.1.25)$$

$$\phi_1(x, y) = 0 \qquad \text{(upstream)} \qquad (2.1.26)$$

Order ϵ^2

$$\phi_{2yy} - B^2 \phi_{2xx} = M^2[(\gamma + 1)\phi_{1x}\phi_{1xx} + (\gamma - 1)\phi_{1x}\phi_{1yy} + 2\phi_{1y}\phi_{1xy}] \qquad (2.1.27)$$

$$\phi_{2y} = \phi_{1x}T' - \phi_{1yy}T \quad \text{at} \quad y = 0 \quad \text{and} \quad 0 \le x \le l \qquad (2.1.28)$$

$$\phi_2(x, y) = 0 \qquad \text{(upstream)} \qquad (2.1.29)$$

The general solution of (2.1.24) is

$$\phi_1 = f(\xi) + g(\eta) \qquad (2.1.30)$$

where

$$\xi = x - By, \qquad \eta = x + By$$

The upstream condition (2.1.26) demands that $g = 0$, while the condition

(2.1.25) demands that $f = -T(\xi)/B$. Therefore

$$\phi_1 = -T(\xi)/B \qquad (2.1.31)$$

Substituting for ϕ_1 into (2.1.27) gives

$$\phi_{2yy} - B^2\phi_{2xx} = M^4(\gamma + 1)f'f'' \qquad (2.1.32)$$

Transforming the left-hand side of (2.1.32) into the ξ and η coordinates, we have

$$\frac{\partial^2 \phi_2}{\partial \xi \, \partial \eta} = -\frac{M^4(\gamma + 1)}{4B^2}f'f'' \qquad (2.1.33)$$

whose solution is

$$\phi_2 = -\frac{M^4(\gamma + 1)}{8B^2}f'^2\eta + h(\xi) \qquad (2.1.34)$$

The function $h'(\xi)$ can be determined from (2.1.28) to be

$$h'(\xi) = \frac{M^4(\gamma + 1)}{4B^4}\xi T'T'' + \frac{1}{B^2}\left[1 - \frac{M^4(\gamma + 1)}{8B^2}\right]T'^2 - TT'' \qquad (2.1.35)$$

Since u (the axial component of velocity) is given by $U(1 + \phi_x)$

$$\frac{u}{U} = 1 - \epsilon\frac{T'}{B} + \epsilon^2\left[\frac{1}{B^2}\left(1 - \frac{M^4(\gamma + 1)}{4B^2}\right)T'^2\right.$$
$$\left. - \frac{\gamma + 1}{2}\frac{M^4}{B^3}yT'T'' - TT''\right] + O(\epsilon^3) \qquad (2.1.36)$$

For $y = O(1)$, the third term in (2.1.36) is bounded and thus is a small correction to the second term which is in turn a small correction to the first term as $\epsilon \to 0$. However, as y increases to $O(\epsilon^{-1})$ and larger, the third term becomes of the same order as the second, and then of the same order as the first because of the presence of the term $(1/2)(\gamma + 1)M^4B^{-3}yT'T''$ which makes u_2/u_1 unbounded as $y \to \infty$. Although there are no circular functions in this problem, this term can be considered a secular term.

2.1.4.   SMALL REYNOLDS NUMBER FLOW PAST A SPHERE

The fourth example given here to show the difficulty arising from an infinite domain is the small Reynolds number, incompressible, uniform flow past a sphere. In the spherical coordinate system shown in Figure 2-2, the full Navier–Stokes equations give the following dimensionless equation for the stream function $\psi(r, \theta)$ ($u_r = \psi_\theta/r^2 \sin\theta$, $u_\theta = -\psi_r/r \sin\theta$) for axisymmetric flow

$$\mathscr{D}^4\psi = \frac{R}{r^2 \sin\theta}\left(\psi_\theta\frac{\partial}{\partial r} - \psi_r\frac{\partial}{\partial \theta} + 2\cot\theta\,\psi_r - 2\frac{\psi_\theta}{r}\right)\mathscr{D}^2\psi \qquad (2.1.37)$$

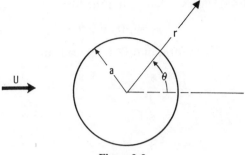

Figure 2-2

where the Reynolds number $R = Ua/\nu$ (ν is kinematic viscosity) and

$$\mathscr{D}^2 = \frac{\partial^2}{\partial r^2} + \frac{\sin\theta}{r^2}\frac{\partial}{\partial\theta}\left(\frac{1}{\sin\theta}\frac{\partial}{\partial\theta}\right) \tag{2.1.38}$$

The boundary conditions at the sphere's surface call for the vanishing of the velocity; that is, in dimensionless form,

$$\psi(1, \theta) = \psi_r(1, \theta) = 0 \tag{2.1.39}$$

The condition that the flow is uniform upstream gives

$$\psi(r, \theta) \to \tfrac{1}{2}r^2 \sin^2\theta \quad \text{as} \quad r \to \infty \tag{2.1.40}$$

Equations (2.1.37) through (2.1.40) form a well-posed problem for the stream function ψ.

Let us seek a formal expansion of the Poincaré type valid for small R

$$\psi(r, \theta; R) = \sum_{m=0}^{\infty} R^m \psi_m(r, \theta) \tag{2.1.41}$$

Substituting (2.1.41) into (2.1.37) through (2.1.40), expanding for small R, and equating coefficients of equal powers of R lead to

Order R^0

$$\mathscr{D}^4\psi_0 = 0 \tag{2.1.42}$$

$$\psi_0(1, \theta) = \psi_{0r}(1, \theta) = 0 \tag{2.1.43}$$

$$\psi_0(r, \theta) \to \tfrac{1}{2}r^2 \sin^2\theta \quad \text{as} \quad r \to \infty \tag{2.1.44}$$

Order R

$$\mathscr{D}^4\psi_1 = \frac{1}{r^2\sin\theta}\left(\psi_{0\theta}\frac{\partial}{\partial r} - \psi_{0r}\frac{\partial}{\partial\theta} + 2\cot\theta\,\psi_{0r} - 2\frac{\psi_{0\theta}}{r}\right)\mathscr{D}^2\psi_0 \tag{2.1.45}$$

$$\psi_1(1, \theta) = \psi_{1r}(1, \theta) = 0 \tag{2.1.46}$$

$$\psi_1(r, \theta) = o(r^2) \quad \text{as} \quad r \to \infty \tag{2.1.47}$$

Equation (2.1.44) suggests that ψ_0 has the form

$$\psi_0 = f(r)\sin^2\theta \tag{2.1.48}$$

Substituting this assumed form into (2.1.42) leads to

$$f^{iv} - \frac{4f''}{r^2} + \frac{8f'}{r^3} - \frac{8f}{r^4} = 0 \tag{2.1.49}$$

whose general solution is

$$f = c_4 r^4 + c_2 r^2 + c_1 r + c_{-1} r^{-1} \tag{2.1.50}$$

The boundary condition (2.1.44) demands that $c_4 = 0$, and $c_2 = 1/2$, while the boundary conditions (2.1.43) demand that $c_1 = -3/4$ and $c_{-1} = 1/4$. Therefore

$$\psi_0 = \frac{1}{4}\left(2r^2 - 3r + \frac{1}{r}\right)\sin^2\theta \tag{2.1.51}$$

This solution was obtained by Stokes (1851).

Substituting for ψ_0 from (2.1.51) into (2.1.45) gives

$$\mathscr{D}^4\psi_1 = -\frac{9}{4}\left(\frac{2}{r^2} - \frac{3}{r^3} + \frac{1}{r^5}\right)\sin^2\theta\cos\theta \tag{2.1.52}$$

Equation (2.1.52) and the boundary conditions (2.1.46) and (2.1.47) suggest that a particular solution for ψ_1 has the form

$$\psi_1 = g(r)\sin^2\theta\cos\theta \tag{2.1.53}$$

With this form for ψ_1, g should satisfy the following equation and boundary conditions

$$g^{iv} - \frac{12g''}{r^2} + \frac{24g'}{r^3} = -\frac{9}{4}\left(\frac{2}{r^2} - \frac{3}{r^3} + \frac{1}{r^5}\right) \tag{2.1.54}$$

$$g(1) = g'(1) = 0 \tag{2.1.55}$$

$$g(r) = o(r^2) \quad \text{as} \quad r \to \infty \tag{2.1.56}$$

The general solution of (2.1.54) is

$$g = b_{-2}r^{-2} + b_0 + b_3 r^3 + b_5 r^5 - \frac{3}{16}r^2 + \frac{9}{32}r + \frac{3}{32}\frac{1}{r} \tag{2.1.57}$$

The boundary condition (2.1.56) demands that $b_3 = b_5 = 0$. However, even with this choice for b_3 and b_5, g does not behave properly as $r \to \infty$ because of the presence of the term $-(3/16)r^2$ in the particular solution for g. It is clear that no values for b_0 and b_{-2} can be found that correct this shortcoming. Moreover, no other complementary solution of (2.1.52) can be found which makes ψ_1 behave properly as $r \to \infty$. The boundary conditions (2.1.55) demand that $b_0 = b_{-2} = -3/32$, hence

$$\psi_1 = -\frac{3}{32}\left(2r^2 - 3r + 1 - \frac{1}{r} + \frac{1}{r^2}\right) \sin^2 \theta \cos \theta \qquad (2.1.58)$$

Here again difficulty with the straightforward expansion arose because of the infinite domain. The two-term expansion

$$\psi = \frac{1}{4}\left(2r^2 - 3r + \frac{1}{r}\right) \sin^2 \theta - \frac{3}{32} R\left(2r^2 - 3r + 1 - \frac{1}{r} + \frac{1}{r^2}\right)$$
$$\times \sin^2 \theta \cos \theta + 0(R^2) \quad \text{as} \quad R \to 0 \quad (2.1.59)$$

satisfies the surface boundary conditions, but it does not satisfy the boundary condition at infinity. Thus, this expansion breaks down for large r and this breakdown is called *Whitehead's paradox* because Whitehead (1889) was the first to obtain this solution, although by iteration, and the first to point out its nonuniformity.

2.2. A Small Parameter Multiplying the Highest Derivative

2.2.1. A SECOND-ORDER EXAMPLE

In order to show the difficulty that arises when a small parameter multiplies the highest derivative, we consider the following example after Latta (1964)

$$\epsilon y'' + y' + y = 0, \qquad 0 \leq x \leq 1 \qquad (2.2.1)$$

$$y(0) = a, \qquad y(1) = b \qquad (2.2.2)$$

where ϵ is a small positive number. First we seek a straightforward expansion of the form

$$y = \sum_{n=0}^{\infty} \epsilon^n y_n(x), \qquad \epsilon \ll 1 \qquad (2.2.3)$$

Substituting into (2.2.1) and equating coefficients of equal powers of ϵ lead to

$$y_0' + y_0 = 0 \qquad (2.2.4)$$

$$y_n' + y_n = -y_{n-1}'' \qquad (2.2.5)$$

It can be seen that at any level of approximation n, y_{n-1} is known, hence y_n for any n is given by a first-order differential equation. Consequently, the solutions of (2.2.4) and (2.2.5) cannot satisfy both of the boundary conditions (2.2.2), and one of them must be dropped. It is shown in Section 4.1.2 that the boundary condition at the origin must be dropped. Solving for the first two terms and imposing the boundary condition $y(1) = b$ give

$$y = be^{1-x} + \epsilon be^{1-x}(1 - x) + O(\epsilon^2) \qquad (2.2.6)$$

At the origin $y = be(1 + \epsilon)$, which is in general different from the a in (2.2.2). Hence the error in (2.2.3) is not uniform over $[0, 1]$, and the expansion breaks down at the origin.

To understand further the nature of the nonuniformity, let us look at the exact solution of (2.2.1) and (2.2.2), namely

$$y = \frac{(ae^{s_2} - b)e^{s_1 x} + (b - ae^{s_1})e^{s_2 x}}{e^{s_2} - e^{s_1}} \qquad (2.2.7)$$

where

$$s_{1,2} = \frac{-1 \pm \sqrt{1 - 4\epsilon}}{2\epsilon} \qquad (2.2.8)$$

It can be shown that

$$\lim_{\substack{\epsilon \to 0 \\ x \text{ fixed}}} y(x, \epsilon) = be^{1-x} \qquad (2.2.9)$$

is in agreement with the first term in (2.2.6) and satisfies the boundary condition $y(1) = b$. To understand what happens at the boundary $x = 0$, let us estimate y to order ϵ from the exact solution and denote it by $\tilde{y}$. As $\epsilon \to 0$

$$s_1 = -1 + O(\epsilon), \qquad s_2 = -\frac{1}{\epsilon} + 1 + O(\epsilon) \qquad (2.2.10)$$

Therefore

$$\tilde{y} = be^{1-x} + (a - be)e^{-(x/\epsilon)+x} + O(\epsilon) \qquad (2.2.11)$$

In the above estimation the term proportional to $\exp[-(x/\epsilon) + x]$ was not neglected because an estimate is being made not only when $\epsilon \to 0$, but when $x \to 0$ also. From the manner in which (2.2.11) was constructed, the order of error is uniform on $[0, 1]$. The behavior of $\tilde{y}$ is shown schematically in Figure 2-3 together with the first term of (2.2.6) denoted by $\hat{y}$. It can be seen that, for small ϵ, $\tilde{y}$ agrees with $\hat{y}$ except in a small region near the origin where it changes quickly in order to satisfy the boundary condition there. Thus $y(x; \epsilon)$ is continuous for $\epsilon > 0$ but discontinuous for $\epsilon = 0$. In fact

$$\lim_{\epsilon \to 0} \lim_{x \to 0} y(x; \epsilon) = a \qquad (2.2.12)$$

$$\lim_{x \to 0} \lim_{\epsilon \to 0} y(x; \epsilon) = be \qquad (2.2.13)$$

which shows the nonuniform convergence of the exact solution $y(x; \epsilon)$ to $\hat{y}$ [the first term in (2.2.6)].

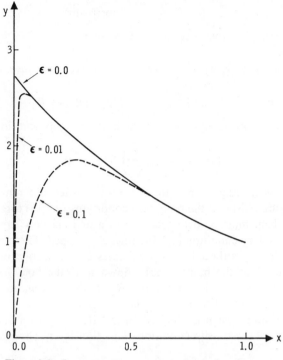

Figure 2-3 IN THIS SPECIAL CASE $a = 0$ AND $b = 1$.

2.2.2. HIGH REYNOLDS NUMBER FLOW PAST A BODY

Consider a two-dimensional viscous incompressible uniform flow past the body shown in Figure 2-1. The complete steady Navier–Stokes equations give for the stream function ψ ($u = \psi_y$, $v = -\psi_x$ with u and v the velocity components along the x and y coordinates)

$$\left(\psi_y \frac{\partial}{\partial x} - \psi_x \frac{\partial}{\partial y} - \frac{1}{R}\nabla^2\right)\nabla^2\psi = 0 \qquad (2.2.14)$$

where the Reynolds number $R = UL/\nu$ with ν the kinematic viscosity of the fluid. Equation (2.2.14) needs to be supplemented by boundary conditions. At the body surface, $y = F(x)$, both components of velocity vanish; thus

$$F'\psi_y[x, F(x)] + \psi_x[x, F(x)] = 0 \qquad (2.2.15)$$

$$\psi[x, F(x)] = 0 \qquad (2.2.16)$$

The second condition represents the vanishing of the velocity tangent to the body (the so-called no-slip condition), while the first condition represents the vanishing of the normal velocity component. The third boundary

condition is
$$\psi(x, y) \to y \qquad \text{(upstream)} \qquad (2.2.17)$$

Seeking a straightforward solution of the form

$$\psi(x, y; R) = \sum_{m=0}^{\infty} \delta_m(R)\psi_m(x, y) \quad \text{as} \quad R \to \infty \qquad (2.2.18)$$

where
$$\delta_0(R) = 1 \quad \text{and} \quad \delta_m(R) = o[\delta_{m-1}(R)] \text{ as } R \to \infty$$

we obtain the following equation for the first term (inviscid flow)

$$\left(\psi_{0y}\frac{\partial}{\partial x} - \psi_{0x}\frac{\partial}{\partial y}\right)\nabla^2\psi_0 = 0 \qquad (2.2.19)$$

which is a third-order rather than a fourth-order differential equation. Thus ψ_0 cannot satisfy all the boundary conditions (2.2.15) through (2.2.17), and one of them must be dropped. Since an inviscid flow can slide on the body, the boundary condition (2.2.16) must be dropped. Therefore the resulting solution for ψ_0, although it approximates the exact solution very well as $R \to \infty$ away from the body, breaks down near the body. Regardless of how small the viscosity is (how large R is), the tangential velocity must vanish at the surface. Therefore for large R the exact solution is close to ψ_0 except in a thin layer near the body where it undergoes a quick change in order to retrieve the no-slip condition. This thin layer is *Prandtl's boundary layer*.

2.2.3. RELAXATION OSCILLATIONS

The problem considered next is that of finding the periodic solutions of equations of the form
$$\epsilon u'' = f(u', u) \qquad (2.2.20)$$

for small ϵ when $f(u', u) = 0$ has no periodic solutions. Van der Pol (1927) was the first to treat a problem of this kind in connection with explaining the jerky oscillations (relaxation oscillations) of an electronic circuit governed by the following equation which is named after him

$$u'' + u = \alpha(u' - \tfrac{1}{3}u'^3) \qquad (2.2.21)$$

If we let $v = u'$, $x = u/\alpha$, and $\epsilon = \alpha^{-2}$, (2.2.21) becomes

$$\epsilon \frac{dv}{dx} = \frac{v - \dfrac{v^3}{3} - x}{v} \qquad (2.2.22)$$

If $\epsilon = 0$, $x = v - (v^3/3)$, which is shown in Figure 2-4. We assume that ϵ is very small, but different from zero, and consider a solution curve that starts at P. Since P is away from Γ, dv/dx is approximately $-\infty$ down to P_1 where

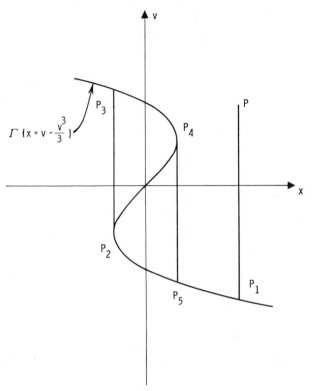

Figure 2-4

it reaches Γ. At this point, $dv/dx = 0$. Since dv/dx is approximately $\pm\infty$ away from Γ, the solution curve tends to follow Γ, staying below it, until it reaches the vicinity of P_2. At this point the solution curve turns almost vertically upward until it intersects Γ at P_3. Since $dv/dx \approx \pm\infty$ away from Γ, the solution curve tends to follow Γ from P_3 clockwise, staying above Γ, until it reaches the vicinity of P_4, where it turns almost vertically downward to intersect Γ at P_5, Then it tends to follow the path from P_5 to P_2. Therefore the limit of the periodic solution as $\epsilon \to 0$ consists of the two segments P_5P_2 and P_3P_4 of Γ, and the two vertical lines P_4P_5 and P_2P_3. Thus the limit solution as $\epsilon \to 0$ satisfies $f(u', u) = 0$ except at certain points where $v = u'$ has jump discontinuities.

2.2.4. UNSYMMETRICAL BENDING OF PRESTRESSED ANNULAR PLATES

The last example in this class is the unsymmetrical bending of a prestressed annular plate introduced by Alzheimer and Davis (1968). The plate

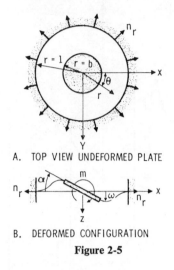

A. TOP VIEW UNDEFORMED PLATE

B. DEFORMED CONFIGURATION

Figure 2-5

is clamped at the outer edge and contains a central rigid inclusion as shown in Figure 2-5. A moment is applied to the rigid inclusion which rotates about a diameter, out of the plane of the plate. For a thin annular plate with in-plane forces and no surface loading, Timoshenko and Woinowsky-Krieger (1959) derived the following equation for the lateral displacement ω

$$\nabla^4 \omega = \frac{r_1^2}{D}\left[n_r \frac{\partial^2 \omega}{\partial r^2} + n_\theta\left(\frac{1}{r}\frac{\partial \omega}{\partial r} + \frac{1}{r^2}\frac{\partial^2 \omega}{\partial \theta^2}\right) + 2n_{r\theta}\frac{\partial}{\partial r}\left(\frac{1}{r}\frac{\partial \omega}{\partial r}\right) \right] \quad (2.2.23)$$

where n_r, n_θ, and $n_{r\theta}$ are the in-plane forces per unit length. In (2.2.23) the radial distance is made dimensionless with respect to the outer radius r_1, and the lateral displacement is made dimensionless with respect to the plate thickness h. The flexural rigidity $D = Eh^3/12(1 - \nu^2)$, where E is Young's modulus and ν is Poisson's ratio.

We assume that the in-plane forces are applied in the form of an initial uniform radial prestressing, and that they are large enough to remain essentially constant during any subsequent lateral motion (i.e., $n_r = n_\theta = n =$ constant, $n_{r\theta} = 0$). Thus (2.2.23) reduces to

$$\epsilon^2 \nabla^4 \omega - \nabla^2 \omega = 0 \quad (2.2.24)$$

where

$$\epsilon^2 = \frac{D}{r_1^2 n}$$

Figure 2-5 yields the following boundary conditions

$$\omega = b\alpha \cos\theta, \qquad \frac{\partial\omega}{\partial r} = \alpha\cos\theta \quad \text{at} \quad r = b \qquad (2.2.25)$$

$$\omega = \frac{\partial\omega}{\partial r} = 0 \quad \text{at} \quad r = 1 \qquad (2.2.26)$$

where α is assumed to be small so that $\sin\alpha \approx \alpha$, and $b = r_2/r_1$ with r_2 the radius of the rigid inclusion.

The boundary conditions (2.2.25) and (2.2.26) suggest a solution of the form

$$\omega = u(r)\cos\theta \qquad (2.2.27)$$

Hence

$$\epsilon^2\left(\frac{d^2}{dr^2} + \frac{1}{r}\frac{d}{dr} - \frac{1}{r^2}\right)\left(\frac{d^2u}{dr^2} + \frac{1}{r}\frac{du}{dr} - \frac{u}{r^2}\right) - \left(\frac{d^2u}{dr^2} + \frac{1}{r}\frac{du}{dr} - \frac{u}{r^2}\right) = 0 \quad (2.2.28)$$

$$u(b) = b\alpha, \qquad \frac{du}{dr}(b) = \alpha \qquad (2.2.29)$$

$$u(1) = 0, \qquad \frac{du}{dr}(1) = 0 \qquad (2.2.30)$$

As $\epsilon \to 0$, (2.2.28) reduces to

$$\frac{d^2u}{dr^2} + \frac{1}{r}\frac{du}{dr} - \frac{u}{r^2} = 0 \qquad (2.2.31)$$

which is of the second order, and its solution cannot satisfy the four boundary conditions (2.2.29) and (2.2.30). Hence two boundary conditions have to be dropped. If we seek a straightforward expansion of the form

$$u = \sum_{n=0}^{\infty} \epsilon^n u_n(r) \qquad (2.2.32)$$

we find that each u_n satisfies (2.2.31). Consequently, (2.2.32) is not valid for all r in $[b, 1]$. A uniformly valid expansion is obtained in Section 4.1.5 by using the method of matched asymptotic expansions.

2.3. Type Change of a Partial Differential Equation

Since the boundary and initial conditions required to make a well-posed problem depend on the type of the partial differential equation, difficulties might arise if the original equation changes, say, from elliptic to parabolic or hyperbolic as the small parameter vanishes. This class can be considered a subclass of that discussed in Sections 2.2.1 through 2.2.4. In the following discussion we describe two examples and the difficulties encountered in the expansion of one of them.

2.3.1. A SIMPLE EXAMPLE

Let us consider the following Dirichlet problem for the function $\phi(x, y, \epsilon)$

$$\phi_{xx} + \epsilon\phi_{yy} - \phi_y = 0, \qquad 0 \leq x, y \leq 1 \qquad (2.3.1)$$

$$\phi(0, y) = a(y) \qquad (2.3.2)$$

$$\phi(x, 0) = b(x) \qquad (2.3.3)$$

$$\phi(1, y) = c(y) \qquad (2.3.4)$$

$$\phi(x, 1) = d(x) \qquad (2.3.5)$$

For $\epsilon > 0$ the above problem is a well-posed problem and admits a unique solution. However, if $\epsilon = 0$, (2.3.1) reduces to

$$\phi_{xx} - \phi_y = 0 \qquad (2.3.6)$$

which is parabolic (diffusion equation). The solution of (2.3.6) cannot in general satisfy all the boundary conditions (2.3.2) through (2.3.5) and one of them must be dropped. As discussed in Section 4.1.2, (2.3.5) must be dropped, and thus the resulting solution is not valid near $y = 1$. For small ϵ the solution of the reduced equation is expected to be close to the exact solution except in a narrow region near $y = 1$ where the latter changes rapidly so that it satisfies the boundary condition which is about to be lost.

It should be noted that the singular nature of the problem depends not only on the change in type of equation but also on the given region in which the solution is obtained. Although the solution of (2.3.1) in the region $0 \leq x, y \leq 1$ does not tend to that of (2.3.6) uniformly, the solution of (2.3.1) tends uniformly to the solution of (2.3.6) in the upper half-plane.

Next we discuss an example in which the change in type of equation does not lead to nonuniformities.

2.3.2. LONG WAVES ON LIQUIDS FLOWING DOWN INCLINED PLANES

In this section we consider the characteristics of waves on the surface of a liquid film flowing down an inclined plane (Figure 2-6). This is a rather involved example, and it is discussed here because it illustrates a general technique for long nonlinear dispersive waves. The flow is governed by the

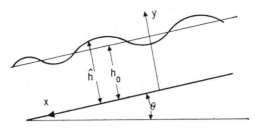

Figure 2-6

Navier–Stokes equations

$$\frac{\partial \hat{u}}{\partial \hat{x}} + \frac{\partial \hat{v}}{\partial \hat{y}} = 0 \qquad (2.3.7)$$

$$\frac{\partial \hat{u}}{\partial \hat{t}} + \hat{u}\frac{\partial \hat{u}}{\partial \hat{x}} + \hat{v}\frac{\partial \hat{u}}{\partial \hat{y}} = -\frac{1}{\rho}\frac{\partial \hat{p}}{\partial \hat{x}} + g\sin\theta + \nu\,\nabla^2\hat{u} \qquad (2.3.8)$$

$$\frac{\partial \hat{v}}{\partial \hat{t}} + \hat{u}\frac{\partial \hat{v}}{\partial \hat{x}} + \hat{v}\frac{\partial \hat{v}}{\partial \hat{y}} = -\frac{1}{\rho}\frac{\partial \hat{p}}{\partial \hat{y}} - g\cos\theta + \nu\,\nabla^2\hat{v} \qquad (2.3.9)$$

where $\hat{u}$ and $\hat{v}$ are the components of the velocity in the x and y directions, $\hat{p}$ is liquid pressure, t is time, and ρ and ν are the liquid's density and kinematic viscosity, respectively. At the liquid/solid interface, both components of velocity vanish; that is

$$\hat{u} = \hat{v} = 0 \quad \text{at} \quad \hat{y} = 0 \qquad (2.3.10)$$

If the liquid surface is flat (i.e., no waves), the following laminar steady solution exists

$$\hat{U} = \frac{g\sin\theta}{2\nu}(2h_0\hat{y} - \hat{y}^2), \qquad \hat{V} = 0$$

$$\hat{P} = \hat{p}_0 - \rho g\cos\theta(\hat{y} - h_0) \qquad (2.3.11)$$

In this solution we used the boundary condition $\partial\hat{u}/\partial\hat{y} = 0$ at $\hat{y} = h_0$ (i.e., no shear).

We next consider fluctuations in this steady-flow configuration. We introduce dimensionless quantities according to

$$\alpha = h_0/l, \qquad h = \hat{h}/h_0, \qquad y = \hat{y}/h_0, \qquad x = \hat{x}/l$$

$$U = \hat{U}/U_L, \qquad u + U = \hat{u}/U_L, \qquad v = \hat{v}/\alpha U_L \qquad (2.3.12)$$

$$t = \hat{t}U_L/l, \qquad P = \hat{P}/\rho g h_0\sin\theta, \qquad p + P = \hat{p}/\rho g h_0\sin\theta$$

where $U_L = gh_0^2 \sin \theta / 2\nu$, l is a characteristic length of the waves, and α is a dimensionless quantity measuring the shallowness of the liquid. Substituting (2.3.12) into (2.3.7) through (2.3.9) and using (2.3.11), we obtain the following equations for the dimensionless fluctuations

$$u_x + v_y = 0 \tag{2.3.13}$$

$$u_t + (u + U)u_x + v(u_y + U') = -\frac{2p_x}{R} + \frac{1}{\alpha R}(u_{yy} + \alpha^2 u_{xx}) \tag{2.3.14}$$

$$v_t + (u + U)v_x + vv_y = -\frac{2}{R\alpha^2}p_y + \frac{1}{\alpha R}(v_{yy} + \alpha^2 v_{xx}) \tag{2.3.15}$$

where $R = U_L h_0/\nu$ is a Reynolds number for the liquid film, $U = 2y - y^2$, and primes denote differentiation with respect to y.

Equation (2.3.13) can be solved by introducing a stream function $\psi(x, y, t)$ such that

$$(u, v) = (\psi_y, -\psi_x)$$

Then (2.3.14) and (2.3.15) can be combined into

$$\psi_{yyyy} = \alpha R[\psi_{yyt} + (U + \psi_y)\psi_{xyy} - (U'' + \psi_{yyy})\psi_x] - 2\alpha^2\psi_{xxyy}$$
$$+ \alpha^3 R[\psi_{xxt} + (U + \psi_y)\psi_{xxx} - \psi_x\psi_{xxy}] - \alpha^4\psi_{xxxx} \tag{2.3.16}$$

Equation (2.3.16) must be supplemented by boundary conditions. At the solid/liquid interface, (2.3.10) gives

$$\psi_y(x, 0, t) = \psi(x, 0, t) = 0 \tag{2.3.17}$$

At the free surface the normal velocity of the liquid is equal to the velocity of the interface; that is

$$h_t + (U + \psi_y)h_x + \psi_x = 0 \quad \text{at} \quad y = h(x) \tag{2.3.18}$$

Moreover, the condition that the tangential stress vanishes at this free surface gives

$$(U' + \psi_{yy} - \alpha^2\psi_{xx})(1 - \alpha^2 h_x^2) - 4\alpha^2\psi_{xy}h_x = 0 \quad \text{at} \quad y = h \tag{2.3.19}$$

Finally, the continuity of the normal stress across this free surface demands that

$$-p + (h - 1)\cot\theta - \frac{Th_{xx}\csc\theta}{(1 + \alpha^2 h_x^2)^{3/2}} - 2\alpha\frac{(\psi_{yy} - \alpha^2\psi_{xx})h_x}{1 + \alpha^2 h_x^2}$$
$$- 2\alpha\psi_{xy}\frac{1 - \alpha^2 h_x^2}{1 + \alpha^2 h_x^2} = C \quad \text{at} \quad y = h \tag{2.3.20}$$

where from (2.3.14)

$$p_x = \tfrac{1}{2}\alpha^{-1}\psi_{yyy} - \tfrac{1}{2}R[\psi_{yt} + (U + \psi_y)\psi_{xy} - (U' + \psi_{yy})\psi_x] + \tfrac{1}{2}\alpha\psi_{xxy} \quad (2.3.21)$$

and

$$T = \frac{\sigma}{\rho g l^2}$$

with σ the liquid surface tension.

To determine an equation describing the dimensionless elevation of the disturbed surface, we follow Benney (1966a) by first finding a perturbation solution for (2.3.16), (2.3.17), and (2.3.19) through (2.3.21) in powers of α and substituting the expansion for ψ into (2.3.18). Thus we let

$$\psi = \psi_0 + \alpha\psi_1 + \cdots$$
$$p = \alpha^{-1}p_{-1} + p_0 + \alpha p_1 \quad (2.3.22)$$

in the former equations and equate coefficients of like powers of α to obtain

Order α^0

$$\psi_{0yyyy} = 0 \quad (2.3.23)$$

$$\psi_0 = \psi_{0y} = 0 \quad \text{at} \quad y = 0 \quad (2.3.24)$$

$$\psi_{0yy} = 2(y - 1) \quad \text{at} \quad y = h \quad (2.3.25)$$

$$\psi_{0yyy} = 0 \quad \text{at} \quad y = h \quad (2.3.26)$$

Order α

$$\psi_{1yyyy} = R[\psi_{0yyt} + (U + \psi_{0y})\psi_{0xyy} - (U'' + \psi_{0yyy})\psi_{0x}] \quad (2.3.27)$$

$$\psi_1 = \psi_{1y} = 0 \quad \text{at} \quad y = 0 \quad (2.3.28)$$

$$\psi_{1yy} = 0 \quad \text{at} \quad y = h \quad (2.3.29)$$

$$p_0 = (h - 1)\cot\theta - T\operatorname{cosec}\theta\, h_{xx} \quad \text{at} \quad y = h \quad (2.3.30)$$

where

$$p_{0x} = \tfrac{1}{2}\psi_{1yyy} - \tfrac{1}{2}R[\psi_{0yt} + (U + \psi_{0y})\psi_{0xy} - (U' + \psi_{0yy})\psi_{0x}] \quad (2.3.31)$$

The solution of (2.3.23) through (2.3.26) is

$$\psi_0 = (h - 1)y^2 \quad (2.3.32)$$

Then (2.3.27) becomes

$$\psi_{1yyyy} = 2R(h_t + 2hh_x y)$$

whose solution subject to (2.3.28) is

$$\psi_1 = A(x, t)y^2 + B(x, t)y^3 + \frac{R}{12}(h_t y^4 + \tfrac{2}{5}hh_x y^5) \quad (2.3.33)$$

Substituting for ψ_1 into (2.3.29) through (2.3.31) and solving the resulting equations for A and B, we obtain

$$B = \tfrac{1}{3}(h_x \cot \theta - T \csc \theta \, h_{xxx})$$
$$A = -h(h_x \cot \theta - T \csc \theta \, h_{xxx}) - \tfrac{1}{2}Rh^2(h_t + \tfrac{2}{3}h^2h_x) \qquad (2.3.34)$$

Since

$$\frac{\partial}{\partial x}\left[\psi(x, y, t)\big|_{y=h}\right] = (\psi_y h_x + \psi_x)\big|_{y=h}$$

we can rewrite (2.3.18) as

$$h_t + (2h - h^2)h_x + \frac{\partial}{\partial x}\left[\psi(x, y, t)\big|_{y=h}\right] = 0 \qquad (2.3.35)$$

Substituting $\psi = \psi_0 + O(\alpha)$ into this equation, we have

$$h_t + 2h^2h_x = 0 \qquad (2.3.36)$$

Then

$$\psi_1(x, h, t) = -\tfrac{2}{3}h^3(h_x \cot \theta - T \csc \theta \, h_{xxx}) + \tfrac{8}{15}Rh^6h_x$$

Letting $\psi = \psi_0 + \alpha\psi_1 + O(\alpha^2)$ in (2.3.35), we obtain

$$h_t + 2h^2h_x + \alpha[-\tfrac{2}{3}h^3(\cot \theta - \tfrac{4}{5}Rh^3)h_{xx} + \tfrac{2}{3}T \csc \theta \, h^3h_{xxxx}$$
$$- 2h^2(\cot \theta - \tfrac{8}{5}Rh^3)h_x{}^2 + 2T \csc \theta \, h^2h_xh_{xxx}] + O(\alpha^2) = 0 \qquad (2.3.37)$$

Let us summarize what we have accomplished in this section. We started with the elliptic equation (2.3.16) which was replaced by (2.3.23) and (2.3.27) which are clearly not elliptic. From these perturbation equations we arrived at (2.3.37) which is clearly hyperbolic. Thus nonuniformities due to the type change of equation did not arise because of the unboundedness of the domain.

2.4. The Presence of Singularities

In this class the expansions have singularities within the region of interest, which are not part of the exact solution. Moreover, in the higher-order terms, the singularities are not only preserved, but they become even more pronounced.

2.4.1. SHIFT IN SINGULARITY

As a first example in this class, we consider the problem discussed by Lighthill (1949a)

$$(x + \epsilon y)\frac{dy}{dx} + (2 + x)y = 0 \quad \text{with} \quad y(1) = e^{-1} \qquad (2.4.1)$$

This equation is singular along the line $x = -\epsilon y$; the boundary condition makes the exact solution $y(x)$ positive for $x \geq 0$, hence $y(x)$ is regular in $0 \leq x < \infty$.

To determine a straightforward expansion, we let

$$y = y_0(x) + \epsilon y_1(x) + \cdots \tag{2.4.2}$$

Substituting (2.4.2) into (2.4.1), expanding, and equating coefficients of ϵ^0 and ϵ to zero lead to

$$x\frac{dy_0}{dx} + (2 + x)y_0 = 0, \qquad y_0(1) = e^{-1} \tag{2.4.3}$$

$$x\frac{dy_1}{dx} + (2 + x)y_1 = -y_0\frac{dy_0}{dx}, \qquad y_1(1) = 0 \tag{2.4.4}$$

The solution of the zeroth-order problem is

$$y_0 = x^{-2}e^{-x} \tag{2.4.5}$$

Substituting for y_0 into (2.4.4) and solving the resultant equation, we have

$$y_1 = x^{-2}e^{-x}\int_1^x e^{-t}t^{-3}(1 + 2t^{-1})\,dt \tag{2.4.6}$$

As $x \to 0$, $y_0 = 0(x^{-2})$, while $y_1 = 0(x^{-5})$. Thus although the exact solution is regular at $x = 0$, the zeroth-order solution is singular at $x = 0$, and the singularity grows stronger.

2.4.2. THE EARTH-MOON-SPACESHIP PROBLEM

We consider next the motion of a spaceship of mass m which is moving in the gravitational field of two fixed mass-centers. The mass M_e of the earth is much larger than the mass M_m of the moon. With respect to the rectangular Cartesian coordinate system shown in Figure 2-7, the equations of motion in dimensionless form are

$$\frac{d^2x}{dt^2} = -(1 - \mu)\frac{x}{r_e^3} - \mu\frac{x - 1}{r_m^3} \tag{2.4.7}$$

$$\frac{d^2y}{dt^2} = -(1 - \mu)\frac{y}{r_e^3} - \mu\frac{y}{r_m^3} \tag{2.4.8}$$

$$r_e^2 = x^2 + y^2, \qquad r_m^2 = (x - 1)^2 + y^2 \tag{2.4.9}$$

where

$$\mu = M_m/(M_m + M_e)$$

Distances and time were made dimensionless using, respectively, the distance

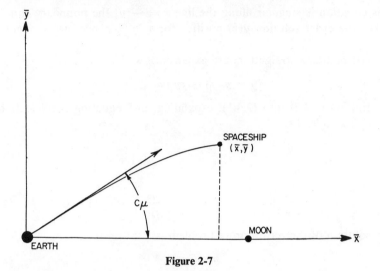

Figure 2-7

d between the mass centers, and

$$\left[\frac{d^3}{G(M_m + M_e)}\right]^{1/2}$$

where G is the universal gravitational constant. This problem has been studied by Lagerstrom and Kevorkian (1963b) for the initial conditions

$$x = 0, \qquad y = 0, \qquad \frac{dy}{dx} = -\mu c \quad \text{at} \quad t = 0 \qquad (2.4.10)$$

$$h = -\rho^2, \qquad \rho \neq 1 \qquad (2.4.11)$$

where h is the total energy of the spaceship.

We interchange the roles of x and t and assume the following straightforward expansions for small μ

$$t = t_0(x) + \mu t_1(x) + \cdots \qquad (2.4.12)$$

$$y = \mu y_1 + \cdots \qquad (2.4.13)$$

Substituting (2.4.12) and (2.4.13) into (2.4.7) through (2.4.9) and equating coefficients of equal powers of μ lead to

$$\frac{t_0''}{t_0'^3} = \frac{1}{x^2} \qquad (2.4.14)$$

$$-\frac{t_1''}{t_0'^3} + 3\frac{t_0''t_1'}{t_0'^4} = \frac{1}{x^2} + \frac{1}{(1-x)^2} \qquad (2.4.15)$$

$$\frac{y_1''}{t_0'^2} - \frac{t_0''}{t_0'^3}y_1' + \frac{y_1}{x^2} = 0 \qquad (2.4.16)$$

The solutions of these equations subject to the initial conditions (2.4.10) and (2.4.11) are

$$\sqrt{2}\, t_0 = \frac{1}{\rho^3} \sin^{-1} \rho\sqrt{x} - \frac{1}{\rho^2} \sqrt{x(1 - \rho^2 x)} \qquad (2.4.17)$$

$$\sqrt{2}\, t_1 = -\frac{2}{\rho^3} \sin^{-1} \rho\sqrt{x} + \frac{2 - \rho^2}{\rho^2(1 - \rho^2)} \sqrt{\frac{x}{1 - \rho^2 x}} - \frac{1}{2(1 - \rho^2)^{3/2}}$$

$$\times \ln \frac{1 + (1 - 2\rho^2)x + 2[(1 - \rho^2)(1 - \rho^2 x)x]^{1/2}}{1 - x} \qquad (2.4.18)$$

$$y_1 = -cx \qquad (2.4.19)$$

Thus the above expansion breaks down as $x \to 1$ because t_1 has a logarithmic singularity. We find that the higher approximations are increasingly more singular near $x = 1$. In fact

$$t_2 = O[(1 - x)^{-1}] \quad \text{as} \quad x \to 1 \qquad (2.4.20)$$

2.4.3. THERMOELASTIC SURFACE WAVES

We consider next the effect of heat conduction on waves propagating along the free surface of an isotropic elastic half-space. Maxwell's modified heat conduction law is used rather than that of Fourier in order to take into account the short time required to establish a steady-state heat conduction when a temperature gradient is suddenly produced in the solid. Thus the heat flux $\mathbf{h}$ is assumed to be given by

$$\tau_0 \frac{\partial \mathbf{h}}{\partial t} + \mathbf{h} = -k \,\mathbf{grad}\, \theta \qquad (2.4.21)$$

where θ denotes the change in the absolute basic temperature θ_0, k is the coefficient of thermal conductivity, and τ_0 is the thermal relaxation time. Equation (2.4.21) implies that thermal signals have finite rather than infinite speed of propagation. Our discussion here follows that of Nayfeh and Nemat–Nasser (1971).

Since the material is assumed to be isotropic, we consider two-dimensional motions in the x-y plane, and we denote the corresponding displacements by u and v, respectively. The x axis is taken to coincide with the free surface and the y axis normal to it and directed toward the body. If $\beta^2 = (\lambda + 2\mu)/\mu$ with λ and μ the Lamé coefficients of elasticity, $b = [2 + (3\lambda/\mu)]\alpha\theta_0$ with α the coefficient of linear thermal expansion, and $g = \alpha(3\lambda + 2\mu)/\rho c_v$ with ρ the material density and c_v the specific heat at constant volume, then the coupled elastic and heat conduction equations can be written as

$$\beta^2 \frac{\partial^2 u}{\partial t^2} = \beta^2 \frac{\partial^2 u}{\partial x^2} + (\beta^2 - 1)\frac{\partial^2 v}{\partial x \, \partial y} + \frac{\partial^2 u}{\partial y^2} - b\frac{\partial \theta}{\partial x} \qquad (2.4.22)$$

$$\beta^2 \frac{\partial^2 v}{\partial t^2} = \beta^2 \frac{\partial^2 v}{\partial y^2} + (\beta^2 - 1)\frac{\partial^2 u}{\partial x \, \partial y} + \frac{\partial^2 v}{\partial x^2} - b\frac{\partial \theta}{\partial y} \qquad (2.4.23)$$

$$\frac{\partial^2 \theta}{\partial x^2} + \frac{\partial^2 \theta}{\partial y^2} - \frac{\partial \theta}{\partial t} - \tau\frac{\partial^2 \theta}{\partial t^2} = g\left(\frac{\partial^2 u}{\partial x \, \partial t} + \frac{\partial^2 v}{\partial y \, \partial t}\right) + g\tau\left(\frac{\partial^3 u}{\partial x \, \partial t^2} + \frac{\partial^3 v}{\partial y \, \partial t^2}\right) \qquad (2.4.24)$$

In the above equations time was made dimensionless by using $1/\omega^*$ and lengths were made dimensionless by using v_p/ω^* where

$$\omega^* = \frac{\rho c_v v_p^2}{k}, \qquad v_p^2 = \frac{\lambda + 2\mu}{\rho} \qquad (2.4.25)$$

At the free surface the normal and tangential stresses as well as the temperature gradient vanish; that is, at $y = 0$

$$\frac{\partial u}{\partial y} + \frac{\partial v}{\partial x} = 0 \qquad (2.4.26)$$

$$(\beta^2 - 2)\frac{\partial u}{\partial x} + \beta^2 \frac{\partial v}{\partial y} - b\theta = 0 \qquad (2.4.27)$$

$$\frac{\partial \theta}{\partial y} = 0 \qquad (2.4.28)$$

The solution of (2.4.22) through (2.4.28) is assumed to be of the form

$$(u, v, \theta) = (a_1, a_2, a_3)\exp(-\alpha y + i\omega t + iqx) \qquad (2.4.29)$$

where the real part of α is positive. The wave speed is then given by $c = \omega/\text{Real}(q)$ and the attenuation constant by $s = \text{Imaginary}(q)$. Substituting (2.4.29) into (2.4.22) through (2.4.25) and setting the determinant of the resulting linear system of equations for a_k to zero lead to the following three solutions for α in terms of q and ω

$$\alpha_1^2 = q^2 - \beta^2\omega^2 \qquad (2.4.30)$$

$$\alpha_2^2 + \alpha_3^2 = 2q^2 - \tau\omega^2 - \omega^2(1 + \tau\epsilon) + i\omega(1 + \epsilon) \qquad (2.4.31)$$

$$\alpha_2^2 \alpha_3^2 = \omega^4\tau - \omega^2 q^2 - i\omega^3 + i\omega q^2(1 + \epsilon) + q^4 - q^2\omega^2\tau(1 + \epsilon) \qquad (2.4.32)$$

where $\epsilon = bg/\beta^2$. For each α_k there exists an eigenvector which is given by

$$\left[1, i\frac{\alpha_k}{q}, i\frac{\beta^2(q^2 - \omega^2 - \alpha_k^2)}{bq}\right]a_{1k} \quad \text{for} \quad k = 2 \text{ and } 3 \qquad (2.4.33)$$

and

$$\left(1,\ i\ \frac{\alpha_1 q}{q^2 - \beta^2 \omega^2},\ 0\right) a_{11} \quad \text{for} \quad k = 1 \qquad (2.4.34)$$

Substituting the eigenfunctions corresponding to the eigenvectors (2.4.33) and (2.4.34) into (2.4.26) through (2.4.28) and setting the determinant of the set of equations for a_{1k} to zero lead to, as $\omega \to \infty$

$$G(c^2) = \frac{A^2[2 - c^2 - c^2\tau(1 + \epsilon) + 2A]}{[1 - c^2\tau(1 + \epsilon) + A]^2} \qquad (2.4.35)$$

where

$$A^2 = (1 - c^2)(1 - c^2\tau) - c^2\tau\epsilon, \qquad G(c^2) = \frac{(1 - \frac{1}{2}\beta^2 c^2)^4}{1 - \beta^2 c^2} \qquad (2.4.36)$$

The classic Rayleigh wave speed can be obtained from (2.4.35) and (2.4.36) by letting $\epsilon = 0$; if $\beta^2 = 3$ (corresponding to a Poisson's ratio of $1/4$), $c^2 = 0.2817$. For small ϵ we may attempt to determine an expansion for c^2 whose first term is the Rayleigh solution. Thus we let

$$c^2 = c_R^2(1 + \epsilon c_1 + \epsilon^2 c_2 + \cdots) \qquad (2.4.37)$$

Substituting (2.4.37) into (2.4.35) and (2.4.36), expanding, and equating coefficients of equal powers of ϵ on both sides, we obtain

$$c_1 = -\frac{\tau c_R^2\left[1 - \dfrac{1 - c_R^2}{F}\right]}{(1 - \tau c_R^2)\left[1 + \dfrac{dG}{dc^2}(c_R^2)\right]} \qquad (2.4.38)$$

where

$$F = 2 - c_R^2(1 + \tau) + 2\sqrt{(1 - c_R^2)(1 - \tau c_R^2)} \qquad (2.4.39)$$

Equation (2.4.38) shows that the expansion (2.4.37) breaks down as τ increases toward the value c_R^{-2} where c_1 becomes unbounded. Higher approximation can be shown to be more singular than the second terms at $\tau = c_R^{-2}$.

If $\tau > c_R^{-2}$ the expansion (2.4.37) must be modified; otherwise A^2 is negative, hence from (2.4.32) a real part of one of the α terms becomes negative. The modified expansion is of the form

$$c^2 = \frac{1}{\tau}(1 + \epsilon c_1 + \epsilon^2 c_2 + \cdots) \qquad (2.4.40)$$

The functions c_1 and c_2 can be determined by substituting (2.4.40) into (2.4.35) and equating coefficients of ϵ and ϵ^2 on both sides. The result is

$$c_1 = -\frac{\tau}{\tau - 1}, \qquad c_2 = -\frac{\tau^3}{(1 - \tau)^3} - \frac{\tau}{\tau - 1} M^2 \qquad (2.4.41)$$

where M is a root of

$$\left[G^{-1}(\tau^{-1}) + \frac{\tau}{1 - \tau}\right]M^2 - \frac{2\tau^2}{(1 - \tau)^2} M + \frac{\tau}{(1 - \tau)^2} = 0 \quad (2.4.42)$$

The coefficient of M^2 vanishes at $\tau = c_R^{-2}$, hence the expansion (2.4.40) is singular at this point. It can also be shown that this singularity is compounded in the higher approximation. Thus the first of the above expansions is valid for $\tau < c_R^{-2}$ and the second is valid for $\tau > c_R^{-2}$; both break down near $\tau = c_R^{-2}$. An expansion valid near this singularity is determined in Section 4.1.6 by using the method of matched asymptotic expansions.

2.4.4. TURNING POINT PROBLEMS

As a last example, we consider the asymptotic expansions of the solutions of

$$y'' + \lambda^2(1 - x^2)y = 0 \quad (2.4.43)$$

for large λ. For $|x| < 1$, the solutions of this equation are oscillatory, while for $|x| > 1$ they behave as exponentials. This suggests an expansion of the form

$$y = e^{\lambda\phi(x;\lambda)} \quad (2.4.44)$$

where

$$\phi = \phi_0(x) + \lambda^{-1}\phi_1(x) + \cdots \quad (2.4.45)$$

Substituting this expansion into (2.4.43) and equating coefficients of like powers of λ, we obtain

$$\phi_0'^2 = -(1 - x^2) \quad (2.4.46)$$

$$2\phi_0'\phi_1' + \phi_0'' = 0 \quad (2.4.47)$$

The solutions of these equations are

$$\phi_0 = \begin{cases} \pm i \int^x \sqrt{1 - \tau^2}\, d\tau & \text{for} \quad |x| < 1 \\[3mm] \pm \int^x \sqrt{\tau^2 - 1}\, d\tau & \text{for} \quad |x| > 1 \end{cases} \quad (2.4.48)$$

$$\phi_1 = -\tfrac{1}{2}\ln \phi_0' + \text{a constant} \quad (2.4.49)$$

Hence

$$y = \frac{1}{(1 - x^2)^{1/4}}\left[a_1 \cos\left(\lambda \int^x \sqrt{1 - \tau^2}\, d\tau\right) + b_1 \sin\left(\lambda \int^x \sqrt{1 - \tau^2}\, d\tau\right)\right]$$

$$\text{for} \quad |x| < 1 \quad (2.4.50)$$

and

$$y = \frac{1}{(x^2 - 1)^{1/4}} \left[a_2 \exp\left(\lambda \int^x \sqrt{\tau^2 - 1}\, d\tau\right) + b_2 \exp\left(-\lambda \int^x \sqrt{\tau^2 - 1}\, d\tau\right) \right]$$

$$\text{for } |x| > 1 \quad (2.4.51)$$

where a_i and b_i are constants.

The expansions (2.4.50) and (2.4.51) are called the Liouville–Green or WKB approximation (Section 7.1.3). These expansions are singular at $x = \pm 1$, hence they are not uniformly valid. The points $x = \pm 1$ are called *turning points*. The nonuniformity in this example arose because of the representation of the solutions in terms of elementary functions (namely, exponential and circular functions). The expansions show that the behavior of the solution changes across $|x| = 1$ from oscillatory to exponential, hence we need to represent the solutions by functions that exhibit this qualitative behavior. In this case the appropriate functions are the Airy functions (Section 7.3.1).

2.5. The Role of Coordinate Systems

In obtaining a parameter perturbation for a quantity such as $u(x; \epsilon)$, we first choose an independent variable which need not be the physical independent variable x but a function ζ of x and the small parameter ϵ. Then we assume

$$u = \sum_{m=0}^{\infty} \delta_m(\epsilon) u_m[\zeta(x; \epsilon)] \quad \text{as} \quad \epsilon \to 0 \quad (2.5.1)$$

where $\delta_m(\epsilon)$ is an asymptotic sequence. We substitute this expansion into the governing equations, expand for small ϵ keeping now ζ fixed, and then equate the coefficient of each δ_m to zero. Thus

$$u_0 = \lim_{\substack{\epsilon \to 0 \\ \zeta \text{ fixed}}} \left[\frac{u}{\delta_0(\epsilon)} \right]$$

$$u_n = \lim_{\substack{\epsilon \to 0 \\ \zeta \text{ fixed}}} \left[\frac{u - \sum_{m=0}^{n-1} \delta_m(\epsilon) u_m(\zeta)}{\delta_n(\epsilon)} \right] \quad (2.5.2)$$

It is clear that, for a given sequence δ_n, u_m depends on the choice of $\zeta(x; \epsilon)$.

Some of these choices ζ lead to nonuniform expansions, whereas others lead to uniform expansions. For example, choosing $\zeta = t$ in (2.1.1), we obtain the nonuniform expansion

$$u(t; \epsilon) = a \cos t + \epsilon a^3 [-\tfrac{3}{8} t \sin t + \tfrac{1}{32}(\cos 3t - \cos t)] + O(\epsilon^2) \quad (2.5.3)$$

for Duffing's equation (Section 2.1.1). Had we chosen $\zeta = [1 + (3/8)\epsilon a^2]t$, we would have obtained

$$u(t; \epsilon) = a \cos \zeta + \frac{\epsilon a^3}{32}\left[\cos 3\zeta - \cos \zeta\right] + O(\epsilon^2) \tag{2.5.4}$$

which is uniformly valid. Coordinates such as $\zeta = [1 + (3/8)\epsilon a^2]t$, which lead to uniform expansions, are called *optimal coordinates* (Kaplun, 1954).

As a second example, we consider the model for the weak nonlinear instability problem of Section 2.1.2. Equation (2.1.18) gives the following expansion

$$u = \epsilon \cos \sigma_1 t \cos kx$$

$$+ \epsilon^3\left[\frac{9}{32\sigma_1} t \sin \sigma_1 t \cos kx + \text{terms bounded as } t \to \infty\right] \tag{2.5.5}$$

This nonuniform expansion corresponds to the choice $\zeta = t$. Had we chosen $\zeta = \sigma_1[1 - (9/32\sigma_1^2)]t$, we would have obtained

$$u = \epsilon \cos \zeta \cos kx + O(\epsilon^3) \tag{2.5.6}$$

which is uniformly valid for all t. Hence this latter ζ is an optimal coordinate.

As a third example, we consider supersonic flow past a thin airfoil (Section 2.1.3). Equation (2.1.36) gives the following expansion for the axial velocity component

$$\frac{u}{U} = 1 - \epsilon \frac{T'}{B} + \epsilon^2\left[\frac{1}{B^2}\left(1 - \frac{M^4(\gamma + 1)}{4B^2}\right)T'^2\right.$$

$$\left. - \frac{\gamma + 1}{2}\frac{M^4}{B^3} yT'T'' - TT''\right] + O(\epsilon^3) \tag{2.5.7}$$

where $T = T(x - By)$; $y = \epsilon T(x)$ (airfoil shape).

This nonuniform expansion was obtained holding x and y fixed. Had we chosen

$$\xi = y \quad \text{and} \quad x - By = \zeta - \epsilon \frac{\gamma + 1}{2}\frac{M^4}{B^2} \xi T'(\zeta) \tag{2.5.8}$$

to be fixed, we would have obtained

$$\frac{u}{U} = 1 - \epsilon \frac{T'(\zeta)}{B} + \epsilon^2\left[\frac{1}{B^2}\left(1 - \frac{M^4(\gamma + 1)}{4B^2}\right)T'^2 - TT''\right] + O(\epsilon^3) \tag{2.5.9}$$

which is uniformly valid. Hence (2.5.8) are optimal coordinates.

It should be mentioned that a coordinate may be optimal for $O(\epsilon)$ but not optimal for $O(\epsilon^2)$. For example

$$\zeta = (1 + \tfrac{3}{8}\epsilon a^2)t \tag{2.5.10}$$

is an optimal coordinate for $O(\epsilon)$ for the Duffing equation (2.1.1), whereas it is not optimal for any higher order. However,

$$\zeta = (1 + \tfrac{3}{8}\epsilon a^2 - \tfrac{15}{256}\epsilon^2 a^4)t \tag{2.5.11}$$

is optimal for $O(\epsilon^2)$ for (2.1.1).

Since most straightforward perturbation expansions (obtained by holding the physical coordinates fixed) are nonuniform, perturbation techniques have been devised to render these expansions uniformly valid. In the method of strained coordinates (Chapter 3), some of these expansions are rendered uniformly valid by determination of the optimal coordinates as near-identity transformations.

In some of the problems considered, such as (2.2.1) and (2.2.2), a uniformly valid expansion was given by (2.2.11) as

$$y = b \exp(1 - x) + (a - be) \exp\left(x - \frac{x}{\epsilon}\right) + O(\epsilon) \tag{2.5.12}$$

This expansion could not be obtained by keeping x fixed or by keeping x/ϵ fixed. Had we kept x fixed, we would have obtained

$$y = b \exp(1 - x) + O(\epsilon) \tag{2.5.13}$$

which is not valid near $x = 0$ because $y(0) = a \neq be$ in general. However, had we kept x/ϵ fixed, we would have obtained

$$y = be + (a - be) \exp\left(-\frac{x}{\epsilon}\right) \tag{2.5.14}$$

which is not uniformly valid because $y(1) = b \neq be$. Thus the solution is represented by two different expansions using the coordinates (scales) x and x/ϵ. Since they are different asymptotic representations of the same function, they can be related to each other by what is called the matching principle (Chapter 4). This suggests that uniformly valid expansions can be obtained by determining different expansions using different scales, relating these expansions by matching, and then combining the expansions. This is the method of matched asymptotic expansions described in Chapter 4.

Rather than obtaining, say, two expansions using two different scales to represent the asymptotic solutions of (2.2.1), we keep both x and x/ϵ or some functions of them fixed in carrying out the expansions. This means that we increase the number of independent variables to two and transform our original ordinary differential equation into a partial differential equation. This is the method of multiple scales described in Chapter 6.

In oscillation problems such as

$$\ddot{u} + u = \epsilon f(u, \dot{u})$$

the unperturbed solution (i.e., $\epsilon = 0$) is

$$u = a \cos \phi, \qquad \phi = t + \theta \qquad (2.5.15)$$

where a and θ are constants. If $\epsilon \neq 0$ the solution can still be expressed in the above form provided that a and θ are taken to be time-dependent. The method of variation of parameters (Section 5.1.1) can be used to find the following equations for a and ϕ

$$\frac{da}{dt} = -\epsilon \sin \phi f[a \cos \phi, -a \sin \phi] \qquad (2.5.16)$$

$$\frac{d\phi}{dt} = 1 - \frac{\epsilon}{a} \cos \phi f[a \cos \phi, -a \sin \phi] \qquad (2.5.17)$$

To determine a uniformly valid expansion of the solution of these equations, we can introduce near-identity transformations for both dependent variables a and ϕ rather than a near-identity transformation for the independent variable as in the method of strained coordinates. This is the method of averaging described in Chapter 5.

Exercises

2.1. Determine a three-term expansion for the solution near unity of

$$(x - 1)(x - \tau) + \epsilon = 0, \qquad \epsilon \ll 1$$

Is it valid for all values of τ?

2.2. Calculate three terms in the asymptotic expansion of

$$\epsilon y' + xy = -1, \qquad y(0) = 1$$

What is its region of nonuniformity?

2.3. The problem of isoenergetic cylindrical shock waves can be reduced to (Levey, 1959)

$$\alpha w^2 g \frac{dg}{dw} = g(1 - w^2) - w(1 - \beta w^2)$$

where β and α are constants. Determine a second-order expansion for g in terms of α when $\alpha \ll 1$ and discuss its uniformity.

2.4. Determine a first-order (two-term) expansion for small ϵ for

$$\ddot{x} + x = \epsilon(\dot{x} - \tfrac{1}{3}\dot{x}^3), \qquad x(0) = a, \qquad \dot{x}(0) = 0$$

Is this expansion uniformly valid?

2.5. Consider the problem

$$(x + \epsilon y)y' + y = 0, \qquad y(1) = 1$$

(a) Determine a second-order (three-term) expansion for this problem assuming that $\epsilon \ll 1$.

(b) What is its region of nonuniformity?

(c) Show that the exact solution to this problem is

$$y = -\frac{x}{\epsilon} + \sqrt{\frac{x^2}{\epsilon^2} + \frac{2}{\epsilon} + 1}$$

(d) Expand this exact solution for small ϵ and compare the result with (a). Can you conclude anything about the source of nonuniformity?

2.6. Find a first-order (two-term) expansion for small ϵ of

$$(x + \epsilon y)y' - \tfrac{1}{2}y = 1 + x^2, \qquad y(1) = 1$$

What is its region of nonuniformity?

2.7. Determine a first-order expansion for small ϵ of

$$(x + \epsilon y)y' + xy = be^{-x}, \qquad y(1) = e^{-1}$$

What is its region of nonuniformity?

2.8. Determine a two-term expansion for the particular solution of

$$\epsilon u'' + (1 - x^2)u = f(x)$$

What conditions must be imposed on f for this expansion to be uniform?

2.9. Determine an expansion for large λ for the solution of

$$xy'' + y' + \lambda^2 x(1 - x^2)y = 0$$

in the form $y = \exp[\lambda\phi_1(x) + \phi_0(x) + \cdots]$. Where does this expansion break down?

2.10. Determine a two-term expansion of

$$\ddot{u} + \omega_0^2 u + k \cos \omega t = \epsilon u^2$$

Is this expansion valid for all values of ω?

2.11. Determine a second-order (three-term) expansion for small ϵ for the solution of

$$\ddot{u} + (\delta + \epsilon \cos 2t)u = 0$$

$$u(0) = a, \qquad \dot{u}(0) = 0$$

For what values of δ is this expansion nonuniform?

2.12. Show that a first-order expansion for small μ of

$$\frac{1}{2}\left(\frac{dx}{dt}\right)^2 = \frac{1 - \mu}{x} + \frac{\mu}{1 - x}, \qquad t(0) = 0$$

is

$$\sqrt{2}t = \tfrac{2}{3}x^{3/2} + \mu\left(\tfrac{2}{3}x^{3/2} + \sqrt{x} - \tfrac{1}{2}\ln\frac{1 + \sqrt{x}}{1 - \sqrt{x}}\right) + O[\mu^2(1 - x)^{-1}]$$

What is its region of nonuniformity?

2.13. For what values of $a > 0$ is the expansion

$$u = a \cos t - \frac{\epsilon}{8} [(a^3 - 4a)t \cos t + \tfrac{1}{4}a^3 \cos 3t] + \cdots$$

uniformly valid? Is there an $a > 0$ for which the expansion

$$u = a \cos t + \frac{\epsilon a^3}{8} (\tfrac{1}{4} \cos 3t - 3t \sin t) + \cdots$$

is uniformly valid.

2.14. Let $t = (1 + \epsilon\sigma)s$ in the second expansion in Exercise 2.13 and expand the result to $O(\epsilon)$ keeping s fixed. Can you choose σ to make this expansion uniformly valid?

2.15. Introduce the new variable

$$\zeta = (1 + \tfrac{3}{8}\epsilon a^2)t$$

in

$$\ddot{u} + u + \epsilon u^3 = 0$$

Determine a first-order expansion for the resulting problem. Is it uniformly valid? Can you conclude anything about the role of the independent variables in rendering the expansions uniform?

2.16. Consider the problem

$$\frac{d}{dx}\left(h^3 p \frac{dp}{dx}\right) = \Lambda \frac{d}{dx}(ph)$$

$$p(0) = p(1) = 1$$

where $h = h(x)$ is a known function. Determine an expansion for large Λ. Discuss the nonuniformity of this expansion. Calculate two terms if $p(1) = 1$ is dropped.

2.17. Consider the problem

$$\frac{d^2\beta}{d\xi^2} = \alpha^2 f \sin(\xi + \beta)$$

$$\frac{d^2 f}{d\xi^2} = \cos(\xi + \beta)$$

$$f'(0) = \beta(0) = f\left(\frac{\pi}{2}\right) = \beta\left(\frac{\pi}{2}\right) = 0$$

which arises in the bending of circular cylindrical tubes (Reissner and Weinitschke, 1963). Determine an expansion to $O(\alpha^2)$ for small α and discuss its uniformity.

2.18. The laminar flow through a channel with uniformly porous walls of different permeabilities can be reduced to (Proudman, 1960; Terrill and Shrestha, 1965)

$$f''' + R(ff'' - f'^2) = c$$

$$f(0) = 1, \quad f'(0) = 0$$

$$f(1) = 1 - \alpha, \quad f'(1) = 0$$

Show that for small α

$$f = 1 + \alpha A[2(e^{-Rx} + Rx - 1) - R(1 - e^{-R})x^2] + O(\alpha^2)$$

$$c = 2\alpha R^2 A(e^{-R} - 1) + O(\alpha^2)$$

and determine A.

2.19. Determine a first-order expansion for

$$u_{tt} - u_{xx} + u = \epsilon u^3, \qquad \epsilon \ll 1$$

$$u(x, 0) = a \cos kx, \qquad u_t(x, 0) = 0$$

and discuss its uniformity.

2.20. Determine a first-order straightforward expansion for small ϵ of

$$u_{tt} - c^2 u_{xx} = \epsilon u u_x$$

$$u(x, 0) = f(x) + g(x), \qquad u_t(x, 0) = c[g'(x) - f'(x)]$$

where $f(x)$ and $g(x)$ are bounded functions of x. Discuss its uniformity.

CHAPTER 3

The Method of Strained Coordinates

In this chapter we describe techniques of rendering the approximate solutions to some of the differential equations discussed in the previous chapters uniformly valid by introducing near-identity transformations of the independent variables. This technique goes back to the nineteenth century when astronomers, such as Lindstedt (1882), Bohlin (1889), and Gyldén (1893), devised techniques to avoid the appearance of secular terms in perturbation solutions of equations such as

$$\ddot{u} + \omega_0^2 u = \epsilon f(u, \dot{u}), \quad \epsilon \ll 1$$

The fundamental idea in Lindstedt's technique is based on the observation that the nonlinearities alter the frequency of the system from the linear one ω_0 to $\omega(\epsilon)$. To account for this change in frequency, he introduced a new variable $\tau = \omega t$ and expanded ω and u in powers of ϵ as

$$u = u_0(\tau) + \epsilon u_1(\tau) + \epsilon^2 u_2(\tau) + \cdots$$

$$\omega = \omega_0 + \epsilon \omega_1 + \epsilon^2 \omega_2 + \cdots$$

Then he chose the parameters ω_i, $i \geq 1$, to prevent the appearance of secular terms. Poincaré (1892) proved that the expansions obtained by Lindstedt's technique are asymptotic.

Various forms of this idea have been utilized to obtain approximate solutions to problems in physics and engineering. The idea is to find a parameter in the problem (such as frequency, wave number, wave speed, eigenvalue, or energy level) that is altered by the perturbations and then expand both the dependent variables as well as this parameter in, say, powers of the strength of these perturbations. The perturbations in the parameter are then chosen to render the expansion uniformly valid. Thus we call this technique the *method of strained parameters*.

This idea is the basis of the Rayleigh–Schrödinger method of obtaining approximate stationary solutions to the Schrödinger equation, in which one expands not only the wave function but also the energy level (Schrödinger,

56

1926). It is also the basis of Stoker's method of treating finite-amplitude water waves in which the stream function and the wave speed are expanded in powers of the steepness ratio of the waves (Stoker, 1957).

If we interpret this parameter expansion as a near-identity transformation, then Lighthill's technique of rendering approximate solutions uniformly valid is a generalization of this technique. According to Lighthill (1949a, 1961), if we encounter a nonuniformity in expanding a function such as $u(x_1, x_2, \ldots, x_n; \epsilon)$ in powers of ϵ, we expand not only the dependent variable u but also the independent variable exhibiting the nonuniformity, say x_1, in powers of ϵ in terms of a new independent variable as

$$u = \sum_{m=0}^{N-1} \epsilon^m u_m(s, x_2, x_3, \ldots, x_n) + O(\epsilon^N)$$

$$x_1 = s + \sum_{m=1}^{N} \epsilon^m \xi_m(s, x_2, x_3, \ldots, x_n) + O(\epsilon^{N+1})$$

The last expansion can be viewed as a near-identity transformation from x_1 to s. The functions ξ_m are called *straining functions*, and they are determined such that the expansion for u is uniformly valid. In other words, $u_m/u_{m-1} < \infty$ for all values of x_1 of interest, or equivalently higher approximations are no more singular than the first. Note that if $\xi_m = \omega_m s$ with ω_m constant, Lighthill's technique becomes the Lindstedt–Poincaré technique. Since Lighthill's transformation strains a coordinate rather than a parameter, his technique is called the *method of strained coordinates*.

For hyperbolic differential equations, Lighthill's technique is equivalent to expanding the dependent and independent variables in terms of some, or all, of the exact characteristics of the equations (Whitham, 1952, 1953; Lin, 1954; Fox, 1955).

Rather than introducing the transformation into the differential equations and then carrying out the expansion in terms of the new variable, Pritulo (1962) suggested introducing the transformation into the nonuniform straightforward expansion. The transformation can then be found directly therefrom by solving algebraic rather than differential equations. This is another form of the method of renormalization (Section 7.4.2) introduced by Rayleigh in his analysis of scattering. Rayleigh carried out an expansion $u = u_0 + \epsilon u_1$ for scattering from a thin slab and then recast this expansion in the form $u = u_0 \exp(\epsilon u_1/u_0)$ to make it valid for many slabs.

The method of strained parameters is described by applying it to several physical examples in the following section. In Section 3.2, Lighthill's technique is applied first to ordinary differential equations and then to partial differential equations. This is followed by a description of Temple's method of linearization. The method of renormalization is then taken up in Section 3.4, while limitations of the method of strained coordinates are discussed in Section 3.5.

3.1. The Method of Strained Parameters

As discussed above, this method is based on the presence of a parameter in the problem which is altered by the perturbations. As shown in the following sections, this parameter may be the frequency of a weakly nonlinear system, the energy level in a quantum mechanical problem, the characteristic exponent in the normal solution of a linear problem with periodic coefficients, the wave number or frequency in plasma oscillations, and the wave speed or frequency in finite-amplitude surface waves.

3.1.1. THE LINDSTEDT–POINCARÉ METHOD

The simple examples of Sections 1.1.2 and 2.1.1 show that truncated straightforward expansions in powers of ϵ of equations of the form

$$\ddot{u} + \omega_0{}^2 u = \epsilon f(u, \dot{u}) \qquad (3.1.1)$$

are valid only for short intervals of time because of the presence of secular terms. The essence of the Lindstedt–Poincaré technique is to prevent the appearance of these secular terms by introducing a new variable

$$t = s(1 + \epsilon\omega_1 + \epsilon^2\omega_2 + \cdots) \qquad (3.1.2)$$

in (3.1.1) to obtain

$$(1 + \epsilon\omega_1 + \epsilon^2\omega_2 + \cdots)^{-2}\frac{d^2 u}{ds^2} + \omega_0{}^2 u$$

$$= \epsilon f\left[u, (1 + \epsilon\omega_1 + \epsilon^2\omega_2 + \cdots)^{-1}\frac{du}{ds}\right] \qquad (3.1.3)$$

Letting

$$u = \sum_{n=0}^{\infty} \epsilon^n u_n(s) \qquad (3.1.4)$$

in (3.1.3) and equating coefficients of like powers of ϵ, we obtain equations to determine u_m in succession. The solutions for u_m contain secular terms unless the ω_m have specific values.

This technique was applied to a wide range of physical and mathematical problems. For example, Keller (1968) adapted this technique to boundary value problems for systems of ordinary differential equations. A combination of this technique and the Ritz-Galërkin procedure is frequently used in the dynamic response of elastic bodies (e.g., Han, 1965; Bauer, 1968; Sweet, 1971). As an example, we consider Duffing's equation

$$\frac{d^2 u}{dt^2} + u + \epsilon u^3 = 0 \qquad (3.1.5)$$

Under the transformation (3.1.2) it becomes

$$\frac{d^2u}{ds^2} + (1 + \epsilon\omega_1 + \epsilon^2\omega_2 + \cdots)^2(u + \epsilon u^3) = 0 \qquad (3.1.6)$$

We substitute (3.1.4) into (3.1.6) and equate coefficients of like powers of ϵ to obtain

$$\frac{d^2u_0}{ds^2} + u_0 = 0 \qquad (3.1.7)$$

$$\frac{d^2u_1}{ds^2} + u_1 = -u_0^3 - 2\omega_1 u_0 \qquad (3.1.8)$$

$$\frac{d^2u_2}{ds^2} + u_2 = -3u_0^2 u_1 - 2\omega_1(u_1 + u_0^3) - (\omega_1^2 + 2\omega_2)u_0 \qquad (3.1.9)$$

The general solution of (3.1.7) is

$$u_0 = a \cos(s + \phi) \qquad (3.1.10)$$

where a and ϕ are constants of integration. With (3.1.10), (3.1.8) becomes

$$\frac{d^2u_1}{ds^2} + u_1 = -\tfrac{1}{4}a^3 \cos 3(s + \phi) - (\tfrac{3}{4}a^2 + 2\omega_1)a \cos(s + \phi) \qquad (3.1.11)$$

If a straightforward perturbation expansion is used, $\omega_n \equiv 0$, and (3.1.11) reduces to (2.1.7) whose particular solution contains a secular term. In order to avoid this secular term, ω_1 is chosen to eliminate the coefficient of $\cos(s + \phi)$ on the right-hand side of (3.1.11). This condition determines ω_1 to be

$$\omega_1 = -\tfrac{3}{8}a^2 \qquad (3.1.12a)$$

Then the solution of (3.1.11) becomes

$$u_1 = \tfrac{1}{32}a^3 \cos 3(s + \phi) \qquad (3.1.12b)$$

Substituting for u_0, u_1, and ω_1 into (3.1.9), we obtain

$$\frac{d^2u_2}{ds^2} + u_2 = (\tfrac{51}{128}a^4 - 2\omega_2)a \cos(s + \phi) + \text{NST} \qquad (3.1.13)$$

where NST stands for terms that do not produce secular terms. Secular terms are eliminated if

$$\omega_2 = \tfrac{51}{256}a^4 \qquad (3.1.14)$$

Therefore

$$u = a \cos(\omega t + \phi) + \frac{\epsilon}{32} a^3 \cos 3(\omega t + \phi) + O(\epsilon^2) \qquad (3.1.15)$$

where a and θ are constants of integration, and

$$\omega = (1 - \tfrac{3}{8}a^2\epsilon + \tfrac{51}{256}a^4\epsilon^2 + \cdots)^{-1}$$
$$= 1 + \tfrac{3}{8}a^2\epsilon - \tfrac{15}{256}a^4\epsilon^2 + O(\epsilon^3) \qquad (3.1.16)$$

3.1.2. TRANSITION CURVES FOR THE MATHIEU EQUATION

As a second example, we determine the transition curves that separate stable from unstable solutions of the Mathieu equation

$$\ddot{u} + (\delta + \epsilon \cos 2t)u = 0 \qquad (3.1.17)$$

which has been studied extensively. This equation is a special case of Hill's equation which is a linear differential equation with periodic coefficients. Equations similar to this appear in many problems in applied mathematics such as stability of a transverse column subjected to a periodic longitudinal load, stability of periodic solutions of nonlinear conservative systems, electromagnetic wave propagation in a medium with periodic structure, the lunar motion, and the excitation of certain electrical systems.

The qualitative nature of the solutions of (3.1.17) can be described by using Floquet theory (see, e.g., Coddington and Levinson, 1955, Chapter 3). This equation has normal solutions of the form

$$u = e^{\gamma t}\phi(t) \qquad (3.1.18)$$

where ϕ is a periodic function of t having a period π or 2π, and γ may be real or complex depending on the values of the parameters δ and ϵ. Floquet theory shows that the transition curves, in the δ-ϵ plane, separating stable from unstable solutions, correspond to periodic solutions of (3.1.17). Some of these curves are determined below by expanding both δ and u as functions of ϵ. Thus we let

$$\delta = n^2 + \epsilon\delta_1 + \epsilon^2\delta_2 + \cdots \qquad (3.1.19)$$

$$u(t) = u_0 + \epsilon u_1 + \epsilon^2 u_2 + \cdots \qquad (3.1.20)$$

where n is an integer including zero, and u_m/u_0 is bounded for all m in order that (3.1.20) be a uniformly valid asymptotic expansion.

Substituting (3.1.19) and (3.1.20) into (3.1.17), expanding, and equating coefficients of equal powers of ϵ, we obtain

$$\ddot{u}_0 + n^2 u_0 = 0 \qquad (3.1.21)$$

$$\ddot{u}_1 + n^2 u_1 = -(\delta_1 + \cos 2t)u_0 \qquad (3.1.22)$$

$$\ddot{u}_2 + n^2 u_2 = -(\delta_1 + \cos 2t)u_1 - \delta_2 u_0 \qquad (3.1.23)$$

The solution of the zeroth-order equation is

$$u_0 = \begin{cases} \cos nt \\ \sin nt \end{cases} \qquad n = 0, 1, 2, \ldots \qquad (3.1.24)$$

Next, we determine the higher approximations for the cases $n = 0, 1$, and 2.

The Case of $n = 0$

In this case $u_0 = 1$, and (3.1.22) becomes

$$\ddot{u}_1 = -\delta_1 - \cos 2t \qquad (3.1.25)$$

In order for (3.1.20) to be a uniformly valid asymptotic expansion, δ_1 must vanish; and

$$u_1 = \tfrac{1}{4} \cos 2t + c \qquad (3.1.26)$$

where c is a constant. With u_0 and u_1 known (3.1.23) becomes

$$\ddot{u}_2 = -\delta_2 - \tfrac{1}{8} - c \cos 2t - \tfrac{1}{8} \cos 4t \qquad (3.1.27)$$

In order for u_2/u_0 to be bounded, δ_2 must be equal to $-1/8$, hence

$$\delta = -\tfrac{1}{8}\epsilon^2 + O(\epsilon^3) \qquad (3.1.28)$$

The Case of $n = 1$

In this case $u_0 = \cos t$ or $\sin t$. Taking $u_0 = \cos t$, we find that (3.1.22) becomes

$$\ddot{u}_1 + u_1 = -(\delta_1 + \tfrac{1}{2}) \cos t - \tfrac{1}{2} \cos 3t \qquad (3.1.29)$$

In order that u_1/u_0 be bounded, δ_1 must be equal to $-1/2$, and then

$$u_1 = \tfrac{1}{16} \cos 3t \qquad (3.1.30)$$

Equation (3.1.23) then becomes

$$\ddot{u}_2 + u_2 = -(\tfrac{1}{32} + \delta_2) \cos t + \tfrac{1}{32} \cos 3t - \tfrac{1}{32} \cos 5t \qquad (3.1.31)$$

The condition that u_2/u_0 be bounded demands that $\delta_2 = -1/32$, hence

$$\delta = 1 - \tfrac{1}{2}\epsilon - \tfrac{1}{32}\epsilon^2 + O(\epsilon^3) \qquad (3.1.32)$$

Had we used $u_0 = \sin t$, we would have obtained the transition curve

$$\delta = 1 + \tfrac{1}{2}\epsilon - \tfrac{1}{32}\epsilon^2 + O(\epsilon^3) \qquad (3.1.33)$$

The Case of $n = 2$

In this case $u_0 = \cos 2t$ or $\sin 2t$, and (3.1.22) becomes, in the former case,

$$\ddot{u}_1 + 4u_1 = -\tfrac{1}{2} - \delta_1 \cos 2t - \tfrac{1}{2} \cos 4t \qquad (3.1.34)$$

Since u_1/u_0 must be bounded, δ_1 must vanish, hence

$$u_1 = -\tfrac{1}{8} + \tfrac{1}{24} \cos 4t \qquad (3.1.35)$$

Substituting for u_0 and u_1 into (3.1.23) yields

$$\ddot{u}_2 + 4u_2 = -(\delta_2 - \tfrac{5}{48}) \cos 2t - \tfrac{1}{48} \cos 6t \qquad (3.1.36)$$

Because u_2/u_0 must be bounded, we require that $\delta_2 = 5/48$, hence

$$\delta = 4 + \tfrac{5}{48}\epsilon^2 + O(\epsilon^3) \qquad (3.1.37)$$

Taking $u_0 = \sin 2t$ leads to

$$\delta = 4 - \tfrac{1}{48}\epsilon^2 + O(\epsilon^3) \qquad (3.1.38)$$

3.1.3. CHARACTERISTIC EXPONENTS FOR THE MATHIEU EQUATION (WHITTAKER'S METHOD)

Floquet theory (see, e.g., Coddington and Levinson 1955, Chapter 3) shows that (3.1.17) has solutions of the form indicated in (3.1.18), where ϕ is periodic (having a period π or 2π) and γ is a real or a complex constant. Substituting (3.1.18) into (3.1.17), we transform the latter into

$$\ddot{\phi} + 2\gamma\dot{\phi} + (\delta + \gamma^2 + \epsilon \cos 2t)\phi = 0 \qquad (3.1.39)$$

Since the transition curves correspond to $\gamma = 0$, Whittaker (1914) obtained an approximation to ϕ near the transition curves by assuming the following expansions for ϕ, δ, and γ

$$\phi = \phi_0 + \epsilon\phi_1 + \epsilon^2\phi_2 + \cdots \qquad (3.1.40)$$

$$\delta = \delta_0 + \epsilon\delta_1 + \epsilon^2\delta_2 + \cdots \qquad (3.1.41)$$

$$\gamma = \epsilon\gamma_1 + \epsilon^2\gamma_2 + \cdots \qquad (3.1.42)$$

Next we carry out the solution for the case $\delta_0 = 1$.

Substituting (3.1.40) through (3.1.42) into (3.1.39) and equating like powers of ϵ, we obtain

$$\ddot{\phi}_0 + \phi_0 = 0 \qquad (3.1.43)$$

$$\ddot{\phi}_1 + \phi_1 = -2\gamma_1\dot{\phi}_0 - (\delta_1 + \cos 2t)\phi_0 \qquad (3.1.44)$$

$$\ddot{\phi}_2 + \phi_2 = -2\gamma_1\dot{\phi}_1 - 2\gamma_2\dot{\phi}_0 - (\gamma_1^2 + \delta_2)\phi_0 - (\delta_1 + \cos 2t)\phi_1 \qquad (3.1.45)$$

The general solution of (3.1.43) is

$$\phi_0 = a \cos t + b \sin t \qquad (3.1.46)$$

where a and b are constants. Equation (3.1.44) then becomes

$$\ddot{\phi}_1 + \phi_1 = [2\gamma_1 a + (\tfrac{1}{2} - \delta_1)b] \sin t - [2\gamma_1 b + (\tfrac{1}{2} + \delta_1)a] \cos t$$
$$- \tfrac{1}{2}a \cos 3t - \tfrac{1}{2}b \sin 3t \qquad (3.1.47)$$

Since ϕ is periodic, the terms that produce secular terms must vanish; that is

$$2\gamma_1 a + (\tfrac{1}{2} - \delta_1)b = 0 \qquad (3.1.48)$$

$$(\tfrac{1}{2} + \delta_1)a + 2\gamma_1 b = 0 \qquad (3.1.49)$$

For a nontrivial solution for a and b, the determinant of the coefficients of a and b in (3.1.48) and (3.1.49) must be zero. Hence

$$\gamma_1{}^2 = -\tfrac{1}{4}(\delta_1{}^2 - \tfrac{1}{4}) \qquad (3.1.50)$$

Then

$$b = \frac{2\gamma_1}{\delta_1 - \tfrac{1}{2}}\, a \qquad (3.1.51)$$

The solution of (3.1.47) subject to the conditions (3.1.48) and (3.1.49) becomes

$$\phi_1 = \tfrac{1}{16}a \cos 3t + \tfrac{1}{16}b \sin 3t \qquad (3.1.52)$$

With the above results (3.1.45) becomes

$$\ddot{\phi}_2 + \phi_2 = [2\gamma_2 a - (\delta_2 + \gamma_1{}^2 + \tfrac{1}{32})b] \sin t - [(\delta_2 + \gamma_1{}^2 + \tfrac{1}{32})a$$
$$+ 2\gamma_2 b] \cos t + \text{NST} \qquad (3.1.53)$$

Terms that produce secular terms will be eliminated if

$$2\gamma_2 a - (\delta_2 + \gamma_1{}^2 + \tfrac{1}{32})b = 0 \qquad (3.1.54)$$

$$(\delta_2 + \gamma_1{}^2 + \tfrac{1}{32})a + 2\gamma_2 b = 0 \qquad (3.1.55)$$

Since b is related to a by (3.1.51), (3.1.54) and (3.1.55) are satisfied simultaneously if and only if

$$\gamma_2 = 0 \quad \text{and} \quad \delta_2 = -\gamma_1{}^2 - \tfrac{1}{32} \qquad (3.1.56)$$

Therefore, to first approximation

$$u = ae^{\pm(1/2)\epsilon t\sqrt{(1/4)-\delta_1{}^2}}\Bigg[(\cos t + \tfrac{1}{16}\epsilon \cos 3t)$$

$$+ \frac{2\gamma_1}{\delta_1 - \tfrac{1}{2}}(\sin t + \tfrac{1}{16}\epsilon \sin 3t)\Bigg] + O(\epsilon^2) \qquad (3.1.57)$$

$$\delta = 1 + \epsilon\delta_1 + \tfrac{1}{4}\epsilon^2(\delta_1{}^2 - \tfrac{3}{8}) + O(\epsilon^3) \qquad (3.1.58)$$

If we let

$$\delta_1 = \tfrac{1}{2}\cos 2\sigma \qquad (3.1.59)$$

then

$$\gamma_1 = \tfrac{1}{4}\sin 2\sigma, \qquad \delta_2 = \tfrac{1}{32}(\cos 4\sigma - 2)$$

$$\frac{b}{a} = \frac{\sin 2\sigma}{\cos 2\sigma - 1} = -\cot \sigma \qquad (3.1.60)$$

Hence (3.1.57) and (3.1.58) become

$$u = \tilde{a}e^{(1/4)(\sin 2\sigma)\epsilon t}[\sin(t - \sigma) + \tfrac{1}{16}\epsilon \sin(3t - \sigma)] + O(\epsilon^2) \quad (3.1.61)$$

$$\delta = 1 + \tfrac{1}{2}\epsilon \cos 2\sigma + \tfrac{1}{32}\epsilon^2(\cos 4\sigma - 2) + O(\epsilon^3) \quad (3.1.62)$$

where $\tilde{a}$ is a constant.

3.1.4. THE STABILITY OF THE TRIANGULAR POINTS IN THE ELLIPTIC RESTRICTED PROBLEM OF THREE BODIES

We consider next a fourth-order system with periodic coefficients involving the stability of the triangular points in the elliptic restricted problem of three bodies. The problem is governed mathematically by

$$x'' - 2y' - \frac{h_2 x}{1 + e \cos f} = 0 \quad (3.1.63)$$

$$y'' + 2x' - \frac{h_1 y}{1 + e \cos f} = 0 \quad (3.1.64)$$

where primes denote differentiation with respect to f, and

$$h_{1,2} = \tfrac{3}{2}[1 \pm \sqrt{1 - 3\mu(1 - \mu)}] \quad (3.1.65)$$

Equations (3.1.63) through (3.1.65) describe the linearized motion of a particle near the triangular points in the restricted problem of three bodies. Here e is the eccentricity of the orbit of the two primaries and μ is the ratio of the smaller primary to the sum of the two primaries. If $e = 0$, it is known that the motion is stable if $0 \leq \mu < \bar{\mu} \approx 0.03852$ and unstable for $\mu \geq \bar{\mu}$. Therefore $\bar{\mu}$ is the intersection of a transition curve (from stable to unstable motion) with the μ axis in the e-μ plane. Also, it is known from Floquet theory that periodic solutions with periods 2π and 4π correspond to transition curves. In the interval $0 \leq \mu < \bar{\mu}$, the period 2π corresponds to $\mu = 0$, while 4π corresponds to $\mu_0 = (1 - 2\sqrt{2}/3)/2$; $\mu = 0$ corresponds to a transition curve that coincides with the e axis. In the following discussion we present an analysis following Nayfeh and Kamel (1970a) for the determination of the transition curves that intersect the μ axis at μ_0.

We assume that

$$x = \sum_{n=0}^{\infty} e^n x_n(f) \quad (3.1.66)$$

$$y = \sum_{n=0}^{\infty} e^n y_n(f) \quad (3.1.67)$$

$$\mu = \sum_{n=0}^{\infty} e^n \mu_n \quad (3.1.68)$$

Substituting (3.1.68) into (3.1.65) and expanding for small e, we have

$$h_1 = \sum_{n=0}^{\infty} a_n(\mu_0, \mu_1, \ldots, \mu_n)e^n \qquad (3.1.69)$$

$$h_2 = \sum_{n=0}^{\infty} b_n(\mu_0, \mu_1, \ldots, \mu_n)e^n \qquad (3.1.70)$$

where

$$a_0, b_0 = [\tfrac{3}{2}(1 + k), \tfrac{3}{2}(1 - k)], \qquad k = \sqrt{\tfrac{11}{12}} \qquad (3.1.71)$$

$$b_1 = -a_1 = 3\sqrt{\tfrac{6}{11}}\,\mu_1 \qquad (3.1.72)$$

Substituting (3.1.66) through (3.1.70) into (3.1.63) and (3.1.64) and equating coefficients of like powers of ϵ, we obtain

$$x_n'' - 2y_n' = \sum_{\substack{t=0,s=0,r=0 \\ n=s+r+t}} (-1)^t x_r b_s \cos^t f \qquad (3.1.73)$$

$$y_n'' + 2x_n' = \sum_{\substack{t=0,s=0,r=0 \\ n=r+s+t}} (-1)^t y_r a_s \cos^t f \qquad (3.1.74)$$

The zeroth-order equations admit the following periodic solutions of period 4π

$$x_0 = \cos \tau, \qquad y_0 = -\alpha \sin \tau \qquad (3.1.75)$$

$$x_0 = \sin \tau, \qquad y_0 = \alpha \cos \tau \qquad (3.1.76)$$

where

$$\tau = \frac{f}{2}, \qquad \alpha = b_0 + \tfrac{1}{4} = (a_0 + \tfrac{1}{4})^{-1} = \tfrac{1}{4}(7 - \sqrt{33}) \qquad (3.1.77)$$

There are two transition curves intersecting the μ axis at $\mu = \mu_0$ corresponding to the above two independent solutions. If we take (3.1.75) we find that the first-order problem becomes

$$x_1'' - 2y_1' - b_0 x_1 = (b_1 - \tfrac{1}{2}b_0)\cos \tau - \tfrac{1}{2}b_0 \cos 3\tau \qquad (3.1.78)$$

$$y_1'' + 2x_1' - a_0 y_1 = -\alpha(a_1 + \tfrac{1}{2}a_0)\sin \tau + \tfrac{1}{2}\alpha a_0 \sin 3\tau \qquad (3.1.79)$$

The terms proportional to $\cos \tau$ and $\sin \tau$ lead to secular terms in the particular solutions for x_1 and y_1. To determine the condition necessary for the removal of these secular terms, we assume a particular solution of the form

$$x_p = 0, \qquad y_p = c \sin \tau \qquad (3.1.80)$$

Substituting (3.1.80) into (3.1.78) and (3.1.79) and equating the coefficients of $\cos \tau$ and $\sin \tau$ on both sides, we obtain

$$c = -\left(b_1 - \frac{b_0}{2}\right), \qquad c(a_0 + \tfrac{1}{4}) = \alpha\left(a_1 + \frac{a_0}{2}\right) \qquad (3.1.81)$$

Elimination of c leads to

$$b_1 - \tfrac{1}{2}b_0 = -\alpha^2(a_1 + \tfrac{1}{2}a_0) \qquad (3.1.82)$$

Since $b_1 = -a_1$

$$b_1 = \frac{b_0 - a_0\alpha^2}{2(1 - \alpha^2)} \approx -0.1250 \qquad (3.1.83)$$

Hence from (3.1.72)

$$\mu_1 \approx -0.05641 \qquad (3.1.84)$$

Therefore the transition curve to first order is given by

$$\mu = 0.02859 - 0.05641e + O(e^2) \qquad (3.1.85)$$

Had we used the solution given by (3.1.76) for the zeroth-order problem, we would have obtained the second branch

$$\mu = 0.02859 + 0.05641e + O(e^2) \qquad (3.1.86)$$

The above analysis could be continued to higher orders in a straightforward manner and has been carried out to fourth order by Nayfeh and Kamel (1970a).

3.1.5. CHARACTERISTIC EXPONENTS FOR THE TRIANGULAR POINTS IN THE ELLIPTIC RESTRICTED PROBLEM OF THREE BODIES

It is known from Floquet theory that (3.1.63) and (3.1.64) have normal solutions of the form

$$x, y = e^{\gamma f}[\phi(f), \psi(f)] \qquad (3.1.87)$$

where ϕ and ψ are periodic, having a period of either 2π or 4π, and γ is a real or complex number. Substitution of (3.1.87) into (3.1.63) and (3.1.64) transforms the latter into

$$\phi'' + 2\gamma\phi' - 2\psi' + \gamma^2\phi - 2\gamma\psi - \frac{h_2\phi}{1 + e\cos f} = 0 \qquad (3.1.88)$$

$$\psi'' + 2\gamma\psi' + 2\phi' + \gamma^2\psi + 2\gamma\phi - \frac{h_1\psi}{1 + e\cos f} = 0 \qquad (3.1.89)$$

The transition curves correspond to $\gamma = 0$, hence near these transition curves γ is small. Thus to obtain expansions for ϕ and ψ valid near the transition curves intersecting the μ axis at μ_0, we let

$$\phi = \phi_0 + e\phi_1 + \cdots \qquad (3.1.90)$$

$$\psi = \psi_0 + e\psi_1 + \cdots \qquad (3.1.91)$$

$$\gamma = e\gamma_1 + \cdots \qquad (3.1.92)$$

$$\mu = \mu_0 + e\mu_1 + \cdots \qquad (3.1.93)$$

Substituting (3.1.90) through (3.1.93) into (3.1.88), (3.1.89), and (3.1.65) and equating like powers of e, we obtain

Order e^0

$$\phi_0'' - 2\psi_0' - b_0\phi_0 = 0 \qquad (3.1.94)$$

$$\psi_0'' + 2\phi_0' - a_0\psi_0 = 0 \qquad (3.1.95)$$

Order e

$$\phi_1'' - 2\psi_1' - b_0\phi_1 = -2\gamma_1\phi_0' + 2\gamma_1\psi_0 + b_1\phi_0 - b_0\phi_0\cos f \qquad (3.1.96)$$

$$\psi_1'' + 2\phi_1' - a_0\psi_1 = -2\gamma_1\psi_0' - 2\gamma_1\phi_0 + a_1\psi_0 - a_0\psi_0\cos f \qquad (3.1.97)$$

The general solution of (3.1.94) and (3.1.95) is

$$\phi_0 = A\cos\tau + B\sin\tau \qquad (3.1.98)$$

$$\psi_0 = \alpha B\cos\tau - \alpha A\sin\tau \qquad (3.1.99)$$

This solution determines the right-hand sides of (3.1.96) and (3.1.97). Thus

$$\phi_1'' - 2\psi_1' - b_0\phi_1 = P_{11}\cos\tau + Q_{11}\sin\tau$$
$$- \tfrac{1}{2}b_0 A\cos 3\tau - \tfrac{1}{2}b_0 B\sin 3\tau \qquad (3.1.100)$$

$$\psi_1'' + 2\phi_1' - a_0\psi_1 = P_{12}\cos\tau + Q_{12}\sin\tau$$
$$- \tfrac{1}{2}a_0\alpha B\cos 3\tau + \tfrac{1}{2}a_0\alpha A\sin 3\tau \qquad (3.1.101)$$

where

$$P_{11} = \gamma_1(2\alpha - 1)B + \left(b_1 - \frac{b_0}{2}\right)A$$

$$P_{12} = \gamma_1(\alpha - 2)A + \alpha\left(a_1 - \frac{a_0}{2}\right)B$$

$$Q_{11} = -\gamma_1(2\alpha - 1)A + \left(b_1 + \frac{b_0}{2}\right)B$$

$$Q_{12} = \gamma_1(\alpha - 2)B - \alpha\left(a_1 + \frac{a_0}{2}\right)A$$

Since ϕ and ψ are periodic, the secular terms in the solution of ϕ_1 and ψ_1 must vanish. To eliminate these secular terms, we let

$$\phi_1 = 0, \qquad \psi_1 = c_1\cos\tau + c_2\sin\tau \qquad (3.1.102)$$

Substituting (3.1.102) into (3.1.100) and (3.1.101) and equating the coefficients of $\cos \tau$ and $\sin \tau$ on both sides, we obtain

$$c_1 = Q_{11}, \qquad\qquad -c_2 = P_{11}$$
$$-(a_0 + \tfrac{1}{4})c_1 = P_{12}, \qquad -(a_0 + \tfrac{1}{4})c_2 = Q_{12}$$
(3.1.103)

Elimination of c_1 and c_2 from (3.1.103) results in

$$P_{11} = \alpha Q_{12}, \qquad Q_{11} = -\alpha P_{12} \qquad (3.1.104)$$

Substituting the above expressions for P and Q into (3.1.104) and rearranging, we obtain

$$\left[b_1 - \frac{b_0}{2} + \alpha^2\left(a_1 + \frac{a_0}{2}\right) \right] A - \gamma_1(1 - 4\alpha + \alpha^2)B = 0 \quad (3.1.105)$$

$$\gamma_1(1 - 4\alpha + \alpha^2)A + \left[b_1 + \frac{b_0}{2} + \alpha^2\left(a_1 - \frac{a_0}{2}\right) \right] B = 0 \quad (3.1.106)$$

For a nontrivial solution the determinant of the coefficients of A and B in (3.1.105) and (3.1.106) must vanish. This condition leads to

$$\gamma_1^2 = - \frac{[b_1 - \tfrac{1}{2}b_0 + \alpha^2(a_1 + \tfrac{1}{2}a_0)][b_1 + \tfrac{1}{2}b_0 + \alpha^2(a_1 - \tfrac{1}{2}a_0)]}{(1 - 4\alpha + \alpha^2)^2} \quad (3.1.107)$$

Then

$$\frac{B}{A} = \frac{b_1 - \tfrac{1}{2}b_0 + \alpha^2(a_1 + \tfrac{1}{2}a_0)}{\gamma_1(1 - 4\alpha + \alpha^2)} = \tan \sigma \qquad (3.1.108)$$

Therefore, to first approximation

$$x, y = e^{e^{\gamma_1 t}}[\cos (\tfrac{1}{2}f - \sigma), -\alpha \sin (\tfrac{1}{2}f - \sigma)] + O(e) \qquad (3.1.109)$$

The transition curves (3.1.85) and (3.1.86) correspond to $\gamma_1 = 0$, while (3.1.75) and (3.1.76) can be obtained from (3.1.109) by letting $\gamma_1 = 0$ and $\sigma = 0$ or $\pi/2$.

The present analysis can be continued to higher order in a straightforward manner even though the algebra is involved. The expansion has been carried out to second order by Nayfeh (1970a).

3.1.6. A SIMPLE LINEAR EIGENVALUE PROBLEM

Let us now consider the problem of determining the eigenvalue λ and the eigenfunction u where

$$u'' + [\lambda + \epsilon f(x)]u = 0, \qquad f(x) = f(-x) \qquad (3.1.110)$$
$$u(0) = u(1) = 0 \qquad (3.1.111)$$

and ϵ is a small quantity. If $\epsilon = 0$, the eigenfunctions and eigenvalues are given, respectively, by

$$u_n = \sqrt{2} \sin n\pi x, \qquad n = 1, 2, 3, \ldots \qquad (3.1.112)$$

$$\lambda_n = n^2 \pi^2 \qquad (3.1.113)$$

The above eigenfunctions are orthonormal; that is

$$\int_0^1 u_n(x) u_m(x)\, dx = \delta_{mn} \qquad (3.1.114)$$

where δ_{mn}, the Kronecker delta function, is defined as follows

$$\delta_{mn} = \begin{cases} 0 & m \neq n \\ 1 & m = n \end{cases}$$

For a small but nonzero ϵ, we obtain an approximate solution to u_n and λ_n by letting

$$u_n = \sqrt{2} \sin n\pi x + \epsilon u_{n1} + \epsilon^2 u_{n2} + \cdots \qquad (3.1.115)$$

$$\lambda_n = n^2 \pi^2 + \epsilon \lambda_{n1} + \epsilon^2 \lambda_{n2} + \cdots \qquad (3.1.116)$$

Substituting (3.1.115) and (3.1.116) into (3.1.110) and (3.1.111) and equating coefficients of like powers of ϵ, we obtain

$$u_{n1}'' + n^2 \pi^2 u_{n1} = -f(x) u_{n0} - \lambda_{n1} u_{n0}$$
$$u_{n1}(0) = u_{n1}(1) = 0 \qquad (3.1.117)$$

$$u_{n2}'' + n^2 \pi^2 u_{n2} = -f(x) u_{n1} - \lambda_{n1} u_{n1} - \lambda_{n2} u_{n0}$$
$$u_{n2}(0) = u_{n2}(1) = 0 \qquad (3.1.118)$$

where the zeroth-order problem is satisfied identically, and $u_{n0} = \sqrt{2} \sin n\pi x$.

We assume that u_{n1} can be expressed as a linear combination of the zeroth-order eigenfunctions u_{n0}; that is

$$u_{n1} = \sum_{m=1}^{\infty} a_{nm} \sqrt{2} \sin m\pi x \qquad (3.1.119)$$

This solution satisfies the boundary conditions on u_{n1}. Substituting (3.1.119) into (3.1.117), we obtain

$$\sum_{m=1}^{\infty} \sqrt{2}\, \pi^2 (n^2 - m^2) a_{nm} \sin m\pi x = -\sqrt{2} f(x) \sin n\pi x - \sqrt{2}\, \lambda_{n1} \sin n\pi x \qquad (3.1.120)$$

Multiplying (3.1.120) by $\sqrt{2} \sin k\pi x$, integrating from 0 to 1, and using the orthonormality property (3.1.114) of the eigenfunctions (3.1.112), we obtain

$$\pi^2(n^2 - k^2)a_{nk} = -F_{nk} - \lambda_{n1}\delta_{nk} \tag{3.1.121}$$

where

$$F_{nk} = 2\int_0^1 f(x) \sin n\pi x \sin k\pi x \, dx \tag{3.1.122}$$

If $k = n$, the left-hand side of (3.1.121) vanishes, hence

$$\lambda_{n1} = -F_{nn} = -2\int_0^1 f(x) \sin^2 n\pi x \, dx \tag{3.1.123}$$

However, if $k \neq n$

$$a_{nk} = -\frac{F_{nk}}{\pi^2(n^2 - k^2)} \tag{3.1.124}$$

Condition (3.1.123) is equivalent to the removal of secular terms. The function u_{n1} is then

$$u_{n1} = -\sum_{k \neq n} \frac{F_{nk}}{\pi^2(n^2 - k^2)} \sqrt{2} \sin k\pi x + a_{nn}\sqrt{2} \sin n\pi x \tag{3.1.125}$$

Note that a_{nn} is still undetermined. It is determined in the final solution by normalization of u_n.

Proceeding to second order, we assume that

$$u_{n2} = \sum_{r=1}^{\infty} b_{nr}\sqrt{2} \sin r\pi x \tag{3.1.126}$$

Substituting for u_{n2}, u_{n1}, and u_{n0} into (3.1.118), we obtain

$$\pi^2\sum_{r=1}^{\infty}(n^2 - r^2)\sqrt{2} \, b_{nr} \sin r\pi x = -\sum_{k=1}^{\infty} a_{nk}\sqrt{2} f(x) \sin k\pi x$$
$$-\sum_{k=1}^{\infty} a_{nk}\lambda_{n1}\sqrt{2} \sin k\pi x - \lambda_{n2}\sqrt{2} \sin n\pi x \tag{3.1.127}$$

Multiplying (3.1.127) by $\sqrt{2} \sin s\pi x$, integrating from 0 to 1, and using the orthonormality property (3.1.114), we obtain

$$\pi^2(n^2 - s^2)b_{ns} = -\sum_{k=1}^{\infty} a_{nk}F_{ks} - a_{ns}\lambda_{n1} - \lambda_{n2}\delta_{ns} \tag{3.1.128}$$

If $s = n$, the left-hand side of (3.1.128) vanishes, while the right-hand side gives

$$\lambda_{n2} = -\sum_{k \neq n} a_{nk}F_{kn} = \sum_{k \neq n} \frac{F_{nk}^2}{\pi^2(n^2 - k^2)} \tag{3.11.29}$$

If $s \neq n$, (3.1.128) gives

$$b_{ns} = \sum_{k \neq n} \frac{F_{nk}F_{ks}}{\pi^4(n^2 - k^2)(n^2 - s^2)} - \frac{a_{nn}F_{ns}}{\pi^2(n^2 - s^2)} - \frac{F_{nn}F_{ns}}{\pi^4(n^2 - s^2)^2} \qquad (3.1.130)$$

Here again b_{nn} is still undetermined; it will be determined from the normalization of u_n.

To normalize u_n we require that

$$\int_0^1 (u_{n0} + \epsilon u_{n1} + \epsilon^2 u_{n2})^2 \, dx = 1 \qquad (3.1.131)$$

Since u_{n0} is normalized, we obtain

$$\int_0^1 u_{n0}u_{n1} \, dx = 0 \qquad (3.1.132)$$

$$\int_0^1 (2u_{n0}u_{n2} + u_{n1}^2) \, dx = 0 \qquad (3.1.133)$$

Condition (3.1.132) gives $a_{nn} = 0$, while condition (3.1.133) gives

$$b_{nn} = -\tfrac{1}{2} \sum_{k \neq n} a_{nk}^2 \qquad (3.1.134)$$

Therefore, to second order

$$u_n = \sqrt{2} \sin n\pi x - \epsilon \sum_{k \neq n} \frac{F_{nk}}{\pi^2(n^2 - k^2)} \sqrt{2} \sin k\pi x$$

$$+ \epsilon^2 \sum_{k \neq n} \left\{ \left[\sum_{s \neq n} \frac{F_{ns}F_{ks}}{\pi^4(n^2 - s^2)(n^2 - k^2)} - \frac{F_{nn}F_{nk}}{\pi^4(n^2 - k^2)^2} \right] \sqrt{2} \sin k\pi x \right.$$

$$\left. - \frac{1}{2} \frac{F_{nk}^2}{\pi^4(n^2 - k^2)^2} \sqrt{2} \sin n\pi x \right\} + O(\epsilon^3) \qquad (3.1.135)$$

$$\lambda = n^2\pi^2 - \epsilon F_{nn} + \epsilon^2 \sum_{k \neq n} \frac{F_{nk}^2}{\pi^2(n^2 - k^2)} + O(\epsilon^3) \qquad (3.1.136)$$

The expansion method described in this section is called the Rayleigh–Schrödinger method; it was developed by Schrödinger (1926) to treat stationary solutions of the Schrödinger equation. For more references and more complete treatment, we refer the reader to the book edited by Wilcox (1966) and to the article by Hirschfelder (1969).

3.1.7. A QUASI-LINEAR EIGENVALUE PROBLEM

Let us next consider the eigenvalue problem

$$H\phi + \lambda\phi = \epsilon F(\phi) \qquad (3.1.137)$$

subject to the linear homogeneous boundary condition

$$B(\phi) = 0 \qquad (3.1.138)$$

where H is a linear operator and F is a nonlinear operator of ϕ. We seek an approximate solution for small ϵ by letting

$$\phi = \phi_0 + \epsilon\phi_1 + \cdots \qquad (3.1.139)$$

$$\lambda = \lambda_0 + \epsilon\lambda_1 + \cdots \qquad (3.1.140)$$

Substituting (3.1.139) and (3.1.140) into (3.1.137) and (3.1.138) and equating coefficients of like powers of ϵ, we obtain

$$H\phi_0 + \lambda_0\phi_0 = 0, \qquad B(\phi_0) = 0 \qquad (3.1.141)$$

$$H\phi_1 + \lambda_0\phi_1 = -\lambda_1\phi_0 + F(\phi_0), \qquad B(\phi_1) = 0 \qquad (3.1.142)$$

We must distinguish between two cases depending on whether the eigenvalues of (3.1.141) are distinct or not. The first case is called the *nondegenerate* case, while the second case is called the *degenerate* case because more than one eigenfunction corresponds to a repeated eigenvalue. Both of these cases are treated in order below.

The Nondegenerate Case. Let us assume that (3.1.141) is solvable and that its solution yields the eigenfunctions u_n corresponding to the eigenvalues $\mu_n, n = 1, 2, \ldots$. We assume further that $\mu_m \neq \mu_n$ if $m \neq n$, and that the eigenfunctions $\{u_n\}$ form an orthonormal set so that

$$\int_D u_n \bar{u}_m \, d\mathbf{x} = \delta_{mn} \qquad (3.1.143)$$

where $\mathbf{x}$ is the vector representing the coordinates, $\bar{u}$ is the complex conjugate of u, and the integration is over the domain D of interest. To solve (3.1.142) we expand ϕ_1 in terms of the orthonormal set $\{u_n\}$ as in the previous section; that is

$$\phi_1 = \sum_{m=1}^{\infty} a_m u_m \qquad (3.1.144)$$

Thus ϕ_1 satisfies $B(\phi_1) = 0$ because $B(u_m) = 0$ for each m. Letting $\phi_0 = u_n$ and $\lambda_0 = \mu_n$, and substituting (3.1.144) into (3.1.142), we obtain

$$\sum_{m=1}^{\infty} (\mu_n - \mu_m)a_m u_m = -\lambda_1 u_n + F(u_n) \qquad (3.1.145)$$

Multiplying (3.1.145) by $\bar{u}_s$, integrating over D, and using the orthonormality condition, we obtain

$$(\mu_n - \mu_s)a_s = -\lambda_1 \delta_{ns} + F_{ns} \qquad (3.1.146)$$

where

$$F_{ns} = \int_D F(u_n) \bar{u}_s \, d\mathbf{x} \tag{3.1.147}$$

If $n = s$, (3.1.146) gives

$$\lambda_1 = F_{nn} \tag{3.1.148}$$

If $n \neq s$

$$a_s = \frac{F_{ns}}{\mu_n - \mu_s} \tag{3.1.149}$$

Thus

$$\phi_1 = \sum_{m \neq n} \frac{F_{nm}}{\mu_n - \mu_m} u_m + a_{nn} u_n \tag{3.1.150}$$

The coefficient a_{nn} can be determined to be zero if we assume that $\phi = \phi_0 + \epsilon\phi_1 + O(\epsilon^2)$ is normalized as in the previous section.
Therefore, to first order

$$\phi = u_n + \epsilon \sum_{m \neq n} \frac{F_{nm}}{\mu_n - \mu_m} u_m + O(\epsilon^2) \tag{3.1.151}$$

$$\lambda = \mu_n + \epsilon F_{nn} + O(\epsilon^2) \tag{3.1.152}$$

As an example, let us consider the problem

$$\frac{d^2\phi}{dx^2} + \lambda\phi = -\epsilon\phi^3 \tag{3.1.153}$$

$$\phi(0) = \phi(1) = 0 \tag{3.1.154}$$

Here D is the interval $[0, 1]$, and

$$u_n = \sqrt{2} \sin n\pi x, \qquad \mu_n = n^2\pi^2 \tag{3.1.155}$$

Since $F(\phi) = -\phi^3$ and $\bar{u}_m = u_m$

$$\begin{aligned} F_{nm} &= -4 \int_0^1 \sin^3 n\pi x \sin m\pi x \, dx \\ &= \int_0^1 (\sin 3n\pi x - 3 \sin n\pi x) \sin m\pi x \, dx \\ &= \tfrac{1}{2}\delta_{m,3n} - \tfrac{3}{2}\delta_{nm} \end{aligned} \tag{3.1.156}$$

Therefore (3.1.151) and (3.1.152) become

$$\phi = \sqrt{2} \sin n\pi x - \frac{\epsilon\sqrt{2}}{16n^2\pi^2} \sin 3n\pi x + O(\epsilon^2) \tag{3.1.157}$$

$$\lambda = \dot{n}^2\pi^2 - \tfrac{3}{2}\epsilon + O(\epsilon^2) \tag{3.1.158}$$

The Degenerate Case.   In this case let $\mu_{n+k} = \mu_n$ for $k = 0, 1, 2, \ldots, M$. Then

$$\phi_0 = \sum_{k=0}^{M} b_k u_{n+k} \tag{3.1.159}$$

We substitute for ϕ_1 and ϕ_0 from (3.1.144) and (3.1.159) into (3.1.142) and take $\lambda_0 = \mu_n$ to obtain

$$\sum_{m=1}^{\infty} (\mu_n - \mu_m) a_m u_m = -\lambda_1 \sum_{k=0}^{M} b_k u_{n+k} + F\left[\sum_{k=0}^{M} b_k u_{n+k}\right] \tag{3.1.160}$$

We multiply (3.1.160) by $\bar{u}_s$ and integrate over D to obtain

$$(\mu_n - \mu_s) a_s = -\lambda_1 \sum_{k=0}^{M} b_k \delta_{s,n+k} + \mathscr{F}_s(b_0, b_1, \ldots, b_M) \tag{3.1.161}$$

where

$$\mathscr{F}_s = \int_D F\left[\sum_{k=0}^{M} b_k u_{n+k}\right] \bar{u}_s \, d\mathbf{x} \tag{3.1.162}$$

If $s = n + k$ for $k = 0, 1, 2, \ldots, M$, (3.1.161) gives

$$\mathscr{F}_{n+k}(b_0, b, \ldots, b_M) - \lambda_1 b_k = 0, \qquad k = 0, 1, 2, \ldots, M \tag{3.1.163}$$

These constitute $M + 1$ homogeneous algebraic equations for the $M + 1$ unknown b_m terms and the eigenvalue λ_1. If $s \neq n + k$, $k = 0, 1, \ldots, M$,

$$a_s = \frac{\mathscr{F}_s(b_0, b_1, \ldots, b_M)}{\mu_n - \mu_s} \tag{3.1.164}$$

As an example, let us consider the problem

$$\frac{d^4\phi}{dx^4} + 5\pi^2 \frac{d^2\phi}{dx^2} + \lambda\phi = \epsilon\phi \frac{d\phi}{dx} \tag{3.1.165}$$

$$\phi(0) = \phi''(0) = \phi(1) = \phi''(1) = 0 \tag{3.1.166}$$

In this case the solution of the linearized problem is

$$u_n = \sqrt{2} \sin n\pi x, \qquad \mu_n = n^2(5 - n^2)\pi^4 \tag{3.1.167}$$

Thus $\mu_1 = \mu_2 = 4\pi^4$, and we have degeneracy. We assume that, corresponding to the eigenvalue μ_1

$$\phi_0 = b_0\sqrt{2} \sin \pi x + b_1\sqrt{2} \sin 2\pi x \tag{3.1.168}$$

Then

$$F(\phi_0) = \phi_0 \frac{d\phi_0}{dx}$$

$$= \pi[-b_0 b_1 \sin \pi x + b_0{}^2 \sin 2\pi x + 3b_0 b_1 \sin 3\pi x + 2b_1^2 \sin 4\pi x] \tag{3.1.169}$$

Hence

$$\mathscr{F}_s = \sqrt{2} \int_0^1 F(\phi_0) \sin s\pi x \, dx$$

$$= \tfrac{1}{2}\sqrt{2}\,\pi[-b_0 b_1 \delta_{1s} + b_0{}^2 \delta_{2s} + 3b_0 b_1 \delta_{3s} + 2b_1{}^2 \delta_{4s}] \quad (3.1.170)$$

With $\mathscr{F}_s$ known (3.1.163) becomes

$$-\tfrac{1}{2}\sqrt{2}\pi b_0 b_1 - \lambda_1 b_0 = 0 \quad (3.1.171)$$

$$\tfrac{1}{2}\sqrt{2}\pi b_0{}^2 - \lambda_1 b_1 = 0 \quad (3.1.172)$$

While (3.1.164) gives

$$a_3 = \frac{3b_0 b_1}{40\sqrt{2}\,\pi^3}, \qquad a_4 = \frac{b_1{}^2}{90\sqrt{2}\,\pi^3} \quad (3.1.173)$$

Since $b_0 \neq 0$, (3.1.171) yields

$$b_1 = -\frac{\sqrt{2}}{\pi} \lambda_1 \quad (3.1.174)$$

Substituting for b_1 into (3.1.172) and solving for λ_1, we obtain

$$\lambda_1 = \mp \frac{1}{\sqrt{2}} i\pi b_0 \quad (3.1.175)$$

Hence

$$b_1 = \pm i b_0 \quad (3.1.176)$$

Therefore

$$\phi = b_0\sqrt{2} \sin \pi x \pm i b_0 \sqrt{2} \sin 2\pi x + \epsilon \left[a_1\sqrt{2} \sin \pi x + a_2\sqrt{2} \sin 2\pi x \right.$$

$$\left. \pm \frac{3}{40\pi^3} i b_0{}^2 \sin 3\pi x - \frac{1}{90\pi^3} b_0{}^2 \sin 4\pi x \right] + O(\epsilon^2) \quad (3.1.177)$$

$$\lambda = 4 \mp \frac{\pi}{\sqrt{2}} \epsilon i b_0 + O(\epsilon^2) \quad (3.1.178)$$

The constants a_1 and a_2 can be related to b_0 by normalizing ϕ_0. The solutions corresponding to μ_n, $n > 1$ are

$$\phi = \sqrt{2} \sin n\pi x + \epsilon \frac{1}{15n(n^2 - 1)\pi^3} \sin 2n\pi x + O(\epsilon^2) \quad (3.1.179)$$

$$\lambda = n^2(5 - n^2)\pi^4 + O(\epsilon^2) \quad (3.1.180)$$

3.1.8. THE QUASI-LINEAR KLEIN–GORDON EQUATION

We consider in this section the problem of the determination of periodic finite-amplitude traveling waves for the equation

$$u_{tt} - \alpha^2 u_{xx} + \gamma^2 u = \beta u^3 \qquad (3.1.181)$$

If we neglect the nonlinear term βu^3, we obtain the linear traveling harmonic wave

$$u = a \cos (kx - \omega t), \qquad \omega^2 = \alpha^2 k^2 + \gamma^2 \qquad (3.1.182)$$

The phase speed for this wave is ω/k which is independent of the amplitude a. In the nonlinear problem the phase speed is in general a function of the amplitude.

To determine the dependence of the phase speed c on the amplitude, we assume that

$$u = u(\theta), \qquad \theta = x - ct \qquad (3.1.183)$$

so that (3.1.181) becomes

$$(c^2 - \alpha^2)u'' + \gamma^2 u = \beta u^3 \qquad (3.1.184)$$

where primes denote differentiation with respect to θ. We assume the amplitude to be small and expand both u and c as

$$u = au_1 + a^3 u_3 + \cdots$$
$$c = c_0 + a^2 c_2 + \cdots \qquad (3.1.185)$$

Had we included the terms ac_1 and $a^2 u_2$, we would have found that $c_1 = 0$ and u_2 satisfies the same equation as u_1, hence u_2 is not included.

Substituting (3.1.185) into (3.1.184) and equating coefficients of equal powers of a, we obtain

$$(c_0^2 - \alpha^2)u_1'' + \gamma^2 u_1 = 0 \qquad (3.1.186)$$

$$(c_0^2 - \alpha^2)u_3'' + \gamma^2 u_3 = -2c_0 c_2 u_1'' + \beta u_1^3 \qquad (3.1.187)$$

We take the solution of (3.1.186) to be

$$u_1 = \cos k\theta, \qquad c_0^2 = \alpha^2 + \gamma^2 k^{-2} \qquad (3.1.188)$$

so that (3.1.185) coincides with (3.1.183) to $O(a)$. Then (3.1.187) becomes

$$(c_0{}^2 - \alpha^2)u_3'' + \gamma^2 u_3 = (2c_0 c_2 k^2 + \tfrac{3}{4}\beta)\cos k\theta + \tfrac{1}{4}\beta \cos 3k\theta \quad (3.1.189)$$

Secular terms are eliminated if $c_2 = -3\beta/8c_0 k^2$. Then

$$u_3 = -\frac{\beta}{32\gamma^2}\cos 3k\theta$$

Therefore

$$u = a\cos k\theta - \frac{a^3\beta}{32\gamma^2}\cos 3k\theta + \cdots$$

$$c = \sqrt{\alpha^2 + \gamma^2 k^{-2}}\left[1 - \frac{3a^2\beta}{8(\alpha^2 k^2 + \gamma^2)}\right] + \cdots \quad (3.1.190)$$

The technique used in this section was formalized by Stoker (1957) for the treatment of surface waves in liquids. This technique was used to treat the interaction of capillary and gravity waves in deep and finite-depth water by Pierson and Fife (1961) and Barakat and Houston (1968), respectively. It was also used by Maslowe and Kelly (1970) to treat waves in Kelvin–Helmholtz flow.

Instead of expanding the phase speed, we could have expanded the wavenumber to determine the wave number shift or the frequency to determine the frequency shift. Variations of this technique were applied to a variety of problems. For example, Malkus and Veronis (1958) treated the Bénard convection problem. Pedlowsky (1967) determined the response of a bounded ocean to a surface wind oscillating near one of the Rossby wave frequencies. Keller and Ting (1966) and Millman and Keller (1969) obtained periodic solutions for various systems governed by nonlinear partial differential equations, while Keller and Millman (1969) treated nonlinear electromagnetic and acoustic wave propagation. Rajappa (1970) investigated nonlinear Rayleigh–Taylor instability.

3.2. Lighthill's Technique

The essence of Lighthill's technique is to expand not only the dependent variable $u(x_1, x_2, \ldots, x_n; \epsilon)$ in powers of the small parameter ϵ, but also to expand one of the independent variables, say x_1, in powers of ϵ. Lighthill (1949a, 1961) introduced a new independent variable and then expanded both u and x_1 in powers of ϵ with coefficients depending on s. To a first approximation he assumed that x_1 and s are identical. Thus Lighthill assumed the following expansions for u and x_1

$$u = \sum_{m=0}^{\infty} \epsilon^m u_m(s, x_2, x_3, \ldots, x_n) \tag{3.2.1}$$

$$x_1 = s + \sum_{m=1}^{\infty} \epsilon^m \xi_m(s, x_2, x_3, \ldots, x_n) \tag{3.2.2}$$

It is clear that the straightforward expansion (Poincaré type) consists of (3.2.1) alone with s replaced by x_1. Since this straightforward expansion is not uniformly valid, Lighthill introduced (3.2.2) and chose ξ_m (called straining functions) so as to make both of the above expansions uniformly valid; that is, he chose ξ_m so that the resulting approximation is uniformly valid. In some cases this is accomplished by requiring that

$$\frac{u_m}{u_{m-1}} \quad \text{and} \quad \frac{\xi_m}{\xi_{m-1}} \quad \text{be bounded} \tag{3.2.3}$$

In other words, *higher approximations shall be no more singular than the first.*

Comparing (3.2.1) and (3.2.2) with (3.1.2) and (3.1.4), we see that Lighthill's technique is an extension of the method of strained parameters.

This technique was modified by Kuo (1953, 1956) to apply to viscous flows. For this reason Tsien (1956) called it the PLK method, that is, the Poincaré–Lighthill–Kuo method.

This method was applied to a variety of problems, especially wave propagation in nondispersive media. Lighthill (1949b) treated conical shock waves in steady supersonic flow. Whitham (1952) determined the pattern of shock waves on an axisymmetric projectile in steady supersonic flow and treated the propagation of spherical shocks in stars (1953). Legras (1951, 1953) and Lee and Sheppard (1966) applied it to steady supersonic flow past a thin airfoil, while Rao (1956) applied it to sonic booms. Holt and Schwartz (1963), Sakurai (1965), Holt (1967), and Akinsete and Lee (1969) investigated nonsimilar effects in the collapsing of an empty spherical cavity, while Jahsman (1968) treated the collapse of a gas-filled spherical cavity. Sirignano and Crocco (1964) analyzed combustion instability which is driven by chemical kinetics. Savage and Hasegawa (1967) studied the attenuation of pulses in metals, while Sakurai (1968) discussed the effect of plasma impedance in the problem of inverse pinch. Einaudi (1969, 1970) applied it to the propagation of acoustic gravity waves. Lewak (1969) and Zawadzki and Lewak (1971) solved Vlasov's equation. Espedal (1971) used a combination of this technique and the method of matched asymptotic expansions to determine the effect of ion-ion collision on an ion–acoustic plasma pulse.

Asano and Taniuti (1969, 1970) and Asano (1970) extended this technique to nondispersive wave propagation in slightly inhomogeneous media.

Melnik (1965) applied it to the entropy layer in the vicinity of a conical symmetry plane. McIntyre (1966) treated optimal control with discontinuous forcing functions. Ross (1970) applied it to diffusion-coupled biochemical reaction kinetics.

Barua (1954) analyzed secondary flows due to rotation in an unheated tube, Morton (1959) treated laminar convection in a heated tube, and Morris (1965) investigated the case of laminar convection in a vertical tube. Chang, Akins, and Bankoff (1966) analyzed the free convection of a liquid metal from a uniformly heated plate.

Crane (1959) rendered an asymptotic expansion for boundary layers uniformly valid. Goldburg and Cheng (1961) discussed the anomaly arising from the application of this technique and parabolic coordinates to the trailing edge boundary layer. Ockendon (1966) investigated the separation points in the Newtonian theory of hypersonic flow.

Since Lighthill's technique is a generalization of the method of strained parameters, the first technique gives results identical to those obtained using the latter technique whenever it is applicable. Therefore the examples discussed next are problems that cannot be treated by the method of strained parameters.

3.2.1. A FIRST-ORDER DIFFERENTIAL EQUATION

The first example treated by Lighthill is the first-order differential equation

$$(x + \epsilon y) \frac{dy}{dx} + q(x)y = r(x), \qquad y(1) = b > 0 \tag{3.2.4}$$

where $q(x)$ and $r(x)$ are regular functions for all x of interest. Wasow (1955) determined the necessary conditions for the convergence of Lighthill's expansion for this problem; the proof had an error which was corrected by Lighthill. Usher (1971) investigated the necessary conditions for the applicability of this technique to equations of the form

$$y' = f(x, y) + \epsilon g(x, y) + \cdots$$

Comstock (1968) showed that Lighthill's technique may lead to an erroneous expansion (see Exercise 3.28) for

$$(x^n + \epsilon y)y' + nx^{n-1}y = mx^{m-1}, \qquad y(1) = a > 1$$

while Burnside (1970) investigated the uniformity of the expansion obtained by straining $z = x^n$ rather than x.

It is clear that the region of nonuniformity is in the neighborhood of $x = 0$. For $\epsilon = 0$, (3.2.4) has the solution

$$y = \left[\exp \int^x - \frac{q(t)}{t} \, dt \right] \left[\int^x \frac{r(t)}{t} \left(\exp \int^t \frac{q(\tau)}{\tau} \, d\tau \right) + c \right] \tag{3.2.5}$$

Let $q(0) = q_0$, then

$$\exp \int^x \frac{q(t)}{t}\, dt = x^{q_0} R(x) \tag{3.2.6}$$

where $R(x)$ stands for a function regular at $x = 0$. Since $r(x)$ is regular at $x = 0$

$$y = R(x) + O(x^{-q_0}) \quad \text{as} \quad x \to 0 \tag{3.2.7}$$

except when q_0 is a negative integer. In the latter case

$$y = R(x) + O(x^{-q_0} \ln x) \quad \text{as} \quad x \to 0 \tag{3.2.8}$$

Equations (3.2.7) and (3.2.8) show that the zeroth-order solution of (3.2.4) is bounded or unbounded as $x \to 0$ depending on whether $q_0 < 0$ or $q_0 \geq 0$. In order to show the details of the method, we apply it to a specific example for which $q_0 = 2$.

In this case we consider the following problem treated by Lighthill (1949a) and Tsien (1956)

$$(x + \epsilon y)\frac{dy}{dx} + (2 + x)y = 0, \qquad y(1) = Ae^{-1} \tag{3.2.9}$$

where A is a constant. Following Lighthill, we assume that

$$y = \sum_{m=0}^{\infty} \epsilon^m y_m(s) \tag{3.2.10}$$

$$x = s + \sum_{m=1}^{\infty} \epsilon^m x_m(s) \tag{3.2.11}$$

Then

$$\frac{dy}{dx} = \frac{\dfrac{dy}{ds}}{\dfrac{dx}{ds}} = \frac{\displaystyle\sum_{m=0}^{\infty} \epsilon^m y_m'(s)}{1 + \displaystyle\sum_{m=1}^{\infty} \epsilon^m x_m'(s)} \tag{3.2.12}$$

In order to apply the boundary condition, we need to determine the value of s, denote it by $\tilde{s}$, corresponding to $x = 1$; that is, we must solve

$$\tilde{s} = 1 - \sum_{m=1}^{\infty} \epsilon^m x_m(\tilde{s}) \tag{3.2.13}$$

We expand $\tilde{s}$ in powers of ϵ according to

$$\tilde{s} = 1 + \epsilon \tilde{s}_1 + \epsilon^2 \tilde{s}_2 + \cdots \tag{3.2.14}$$

Substituting (3.2.14) into (3.2.13), expanding, and equating coefficients of equal powers of ϵ lead to

$$\tilde{s} = 1 - \epsilon x_1(1) - \epsilon^2[x_2(1) - x_1(1)x_1'(1)] + \cdots \tag{3.2.15}$$

The boundary condition can now be written as

$$Ae^{-1} = y_0(1) + \epsilon[y_1(1) - y_0'(1)x_1(1)] + \cdots \quad (3.2.16)$$

or

$$y_0(1) = Ae^{-1} \quad (3.2.17)$$

$$y_1(1) = y_0'(1)x_1(1) \quad (3.2.18)$$

Substituting (3.2.10) through (3.2.12) into (3.2.9), expanding, and equating the coefficients of ϵ^0 and ϵ to zero yield

$$sy_0' + (2 + s)y_0 = 0 \quad (3.2.19)$$

$$sy_1' + (2 + s)y_1 = -(2 + s)y_0x_1' - (y_0 + y_0')x_1 - y_0y_0' \quad (3.2.20)$$

The solution for y_0 is

$$y_0 = Ae^{-s}s^{-2} \quad (3.2.21)$$

With this solution (3.2.20) becomes

$$\frac{d}{ds}\left(\frac{y_1}{y_0}\right) = \frac{1}{s}\left[-(2 + s)x_1' + \frac{2}{s}x_1 + Ae^{-s}s^{-2}\left(\frac{2}{s} + 1\right)\right] \quad (3.2.22)$$

If $x_1 = 0$, (3.2.22) reduces to the equation for the first-order term in the straightforward expansion, where y_1 is more singular at $x = 0$ than y_0. In fact, $y_0 = O(x^{-2})$, while $y_1 = O(x^{-5})$ as $x \to 0$. To render the above expansion uniformly valid, x_1 can be chosen so that y_1 is no more singular than y_0 by eliminating the right-hand side of (3.2.22). However, Lighthill found that a uniformly valid expansion can be obtained by choosing x_1 to eliminate the worst singularity. Thus he put

$$x_1' - \frac{x_1}{s} = \frac{A}{s^3} \quad (3.2.23)$$

or

$$x_1 = -\frac{A}{3s^2} \quad (3.2.24)$$

Then (3.2.22) becomes

$$\frac{d}{ds}\left[\frac{y_1}{y_0}\right] = -\frac{2A}{3s^3} - \frac{2A}{s^4} + Ae^{-s}\left(\frac{1}{s^3} + \frac{2}{s^4}\right) \quad (3.2.25)$$

Hence

$$y_1 = A^2e^{-s}s^{-2}\left[\frac{2}{3s^3} + \frac{1}{3s^2} - \int_s^1 e^{-\xi}\left(\frac{2}{\xi^4} - \frac{1}{\xi^3}\right)d\xi\right] \quad (3.2.26)$$

The straining function x_2 can be found from the elimination of the worst singularity in y_2 to be

$$x_2 = -\frac{3A^2}{10s^4} \quad (3.2.27)$$

Therefore

$$y = Ae^{-s}s^{-2}\left\{1 + A\epsilon\left[\frac{2}{3s^3} + \frac{1}{3s^2} - \int_s^1 e^{-\xi}\left(\frac{2}{\xi^4} + \frac{1}{\xi^3}\right) d\xi\right]\right\} + O\left(\frac{\epsilon^2}{s^6}\right) \quad (3.2.28)$$

where

$$x = s - \frac{\epsilon A}{3s^2} - \frac{3\epsilon^2 A^2}{10s^4} + O\left(\frac{\epsilon^3}{s^6}\right) \quad (3.2.29)$$

The roughest approximation that is uniformly valid near the origin is

$$y = Ae^{-s}s^{-2} \quad (3.2.30)$$

where s is the root of

$$x = s - \frac{\epsilon A}{3s^2} \quad (3.2.31)$$

which is approximately x when $x \geq 0$ and $\epsilon \ll 1$. This expansion is assumed to start from a positive value of x, and it is required to continue downward and through $x = 0$. For a physical problem this continuation stops if there exists a real branch point of s as a function of x before the origin. The branch point is given by $dx/ds = 0$, or equivalently by $x + \epsilon y = 0$, which is a singularity of the original equation (3.2.9). In this case the branch point is given by $s \approx (-2A\epsilon/3)^{1/3}$ which is positive if and only if $A < 0$. Therefore the above expansion would be valid up to the origin if $A > 0$, and stops at $x \approx (3/2)(-2A\epsilon/3)^{1/3}$ if $A < 0$.

If $A = 1$ then $x = 0$ corresponds to

$$s = \left(\frac{\epsilon}{3}\right)^{1/3} + \frac{9}{10}\left(\frac{\epsilon}{3}\right)^{2/3} + O(\epsilon) \quad (3.2.32)$$

Hence at $x = 0$

$$y = \left(\frac{3}{\epsilon}\right)^{2/3} - \frac{3}{10}\left(\frac{3}{\epsilon}\right)^{1/3} + O(1) \quad (3.2.33)$$

3.2.2.   THE ONE-DIMENSIONAL EARTH-MOON-SPACESHIP PROBLEM

The one-dimensional earth-moon-spaceship problem was studied by Nayfeh (1965a) and can be reduced to (see Section 2.4.2)

$$\frac{1}{2}\left(\frac{dx}{dt}\right)^2 = \frac{1 - \mu}{x} + \frac{\mu}{1 - x}, \qquad t(0) = 0 \quad (3.2.34)$$

We assume that

$$t = t_0(s) + \mu t_1(s) + O(\mu^2) \quad (3.2.35)$$

$$x = s + \mu x_1(s) + O(\mu^2) \quad (3.2.36)$$

Substituting (3.2.35) and (3.2.36) into (3.2.34) and equating coefficients of equal powers of μ, we have

$$2t_0'^2 = s, \qquad t_0(0) = 0 \tag{3.2.37}$$

$$\frac{t_1'}{t_0'^3} = \frac{x_1'}{t_0'^2} + \frac{x_1}{s^2} + \frac{1}{s} - \frac{1}{1-s}, \qquad t_1(0) = t_0'(0)x_1(0) \tag{3.2.38}$$

The solution of (3.2.37) is

$$\sqrt{2}\, t_0 = \tfrac{2}{3}s^{3/2} \tag{3.2.39}$$

If $x_1 = 0$, $t_1 = O[\ln(1-x)]$ as $x \to 1$. The singularity in t_1 can be removed if the right-hand side of (3.2.38) is eliminated; that is

$$\frac{x_1'}{t_0'^2} + \frac{x_1}{s^2} + \frac{1}{s} - \frac{1}{1-s} = 0 \tag{3.2.40}$$

whose solution is

$$x_1 = -1 + \tfrac{1}{2}s^{-1/2} \ln \frac{1 + s^{1/2}}{1 - s^{1/2}} - \tfrac{2}{3}s \tag{3.2.41}$$

Therefore, to first approximation

$$\sqrt{2}\, t = \tfrac{2}{3}s^{3/2} + O(\mu) \tag{3.2.42}$$

where s is the root of

$$x = s - \mu\left[1 - \tfrac{1}{2}s^{-1/2} \ln \frac{1 + s^{1/2}}{1 - s^{1/2}} + \tfrac{2}{3}s\right] + O(\mu^2) \tag{3.2.43}$$

3.2.3. A SOLID CYLINDER EXPANDING UNIFORMLY IN STILL AIR

Let us next solve the problem of a cylindrical shock wave produced by a cylindrical solid body expanding uniformly from zero in inviscid nonconducting still air. This problem was also studied by Lighthill. The radial expansion velocity is assumed to be ϵa_0 where a_0 is the speed of sound in still air and ϵ is a small quantity. The shock propagates with a uniform radial velocity Ma_0 where M is the shock Mach number. The flow between the cylinder and the shock is adiabatic and isentropic, hence it can be represented by a potential function $\varphi(r, t)$ (the radial velocity $q = \varphi_r$) given by

$$a^2 \nabla^2 \varphi = \frac{\partial^2 \varphi}{\partial t^2} + 2 \frac{\partial \varphi}{\partial r} \frac{\partial^2 \varphi}{\partial r \, \partial t} + \left(\frac{\partial \varphi}{\partial r}\right)^2 \frac{\partial^2 \varphi}{\partial r^2} \tag{3.2.44}$$

where a is the local speed of sound and is related to a_0 by Bernoulli's equation; namely

$$a^2 + (\gamma - 1)\left[\frac{\partial \varphi}{\partial t} + \frac{1}{2}\left(\frac{\partial \varphi}{\partial r}\right)^2\right] = a_0^2 \tag{3.2.45}$$

where γ is the ratio of the gas specific heats. The gas is assumed perfect with constant specific heats. Three boundary conditions must be satisfied by φ. The velocity of the air at the cylinder's surface is equal to the velocity of its expansion, that is

$$\frac{\partial \varphi}{\partial r}(\epsilon a_0 t) = \epsilon a_0 \qquad (3.2.46)$$

The second condition is the continuity of φ across the shock. Since $\varphi = 0$ in still air

$$\varphi(Ma_0 t) = 0 \qquad (3.2.47)$$

The third condition is the Rankine–Hugonoit relation between the shock velocity and the velocity of the air behind it; that is

$$\frac{\partial \varphi}{\partial r}(Ma_0 t) = \frac{2a_0(M^2 - 1)}{M(\gamma + 1)} \qquad (3.2.48)$$

Since there is no fundamental length in this problem, all flow quantities are functions only of $r/a_0 t$. Thus we let

$$\varphi = a_0^2 t f(x), \qquad x = \frac{r}{a_0 t} \qquad (3.2.49)$$

Then the problem becomes

$$\left[1 - x^2 + (\gamma + 1)x\frac{df}{dx} - (\gamma - 1)f - \tfrac{1}{2}(\gamma + 1)\left(\frac{df}{dx}\right)^2\right]\frac{d^2f}{dx^2}$$
$$+ \frac{1}{x}\frac{df}{dx}\left\{1 + (\gamma - 1)\left[x\frac{df}{dx} - f - \frac{1}{2}\left(\frac{df}{dx}\right)^2\right]\right\} = 0 \quad (3.2.50)$$

subject to the boundary conditions

$$\frac{df}{dx}(\epsilon) = \epsilon \qquad (3.2.51)$$

$$f(M) = 0 \qquad (3.2.52)$$

$$\frac{df}{dx}(M) = 2\frac{(M^2 - 1)}{M(\gamma + 1)} \qquad (3.2.53)$$

Since there are three boundary conditions imposed on a second-order differential equation, a relationship must exist between M and ϵ.

Since ϵ is small, f is small, hence the zeroth-order term in the straightforward expansion is the solution of the linearized form of (3.2.50); that is

$$(1 - x^2)\frac{d^2f}{dx^2} + \frac{1}{x}\frac{df}{dx} = 0 \qquad (3.2$$

Using the above boundary conditions, we find that

$$f = \epsilon^2 \int_1^x \sqrt{\frac{1}{x^2} - 1} \, dx, \qquad M = 1 \qquad (3.2.55)$$

This approximate solution has no physical meaning for $x > 1$, and the shock Mach number must be greater than 1 for there to be any propagation.

In order to determine a valid solution beyond $x = 1$, and in order to determine by how much M exceeds 1, we find it convenient to transform the second-order equation (3.2.50) into a system of two first-order equations by letting

$$\frac{df}{dx} = g \qquad (3.2.56)$$

Then we assume the following expansions

$$f = \epsilon^2 f_0 + \epsilon^4 f_1 + \cdots \qquad (3.2.57)$$

$$g = \epsilon^2 g_0 + \epsilon^4 g_1 + \cdots \qquad (3.2.58)$$

$$x = s + \epsilon^2 x_1(s) + \epsilon^4 x_2(s) + \cdots \qquad (3.2.59)$$

$$M = 1 + \epsilon^2 M_1 + \epsilon^4 M_2 + \cdots \qquad (3.2.60)$$

The zeroth-order term is given by (3.2.54) if x is replaced by s; that is

$$g_0 = \sqrt{\frac{1}{s^2} - 1}$$

$$f_0 = \int_1^s g_0(\xi) \, d\xi \qquad (3.2.61)$$

The first-order problem is

$$(1 - s^2)g_1' + \frac{g_1}{s} - (1 - s^2)g_0' x_1' + [-2s x_1 + (\gamma + 1)s g_0 - (\gamma - 1)f_0]g_0'$$

$$+ \frac{g_0}{s}(\gamma - 1)(s g_0 - f_0) - \frac{g_0 x_1}{s^2} = 0 \qquad (3.2.62)$$

$$f_1' = g_1 + g_0 x_1' \qquad (3.2.63)$$

As $s \to 1$, $g_0 \to \sqrt{2(1 - s)}$; $f_0 \to -\tfrac{2}{3}\sqrt{2}\,(1 - s)^{3/2}$, hence (3.2.62) becomes

$$g_1 = \frac{-x_1}{\sqrt{2(1 - s)}} + \gamma + 1 + O(\sqrt{1 - s}) \quad \text{as} \quad s \to 1 \qquad (3.2.64)$$

Thus g_1 will have a singularity at $s = 1$ unless $x_1 = 0$, and $g_1(1) = \gamma + 1$ as a consequence.

In order to determine M_1, we use the boundary conditions at the shock. If $\tilde{s}$ corresponds to the position of the shock $x = M$, then to order ϵ^2

$$\tilde{s} = 1 + \epsilon^2 M_1 + \cdots \tag{3.2.65}$$

and the boundary condition (3.2.53) gives

$$\epsilon^2 g_0 (1 + \epsilon^2 M_1 + \cdots) + \cdots = \frac{4}{\gamma + 1} \epsilon^2 M_1 + \cdots \tag{3.2.66}$$

Substituting for g_0 from (3.2.61) and equating the coefficients of ϵ^2 on both sides, we obtain $M_1 = 0$. Hence

$$\tilde{s} = 1 + \epsilon^4 [M_2 - x_2(1)] + \cdots \tag{3.2.67}$$

Then (3.2.52) and (3.2.57) give

$$f_1(1) = 0 \tag{3.2.68}$$

The second-order equation gives for g_2

$$g_2 = -\frac{2x_2 - (\gamma + 1)^2}{2\sqrt{2(1 - s)}} + O(1) \quad \text{as} \quad s \to 1 \tag{3.2.69}$$

In order to remove the singularity from g_2, we set

$$x_2 = \tfrac{1}{2}(\gamma + 1)^2 \tag{3.2.70}$$

Hence $\tilde{s} = 1 + \epsilon^4 [M_2 - (\gamma + 1)^2/2]$ and, to fourth order, the boundary condition (3.2.53) gives

$$\sqrt{(\gamma + 1)^2 - 2M_2} + (\gamma + 1) = \frac{4M_2}{\gamma + 1} \tag{3.2.71}$$

whose solution is $M_2 = 3(\gamma + 1)^2/8$. Consequently

$$M = 1 + \tfrac{3}{8}(\gamma + 1)^2 \epsilon^4 + O(\epsilon^6) \tag{3.2.72}$$

Pandey (1968) treated the case of a solid cylinder expanding uniformly in still water rather than air.

3.2.4. SUPERSONIC FLOW PAST A THIN AIRFOIL

The fourth example in the application of Lighthill's method is the determination of a uniformly valid expansion for supersonic flow past a thin airfoil (Legras, 1951, 1953) discussed in Section 2.1.3. To carry out the expansion, it is more convenient to transform the original second-order differential equation into a system of two first-order equations by letting

$$u = \phi_x, \qquad v = \phi_y$$

in (2.1.19) through (2.1.21) and obtaining

$$v_y - B^2 u_x = M^2[(\gamma + 1)uu_x + (\gamma - 1)uv_y + 2vv_x + \text{cubic terms}] \quad (3.2.73)$$

$$u_y = v_x \quad (3.2.74)$$

$$\frac{v + \epsilon Tv_y + \cdots}{1 + u + \epsilon Tu_y + \cdots} = \epsilon T'(x) \quad \text{at} \quad y = 0 \quad (3.2.75)$$

$$u(x, y) = v(x, y) = 0 \quad \text{(upstream)} \quad (3.2.76)$$

The straightforward expansion of this problem was found in Section 2.1.3 to break down as $y \to \infty$. Since u and v vanish upstream, a uniformly valid expansion can be obtained by straining the linearized outgoing characteristic (i.e., $x - By = $ a constant). Thus we let

$$u = \epsilon u_1(\xi, \eta) + \epsilon^2 u_2(\xi, \eta) + \cdots \quad (3.2.77)$$

$$v = \epsilon v_1(\xi, \eta) + \epsilon^2 v_2(\xi, \eta) + \cdots \quad (3.2.78)$$

where

$$x - By = \xi + \epsilon G_1(\xi, \eta) + \epsilon^2 G_2(\xi, \eta) + \cdots \quad (3.2.79)$$

$$y = \eta \quad (3.2.80)$$

The straining functions G_i can be determined by imposing the condition that (3.2.77) and (3.2.78) be uniformly valid for large distances; that is, u_2/u_1 and v_2/v_1 are bounded. This condition has been shown to be equivalent to requiring ξ to be the outgoing characteristic of the nonlinear equations (Lighthill, 1949a; Whitham, 1952, 1953; Lin, 1954; Fox, 1955).

Since the characteristics of (2.1.19) are given by

$$[1 - M^2(\gamma - 1)\phi_x + \cdots](dx)^2 + [2M^2\phi_y + \cdots] \, dx \, dy$$
$$- [B^2 + M^2(\gamma + 1)\phi_x + \cdots](dy)^2 = 0 \quad (3.2.81)$$

the equation for the outgoing characteristic is

$$\left.\frac{dx}{dy}\right|_{\xi=\text{constant}} = c \quad (3.2.82)$$

where

$$c = B + \frac{M^2}{2B}\{[B^2(\gamma - 1) + (\gamma + 1)]u - 2Bv\} + \cdots \quad (3.2.83)$$

Equation (3.2.82) can be rewritten as

$$\frac{\partial x}{\partial \eta} = c\frac{\partial y}{\partial \eta} \quad (3.2.84)$$

The problem is thus reduced to expanding the dependent variables u and v as well as the independent variable x in terms of ϵ, the variable $\eta = y$, and the outgoing characteristic ξ. Thus (3.2.79) is equivalent to

$$x = x_0(\xi, \eta) + \epsilon x_1(\xi, \eta) + \epsilon^2 x_2(\xi, \eta) + \cdots \qquad (3.2.85)$$

where

$$x_0 = \xi + B\eta \quad \text{and} \quad x_i = G_i \quad \text{for} \quad i \geq 1 \qquad (3.2.86)$$

To fix the parametrization we need to place an initial condition on x. This condition is taken to be

$$x(\xi, 0) = \xi \qquad (3.2.87)$$

which is equivalent to choosing G_i to vanish at $y = 0$.

To transform from the independent variables x and y to ξ and η, we note that

$$\frac{\partial}{\partial \xi} = x_\xi \frac{\partial}{\partial x}$$

$$\frac{\partial}{\partial \eta} = x_\eta \frac{\partial}{\partial x} + y_\eta \frac{\partial}{\partial y} = c\frac{\partial}{\partial x} + \frac{\partial}{\partial y}$$

on account of (3.2.80) and (3.2.84). Hence

$$\frac{\partial}{\partial x} = \frac{1}{x_\xi}\frac{\partial}{\partial \xi}, \qquad \frac{\partial}{\partial y} = \frac{\partial}{\partial \eta} - \frac{c}{x_\xi}\frac{\partial}{\partial \xi} \qquad (3.2.88)$$

Substituting (3.2.77), (3.2.78), and (3.2.85) into (3.2.73) through (3.2.76), (3.2.83), (3.2.84), and (3.2.87), using (3.2.88), and equating coefficients of equal powers of ϵ, we obtain

Order ϵ

$$x_{0\xi}v_{1\eta} - (Bv_{1\xi} + B^2u_{1\xi}) = 0 \qquad (3.2.89)$$

$$x_{0\xi}u_{1\eta} - (Bu_{1\xi} + v_{1\xi}) = 0 \qquad (3.2.90)$$

$$v_1(\xi, 0) = T'(\xi) \qquad (3.2.91)$$

$$x_{1\eta} = \frac{M^2}{2B}\{[B^2(\gamma - 1) + (\gamma + 1)]u_1 - 2Bv_1\} \qquad (3.2.92)$$

$$x_1(\xi, 0) = 0 \qquad (3.2.93)$$

The solution of (3.2.89) through (3.2.91) that vanishes upstream is

$$v_1 = T'(\xi), \qquad u_1 = -B^{-1}T'(\xi) \qquad (3.2.94)$$

which coincides with the linearized solution. Then (3.2.92) becomes

$$x_{1\eta} = -\tfrac{1}{2}M^4(\gamma + 1)B^{-2}T'(\xi) \qquad (3.2.95)$$

whose solution subject to (3.2.93) is

$$x_1 = -\tfrac{1}{2}M^4(\gamma + 1)B^{-2}\eta T'(\xi) \tag{3.2.96}$$

Therefore a first-order uniformly valid expansion is given by the first terms in (3.2.77) and (3.2.78) where

$$x - By = \xi - \tfrac{1}{2}\epsilon M^4(\gamma + 1)B^{-2}yT'(\xi) + O(\epsilon^2) \tag{3.2.97}$$

on account of (3.2.85), (3.2.86), and (3.2.96). This solution shows that the uniformly valid first-order expansion for a hyperbolic system of equations is simply the linearized solution with the linearized characteristic replaced by the characteristic calculated by including the first-order nonlinear terms.

Higher-order approximations can be obtained in a straightforward manner. A second-order expansion was obtained by Lee and Sheppard (1966).

For general problems the velocity potential ϕ does not vanish upstream. In this case a uniformly valid expansion can be obtained by expanding the dependent as well as both of the independent variables x and y in terms of ϵ, and both characteristics ξ and η of the nonlinear equations. Thus we augment (3.2.83) and (3.2.84) by equations describing the ingoing characteristic η and introduce another expansion for y similar to (3.2.85). We next illustrate such a procedure for a more general system of hyperbolic equations.

3.2.5. EXPANSIONS BY USING EXACT CHARACTERISTICS— NONLINEAR ELASTIC WAVES

In the hyperbolic differential equation discussed above, a uniformly valid expansion was obtained by straining one of the characteristics of the linearized equation. The resulting strained variable was a better approximation to the exact characteristic. Lin (1954) and Fox (1955) generalized Lighthill's technique for problems of hyperbolic differential equations in two independent variables by adopting characteristic parameters as the independent variables, a procedure that amounts to straining two families of characteristics. Thus they were able to treat general waves in fluid flow in which the ingoing and outgoing waves interact.

This method was applied by Verhagen and Van Wijngaarden (1965) to the hydraulic jump problem. Guiraud (1965), Oswatitsch (1965), and Zierep and Heynatz (1965) applied it to gas dynamics waves of finite amplitude. Gretler (1968) devised an indirect method to calculate the plane flow past an airfoil, while Van Wijngaarden (1968) analyzed the oscillation near resonance in open pipes. Thermally driven nonlinear one-dimensional oscillations were treated by Chu and Ying (1963) and Rehm (1968) and by Gundersen (1967) for a conducting fluid. Chu (1963) and Mortell (1971) studied self-sustained oscillations in a pipe. Lick (1969) analyzed propagation of waves in isentropic and chemically reacting compressible fluids, while

Lesser (1970) investigated wave propagation in inhomogeneous media. Parker and Varley (1968) treated the nonlinear interaction of stretching and deflection waves in elastic membranes and strings, while Mortell and Varley (1971) discussed the nonlinear free vibration of an elastic panel. Richmond and Morrison (1968) applied this technique to an axisymmetric plasticity problem, while Davison (1968) obtained a second-order expansion for nonlinear elastic waves in an isotropic medium by using two characteristics as independent variables. Now, we explain the technique by determining a first-order expansion for nonlinear elastic waves in anisotropic materials.

If u and v represent the displacements in the x and y directions, then

$$\rho u_{tt} = \sigma_x \qquad (3.2.98)$$

$$\rho v_{tt} = \tau_x \qquad (3.2.99)$$

where ρ is the material density. We let $P = u_x$ and $Q = v_x$ and assume the stresses σ and τ to be polynomials in P and Q such that

$$\begin{aligned} \sigma_P \to \lambda + 2\mu, \qquad \sigma_Q \to 0 \\ \tau_P \to 0, \qquad \tau_Q \to \mu \end{aligned} \qquad (3.2.100)$$

as P and $Q \to 0$. Here, λ and μ are the Lamé moduli of the linear theory of elasticity. Thus

$$\frac{\sigma}{\rho} = c_p^2 P + \tfrac{1}{2}a_1 P^2 + a_2 PQ + \tfrac{1}{2}a_3 Q^2 + \cdots \qquad (3.2.101)$$

$$\frac{\tau}{\rho} = c_v^2 Q + \tfrac{1}{2}b_1 P^2 + b_2 PQ + \tfrac{1}{2}b_3 Q^2 + \cdots \qquad (3.2.102)$$

where

$$c_p^2 = \frac{\lambda + 2\mu}{\rho}, \qquad c_s^2 = \frac{\mu}{\rho} \qquad (3.2.103)$$

with c_p and c_s the principal and shear speeds of propagation. By letting

$$R = u_t \quad \text{and} \quad S = v_t \qquad (3.2.104)$$

and substituting (3.2.101) and (3.2.102) into (3.2.98) and (3.2.99), we have

$$R_t - c_p^2 P_x = \alpha P_x + \beta Q_x + \cdots \qquad (3.2.105)$$

$$S_t - c_s^2 Q_x = \gamma P_x + \delta Q_x + \cdots \qquad (3.2.106)$$

where

$$\alpha = a_1 P + a_2 Q, \qquad \beta = a_2 P + a_3 Q \qquad (3.2.107)$$

$$\gamma = b_1 P + b_2 Q, \qquad \delta = b_2 P + b_3 Q \qquad (3.2.108)$$

Since $P = u_x$ and $Q = v_x$, (3.2.104) gives

$$P_t = R_x, \qquad Q_t = S_x \tag{3.2.109}$$

Note that we have replaced the system of the two second-order differential equations (3.2.98) and (3.2.99) by a system of four first-order equations, which is more convenient for the application of the method of strained coordinates.

To complete the problem formulation, we need to specify the initial conditions. We consider the case investigated by Davison (1968) in which the material is initially unstressed, at rest, and occupying the half-space $x \geq 0$ when a disturbance is introduced at $x = 0$; that is

$$P(0, t) = \epsilon\phi(t), \qquad Q(0, t) = \epsilon\psi(t) \quad \text{for} \quad t \geq 0$$
$$P(x, 0) = Q(x, 0) = R(x, 0) = S(x, 0) = 0 \quad \text{for} \quad x \geq 0 \tag{3.2.110}$$

with

$$\phi(t) = \psi(t) \equiv 0 \quad \text{for} \quad t \leq 0 \tag{3.2.111}$$

where ϕ and ψ are known functions and ϵ is a small but finite dimensionless quantity. The condition (3.2.111) is equivalent to the vanishing of P and Q along the ingoing characteristics.

To obtain a uniformly valid expansion for the above problem, we expand the dependent as well as the independent variables in terms of ϵ and the outgoing characteristic parameters ξ and η. Thus we let

$$P = \epsilon P_1(\xi, \eta) + \epsilon^2 P_2(\xi, \eta) + \cdots \tag{3.2.112}$$

$$Q = \epsilon Q_1(\xi, \eta) + \epsilon^2 Q_2(\xi, \eta) + \cdots \tag{3.2.113}$$

$$R = \epsilon R_1(\xi, \eta) + \epsilon^2 R_2(\xi, \eta) + \cdots \tag{3.2.114}$$

$$S = \epsilon S_1(\xi, \eta) + \epsilon^2 S_2(\xi, \eta) + \cdots \tag{3.2.115}$$

$$x = x_0(\xi, \eta) + \epsilon x_1(\xi, \eta) + \epsilon^2 x_2(\xi, \eta) + \cdots \tag{3.2.116}$$

$$t = t_0(\xi, \eta) + \epsilon t_1(\xi, \eta) + \epsilon^2 t_2(\xi, \eta) + \cdots \tag{3.2.117}$$

To first-order quantities in P and Q, the characteristic wave speeds of (3.2.105), (3.2.106), and (3.2.109) are given by

$$c = \pm\left(c_p + \frac{\alpha}{2c_p}\right), \pm\left(c_s + \frac{\delta}{2c_s}\right) \tag{3.2.118}$$

Thus to $O(P, Q)$ the outgoing characteristics ξ and η are given by

$$x_\eta = c_1 t_\eta \tag{3.2.119}$$

$$x_\xi = c_2 t_\xi \tag{3.2.120}$$

where c_1 and c_2 are the positive speeds in (3.2.118). To fix the parametrization we need to place initial conditions on x and t. These conditions are taken to be

$$x(\xi, \xi) = 0 \quad \text{and} \quad t(\xi, \xi) = \xi \tag{3.2.121}$$

In terms of these new independent variables, the initial conditions (3.2.110) become

$$P(\xi, \xi) = \epsilon\phi(\xi), \qquad Q(\xi, \xi) = \epsilon\psi(\xi)$$
$$P(0, \eta) = Q(\xi, 0) = R(0, \eta) = S(\xi, 0) = 0 \tag{3.2.122}$$

To transform from the independent variables x and t to ξ and η, we note that

$$\frac{\partial}{\partial \xi} = \frac{\partial x}{\partial \xi}\frac{\partial}{\partial x} + \frac{\partial t}{\partial \xi}\frac{\partial}{\partial t} = \frac{\partial t}{\partial \xi}\left(\frac{\partial}{\partial t} + c_2\frac{\partial}{\partial x}\right)$$

$$\frac{\partial}{\partial \eta} = \frac{\partial t}{\partial \eta}\left(\frac{\partial}{\partial t} + c_1\frac{\partial}{\partial x}\right) \tag{3.2.123}$$

Hence

$$\frac{\partial}{\partial x} = \frac{1}{c_2 - c_1}\left(\frac{1}{t_\xi}\frac{\partial}{\partial \xi} - \frac{1}{t_\eta}\frac{\partial}{\partial \eta}\right)$$

$$\frac{\partial}{\partial t} = \frac{-1}{c_2 - c_1}\left(\frac{c_1}{t_\xi}\frac{\partial}{\partial \xi} - \frac{c_2}{t_\eta}\frac{\partial}{\partial \eta}\right) \tag{3.2.124}$$

Substituting (3.2.112) through (3.2.117) into (3.2.105), (3.2.106),(3.2.109), and (3.2.118) through (3.2.122), using (3.2.124), and equating coefficients of equal powers of ϵ, we obtain

Order ϵ^0

$$x_{0\eta} - c_p t_{0\eta} = 0 \tag{3.2.125}$$

$$x_{0\xi} - c_s t_{0\xi} = 0 \tag{3.2.126}$$

$$x_0(\xi, \xi) = 0, \qquad t_0(\xi, \xi) = \xi \tag{3.2.127}$$

Order ϵ

$$-t_{0\eta}(c_p P_1 + R_1)_\xi + t_{0\xi}(c_s P_1 + R_1)_\eta = 0 \tag{3.2.128}$$

$$-t_{0\eta}(c_p Q_1 + S_1)_\xi + t_{0\xi}(c_s Q_1 + S_1)_\eta = 0 \tag{3.2.129}$$

$$-t_{0\eta}(c_p R_1 + c_p{}^2 P_1)_\xi + t_{0\xi}(c_s R_1 + c_p{}^2 P_1)_\eta = 0 \tag{3.2.130}$$

$$-t_{0\eta}(c_p S_1 + c_s{}^2 Q_1)_\xi + t_{0\xi}(c_s S_1 + c_s{}^2 Q_1)_\eta = 0 \tag{3.2.131}$$

$$(x_1 - c_p t_1)_\eta = \tfrac{1}{2}c_p^{-1}(a_1 P_1 + a_2 Q_1)t_{0\eta} \tag{3.2.132}$$

$$(x_1 - c_s t_1)_\xi = \tfrac{1}{2}c_s^{-1}(b_2 P_1 + b_3 Q_1)t_{0\xi} \tag{3.2.133}$$

$$x_1(\xi, \xi) = t_1(\xi, \xi) = 0 \tag{3.2.134}$$

$$P_1(\xi, \xi) = \phi(\xi), \qquad Q_1(\xi, \xi) = \psi(\xi)$$
$$P_1(0, \eta) = Q_1(\xi, 0) = R_1(0, \eta) = S_1(\xi, 0) = 0 \tag{3.2.135}$$

The solution of (3.2.125) through (3.2.127) is

$$x_0 = \frac{c_p c_s(\xi - \eta)}{c_p - c_s}, \qquad t_0 = \frac{c_p \xi - c_s \eta}{c_p - c_s} \tag{3.2.136}$$

which is simply the linearized characteristics

$$t_0 - \frac{x_0}{c_p} = \xi, \qquad t_0 - \frac{x_0}{c_s} = \eta$$

Substituting for t_0 from (3.2.136) into (3.2.128) through (3.2.131) and solving the resulting equations subject to (3.2.135), we obtain

$$P_1(\xi, \eta) = \phi(\xi), \qquad Q_1(\xi, \eta) = \psi(\eta)$$
$$R_1(\xi, \eta) = -c_p \phi(\xi), \qquad S_1(\xi, \eta) = -c_s \psi(\eta) \tag{3.2.137}$$

which are, except for an ϵ factor, simply the solution of the linearized problem. With this solution (3.2.132) and (3.2.133) become

$$(x_1 - c_p t_1)_\eta = \Gamma_1 a_1 \phi(\xi) + \Gamma_1 a_2 \psi(\eta)$$
$$(x_1 - c_s t_1)_\xi = \Gamma_2 b_2 \phi(\xi) + \Gamma_2 b_3 \psi(\eta) \tag{3.2.138}$$

where $(\Gamma_1, \Gamma_2) = -(1/2)(c_p - c_s)^{-1}(c_s/c_p, -c_p/c_s)$. The solution of (3.2.138) subject to (3.2.134) is

$$x_1 - c_p t_1 = \Gamma_1 a_1(\eta - \xi)\phi(\xi) + \Gamma_1 a_2 \int_\xi^\eta \psi(\zeta)\, d\zeta$$
$$x_1 - c_s t_1 = \Gamma_2 b_2 \int_\eta^\xi \phi(\zeta)\, d\zeta + \Gamma_2 b_3(\xi - \eta)\psi(\eta) \tag{3.2.139}$$

Therefore a first-order uniformly valid expansion is

$$P = \epsilon\phi(\xi) + O(\epsilon^2), \qquad R = -\epsilon c_p \phi(\xi) + O(\epsilon^2)$$
$$Q = \epsilon\psi(\eta) + O(\epsilon^2), \qquad S = -\epsilon c_s \psi(\eta) + O(\epsilon^2) \tag{3.2.140}$$

where ξ and η are given by

$$x - c_p t = -c_p \xi + \epsilon\Gamma_1\left[a_1(\eta - \xi)\phi(\xi) + a_2 \int_\xi^\eta \psi(\zeta)\, d\zeta \right] + O(\epsilon^2)$$
$$x - c_s t = -c_s \eta + \epsilon\Gamma_2\left[b_2 \int_\eta^\xi \phi(\zeta)\, d\zeta + b_3(\xi - \eta)\psi(\eta) \right] + O(\epsilon^2) \tag{3.2.141}$$

As in the problem of supersonic flow past a thin airfoil treated in the previous section, the first-order uniformly valid expansion is simply the linearized solution with the linearized characteristics replaced by the characteristics calculated by including the first-order nonlinear terms.

The solution can be continued in a straightforward manner to higher orders. For the isotropic case the solution was carried to second order by Davison (1968) and Nair and Nemat–Nasser (1971) for homogeneous and inhomogeneous materials, respectively.

3.3. Temple's Technique

In order to determine a uniformly valid expansion for the problem

$$\frac{du}{dx} = F(x, u, \epsilon) \quad \text{with} \quad u(x_0) = u_0 \tag{3.3.1}$$

Temple (1958) introduced a new independent variable s, as Lighthill did, and assumed that

$$u = u(s, \epsilon) \quad \text{and} \quad x = x(s, \epsilon) \tag{3.3.2}$$

Whereas Lighthill assumed that

$$u = u_0(s) + \epsilon u_1(s) + \epsilon^2 u_2(s) + \cdots \tag{3.3.3}$$

$$x = s + \epsilon x_1(s) + \epsilon^2 x_2(s) + \cdots \tag{3.3.4}$$

and chose x_i so that the above two expansions are uniformly valid, Temple replaced the original equation (3.3.1) by two new equivalent equations

$$\frac{du}{ds} = U(x, u, s, \epsilon), \qquad \frac{dx}{ds} = X(x, u, s, \epsilon) \tag{3.3.5}$$

such that U and X are regular in ϵ. Then he determined a straightforward perturbation expansion for u and x. Thus Temple's technique determines x_i in a systematic manner. A similar approach has been used by Whitham, Lighthill, Fox, Lin, and Davison and discussed earlier in the case of hyperbolic equations where the uniformization was achieved by expanding in one or more characteristic parameters.

As an example, let us consider

$$(x + \epsilon y)\frac{dy}{dx} + (2 + x)y = 0, \qquad y(1) = e^{-1} \tag{3.3.6}$$

This example has been discussed by Temple, and it is a special case of the problem given by (3.2.9). Temple replaces the above equation by

$$s\frac{dx}{ds} = x + \epsilon y, \qquad s\frac{dy}{ds} = -(2 + x)y \tag{3.3.7}$$

These equations are analytic in ϵ and possess the following expansions

$$y = s^{-2}e^{-s}\left[1 - \epsilon \int_1^s \phi(t)\, dt\right] + O(\epsilon^2) \tag{3.3.8}$$

$$x = s[1 + \epsilon\phi(s)] + O(\epsilon^2) \tag{3.3.9}$$

where

$$\phi(s) = \int_1^s s^{-4} e^{-s} \, ds \qquad (3.3.10)$$

As $s \to 0$

$$x = s - \tfrac{1}{3}\epsilon s^{-2} + O(\epsilon^2 s^{-4}) \qquad (3.3.11)$$

$$y = s^{-2} - \tfrac{1}{6}\epsilon s^{-4} + O(\epsilon^2 s^{-6}) \qquad (3.3.12)$$

Hence at $x = 0$

$$y = \left(\frac{3}{\epsilon}\right)^{2/3} + O(\epsilon^{-1/3}) \qquad (3.3.13)$$

in agreement with (3.2.33) obtained by Lighthill's technique.

3.4. Renormalization Technique

Pritulo (1962) showed that in order to determine a uniformly valid perturbation expansion for a given problem we need not introduce the transformation (3.2.2) into the differential equations and determine ξ_n. Instead, we carry out the straightforward expansion in terms of the original variables and then introduce the transformation (3.2.2) into this straightforward expansion. In order to render this expansion uniformly valid, we impose the condition of Lighthill that the singularities do not grow stronger as the order of approximation increases. Thus we obtain algebraic equations for the determination of ξ_n rather than differential equations, thereby simplifying the whole procedure. However, Pritulo assumed that the coefficients of the series (3.2.1), with the exception of perhaps u_0, satisfy linear equations, and thus he asserted that under this condition his method becomes effective. This technique was rediscovered by Usher (1968).

This technique is closely related to the method described in Section 7.4.2, which was originated in the work of Rayleigh on scattering. After determining the scattering from a thin slab, Rayleigh recast it into an exponential to render it valid for scattering from many slabs.

We apply this technique to several examples treated earlier in this book and render them uniformly valid.

3.4.1. THE DUFFING EQUATION
This problem was introduced in Section 2.1.1 where we determined the straightforward perturbation expansion

$$u = a \cos t + \epsilon a^3 [-\tfrac{3}{8} t \sin t + \tfrac{1}{32}(\cos 3t - \cos t)] + O(\epsilon^2) \qquad (3.4.1)$$

A uniformly valid expansion was obtained in Section 3.1.1 by using the Lindstedt–Poincaré procedure.

To determine a uniformly valid expansion from (3.4.1), we introduce the transformation (3.1.2) into the above series. Expanding and collecting coefficients of equal powers of ϵ, we have

$$u = a \cos s - \epsilon[a(\omega_1 + \tfrac{3}{8}a^2)s \sin s - \tfrac{1}{32}a^3(\cos 3s - \cos s)] + O(\epsilon^2) \quad (3.4.2)$$

The secular terms in (3.4.2) are eliminated if

$$\omega_1 = -\tfrac{3}{8}a^2 \quad (3.4.3)$$

Therefore a uniformly valid expansion is

$$u = a \cos s + \tfrac{1}{32}\epsilon a^3(\cos 3s - \cos s) + O(\epsilon^2) \quad (3.4.4)$$

where

$$t = s(1 - \tfrac{3}{8}\epsilon a^2) + O(\epsilon^2) \quad (3.4.5)$$

which is in agreement with (3.1.15) and (3.1.16) obtained by using the Lindstedt–Poincaré procedure.

3.4.2. A MODEL FOR WEAK NONLINEAR INSTABILITY

As a second example, we render uniformly valid the straightforward expansion

$$u = \epsilon \cos \sigma_1 t \cos kx + \epsilon^3 \left(\frac{9}{32\sigma_1} t \sin \sigma_1 t \cos kx + \text{terms bounded as } t \to \infty \right) \quad (3.4.6)$$

obtained in Section 2.1.2 for the model problem (2.1.10) and (2.1.11).

We let

$$t = s(1 + \epsilon^2 \omega_2 + \cdots) \quad (3.4.7)$$

in (3.4.6) and expand for small ϵ to obtain

$$u = \epsilon \cos \sigma_1 s \cos kx + \epsilon^3 \left[\left(\frac{9}{32\sigma_1} - \sigma_1 \omega_2 \right) s \sin \sigma_1 s \cos kx \right.$$
$$\left. + \text{terms bounded as } s \to \infty \right]$$

Secular terms will be eliminated if $\omega_2 = 9/32\sigma_1^2$. Therefore a uniformly valid expansion is

$$u = \epsilon \cos \sigma t \cos kx + O(\epsilon^3) \quad (3.4.8)$$

where

$$\sigma = \sqrt{k^2 - 1} \left[1 - \frac{9\epsilon^2}{32(k^2 - 1)} \right] + O(\epsilon^3) \quad (3.4.9)$$

If $k > 1$, σ is real and (3.4.8) is valid for times as large as $O(\epsilon^{-2})$ and it represents oscillatory standing waves with amplitude-dependent frequencies.

However, if $k < 1$, σ is imaginary and (3.4.8) represents growing waves. Since after a short time $\cosh 3\tilde{\sigma}t$, where $\tilde{\sigma}$ is real, dominates $\cosh \tilde{\sigma}t$, (3.4.8) is valid only for short times. Equation (3.4.9) shows that $\sigma \to \infty$ as $k \to 1$, and the second term on the right-hand side is of the same order as the first term when $k - 1 = O(\epsilon^2)$. Therefore although the above expansion is valid for a wide range of k, it breaks down when $k - 1 = O(\epsilon^2)$. We show in Section 3.5.1 that the application of the method of strained parameters to determine an expansion valid near $k = 1$ leads to erroneous results. An expansion valid near $k = 1$ is obtained by using the method of multiple scales in Section 6.2.8.

3.4.3. SUPERSONIC FLOW PAST A THIN AIRFOIL

As a third example for the application of the renormalization technique, we render uniformly valid the straightforward expansion for the velocity component in the axial direction obtained in Section 2.1.3 for flow past a thin airfoil. From (2.1.36) the straightforward expansion is

$$\frac{u}{U} = 1 - \epsilon \frac{T'(\xi)}{B} + \epsilon^2\left[\frac{1}{B^2}\left(1 - \frac{M^4(\gamma + 1)}{4B^2}\right)T'^{\,2}(\xi)\right.$$

$$\left. - \frac{\gamma + 1}{2}\frac{M^4}{B^3}\, yT'(\xi)T''(\xi) - T(\xi)T''(\xi)\right] + O(\epsilon^3) \quad (3.4.10)$$

In order to render this expansion uniformly valid, we let

$$\xi = s + \epsilon\xi_1(s, y) + O(\epsilon^2) \quad (3.4.11)$$

in (3.4.10), expand, and collect coefficients of equal powers of ϵ to obtain

$$\frac{u}{U} = 1 - \frac{\epsilon T'(s)}{B} + \epsilon^2\left[\frac{1}{B^2}\left(1 - \frac{M^4(\gamma + 1)}{4B^2}\right)T'^{\,2}(s) - T(s)T''(s)\right.$$

$$\left. - \left(\xi_1(s, y) + \frac{\gamma + 1}{2B^2}M^4yT'(s)\right)\frac{T''(s)}{B}\right] + O(\epsilon^3) \quad (3.4.12)$$

This expansion can be rendered uniformly valid for all y by choosing

$$\xi_1(s, y) = -\frac{\gamma + 1}{2B^2}M^4yT'(s) \quad (3.4.13)$$

Therefore a uniformly valid expansion is given by

$$\frac{u}{U} = 1 - \epsilon\frac{T'(s)}{B} + O(\epsilon^2) \quad (3.4.14)$$

where

$$\xi = s - \epsilon \frac{\gamma + 1}{2B^2} M^4 y T'(s) + O(\epsilon^2) \tag{3.4.15}$$

which is in full agreement with (3.2.94) and (3.2.97) obtained by using Lighthill's technique.

3.4.4. SHIFT IN SINGULARITY

As a fourth example, we consider the problem given in (2.4.1). The straightforward expansion was obtained in Section 2.4.1 to be

$$y = x^{-2} e^{-x} \left[1 + \epsilon \int_1^x e^{-t} t^{-3} (1 + 2t^{-1}) \, dt \right] + O(\epsilon^2) \tag{3.4.16}$$

To render this expansion uniformly valid, we let

$$x = s + \epsilon x_1(s) + \cdots \tag{3.4.17}$$

in (3.4.16) and collect coefficients of equal powers of ϵ to obtain

$$y = s^{-2} \left[1 - \frac{2\epsilon}{s} (x_1 + \tfrac{1}{3} s^{-2}) \right] + O(\epsilon^2) \quad \text{as} \quad s \to 0 \tag{3.4.18}$$

The above expansion can be rendered uniformly valid by choosing

$$x_1 = -\tfrac{1}{3} s^{-2} \tag{3.4.19}$$

to remove the worst singularity. Therefore a uniformly valid expansion is given by

$$y = s^{-2} e^{-s} + O(\epsilon) \tag{3.4.20}$$

where

$$x = s - \tfrac{1}{3} \epsilon s^{-2} + O(\epsilon^2) \tag{3.4.21}$$

which is in full agreement with (3.2.30) and (3.2.31) obtained by Lighthill's technique.

3.5. Limitations of the Method of Strained Coordinates

In the preceding sections it was evident that the method of strained coordinates is a powerful technique for determining uniformly valid expansions for widely varying physical problems. However, in spite of the successful treatment of hyperbolic differential equations for waves traveling in one or two directions, the method may not yield uniformly valid expansions for elliptic differential equations. Although Lighthill (1951) obtained a uniformly valid second-order expansion for incompressible flow past a rounded thin airfoil, Fox (1953) found higher-order expansions are not uniformly valid.

She also found that even a second-order uniformly valid expansion could not be obtained for the case of compressible flow past a thin airfoil. Thus Lighthill (1961) advised in a later article that his method be used only for hyperbolic differential equations. In spite of this, Vaglio-Laurin (1962) successfully applied this technique in conjunction with the method of integral relations to the blunt body problem (a mixed boundary value problem). Moreover, Emanuel (1966) and Kuiken (1970) successively applied this technique to parabolic problems involving, respectively, unsteady, diffusing, reacting tubular flow, and flow down an inclined surface originated by strong fluid injection.

It should be mentioned that Hoogstraten (1967) modified this technique to treat subsonic thin airfoil problems by introducing a function approximating uniformly a mapping of the physical plane onto a plane in which the airfoil is represented by its chord.

It was suggested by Tsien (1956) that the failure of the method of strained coordinates in treating the thin airfoil problem is the expansion of a function near an irregular point. Fortunately, one can see that the singularities are transferred from the dependent variables to the straining functions and thus realizes the nonuniformity of the resulting expansion. Yuen (1968) expanded a function near an irregular singular point to obtain an expansion valid near the cutoff wave number in the nonlinear stability of a cylindrical liquid jet. However, the resulting expansion did not have any singularity although it fails at the cutoff wave number as shown by Nayfeh (1970c). We show the difficulties encountered by Yuen by using a model problem for weak non-linear instability of a standing wave (Section 3.5.1).

The method of strained coordinates has been shown by Levey (1959) to fail for the class of singular perturbations in which the small parameter multiplies the highest derivative (Section 3.5.2). He showed that this method yields erroneous results for the problem of cylindrical shock waves. However, it can be shown that straining the dependent rather than the independent variable leads to a uniformly valid expansion (Exercise 3.33).

Although this technique yields uniformly valid expansions for the periodic orbits of weakly nonlinear oscillations, it was shown by Nayfeh (1966) that it does not yield any information about any solution besides the limit cycle and limit points. In general, if the amplitude is varying, then the method of strained coordinates does not apply.

We show the difficulties of the method of strained coordinates with the following examples.

3.5.1. A MODEL FOR WEAK NONLINEAR INSTABILITY

As discussed in Section 3.4.2, the expansion (3.4.9) for σ breaks down when $k - 1 = O(\epsilon^2)$. To apply the method of strained parameters to this expansion, we let

$$k = \alpha + \epsilon^2 k_2 \qquad (3.5.1)$$

in (3.4.9), expand for small ϵ, collect coefficients of powers of ϵ, and obtain

$$\sigma = \sqrt{\alpha^2 - 1}\left[1 - \epsilon^2 \frac{\frac{9}{32} - \alpha k_2}{\alpha^2 - 1}\right] + O(\epsilon^3) \qquad (3.5.2)$$

In order that the coefficient of ϵ^2 be no more singular than the first term as $\alpha \to 1$, we choose k_2 to be 9/32. Then (3.5.2) becomes

$$\sigma = \sqrt{\alpha^2 - 1}\left[1 + \frac{9\epsilon^2}{32(\alpha + 1)}\right] + \cdots \qquad (3.5.3)$$

which is bounded as $\alpha \to 1$. Neutral stability corresponds to $\sigma = 0$; that is, $\alpha = 1$, or from (3.5.1)

$$k = 1 + \tfrac{9}{32}\epsilon^2 + \cdots \qquad (3.5.4)$$

To show the invalidity of the expansion (3.5.3), it is enough to show the invalidity of the neutral stability condition (3.5.4). The neutral stability configuration is by definition independent of t, hence it is governed by the equation

$$u_{xx} + u = -u^3 \qquad (3.5.5)$$

Letting

$$u = \epsilon \cos kx + \sum_{n=2}^{\infty} A_n \cos nkx, \qquad A_n = O(\epsilon^2)$$

in (3.5.5) and equating the coefficient of $\cos kx$ to zero, we obtain

$$k^2 = 1 + \tfrac{3}{4}\epsilon^2 + \cdots$$

Hence

$$k = 1 + \tfrac{3}{8}\epsilon^2 + \cdots \qquad (3.5.6)$$

which is different from (3.5.4). Therefore the expansions (3.5.3) and (3.5.4) are incorrect. A correct expansion valid near $k = 1$ is obtained in Section 6.2.8 by using the method of multiple scales.

3.5.2.   A SMALL PARAMETER MULTIPLYING THE HIGHEST DERIVATIVE

Levey (1959) showed that the application of the method of strained coordinates to the problem of cylindrical shock waves (introduced in Exercise 2.3) leads to incorrect results. The thickness of the shock was found by Wu (1956) to be independent of its strength, contrary to the result obtained by Levey using topology analysis. Rather than showing the invalidity of the expansion by discussing the problem of cylindrical shock waves, we follow

Levey and discuss a simpler problem that has the same features and has an exact solution for comparison. The problem is

$$\epsilon \frac{dy}{dx} = -xy - 1 \tag{3.5.7}$$

where ϵ is a small positive number. Its exact solution which passes through the point (x_0, y_0) is

$$y = y_0 e^{(x_0^2 - x^2)/2\epsilon} - \frac{1}{\epsilon} e^{-x^2/2\epsilon} \int_{x_0}^x e^{t^2/2\epsilon} \, dt \tag{3.5.8}$$

The straightforward perturbation expansion of (3.5.7) can be obtained by letting

$$y = \sum_{n=0}^{\infty} \epsilon^n y_n(x) \tag{3.5.9}$$

Substituting (3.5.9) into (3.5.7), equating coefficients of equal powers of ϵ, and solving the resulting equations, we obtain

$$y = -x^{-1} - \epsilon x^{-3} - \cdots - 1.3.5 \cdots (2n-1)\epsilon^n x^{-2n-1} \cdots \tag{3.5.10}$$

It can be verified that (3.5.10) is the asymptotic expansion for large x of the exact solution (3.5.8). It can be seen that the above expansion breaks down near $x = 0$ because the first term is singular and higher-order terms are more and more singular. When $x = O(\epsilon^{1/2})$, all terms in the above expansion are of $O(\epsilon^{-1/2})$. Thus the above expansion is never an adequate expansion in the region $x = O(\epsilon^{1/2})$.

To apply the method of strained coordinates to this problem, we let

$$x = s + \epsilon x_1(s) + \epsilon^2 x_2(s) + \cdots \tag{3.5.11}$$

in (3.5.10), expand for small ϵ, and collect coefficients of equal powers of ϵ. The straining functions x_n are then chosen so that the higher-order terms are no more singular than the first. In this problem this amounts to eliminating all the terms except the first. The result is

$$y = -\frac{1}{s} \tag{3.5.12}$$

where

$$x = s + \frac{\epsilon}{s} + \frac{\epsilon^2}{s^3} + \cdots + \frac{a_n \epsilon^n}{s^{2n-1}} + \cdots \tag{3.5.13}$$

$$a_{2m} = (4m - 2)\left[\sum_{r=0}^{m-2} a_{2m-1-r}a_{r+1} + \tfrac{1}{2}a_m^2\right], \quad m \geq 2$$

$$a_{2m+1} = 4m \sum_{r=0}^{m-1} a_{2m-r}a_{r+1}, \quad m \geq 1 \tag{3.5.14}$$

It follows from the last relations that

$$a_n > 2^{n-2}(n-1)!, \qquad n > 2 \tag{3.5.15}$$

Therefore the expansion (3.5.13) is divergent; in fact, it is "more" divergent than (3.5.10), and the expansion breaks down as x approaches $O(\epsilon^{1/2})$. All that has been achieved is the exchange of an invalid expansion of one variable for an invalid expansion of the other. The reason for the breakdown of the expansion is the dropping of the highest derivative whose effect is small for large x but becomes significant as x approaches the region $O(\epsilon^{1/2})$.

3.5.3.   THE EARTH-MOON-SPACESHIP PROBLEM

Next we show that the application of the method of strained coordinates to the two-dimensional earth-moon-space problem (introduced in Section 2.4.2) leads to an invalid expansion. To render the expansion (2.4.17) and (2.4.18) uniformly valid using the method of strained coordinates, we substitute

$$x = s + \mu x_1(s) + \cdots \tag{3.5.16}$$

into this expansion and obtain

$$
\sqrt{2}\,t = \frac{1}{\rho^3} \sin^{-1} \rho\sqrt{s} - \frac{1}{\rho^2} \sqrt{s(1 - \rho^2 s)}
$$
$$
+ \mu \left[\sqrt{\frac{s}{1 - \rho^2 s}}\, x_1 - \frac{2}{\rho^3} \sin^{-1} \rho\sqrt{s} + \frac{2 - \rho^2}{\rho^2(1 - \rho^2)} \sqrt{\frac{s}{1 - \rho^2 s}} \right.
$$
$$
\left. - \frac{1}{2(1 - \rho^2)^{3/2}} \ln \frac{1 + (1 - 2\rho^2)s + 2\sqrt{(1 - \rho^2)(1 - \rho^2 s)s}}{1 - s} \right] + O(\mu^2) \tag{3.5.17}
$$

The straining function x_1 is chosen so that the term between the brackets is no more singular than the first term as $s \to 1$; that is

$$x_1 = -\frac{1}{2(1 - \rho^2)} \ln (1 - s) \tag{3.5.18}$$

The straining function could have been chosen to eliminate the term in the brackets. The resulting first-order expansion is then

$$\sqrt{2}\,t = \frac{1}{\rho^3} \sin^{-1} \rho\sqrt{s} - \frac{1}{\rho^2} \sqrt{s(1 - \rho^2 s)} + O(\mu^2) \tag{3.5.19}$$

$$y = -\mu c s + O(\mu^2) \tag{3.5.20}$$

where

$$x = s - \frac{1}{2}\frac{\mu}{1 - \rho^2} \ln (1 - s) + O(\mu^2) \qquad (3.5.21)$$

Comparing this expansion with the exact solution, Nayfeh (1965a) showed that this expansion gives a divergent trajectory near the moon although the expansion obtained in the one-dimensional case agrees fairly well with the exact solution. In the one-dimensional case there is a singularity at $x = 1 + \mu/(1 - \rho^2) + O(\mu^2)$ which is outside the range of interest $0 \le x \le 1$. In the straightforward expansion the singularity is shifted to $x = 1$, and the straining of x moves the singularity from $x = 1$ toward its right position. However, in the two-dimensional case there is a sharp change in the direction of the spacecraft in the neighborhood of the moon, and the straining which is $O(\mu)$ cannot cope with such a sharp change.

Exercises

3.1. Consider the problem

$$\ddot{u} + u = \epsilon u^2, \qquad u(0) = a, \qquad \dot{u}(0) = 0$$

(a) Determine a second-order (three-term) straightforward expansion and discuss its uniformity.

(b) Render this expansion uniformly valid using the method of renormalization.

(c) Determine a first-order (two-term) uniformly valid expansion using the method of strained parameters and compare the result with (b).

3.2. (a) Show that the motion of a point mass that moves freely along a parabola $x^2 = 2pz$ rotating about its axis with an angular velocity ω is given by

$$\left(1 + \frac{x^2}{p^2}\right)\ddot{x} + \frac{x\dot{x}^2}{p^2} + \left(\frac{g}{p} - \omega^2\right)x = 0$$

(b) Determine a two-term straightforward expansion for small amplitude and discuss its uniformity.

(c) Render this expansion uniformly valid using the method of renormalization.

(d) Determine a one-term uniformly valid expansion using the method of strained parameters and compare the result with (c).

3.3. Determine a two-term uniformly valid expansion for small amplitudes for the solution of

$$\ddot{\theta} + \frac{g}{l} \sin \theta = 0$$

which describes the oscillations of a pendulum.

3.4. Determine a second-order uniformly valid expansion for the periodic solution of

$$\ddot{u} + u = \epsilon(1 - u^2)\dot{u}$$

Note that the amplitude is not arbitrary.

3.5. Determine a first-order uniformly valid expansion for the periodic solution of

$$\ddot{u} + u = \epsilon(1 - z)\dot{u}$$

$$\tau\dot{z} + z = u^2$$

where τ is a constant.

3.6. Consider the equation

$$\ddot{u} + \omega_0^2 u = \epsilon u^2 + k \cos \omega t$$

Determine first-order uniformly valid expansions for the periodic solutions when
(a) $\omega_0 \approx 2\omega$ [Hint: let $\omega_0 = 2\omega + \epsilon\sigma$ and $u = u_0 + \epsilon u_1 + \cdots$ where $u_0 = a \cos (2\omega t + \beta) + (1/3)k\omega^{-2} \cos \omega t$, then determine a and β from the equation for u_1], and
(b) $\omega_0 \approx \omega/2$ (a is arbitrary in this case).

3.7. Consider the equation

$$\ddot{u} + \omega_0^2 u = \epsilon u^3 + k \cos \omega t$$

Determine first-order uniformly valid expansions for the periodic solutions when
(a) $\omega_0 \approx 3\omega$ and (b) $\omega_0 \approx \omega/3$.

3.8. Determine second-order expansions for the odd solutions corresponding to the transition curves of

$$\ddot{u} + (\delta + \epsilon \cos 2t)u = 0$$

when δ is near 1 and 4.

3.9. Consider the equation

$$\ddot{u} + \frac{\delta u}{1 + \epsilon \cos 2t} = 0$$

(a) Determine second-order expansions for the transition curves near $\delta = 0$, 1, and 4 (Shen, 1959).
(b) Use Whittaker's technique to determine second-order expansions for u near these curves.

3.10. Consider the equation

$$\ddot{u} + \frac{\delta - \epsilon \cos^2 t}{1 - \epsilon \cos^2 t} u = 0$$

(a) Determine second-order expansions for the first three transition curves (Rand and Tseng, 1969) (i.e., near $\delta = 0$, 1, and 4).
(b) Use Whittaker's technique to determine u near these curves.

3.11. Consider the equation

$$\ddot{u} + (\delta + \epsilon \cos^3 t)u = 0$$

Determine second-order expansions for the first three transition curves using both the method of strained parameters and Whittaker's technique.

3.12. Determine a periodic solution to $O(\epsilon)$ of the problem (Mulholland, 1971)

$$\dddot{u} + \ddot{u} + \dot{u} + u = (1 - u^2 - \dot{u}^2 - \ddot{u}^2)(\ddot{u} + \dot{u})$$

3.13. Determine first-order expansions for

$$\ddot{u} + \lambda u = \epsilon u^3$$

subject to (a) $u(0) = u(\pi) = 0$, and (b) $u(t) = u(t + 2\pi)$.

3.14. Determine a first-order expansion for

$$\ddot{u} + \lambda u = \epsilon(\sin 2t + u^2)u$$

$$u(t) = u(t + 2\pi)$$

3.15. Determine first-order expansions for

(a) $\ddot{u} + \lambda u = \epsilon t u$

$$u(t) = u(t + 2\pi)$$

(b) $\ddot{u} + \lambda u = \epsilon(\alpha \cos 2t + \beta \sin 2t)u$

$$u(t) = u(t + 2\pi)$$

3.16. Determine a first-order expansion for small amplitudes of the periodic solution of

$$\ddot{x} + \frac{k}{m}x + g(1 - \cos\theta) - (l + x)\dot{\theta}^2 = 0$$

$$\ddot{\theta} + \frac{g}{l + x}\sin\theta + \frac{2}{l + x}\dot{x}\dot{\theta} = 0$$

which describe the oscillation of a swinging spring of length l and constant k when $\omega_1^2 = k/m \approx 4\omega_2^2 = 4g/l$.

3.17. The free vibrations of a simply supported beam on an elastic foundation are given by

$$u_{xxxx} + \gamma u + \epsilon\gamma u^3 + u_{tt} = 0$$

$$u(0, t) = u(\pi, t) = u_{xx}(0, t) = u_{xx}(\pi, t) = 0$$

$$u(x, 0) = a\sin x, \qquad u_t(x, 0) = 0$$

where γ, a, and ϵ are constants. Determine an expansion to $O(\epsilon)$ of the frequency of oscillation (Han, 1965).

3.18. Carry out the expansions of Sections 3.1.4 and 3.1.5 to second order.

3.19. Consider uniform traveling wave solutions of

$$u_{tt} - u_{xx} + u = \epsilon u^3$$

of the form $u = a \exp i(kx - \omega t) +$ higher harmonics and determine the frequency and wave number shifts.

3.20. Consider the problem

$$u_{tt} + u_{xx} + u_{xxxx} = \epsilon u^3$$

$$u(x, 0) = a \cos kx, \qquad u_t(x, 0) = 0$$

(a) Determine a first-order straightforward expansion.

(b) Render this expansion valid using the method of renormalization.

(c) Determine an expansion valid for $t = O(\epsilon^{-1})$ using the method of strained parameters.

(d) Show that the frequency is invalid near $k = 1$.

(e) Apply the method of renormalization to this frequency to remove the singularity.

(f) Show that the resulting expansion is erroneous.

3.21. Consider the eigenvalue problem

$$\phi_{xx} + \phi_{yy} + \lambda\phi = \epsilon x^2\phi$$

$$\phi(x, 0) = \phi(x, \pi) = \phi(0, y) = \phi(\pi, y) = 0$$

Determine first-order expansions when λ is near 2 and 5.

3.22. Consider the problem

$$\nabla^2\phi + \lambda\phi = \epsilon f(x, y, z)\phi$$

with ϕ vanishing on the surfaces of a cube of length π. Determine first-order expansions when $\lambda \approx 3$ and 6 if (a) $f = x^2$ and (b) $f = x^2y$.

3.23. The free transverse vibrations of a simply supported beam are given by

$$EIw_{xxxx} - Tw_{xx} + \rho w_{tt} = 0$$

$$T = \frac{ES}{2l}\int_0^l (w_x)^2\,dx$$

$$w(0, t) = w(l, t) = w_{xx}(0, t) = w_{xx}(l, t) = 0$$

$$w(x, 0) = a\sin\frac{\pi x}{l}, \qquad w_t(x, 0) = 0$$

where E, I, ρ, S, a, and l are constant. Determine a first-order expansion for small amplitude (Evensen, 1968).

3.24. Consider the problem

$$(x + \epsilon y)y' + y = 0, \qquad y(1) = 1$$

(a) Determine a second-order straightforward expansion. What is its region of nonuniformity?

(b) Render this expansion uniformly valid using the method of renormalization.

(c) Determine a first-order expansion (two terms in y and three terms in x) using Lighthill's technique and compare it with (b).

(d) Determine the exact solution by interchanging the roles of dependent and independent variables and compare it with (b) and (c).

3.25. Show that a uniformly valid expansion of

$$(x + \epsilon y)y' + xy = be^{-x}, \qquad y(1) = e^{-1}$$

is

$$y = e^{-\xi}(b \ln \xi + 1) + O(\epsilon)$$

where $x = \xi - \epsilon(b \ln \xi + b + 1) + O(\epsilon^2)$.

3.26. Consider the problem

$$(x + \epsilon y)y' - \tfrac{1}{2}y = 1 + x^2, \qquad y(1) = 1$$

(a) Determine a second-order straightforward expansion and discuss its uniformity.

(b) Render this expansion uniformly valid using the method of renormalization.

(c) Obtain a first-order expansion using Lighthill's technique and compare the result with (b).

3.27. Use the method of renormalization to render the expansion of Exercise 2.12 uniformly valid.

3.28. Consider the problem

$$(x^n + \epsilon y)y' + nx^{n-1}y - mx^{m-1} = 0$$

$$y(1) = a > 1$$

(a) Show that its exact solution is given by

$$x^n y + \tfrac{1}{2}\epsilon y^2 = x^m + (a + \tfrac{1}{2}\epsilon a^2 - 1)$$

(b) Show that application of Lighthill's technique gives

$$y \sim y_0 = (\xi^m + a - 1)\xi^{-n}$$

$$x = \xi - \frac{1}{2}\frac{\epsilon\xi(y_0^2 - a^2)}{(n - m)\xi^m + n(a - 1)} + O(\epsilon^2)$$

(c) Show that (Comstock, 1968) the approximate solution is erroneous near ($x = 0$) except for special values of m and n.

(d) Introduce a new variable $z = x^n$ in the original problem and then strain the variable z to determine an approximate solution to y. Determine the conditions under which this new expansion is valid near the origin (Burnside, 1970), hence determine the role of changing the independent variable on rendering the approximate solution uniformly valid.

3.29. Consider the problem

$$(1 + \epsilon u)\frac{\partial u}{\partial x} + \frac{\partial u}{\partial y} = 0, \qquad u(x, 0) = \epsilon\phi(x)$$

(a) Determine a first-order straightforward expansion for $\epsilon \ll 1$ and discuss its uniformity.

(b) Render this expansion uniformly valid using the method of renormalization.

(c) Determine a first-order expansion using Lighthill's technique and compare the result with (b).

3.30. Consider the problem

$$u_{tt} - c^2 u_{xx} = u_x u_{xx}$$
$$u(0, t) = \epsilon\phi(t), \qquad \phi(t) = 0 \quad \text{for} \quad t \leq 0$$
$$u(x, 0) = 0 \quad \text{for} \quad x \geq 0$$

(a) Determine a first-order straightforward expansion and render it "uniformly valid" using the method of renormalization.

(b) Determine a first-order expansion using Lighthill's technique and compare it with (a). Show that renormalizing u rather than u_x leads to an erroneous result.

3.31. Consider the problem (Lighthill, 1949a)

$$\frac{\partial u}{\partial x} + \frac{n}{x + y} u = u\left(\frac{\partial^2 v}{\partial x^2} + \frac{\partial u}{\partial y}\right), \qquad u = v_y$$
$$u(x, 0) = v(x, 0) = 0$$
$$u(0, y) = \epsilon\phi(y)y^{-n}, \qquad 0 < n < 1$$

where $\phi(0) = 0$. Show that a first-order uniformly valid expansion is

$$u = \epsilon\phi(\eta)(x + \eta)^{-n} + O(\epsilon^2)$$
$$y = \eta - \frac{\epsilon\phi(\eta)(x + \eta)^{1-n}}{1 - n} + O(\epsilon^2)$$

3.32. Consider the problem

$$u_{tt} - c^2 u_{xx} = \epsilon u_x u_{xx}$$
$$u(x, 0) = f(x) + g(x), \qquad u_t(x, 0) = c\big(g'(x) - f'(x)\big)$$

where $f(x)$ and $g(x)$ are bounded functions of x.

(a) Determine a first-order straightforward expansion. Can you render it uniformly valid using the method of renormalizaion?

(b) Obtain a first-order expansion using the method of strained coordinates.

3.33. Consider the problem

$$\epsilon y' + y = 1, \qquad y(1) = 1$$

(a) Show that Lighthill's technique fails to give a uniformly valid expansion.

(b) Show that straining y rather than x gives a uniformly valid expansion.

(c) Investigate whether straining y' would yield a uniformly valid expansion for

$$\epsilon y'' + y' + y = 0$$
$$y(0) = \alpha, \qquad y(1) = \beta$$

3.34. Consider the problem

$$\ddot{u} + u = \epsilon f(u, \dot{u})$$

(a) Show that the method of strained coordinates (MSC) leads to

$$u = a \sin \phi + O(\epsilon), \qquad \phi = s + c$$

where

$$t = s + \epsilon t_1(s) + \cdots$$

$$at_1'' = \alpha = -\frac{1}{\pi} \int_0^{2\pi} f[a \sin \phi, a \cos \phi] \cos \phi \, d\phi$$

$$2at_1' = \beta = \frac{1}{\pi} \int_0^{2\pi} f[a \sin \phi, a \cos \phi] \sin \phi \, d\phi$$

(b) Show that $\alpha = 0$ and β is a constant so that $t_1 = (1/2)\beta a^{-1}s + a$ constant.

(c) Hence show that the MSC yields only the limit cycles or limit points for this problem (Nayfeh, 1966).

CHAPTER 4

The Methods of Matched and Composite Asymptotic Expansions

The results of Section 3.5 show that the method of strained coordinates is not capable of yielding uniformly valid expansions in cases in which sharp changes in dependent variables take place in some regions of the domain of the independent variables. In these cases straightforward expansions generally break down in these regions, and near-identity transformations of the independent variables (strained coordinates) cannot cope with such sharp changes. To obtain uniformly valid expansions, we must recognize and utilize the fact that the sharp changes are characterized by magnified scales which are different from the scale characterizing the behavior of the dependent variables outside the sharp-change regions.

One technique of dealing with this problem is to determine straightforward expansions (called outer expansions) using the original variables and to determine expansions (called inner expansions) describing the sharp changes using magnified scales. The outer expansions break down in the inner regions (sharp-change regions), while the inner expansions break down away from these regions. To relate these expansions a so-called matching procedure is used. This technique is called the method of inner and outer expansions or, after Bretherton (1962), the method of matched asymptotic expansions.

A second technique for determining a uniformly valid expansion is to assume that each dependent variable is the sum of (1) a part characterized by the original independent variable and (2) parts characterized by magnified independent variables, one for each sharp-change region. This is the method of composite expansions in its simplest form.

In the next section we describe the method of matched asymptotic expansions. For more references and applications, we refer the reader to Van Dyke (1964), Wasow (1965), Cole (1968), and O'Malley (1968b). Carrier (1970) reviewed the application of this technique in geophysics, and Germain (1967) reviewed its application in aerodynamics.

4.1. The Method of Matched Asymptotic Expansions

4.1.1. INTRODUCTION—PRANDTL'S TECHNIQUE

To describe the method of matched asymptotic expansions, we discuss the simple boundary value problem

$$\epsilon y'' + y' + y = 0 \tag{4.1.1}$$

$$y(0) = \alpha, \qquad y(1) = \beta \tag{4.1.2}$$

introduced in Section 2.2.1. As $\epsilon \to 0$, (4.1.1) reduces to

$$y' + y = 0 \tag{4.1.3}$$

which is a first-order equation that cannot satisfy in general both of the boundary conditions (4.1.2). Hence one of these boundary conditions must be dropped. As shown in Section 4.1.2, the boundary condition $y(0) = \alpha$ must be dropped. This can also be seen from the exact solution (2.2.7).
   As $\epsilon \to 0$ and for fixed $x \neq 0$

$$y \to \beta e^{1-x} \tag{4.1.4}$$

which is the solution of the reduced equation (4.1.3) subject to $y(1) = \beta$. The solution of the reduced equation is denoted by y^o and it is called the *outer solution*. For small ϵ the solution of the reduced equation is close to the exact solution (2.2.7) except in a small interval at the end point $x = 0$ where the exact solution changes quickly in order to retrieve the boundary condition $y(0) = \alpha$ which is about to be lost. This small interval across which y changes very rapidly is called the *boundary layer* in fluid mechanics, the *edge layer* in solid mechanics, and the *skin layer* in electrodynamics.
   To determine an expansion valid in the boundary layer, we magnify this layer using the stretching transformation

$$\zeta = \frac{x}{\epsilon} \tag{4.1.5}$$

Determination of appropriate stretching transformations is discussed later in this chapter. With this transformation (4.1.1) becomes

$$\frac{d^2 y}{d\zeta^2} + \frac{dy}{d\zeta} + \epsilon y = 0 \tag{4.1.6}$$

which for a fixed ζ, reduces to

$$\frac{d^2 y}{d\zeta^2} + \frac{dy}{d\zeta} = 0 \tag{4.1.7}$$

as $\epsilon \to 0$. Its general solution is

$$y = A + Be^{-\zeta} \tag{4.1.8}$$

where A and B are constants. Since this solution is valid in the boundary layer, it is valid at the origin; hence it must satisfy the boundary condition $y(x = 0) = \alpha$. Since $\zeta = 0$ corresponds to $x = 0$, $y(\zeta = 0) = \alpha$; hence $B = \alpha - A$ and (4.1.8) becomes

$$y = A + (\alpha - A)e^{-\zeta} \tag{4.1.9}$$

which has an arbitrary constant A. We denote this solution by y^i and call it the *inner solution* or *inner expansion*.

To determine A we note that

$$\lim_{x \to 0} y^o = \beta e \tag{4.1.10}$$

Moreover, from (4.1.5) any small fixed value x_0 corresponds to $\zeta \to \infty$ as $\epsilon \to 0$, and

$$\lim_{\zeta \to \infty} y^i = A \tag{4.1.11}$$

Thus these limits must represent the same value of y at a very small value of $x = x_0 \neq 0$, hence

$$A = \beta e \tag{4.1.12}$$

Therefore

$$y^i = \beta e + (\alpha - \beta e)e^{-\zeta} \tag{4.1.13}$$

In determining the outer and inner expansions, we used two different limit processes—an *outer limit* defined by

$$y^o = \lim_{\substack{\epsilon \to 0 \\ x \text{ fixed}}} y(x; \epsilon) \tag{4.1.14}$$

and an *inner limit* defined by

$$y^i = \lim_{\substack{\epsilon \to 0 \\ \zeta \text{ fixed}}} y(\epsilon\zeta; \epsilon) \tag{4.1.15}$$

The process of determining A is called matching, and we have used the matching principle

$$\lim_{x \to 0} y^o(x; \epsilon) = \lim_{\zeta \to \infty} y^i(x; \epsilon) \tag{4.1.16}$$

which is equivalent to equating

The inner limit of (the outer solution) denoted by $(y^o)^i$ to the outer limit of (the inner solution) denoted by $(y^i)^o$ (4.1.17)

An approximate solution to our original problem is given by (4.1.4) for

x not near zero and by (4.1.13) near $x = 0$. To compute y as a function of all x, one must switch from one solution to the other as x increases at some small value of x such as the value where the solutions may intersect. This switching is not convenient, and we form from these solutions a single uniformly valid solution called the *composite solution* and denote it by y^c according to (Erdélyi, 1961)

$$y^c = y^o + y^i - (y^o)^i = y^o + y^i - (y^i)^o \qquad (4.1.18)$$

Since

$$((y^o)^i)^o = (y^o)^i = (y^i)^o = ((y^o)^i)^i$$

therefore

$$(y^c)^o = y^o + (y^i)^o - (y^o)^i = y^o$$

$$(y^c)^i = (y^o)^i + y^i - (y^o)^i = y^i \qquad (4.1.19)$$

Thus the composite solution is as good an approximation in the outer region as the outer solution, and it is as good an approximation in the inner region as the inner solution. This suggests that the composite solution is a uniform approximation over the whole interval of x including the gap between the outer and inner regions. The success of the matching may be due to the presence of an overlapping region in which both the outer and inner solutions are valid, hence there is no gap between the two regions.

Adding (4.1.4) and (4.1.13) and subtracting βe, which is equal to $(y^o)^i = (y^i)^o$ from (4.1.17), we have

$$y^c = \beta e^{1-x} + (\alpha - \beta e)e^{-x/\epsilon} + O(\epsilon) \qquad (4.1.20)$$

The technique presented in this section was developed by Prandtl (1905) to solve the problem of high speed viscous flow past a body. The stream function representing the two-dimensional flow past a body is governed by a fourth-order partial differential equation. For a viscous fluid the components of fluid velocity normal and tangential to the body must vanish on the body. The latter condition is called the no-slip condition because any slight viscosity makes the fluid stick to the body. If the viscosity vanishes, the equation for the stream function reduces to an equation of third order (Section 2.2.2), hence it cannot satisfy all the boundary conditions. Since an inviscid fluid can slide, the no-slip condition is dropped, and the resulting solution represents the flow field everywhere except in a small region near the body called *Prandtl's boundary layer* across which the tangential velocity changes very rapidly from the value given by the reduced equation (vanishing viscosity) to zero in order to retrieve the no-slip condition which is about to be lost. To describe the flow in this region, Prandtl magnified it by introducing a stretching transformation and appraised the order of magnitude of the various terms in the original differential equation, rejecting those he judged to be small. The resulting simplified equation was then solved and its solution was matched with the solution of the inviscid problem by using the matching principle (4.1.16).

A similar matching procedure was used by Rayleigh (1912), Gans (1915), Jeffreys (1924), Wentzel (1926), Kramers (1926), and Brillouin (1926) to connect the approximate expansions on the different sides of a turning point (cf. Section 7.3.1).

Similar techniques were employed in the nineteenth century by (1) Laplace (1805) to solve the problems of a large nonwetting drop on a plane and a wide meniscus, (2) Maxwell (1866) to solve the torsional vibration of circular disks rotating between nearby fixed disks, and (3) Kirchhoff (1877) to solve the problem of a condenser consisting of two differentially charged finite circular disks.

4.1.2. HIGHER APPROXIMATIONS AND REFINED MATCHING PROCEDURES

Prandtl's method has been extended and generalized over the years by many researchers including Weyl (1942), Friedrichs (1942), Dorodnicyn (1947), Latta (1951), Kaplun (1954, 1957, 1967), Kaplun and Lagerstrom (1957), Proudman and Pearson (1957), Višik and Lyusternik (1957), Vasil'eva (1959, 1963), and Van Dyke (1964). The matching procedures have been formalized by Vasil'eva (1959, 1963), Van Dyke (1964), and Kaplun and Lagerstrom (1967). Carrier (1953, 1954), using specific examples, compared the method of strained coordinates with the method of matched asymptotic expansions.

In this section we determine higher approximations to the problem given by (4.1.1) and (4.1.2). We start by determining the boundary condition that must be dropped and determine the stretching transformation as a by-product. Then, we determine second-order inner and outer expansions and match them by using Van Dyke's principle. Finally, we form a uniformly valid composite expansion.

Which Boundary Condition Must be Dropped? As mentioned in the previous section, if ϵ vanishes, (4.1.1) reduces to the first-order equation (4.1.3) whose solution cannot satisfy both boundary conditions $y(0) = \alpha$ and $y(1) = \beta$ and, consequently, one of them must be dropped. At the end where the boundary condition is dropped, y changes very rapidly from the solution of the reduced equation to the boundary value dropped at that end in a very small region called the *boundary layer* or *region of nonuniformity*.

To determine if the boundary condition $y(1) = \beta$ must be dropped, we introduce the stretching transformation

$$\zeta = (1 - x)\epsilon^{-\lambda}, \qquad \lambda > 0 \qquad (4.1.21)$$

thereby transforming (4.1.1) into

$$\epsilon^{1-2\lambda}\frac{d^2y}{d\zeta^2} - \epsilon^{-\lambda}\frac{dy}{d\zeta} + y = 0 \qquad (4.1.22)$$

As $\epsilon \to 0$, three limiting forms of (4.1.22) arise depending on the value of λ.

The Case of $\lambda > 1$

$$\frac{d^2y}{d\zeta^2} = 0 \quad \text{or} \quad y^i = A + B\zeta \qquad (4.1.23)$$

Since the reduced equation (4.1.3) is assumed to be valid at $x = 0$,

$$y^o = \alpha e^{-x} \qquad \text{(outer solution)} \qquad (4.1.24)$$

The matching principle (4.1.16) demands that

$$\lim_{\zeta \to \infty}(A + B\zeta) = \lim_{x \to 1}\alpha e^{-x}$$

or

$$B = 0 \quad \text{and} \quad A = \alpha e^{-1} \qquad (4.1.25)$$

Hence

$$y^i = \alpha e^{-1} \qquad (4.1.26)$$

Since this solution is valid at $x = 1$, it should satisfy the boundary condition $y(x = 1) = \beta$, hence

$$\beta = \alpha e^{-1} \qquad (4.1.27)$$

which is not true in general. Therefore we discard this case because it does not permit the satisfaction of both boundary conditions.

The Case of $\lambda < 1$

$$\frac{dy}{d\zeta} = 0 \quad \text{or} \quad y^i = A = \beta \qquad (4.1.28)$$

which must be discarded because the matching principle demands (4.1.27).

The Case of $\lambda = 1$

$$\frac{d^2y}{d\zeta^2} - \frac{dy}{d\zeta} = 0 \quad \text{or} \quad y^i = A + Be^\zeta \qquad (4.1.29)$$

Since the matching principle demands that

$$\lim_{\zeta \to \infty}(A + Be^\zeta) = \lim_{x \to 1}(\alpha e^{-x})$$

$$B = 0 \quad \text{and} \quad A = \alpha e^{-1} \qquad (4.1.30)$$

So that this case must be discarded also because the application of the boundary condition $y(x = 1) = \beta$ demands (4.1.27).

Therefore the boundary layer cannot exist at $x = 1$, hence the boundary condition $y(1) = \beta$ must not be dropped.

To investigate whether the boundary condition $y(0) = \alpha$ must be dropped, we introduce the stretching transformation

$$\zeta = x\epsilon^{-\lambda}, \quad \lambda > 0 \tag{4.1.31}$$

thereby transforming (4.1.1) into

$$\epsilon^{1-2\lambda} \frac{d^2y}{d\zeta^2} + \epsilon^{-\lambda} \frac{dy}{d\zeta} + y = 0 \tag{4.1.32}$$

In this case also three different limiting forms of (4.1.32) arise as $\epsilon \to 0$ depending on the value of λ

The Case of $\lambda > 1$

$$\frac{d^2y}{d\zeta^2} = 0 \quad \text{or} \quad y^i = A + B\zeta \tag{4.1.33}$$

The Case of $\lambda < 1$

$$\frac{dy}{d\zeta} = 0 \quad \text{or} \quad y^i = A \tag{4.1.34}$$

The Case of $\lambda = 1$

$$\frac{d^2y}{d\zeta^2} + \frac{dy}{d\zeta} = 0, \quad y^i = A + Be^{-\zeta} \tag{4.1.35}$$

The first two cases must be discarded using arguments similar to those employed above. This leaves the third case which, by utilizing $y(0) = \alpha$, leads to the inner solution

$$y^i = A + (\alpha - A)e^{-\zeta} \tag{4.1.36}$$

Since the matching principle demands

$$\lim_{\zeta \to \infty} [A + (\alpha - A)e^{-\zeta}] = \lim_{x \to 0} (\beta e^{1-x}) \quad \text{or} \quad A = \beta e \tag{4.1.37}$$

then

$$y^i = \beta e + (\alpha - \beta e)e^{-\zeta} \tag{4.1.38}$$

Therefore the boundary layer exists at the end $x = 0$, and the boundary condition $y(0) = \alpha$ cannot be imposed on the reduced equation (4.1.3). We have found as a by-product that the stretching transformation is

$$\zeta = x\epsilon^{-1} \tag{4.1.39}$$

as was chosen in (4.1.5), hence the region of nonuniformity is

$$x = O(\epsilon) \tag{4.1.40}$$

Outer Expansion. We seek an outer expansion for y in the form

$$y^o(x; \epsilon) = \sum_{n=0}^{N-1} \epsilon^n y_n(x) + O(\epsilon^N) \qquad (4.1.41)$$

using the *outer limit process*

$$\epsilon \to 0 \text{ keeping } x \text{ fixed} \qquad (4.1.42)$$

Thus

$$y_0(x) = \lim_{\substack{\epsilon \to 0 \\ x \text{ fixed}}} y(x; \epsilon)$$

and

$$y_m(x) = \lim_{\substack{\epsilon \to 0 \\ x \text{ fixed}}} \frac{y - \sum_{n=0}^{m-1} \epsilon^n y_n(x)}{\epsilon^m} \qquad (4.1.43)$$

To determine this expansion we substitute (4.1.41) into (4.1.1) and equate coefficients of like powers of ϵ to obtain

$$y_0' + y_0 = 0 \qquad (4.1.44)$$

$$y_n' + y_n = -y_{n-1}'', \qquad n \geq 1 \qquad (4.1.45)$$

As mentioned in the previous section, this outer solution is valid everywhere except in the region $x = O(\epsilon)$, hence it must satisfy the boundary condition $y^o(1) = \beta$ which together with (4.1.41) yields

$$y_0(1) = \beta, \qquad y_n(1) = 0 \quad \text{for} \quad n \geq 1 \qquad (4.1.46)$$

The solution of (4.1.44) subject to $y_0(1) = \beta$ is

$$y_0 = \beta e^{1-x} \qquad (4.1.47)$$

while the solution of (4.1.45) for $n = 1$ subject to $y_n(1) = 0$ is

$$y_1 = \beta(1 - x)e^{1-x} \qquad (4.1.48)$$

Therefore

$$y^o = \beta[1 + \epsilon(1 - x)]e^{1-x} + O(\epsilon^2) \qquad (4.1.49)$$

Inner Expansion. To determine an expansion valid near the origin, we use the stretching transformation (4.1.39) to transform (4.1.1) into

$$\frac{d^2 y^i}{d\zeta^2} + \frac{dy^i}{d\zeta} + \epsilon y^i = 0 \qquad (4.1.50)$$

and then seek an inner expansion of y in the form

$$y^i(x; \epsilon) = \sum_{n=0}^{N-1} \epsilon^n Y_n(\zeta) + O(\epsilon^N) \qquad (4.1.51)$$

using the *inner limit process*

$$\epsilon \to 0 \text{ keeping } \zeta = x\epsilon^{-1} \text{ fixed} \tag{4.1.52}$$

Thus

$$Y_0(\zeta) = \lim_{\substack{\epsilon \to 0 \\ \zeta \text{ fixed}}} y(\epsilon\zeta; \epsilon)$$

$$Y_m(\zeta) = \lim_{\substack{\epsilon \to 0 \\ \zeta \text{ fixed}}} \frac{y(\epsilon\zeta; \epsilon) - \sum_{n=0}^{m-1} \epsilon^n Y_n(\zeta)}{\epsilon^m} \tag{4.1.53}$$

To determine this expansion we substitute (4.1.51) into (4.1.50), equate coefficients of like powers of ϵ, keeping in mind that ζ is the independent variable, and obtain

$$Y_0'' + Y_0' = 0 \tag{4.1.54}$$

$$Y_n'' + Y_n' = -Y_{n-1}, \qquad n \geq 1 \tag{4.1.55}$$

While this inner expansion satisfies the boundary condition at $x = 0$, it is not expected to satisfy in general the boundary condition at $x = 1$. Since $x = 0$ corresponds to $\zeta = 0$, the boundary condition $y(x = 0) = \alpha$ together with (4.1.51) gives

$$Y_0(0) = \alpha, \qquad Y_n(0) = 0 \quad \text{for} \quad n \geq 1 \tag{4.1.56}$$

The solution of (4.1.54) subject to $Y_0(0) = \alpha$ is

$$Y_0 = \alpha - A_0(1 - e^{-\zeta}) \tag{4.1.57}$$

While the solution of (4.1.55) and (4.1.56) for $n = 1$ is

$$Y_1 = A_1(1 - e^{-\zeta}) - [\alpha - A_0(1 + e^{-\zeta})]\zeta \tag{4.1.58}$$

Therefore

$$y^i = \alpha - A_0(1 - e^{-\zeta}) + \epsilon\{A_1(1 - e^{-\zeta}) - [\alpha - A_0(1 + e^{-\zeta})]\zeta\} + O(\epsilon^2) \tag{4.1.59}$$

This inner expansion contains the arbitrary constants A_0 and A_1 which must be determined from its matching with the outer expansion (4.1.49).

Refined Matching Procedures. The simplest possible form of matching the inner and outer expansions is that of Prandtl where

$$\lim_{x \to 0} y^o = \lim_{\zeta \to \infty} y^i \tag{4.1.60}$$

This condition leads to the matching of the first terms in both the inner and the outer expansions, thereby giving

$$A_0 = \alpha - \beta e \qquad (4.1.61)$$

It can be easily seen that this matching principle cannot be used to match other than these first terms. In fact

$$\lim_{x \to 0} y^o = \beta e(1 + \epsilon) + O(\epsilon^2) \qquad (4.1.62a)$$

while

$$\lim_{\zeta \to \infty} y^i = \alpha - A_0 + \epsilon[A_1 - (\alpha - A_0)\zeta] + O(\epsilon^2) \qquad (4.1.62b)$$

Since these last two expansions have to be equal according to the matching principle (4.1.60) for all values of ζ

$$A_0 = \alpha, \qquad A_1 = \beta e\left(\frac{1}{\epsilon} + 1\right) \qquad (4.1.63)$$

which violates the assumption that $A_1 = O(1)$ which we used in deriving (4.1.59).

A more general form of the matching condition is

> The inner limit of (the outer limit) equals
> the outer limit of (the inner limit) $\qquad$ (4.1.64)

A still more general form of the matching condition is

> The inner expansion of (the outer expansion) equals
> the outer expansion of (the inner expansion) $\qquad$ (4.1.65)

where these expansions are formed using the outer and the inner limit processes defined by (4.1.41) through (4.1.43) and (4.1.51) through (4.1.53), respectively. Van Dyke (1964) proposed the following matching principle

> The m-term inner expansion of (the n-term outer expansion) equals
> the n-term outer expansion of (the m-term inner expansion) $\qquad$ (4.1.66)

where m and n may be taken to be any two integers which may be equal or unequal. To determine the m-term inner expansion of (the n-term outer expansion), we rewrite the first n terms of the outer expansion in terms of the inner variable, expand it for small ϵ keeping the inner variable fixed, and truncate the resulting expansion after m terms; and conversely for the right-hand side of (4.1.66). Because of its simplicity, Van Dyke's matching principle is widely used. A more general and rigorous matching condition is due to Kaplun (1967) who used intermediate limits. In three articles, Fraenkel (1969) compared these two matching conditions; he concluded that, although Van Dyke's matching condition may be incorrect, it is easier to use than the overlapping principle of Kaplun.

Van Dyke's Matching Principle. To show the application of Van Dyke's matching principles we use it to match the outer expansion (4.1.49) with the inner expansion (4.1.59), taking $m = n = 1$; $m = 1, n = 2$; and $m = n = 2$.

To match the one-term outer expansion with the one-term inner expansion, we proceed systematically as follows

One-term outer expansion: $\quad y \sim \beta e^{1-x}$ $\qquad$ (4.1.67a)

$\quad$ Rewritten in inner variable: $\quad = \beta e^{1-\epsilon\zeta}$ $\qquad$ (4.1.67b)

$\quad$ Expanded for small ϵ: $\qquad = \beta e(1 - \epsilon\zeta + \cdots)$ $\qquad$ (4.1.67c)

$\quad$ One-term inner expansion: $\qquad = \beta e$ $\qquad$ (4.1.67d)

One-term inner expansion: $\quad y \sim \alpha - A_0(1 - e^{-\zeta})$ $\qquad$ (4.1.68a)

$\quad$ Rewritten in outer variable: $\quad = \alpha - A_0(1 - e^{-x/\epsilon})$ $\qquad$ (4.1.68b)

$\quad$ Expanded for small ϵ: $\qquad = \alpha - A_0$ $\qquad$ (4.1.68c)

$\quad$ One-term outer expansion: $\qquad = \alpha - A_0$ $\qquad$ (4.1.68d)

By equating (4.1.67d) and (4.1.68d) according to the matching principle (4.1.66), we obtain

$$\beta e = \alpha - A_0 \quad \text{or} \quad A_0 = \alpha - \beta e \qquad (4.1.69)$$

Now we match the one-term outer expansion with the two-term inner expansion. Taking $m = 1$ and $n = 2$, we have

One-term outer expansion: $\quad y \sim \beta e^{1-x}$ $\qquad$ (4.1.70a)

$\quad$ Rewritten in inner variable: $\quad = \beta e^{1-\epsilon\zeta}$ $\qquad$ (4.1.70b)

$\quad$ Expanded for small ϵ: $\qquad = \beta e(1 - \epsilon\zeta + \tfrac{1}{2}\epsilon^2\zeta^2 + \cdots)$ $\qquad$ (4.1.70c)

$\quad$ Two-term inner expansion: $\qquad = \beta e(1 - \epsilon\zeta)$ $\qquad$ (4.1.70d)

Two-term inner expansion: $\quad y \sim \alpha - A_0(1 - e^{-\zeta}) + \epsilon\{A_1(1 - e^{-\zeta})$
$$- [\alpha - A_0(1 + e^{-\zeta})]\zeta\} \qquad (4.1.71a)$$

$\quad$ Rewritten in outer variable: $\quad = \alpha - A_0(1 - e^{-x/\epsilon}) + \epsilon\{A_1(1 - e^{-x/\epsilon})$
$$- [\alpha - A_0(1 + e^{-x/\epsilon})]\frac{x}{\epsilon}\} \qquad (4.1.71b)$$

$\quad$ Expanded for small ϵ: $\qquad = (\alpha - A_0)(1 - x) + \epsilon A_1$ $\qquad$ (4.1.71c)

$\quad$ One-term outer expansion: $\quad = (\alpha - A_0)(1 - x)$ $\qquad$ (4.1.71d)

Equating (4.1.70d) and (4.1.71d) according to the matching principle (4.1.66) gives

$$\beta e(1 - \epsilon\zeta) = (\alpha - A_0)(1 - x) \qquad (4.1.72a)$$

Since $x = \epsilon\zeta$

$$\alpha - A_0 = \beta e \quad \text{or} \quad A_0 = \alpha - \beta e \qquad (4.1.72b)$$

and no information was obtained about A_1.

Taking $m = n = 2$ in (4.1.66), we obtain

Two-term outer expansion: $\quad y \sim \beta[1 + \epsilon(1 - x)]e^{1-x}$ (4.1.73a)

Rewritten in inner variable: $\quad = \beta[1 + \epsilon(1 - \epsilon\zeta)]e^{1-\epsilon\zeta}$ (4.1.73b)

Expanded for small ϵ: $\quad = \beta e(1 + \epsilon - \epsilon\zeta + \cdots)$ (4.1.73c)

Two-term inner expansion: $\quad = \beta e(1 + \epsilon - \epsilon\zeta)$ (4.1.73d)

Two-term inner expansion: $\quad y \sim \alpha - A_0(1 - e^{-\zeta}) + \epsilon\{A_1(1 - e^{-\zeta})$

$$- [\alpha - A_0(1 + e^{-\zeta})]\zeta\} \qquad (4.1.74a)$$

Rewritten in outer variable: $\quad = \alpha - A_0(1 - e^{-x/\epsilon}) + \epsilon\Big\{A_1(1 - e^{-x/\epsilon})$

$$- [\alpha - A_0(1 + e^{-x/\epsilon})]\frac{x}{\epsilon}\Big\} \qquad (4.1.74b)$$

Expanded for small ϵ: $\quad = (\alpha - A_0)(1 - x) + \epsilon A_1$ (4.1.74c)

Two-term outer expansion: $\quad = (\alpha - A_0)(1 - x) + \epsilon A_1$ (4.1.74d)

Equating (4.1.73d) and (4.1.74d) according to the matching principle (4.1.66), we obtain

$$A_0 = \alpha - \beta e, \qquad A_1 = \beta e \qquad (4.1.75)$$

Therefore

$$y^i = \beta e + (\alpha - \beta e)e^{-\zeta} + \epsilon\{\beta e(1 - e^{-\zeta}) - [\beta e - (\alpha - \beta e)e^{-\zeta}]\zeta\} + O(\epsilon^2) \qquad (4.1.76)$$

Composite Expansion. As discussed previously, the outer expansion is not valid near the origin, while the inner expansion is not valid in general away from the region $x = O(\epsilon)$. To determine an expansion valid over the whole interval, we form a composite expansion y^c according to (Vasil'eva, 1959; Erdélyi, 1961)

$$y^c = y^o + y^i - (y^o)^i = y^o + y^i - (y^i)^o \qquad (4.1.77)$$

These two forms are equivalent since the matching principle (4.1.66) requires that

$$(y^o)^i = (y^i)^o \qquad (4.1.78)$$

Since

$$(y^i)^i = y^i, \qquad (y^o)^o = y^o \qquad (4.1.79)$$

(4.1.77) implies that

$$(y^c)^o = y^o \quad \text{and} \quad (y^c)^i = y^i \tag{4.1.80}$$

Therefore y^c is as good an approximation to y in the outer region as y^o, and it is also as good an approximation to y in the inner region as y^i.

Since $(y^o)^i$ is given by either (4.1.73d) or (4.1.74d), a composite expansion can be formed by adding the outer expansion (4.1.49) and the inner expansion (4.1.76) and subtracting the inner expansion of the outer expansion (4.1.73d). This results in

$$y^c = \beta[1 + \epsilon(1 - x)]e^{1-x} + [(\alpha - \beta e)(1 + x) - \epsilon\beta e]e^{-x/\epsilon} + O(\epsilon^2) \tag{4.1.81}$$

4.1.3.   A SECOND-ORDER EQUATION WITH VARIABLE COEFFICIENTS

We determine in this section a first-order uniformly valid expansion for the solution of

$$\epsilon y'' + a(x)y' + b(x)y = 0 \tag{4.1.82}$$

$$y(0) = \alpha, \quad y(1) = \beta \tag{4.1.83}$$

where $\epsilon \ll 1$, and $a(x)$ and $b(x)$ are analytic functions of x in the interval $[0, 1]$. Equation (4.1.1), discussed in the previous two sections, can be obtained from (4.1.82) by setting $a(x) = b(x) \equiv 1$. If ϵ vanishes, (4.1.82) reduces to the first-order equation

$$a(x)y' + b(x)y = 0 \tag{4.1.84}$$

whose solution cannot satisfy both boundary conditions, and one of them must be dropped as a consequence. As shown below, the boundary condition that must be dropped depends on the sign of $a(x)$ in the interval $[0, 1]$. If $a(x) > 0$, $y(0) = \alpha$ must be dropped, and an inner expansion near $x = 0$ must be developed and matched with the outer solution. If $a(x) < 0$, $y(1) = \beta$ must be dropped, and an inner expansion near $x = 1$ must be obtained and matched with the outer expansion. However, if $a(x)$ changes sign in $[0, 1]$, y may change from oscillatory to exponentially growing or decaying across the zeros of $a(x)$. Such zeros are called *turning or transition points*, and turning point problems are treated in Section 7.3.

Which Boundary Condition Must be Dropped?. To investigate whether $y(0) = \alpha$ must be dropped (i.e., the boundary layer is at the origin), we introduce the stretching transformation

$$\zeta = x\epsilon^{-\lambda}, \quad \lambda > 0 \tag{4.1.85}$$

to transform (4.1.82) into

$$\epsilon^{1-2\lambda}\frac{d^2y}{d\zeta^2} + \epsilon^{-\lambda}a(\epsilon^{\lambda}\zeta)\frac{dy}{d\zeta} + b(\epsilon^{\lambda}\zeta)y = 0 \qquad (4.1.86)$$

As $\epsilon \to 0$, (4.1.86) tends to

$$\frac{d^2y}{d\zeta^2} = 0 \quad \text{if} \quad \lambda > 1$$

$$\frac{dy}{d\zeta} = 0 \quad \text{if} \quad \lambda < 1 \qquad (4.1.87)$$

$$\frac{d^2y}{d\zeta^2} + a(0)\frac{dy}{d\zeta} = 0 \quad \text{if} \quad \lambda = 1$$

In order to be able to match the solutions of (4.1.87) with the outer solution as given by the reduced equation (4.1.84), we need the bounded solutions of (4.1.87) as $\zeta \to \infty$. The bounded solutions of the first two cases are constants, hence they must be discarded because they lead to certain inconsistencies similar to those encountered in Section 4.1.2. Similarly, if $a(0) < 0$, the bounded solution of the last case (i.e., $\lambda = 1$) is a constant and must also be discarded; as a consequence, the boundary layer cannot exist at $x = 0$. If $a(0) > 0$, the general solution for $\lambda = 1$ is

$$y^i = A + Be^{-a(0)\zeta} \qquad (4.1.88)$$

which is bounded as $\zeta \to \infty$ and contains two arbitrary constants; hence it is acceptable as an inner expansion because together with the solution of the reduced equation it can satisfy both boundary conditions.

To investigate whether $y(1) = \beta$ must be dropped, we introduce the stretching transformation

$$\eta = (1 - x)\epsilon^{-\lambda}, \qquad \lambda > 0 \qquad (4.1.89)$$

to transform (4.1.82) into

$$\epsilon^{1-2\lambda}\frac{d^2y}{d\eta^2} - \epsilon^{-\lambda}a(1 - \epsilon^{\lambda}\eta)\frac{dy}{d\eta} + b(1 - \epsilon^{\lambda}\eta) = 0 \qquad (4.1.90)$$

As $\epsilon \to 0$, this equation tends to

$$\frac{d^2y}{d\eta^2} = 0 \quad \text{if} \quad \lambda > 1$$

$$\frac{dy}{d\eta} = 0 \quad \text{if} \quad \lambda < 1 \qquad (4.1.91)$$

$$\frac{d^2y}{d\eta^2} - a(1)\frac{dy}{d\eta} = 0 \quad \text{if} \quad \lambda = 1$$

The bounded solutions of (4.1.91) as $\eta \to \infty$ are constants if $\lambda \neq 1$ or $a(1) > 0$ when $\lambda = 1$, hence they must be discarded. However, if $a(1) < 0$ when $\lambda = 1$, the general solution of (4.1.91) is

$$y^i = A + Be^{a(1)\eta} \qquad (4.1.92)$$

which is bounded as $\eta \to \infty$; hence it is acceptable as an inner expansion because it contains two arbitrary constants which allow, when combined with the outer solution, the satisfaction of both boundary conditions.

In summary, the boundary layer exists at

$$x = 0 \quad \text{if} \quad a(x) > 0$$

and at

$$x = 1 \quad \text{if} \quad a(x) < 0 \qquad (4.1.93)$$

Next we determine a first-order uniformly valid expansion for the former case.

The Case of $a(x) > 0$. In this case the first term of the outer expansion (4.1.41) is governed by

$$a(x)y' + b(x)y = 0, \qquad y(1) = \beta \qquad (4.1.94)$$

which yields

$$y^o = \beta \exp \left[-\int_1^x \frac{b(t)}{a(t)} \, dt \right] \qquad (4.1.95)$$

The general solution for the first term of the inner expansion (4.1.51) is given by (4.1.88). Under the transformation (4.1.85) the boundary condition $y(x = 0) = \alpha$ is transformed into $y(\zeta = 0) = \alpha$, hence

$$y^i = \alpha - B + Be^{-a(0)\zeta} \qquad (4.1.96)$$

To match y^o and y^i, we take $m = n = 1$ in (4.1.66) and proceed as follows

One-term inner expansion of (one-term outer expansion)

$$= \beta \exp \left[-\int_1^0 \frac{b(t)}{a(t)} \, dt \right] \quad (4.1.97)$$

One-term outer expansion of (one-term inner expansion) $= \alpha - B$ (4.1.98)

By equating (4.1.97) and (4.1.98) according to the matching principle (4.1.66), we have

$$B = \alpha - \beta \exp \left[-\int_1^0 \frac{b(t)}{a(t)} \, dt \right] \qquad (4.1.99)$$

Forming a composite expansion by adding (4.1.95) and (4.1.96) and subtracting (4.1.98), we obtain

$$y^c = \beta \exp\left[-\int_1^x \frac{b(t)}{a(t)}\,dt\right] + \left\{\alpha - \beta \exp\left[-\int_1^0 \frac{b(t)}{a(t)}\,dt\right]\right\}$$

$$\times \exp\left[-\frac{a(0)x}{\epsilon}\right] + O(\epsilon) \quad (4.1.100)$$

Letting $a(x) = b(x) = 1$, we have

$$y^c = \beta e^{1-x} + (\alpha - \beta e)e^{-x/\epsilon} + O(\epsilon) \quad (4.1.101)$$

in agreement with (4.1.20). Taking $a(x) = 2x + 1$ and $b(x) = 2$, we obtain

$$y^c = \frac{3\beta}{1 + 2x} + (\alpha - 3\beta)e^{-x/\epsilon} + O(\epsilon) \quad (4.1.102)$$

4.1.4. REYNOLDS' EQUATION FOR A SLIDER BEARING

The pressure distribution in an isothermal, compressible film in an

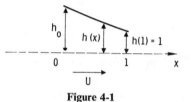

Figure 4-1

infinitely long slider bearing (see Figure 4-1) is given by the Reynolds equation, written in dimensionless quantities

$$\frac{d}{dx}\left(h^3 p \frac{dp}{dx}\right) = \Lambda \frac{d}{dx}(ph) \quad (4.1.103)$$

Here the distance x, the film thickness h, and the pressure p have been made dimensionless with respect to the length L of the bearing in the flow direction, the film thickness T at the·trailing edge, and the ambient pressure p_a, respectively. The bearing number Λ is given by $6\mu LU/p_a T^2$, where μ is the fluid viscosity and U is the velocity of the lower surface. The boundary conditions are

$$p = 1 \quad \text{at} \quad x = 0 \quad \text{and} \quad x = 1 \quad (4.1.104)$$

Following Di Prima (1969) we seek an asymptotic solution to this problem for large Λ using the method of matched asymptotic expansions.

We seek an outer expansion of the form

$$p^o = p_0(x) + \Lambda^{-1}p_1(x) + \cdots \quad (4.1.105)$$

Substituting this expansion into (4.1.103) and equating coefficients of like powers of Λ^{-1}, we obtain

$$\frac{d}{dx}(hp_0) = 0 \tag{4.1.106}$$

$$\frac{d}{dx}(hp_1) = \frac{d}{dx}\left(h^3 p_0 \frac{dp_0}{dx}\right) \tag{4.1.107}$$

Since these equations are of first order, one of the boundary conditions must be dropped, and the outer expansion is not valid near that boundary. Since h and Λ are positive, arguments similar to those used in Sections 4.1.2 and 4.1.3 show that the boundary condition at $x = 1$ must be dropped. Hence $p^o(0) = 1$, which together with (4.1.105) gives

$$p_0(0) = 1, \qquad p_1(0) = 0 \tag{4.1.108}$$

Consequently, the solution for p_0 is

$$p_0 = \frac{h_0}{h(x)} \tag{4.1.109}$$

Hence the solution for p_1 is

$$p_1 = \frac{h_0^2[h'(0) - h'(x)]}{h(x)} \tag{4.1.110}$$

This outer expansion needs to be supplemented by an inner expansion (boundary-layer solution) near $x = 1$. Introducing the stretching transformation

$$\zeta = (1 - x)\Lambda^\sigma, \qquad \sigma > 0 \tag{4.1.111}$$

we transform (4.1.103) into

$$\Lambda^{\sigma-1} \frac{d}{d\zeta}\left[h^3(1 - \zeta\Lambda^{-\sigma})p \frac{dp}{d\zeta}\right] = -\frac{d}{d\zeta}[h(1 - \zeta\Lambda^{-\sigma})p] \tag{4.1.112}$$

As $\Lambda \to \infty$ this equation reduces to

$$\frac{d}{d\zeta}\left(p \frac{dp}{d\zeta}\right) = 0 \qquad \text{if} \quad \sigma > 1$$

$$\frac{dp}{d\zeta} = 0 \qquad \text{if} \quad \sigma < 1 \tag{4.1.113}$$

$$\frac{d}{d\zeta}\left(p \frac{dp}{d\zeta}\right) = -\frac{dp}{d\zeta} \quad \text{if} \quad \sigma = 1$$

The first two cases must be discarded because the solution of the second is a constant while the solution of the first is a constant as $\zeta \to \infty$, hence in neither case can we satisfy the boundary conditions. In the third case a first integral is

$$p \frac{dp}{d\zeta} + p = A \tag{4.1.114}$$

where A is a constant. The solution of (4.1.114) is

$$-\zeta = p + A \ln (p - A) + B \tag{4.1.115}$$

where B is another constant. This solution thus represents the first term P_0 in the inner expansion

$$p^i = P_0(\zeta) + \Lambda^{-1} P_1(\zeta) + \cdots \tag{4.1.116}$$

hence it must satisfy the boundary condition $p(x = 1) = 1$ or $p^i(\zeta = 0) = 1$. This together with (4.1.116) leads to

$$P_0(0) = 1 \tag{4.1.117}$$

which, when substituted into (4.1.115), yields

$$B = -1 - A \ln (1 - A) \tag{4.1.118}$$

Hence P_0 is given implicitly by

$$-\zeta = P_0 - 1 + A \ln \frac{A - P_0}{A - 1} \tag{4.1.119}$$

To determine A we take $m = n = 1$ in the matching principle (4.1.66) and proceed as follows

One-term inner expansion of (one-term outer expansion) = h_0

$$\tag{4.1.120}$$

One-term outer expansion of (one-term inner expansion) = A (4.1.121)

Equation (4.1.121) was obtained by expanding (4.1.119) for large ζ as a result of the implicit functional dependence of P_0 on ζ. Equating (4.1.120) and (4.1.121) according to the matching principle, we have

$$A = h_0 \tag{4.1.122a}$$

hence

$$-\zeta = P_0 - 1 + h_0 \ln \frac{h_0 - P_0}{h_0 - 1} \tag{4.1.122b}$$

A first-order uniformly valid composite expansion is then formed according to

$$y^c = y^o + y^i - (y^o)^i$$

$$= \frac{h_0}{h(x)} + P_0(\zeta) - h_0 + O(\Lambda^{-1}) \qquad (4.1.123)$$

where P_0 is given by (4.1.122b).

4.1.5. UNSYMMETRICAL BENDING OF PRESTRESSED ANNULAR PLATES

We next consider the problem of unsymmetrical bending of prestressed annular plates introduced by Alzheimer and Davis (1968) and discussed in Section 2.2.4. The problem was reduced to a boundary value problem consisting of the fourth-order ordinary differential equation (2.2.28) and the boundary conditions (2.2.29) and (2.2.30). As $\epsilon \to 0$, (2.2.28) reduces from a fourth-order equation to the second-order equation (2.2.31) which can satisfy in general only two of the four boundary conditions. Consequently, its solution is not valid near one or both boundaries where boundary-layer solutions must be introduced.

Using arguments similar to those used in Sections 4.1.2 and 4.1.3, we conclude that boundary layers exist at both ends; that is, at $r = b$ and $r = 1$. The boundary layer at $r = b$ is characterized by the stretching transformation

$$\zeta = (r - b)\epsilon^{-1} \qquad (4.1.124)$$

so that u is governed near $r = b$ by

$$\left[\frac{d^2}{d\zeta^2} + \frac{\epsilon}{b + \epsilon\zeta} \frac{d}{d\zeta} - \frac{\epsilon^2}{(b + \epsilon\zeta)^2} \right]^2 u$$

$$- \left[\frac{d^2u}{d\zeta^2} + \frac{\epsilon}{b + \epsilon\zeta} \frac{du}{d\zeta} - \frac{u\epsilon^2}{(b + \epsilon\zeta)^2} \right] = 0 \quad (4.1.125)$$

However, the boundary layer at $r = 1$ is characterized by the stretching transformation

$$\eta = (1 - r)\epsilon^{-1} \qquad (4.1.126)$$

hence there u is governed by

$$\left[\frac{d^2}{d\eta^2} - \frac{\epsilon}{1 - \epsilon\eta} \frac{d}{d\eta} - \frac{\epsilon^2}{(1 - \epsilon\eta)^2} \right]^2 u$$

$$- \left[\frac{d^2u}{d\eta^2} - \frac{\epsilon}{1 - \epsilon\eta} \frac{du}{d\eta} - \frac{u\epsilon^2}{(1 - \epsilon\eta)^2} \right] = 0 \quad (4.1.127)$$

Outer Expansion. We seek an expansion, under the outer limit process

$\epsilon \to 0$ with r fixed, of the form

$$u^o = \sum_{n=0}^{N=1} \epsilon^n u_n(r) + O(\epsilon^N) \qquad (4.1.128)$$

Substituting this expansion into (2.2.28) and equating coefficients of like powers of ϵ, we have

$$\frac{d^2 u_n}{dr^2} + \frac{1}{r}\frac{du_n}{dr} - \frac{u_n}{r^2} = 0 \qquad \text{for each } n \qquad (4.1.129)$$

Hence

$$u_n = \frac{A_n}{r} + B_n r \qquad (4.1.130)$$

where A_n and B_n are constants. Since the boundary layer exists at both ends, we cannot use the boundary conditions to determine these constants; they are determined from matching the outer solution with the boundary-layer solutions.

Inner Expansion Near $r = b$. In this case we seek an inner expansion, under the inner limit process $\epsilon \to 0$ with $\zeta = (r - b)\epsilon^{-1}$ fixed, of the form

$$u^i = U_0(\zeta) + \epsilon U_1(\zeta) + \cdots \qquad (4.1.131)$$

Substituting this expansion into (4.1.125) and equating coefficients of like powers of ϵ, we obtain

$$\left(\frac{d^2}{d\zeta^2} - 1\right)\frac{d^2 U_0}{d\zeta^2} = 0 \qquad (4.1.132)$$

$$\left(\frac{d^2}{d\zeta^2} - 1\right)\frac{d^2 U_1}{d\zeta^2} = \frac{1}{b}\left(1 - 2\frac{d^2}{d\zeta^2}\right)\frac{dU_0}{d\zeta} \qquad (4.1.133)$$

This expansion must satisfy the boundary conditions (2.2.29); that is

$$u^i(\zeta = 0) = b\alpha, \qquad \frac{du^i}{d\zeta}(\zeta = 0) = \epsilon\alpha \qquad (4.1.134)$$

This together with (4.1.131) leads to

$$U_0(0) = b\alpha, \qquad U_0'(0) = 0 \qquad (4.1.135a)$$

$$U_1(0) = 0, \qquad U_1'(0) = \alpha \qquad (4.1.135b)$$

The general solution of (4.1.132) is

$$U_0 = a_0 + b_0\zeta + c_0 e^{-\zeta} + d_0 e^{\zeta} \qquad (4.1.136)$$

The constant d_0 must be zero; otherwise U_0 would grow exponentially with

ζ and it could not be matched with the outer expansion. The boundary conditions (4.1.135a) demand that

$$b_0 = c_0, \qquad a_0 + c_0 = b\alpha$$

so that U_0 can be written as

$$U_0 = b\alpha + c_0(e^{-\zeta} + \zeta - 1) \qquad (4.1.137)$$

where c_0 is a constant to be determined from matching. To do this, we take $m = n = 1$ in the matching principle (4.1.66) and obtain

$$\frac{A_0}{b} + B_0 b = \begin{cases} \dfrac{c_0(r - b)}{\epsilon} & \text{if } c_0 \neq 0 \\ b\alpha & \text{if } c_0 = 0 \end{cases} \qquad (4.1.138)$$

Hence

$$c_0 = 0 \quad \text{and} \quad \frac{A_0}{b} + B_0 b = b\alpha \qquad (4.1.139)$$

To determine A_0 and B_0, we need another relationship between them which can be provided only from matching the outer expansion with the inner expansion at $r = 1$.

We could have proceeded in this example to higher order and then performed the matching. This would have been at the expense of algebraic inconvenience arising from the terms proportional to c_0. In general, one cannot determine an outer expansion and an inner expansion to any order and then match them to determine the arbitrary constants. For example, one must proceed term by term in the problem of high Reynolds number flow past a body (e.g., Van Dyke, 1964).

With $U_0 = b\alpha$, the solution of (4.1.133) that does not grow exponentially with ζ and satisfies the boundary condition (4.1.135b) is

$$U_1 = c_1(e^{-\zeta} + \zeta - 1) + \alpha\zeta \qquad (4.1.140)$$

To determine c_1 we match the two-term inner expansion with the two-term outer expansion as follows

Two-term inner expansion of (two-term outer expansion)

$$= \frac{A_0}{b} + B_0 b + \epsilon\left[\frac{A_1}{b} + B_1 b + \left(B_0 - \frac{A_0}{b^2}\right)\zeta\right] \qquad (4.1.141)$$

Two-term outer expansion of (two-term inner expansion)

$$= b\alpha + (c_1 + \alpha)(r - b) - \epsilon c_1 \qquad (4.1.142)$$

Equating (4.1.141) and (4.1.142) according to the matching principle, we have

$$\frac{A_0}{b} + B_0 b + \epsilon\left[\frac{A_1}{b} + B_1 b + \left(B_0 - \frac{A_0}{b^2}\right)\zeta\right]$$
$$= b\alpha + (c_1 + \alpha)(r - b) - \epsilon c_1 \quad (4.1.143)$$

Since $\zeta = (r - b)\epsilon^{-1}$

$$\frac{A_1}{b} + B_1 b = -c_1, \qquad B_0 - \frac{A_0}{b^2} = c_1 + \alpha \quad (4.1.144)$$

These provide two equations for A_1, B_1, and c_1; the third relationship is obtained from matching the outer expansion with the inner expansion at $r = 1$.

Inner Expansion Near $r = 1$. Near $r = 1$, we seek an inner expansion u^I, under the inner limit process $\epsilon \to 0$ with $\eta = (1 - r)\epsilon^{-1}$ fixed, of the form

$$u^I = \tilde{U}_0(\eta) + \epsilon\tilde{U}_1(\eta) + \cdots \quad (4.1.145)$$

Substituting this expansion into (4.1.127) and equating coefficients of like powers of ϵ, we obtain

$$\left(\frac{d^2}{d\eta^2} - 1\right)\frac{d^2\tilde{U}_0}{d\eta^2} = 0 \quad (4.1.146)$$

$$\left(\frac{d^2}{d\eta^2} - 1\right)\frac{d^2\tilde{U}_1}{d\eta^2} = \left(2\frac{d^2}{d\eta^2} - 1\right)\frac{d\tilde{U}_0}{d\eta} \quad (4.1.147)$$

Since $r = 1$ corresponds to $\eta = 0$, u^I must satisfy the boundary conditions (2.2.30); that is

$$u^I = \frac{du^I}{d\eta} = 0 \quad \text{at} \quad \eta = 0 \quad (4.1.148)$$

which, when combined with (4.1.145), gives

$$\tilde{U}_0(0) = 0, \qquad \tilde{U}_0'(0) = 0 \quad (4.1.149)$$

$$\tilde{U}_1(0) = 0, \qquad \tilde{U}_1'(0) = 0 \quad (4.1.150)$$

The solution of (4.1.146) and (4.1.149) which does not grow exponentially with η is

$$\tilde{U}_0 = \tilde{c}_0(e^{-\eta} + \eta - 1) \quad (4.1.151)$$

To determine $\tilde{c}_0$ we match the one-term outer expansion with the one-term inner expansion according to

$$A_0 + B_0 = \tilde{c}_0\frac{1 - r}{\epsilon} \quad (4.1.152)$$

Hence

$$\tilde{c}_0 = 0 \quad \text{and} \quad A_0 + B_0 = 0 \qquad (4.1.153)$$

and

$$\tilde{U}_0 = 0 \qquad (4.1.154)$$

Solving (4.1.139) and (4.1.153) for A_0 and B_0, we obtain

$$A_0 = -B_0 = \frac{b^2\alpha}{1 - b^2} \qquad (4.1.155)$$

Hence

$$u_0 = \frac{b^2\alpha}{1 - b^2}\left(\frac{1}{r} - r\right) \qquad (4.1.156)$$

Since $\tilde{U}_0 = 0$, the solution of (4.1.147) and (4.1.150) that does not grow exponentially with η is

$$\tilde{U}_1 = \tilde{c}_1(e^{-\eta} + \eta - 1) \qquad (4.1.157)$$

To determine $\tilde{c}_1$ we match the two-term outer expansion with the two-term inner expansion as follows

Two-term inner expansion of (two-term outer expansion)

$$= 0 + \epsilon(2A_0\eta + A_1 + B_1) \quad (4.1.158)$$

Two-term outer expansion of (two-term inner expansion)

$$= \tilde{c}_1(1 - r) - \epsilon\tilde{c}_1 \quad (4.1.159)$$

Equating (4.1.158) and (4.1.159) according to the matching principle, we have

$$\tilde{c}_1 = 2A_0 = \frac{2b^2\alpha}{1 - b^2}, \qquad A_1 + B_1 = -\tilde{c}_1 = -\frac{2b^2\alpha}{1 - b^2} \quad (4.1.160)$$

With A_0 and B_0 given by (4.1.155), (4.1.144) gives

$$c_1 = -\frac{2\alpha}{1 - b^2} \qquad (4.1.161)$$

while (4.1.144) and (4.1.160) give

$$A_1 = \frac{2\alpha b(1 + b^3)}{(1 - b^2)^2}, \qquad B_1 = -\frac{2\alpha b(1 + b)}{(1 - b^2)^2} \qquad (4.1.162)$$

This completes the determination of the constants of integration in all of the second-order expansions.

Composite Expansion. Let us first summarize the results of the previous three sections

$$u^o = \frac{b^2\alpha}{1 - b^2}\left(\frac{1}{r} - r\right) + \epsilon\frac{2\alpha b}{(1 - b^2)^2}$$

$$\times \left[\frac{1 + b^3}{r} - (1 + b)r\right] + O(\epsilon^2) \qquad (4.1.163)$$

$$u^i = b\alpha - \epsilon\frac{2\alpha}{1 - b^2}\left(e^{-\zeta} + \frac{1 + b^2}{2}\zeta - 1\right) + O(\epsilon^2) \qquad (4.1.164)$$

$$u^I = \epsilon\frac{2b^2\alpha}{1 - b^2}(e^{-\eta} + \eta - 1) + O(\epsilon^2) \qquad (4.1.165)$$

$$(u^o)^i = b\alpha - \frac{2\alpha}{1 - b^2}\epsilon\left(\frac{1 + b^2}{2}\zeta - 1\right) + O(\epsilon^2) \qquad (4.1.166)$$

$$(u^o)^I = \frac{2b^2\alpha}{1 - b^2}\epsilon(\eta - 1) + O(\epsilon^2) \qquad (4.1.167)$$

A uniformly valid composite expansion in the interval $[b, 1]$ is now formed according to

$$u^c = u^o + u^i + u^I - (u^o)^i - (u^o)^I$$

$$= \frac{b^2\alpha}{1 - b^2}\left(\frac{1}{r} - r\right) + \frac{2\epsilon\alpha b^2}{(1 - b^2)^2}\left[\frac{1 + b^3}{r} - (1 + b)r\right]$$

$$- \frac{2\epsilon\alpha}{1 - b^2}e^{-\zeta} + \frac{2\epsilon b^2\alpha}{1 - b^2}e^{-\eta} + O(\epsilon^2) \qquad (4.1.168)$$

4.1.6. THERMOELASTIC SURFACE WAVES
The problem of thermoelastic surface waves was reduced in Section 2.4.3 to the solution of the algebraic equation (2.4.35) for the square of the wave speed $x = c^2$. The straightforward expansion of the solution of this equation for small ϵ can be written as (Section 2.4.3)

$$x = x_R\left\{1 - \epsilon\frac{\tau x_R\left(1 - \frac{1 - x_R}{F}\right)}{(1 - \tau x_R)[1 + G'(x_R)]}\right\} + O(\epsilon^2) \qquad (4.1.169)$$

for small τ, and

$$x = \frac{1}{\tau}\left\{1 - \frac{\epsilon\tau}{\tau - 1} - \frac{\epsilon^2\tau}{\tau - 1}\left[\frac{-\tau^2}{(\tau - 1)^2} + M^2\right]\right\} + O(\epsilon^3) \qquad (4.1.170)$$

for large τ, where $F(c_R{}^2)$ and $M(c_R{}^2)$ are given, respectively, by (2.4.39) and (2.4.42), and

$$x_R = c_R{}^2 \approx 0.2817 \qquad \text{(square of Rayleigh wave speed)}$$

The Rayleigh wave speed is the solution of

$$G(c_R{}^2) = 1 - c_R{}^2 \qquad (4.1.171)$$

It is clear that the expansion (4.1.169) breaks down as $\tau \uparrow 1/x_R$. In fact, $x \to \infty$ as $\tau \to 1/x_R$. The second term in (4.1.170) indicates that this expansion breaks down as $\tau \downarrow 1$. However, the third term shows that this expansion breaks down well before $\tau = 1$. Since $G(x_R) = 1 - x_R$ from (4.1.171), the coefficient of M^2 in (2.4.42) vanishes at $\tau = 1/x_R \approx 3.550$, hence both M and $x \to \infty$ as $\tau \downarrow 1/x_R$. Therefore both of the above expansions break down at or near $\tau_R = 1/x_R$.

In order to determine an expansion valid uniformly for all τ, we consider the above two expansions as outer expansions and supplement them by an inner expansion near $\tau = \tau_R$. In all the examples thus far considered, we needed only to stretch the independent variable. However, in this case we find that it is necessary to stretch both the independent variable τ as well as the dependent variable x. Thus we introduce the transformation

$$T = \frac{\tau - \tau_R}{\tau_R} \epsilon^{-n}, \qquad X = \frac{x_R - x}{x_R} \epsilon^{-m} \qquad 0 < m, n < 1 \quad (4.1.172)$$

Before changing the variables in (2.4.35), we note that

$$[1 - x\tau(1 + \epsilon) + A]^2$$
$$= (1 - x\tau + A)^2 - 2x\tau\epsilon(1 - x\tau + A) + O(\epsilon^2)$$
$$= (1 - x\tau)^2 + 2A(1 - x\tau) + A^2 - 2\epsilon x\tau(1 - x\tau + A) + O(\epsilon^2)$$
$$= (1 - x\tau)\left[2 - x\tau - x + 2A - \epsilon x\tau \frac{3 - 2x\tau + 2A}{1 - x\tau}\right] + O(\epsilon^2)$$

hence we can rewrite (2.4.35) as

$$G(x) = 1 - x - \frac{\epsilon x^2 \tau}{2 - x - x\tau + 2A}\left(1 + \frac{2A}{1 - x\tau}\right) + O(\epsilon^2) \quad (4.1.173)$$

With (4.1.172) this equation becomes

$$X[1 + G'(x_R)] = \frac{2}{\sqrt{1 - x_R}} \frac{\epsilon^{1-3m/2}}{\sqrt{X - \epsilon^{n-m}T}} + O(\epsilon) \qquad (4.1.174)$$

To obtain a nontrivial result as $\epsilon \to 0$, we choose

$$m = n = \tfrac{2}{3}$$

and (4.1.174) tends to

$$X[1 + G'(x_R)] = \frac{2}{\sqrt{1 - x_R}\sqrt{X - T}} \quad \text{as} \quad \epsilon \to 0$$

Letting

$$\xi^2 = X - T \tag{4.1.175}$$

we can rewrite this equation as

$$\xi^3 + T\xi = K \tag{4.1.176}$$

where

$$K = \frac{2}{\sqrt{1 - x_R}\,[1 + G'(x_R)]}$$

Hence the solution of this cubic equation gives the first term in the inner expansion. This cubic equation has one or three real roots depending on whether

$$D = \frac{K^2}{4} + \frac{T^3}{27}$$

is positive definite or negative semidefinite. For positive or small negative T, D is positive, and we have only the following real root

$$\xi = \sqrt[3]{-\tfrac{1}{2}K + \sqrt{D}} - \sqrt[3]{\tfrac{1}{2}K + \sqrt{D}} \tag{4.1.177}$$

For $T < -3\sqrt[3]{K^2/4}$, we have three real roots for ξ, and we choose the one that continues (4.1.177) as D changes sign. We match this inner expansion with the outer expansions (4.1.169) and (4.1.170) and then form uniformly valid composite expansions.

Expansion for $0 \le \tau \le \tau_R$. We match the inner expansion with the outer expansion (4.1.169) by taking $m = n = 2$ in (4.1.66) as follows

Two-term inner expansion of (two-term outer expansion)

$$= x_R\left[1 - \epsilon^{2/3}\frac{K}{\sqrt{-T}}\right] \tag{4.1.178}$$

Two-term outer expansion of (two-term inner expansion)

$$= x_R\left[1 - \epsilon^{2/3}\frac{K}{\sqrt{-T}}\right] \tag{4.1.179}$$

In arriving at (4.1.179) we have found the solution of (4.1.176) for large negative T and picked out the root

$$\xi = \sqrt{-T} - \frac{K}{2T} + \cdots \quad \text{as} \quad T \to -\infty \qquad (4.1.180)$$

since (4.1.177) is not valid for large negative T. Comparing (4.1.178) and (4.1.179), we see that they are equal, hence the outer expansion (4.1.169) indeed matches the inner expansion represented by (4.1.176). Therefore a composite expansion valid for all $0 \leq \tau \leq \tau_R$ is given by

$$x^c = x^o + x^i - (x^o)^i$$

$$= x_R \left\{ 1 - \epsilon \frac{\tau x_R \left(1 - \dfrac{1 - x_R}{F} \right)}{(1 - \tau x_R)[1 + G'(x_R)]} + \epsilon^{2/3} \left[\frac{K}{\sqrt{-T}} - T - \xi^2 \right] \right\} + O(\epsilon^2)$$

$$(4.1.181)$$

Note that this expansion is regular at $\tau = \tau_R$ because the singular contribution of the term multiplying ϵ is exactly canceled by the term $\epsilon^{2/3} K / \sqrt{-T}$.

Expansion for $\tau \geq \tau_R$. In this case T is positive, and we match the outer expansion (4.1.170) with the inner expansion by also taking $m = n = 2$ as follows

Two-term inner expansion of (two-term outer expansion)

$$= x_R(1 - \epsilon^{2/3} T) \qquad (4.1.182)$$

Two-term outer expansion of (two-term inner expansion)

$$= x_R \left[1 - \frac{\tau - \tau_R}{\tau_R} + 0 \cdot \epsilon \right] \qquad (4.1.183)$$

Since (4.1.182) and (4.1.183) are equal, the two expansions indeed match. A composite expansion uniformly valid for $\tau \geq \tau_R$ can be formed according to

$$x^c = x^o + x^i - (x^o)^i$$

$$= \frac{1}{\tau} - \frac{\epsilon}{\tau - 1} - \epsilon^{2/3} x_R \xi^2 + O(\epsilon^2) \qquad (4.1.184)$$

Adnan Nayfeh and Nemat-Nasser (1971) were the first to analyze this problem using the method of matched asymptotic expansion. The results of this section show that the outer and inner expansions are not in general expandable in the same asymptotic sequence, say, powers of ϵ. In the present example the outer expansions are in powers of ϵ, while the inner expansion is in fractional powers of ϵ. Moreover, this example serves to demonstrate that a nonuniformity may arise in the interior of the domain rather than at the boundaries as the earlier examples indicate.

4.1.7. THE EARTH-MOON-SPACESHIP PROBLEM

In the examples discussed so far, the asymptotic sequences contained powers ϵ^n or fractional powers $\epsilon^{m/n}$ of the small parameter. These asymptotic sequences may fail to represent the solution in some instances in which they must be supplemented by log $(1/\epsilon)$ which tends to infinity more slowly than any fractional power of ϵ. The present example serves to demonstrate that one should keep in mind that it may be necessary to supplement the asymptotic sequences by logarithms such as log $(1/\epsilon)$, log [log $(1/\epsilon)$], log {log [log $(1/\epsilon)$]}, or their equivalents.

The problem under consideration is represented mathematically by (2.4.7) through (2.4.9) in Section 2.4.2. The one-dimensional problem was treated in Section 3.2.2 using the method of strained coordinates. It was shown in 3.5.3 that the method of strained coordinates leads to an erroneous solution in the two-dimensional case.

If we take x rather than t as the independent variable, the straightforward expansion (obtained using the outer limit process $\mu \to 0$ with x fixed) is given by (2.4.17) through (2.4.19). It was pointed out that this expansion is invalid near $x = 1$ because the second term (2.4.19) has a logarithmic singularity while the third term has an algebraic singularity at $x = 1$. The size of the nonuniformity, hence the appropriate stretching transformation, may be estimated by taking the ratio of the most singular part of the term $O(\mu^2)$ to that of $O(\mu)$. This estimation suggests that the region of non-uniformity is $1 - x = O(\mu)$. Had we taken the ratio of the singular part in the term $O(\mu)$ to the first term, we would have concluded that the region of nonuniformity is $1 - x = O(e^{-1/\mu})$ which is not correct.

A better way of estimating the size of nonuniformity and determining the stretching transformation is an analysis of the order of magnitude of the various terms in the equations, as we have made in the earlier examples. Thus we introduce the transformation

$$\xi = \frac{1-x}{\mu^\alpha}, \qquad \zeta = \frac{y}{\mu^\alpha}, \qquad \eta = \frac{t-\tau}{\mu^\beta}, \qquad \alpha \text{ and } \beta > 0 \quad (4.1.185)$$

where τ is the time elapsed to reach $x = 1$ and transform (2.4.7) into

$$\frac{d^2\xi}{d\eta^2} = \frac{(1-\mu)\mu^{2\beta-\alpha}(1-\mu^\alpha\xi)}{[(1-\mu^\alpha\xi)^2 + \mu^{2\alpha}\zeta^2]^{3/2}} - \mu^{1-3\alpha+2\beta}\frac{\xi}{(\xi^2+\zeta^2)^{3/2}} \quad (4.1.186)$$

The term on the left-hand side of this equation represents the spaceship's acceleration, the first term on the right-hand side represents the contribution to this acceleration due to the earth's attraction, and the last term represents the contribution to this acceleration due to lunar attraction. If the lunar attraction force is neglected compared to the earth's attraction force, we would arrive at the outer expansion (2.4.12) which is not valid near $x = 1$. Thus in order to obtain an expansion valid near the moon (i.e., $x = 1$), the contribution to the acceleration by the lunar attraction must be of the same order as the acceleration; that is

$$3\alpha - 2\beta = 1 \qquad (4.1.187)$$

This condition can be satisfied by an infinity of α and β values. Since the inner expansion must match the outer expansion, the velocity $d\xi/d\eta$ near the moon must be of the same order as the velocity $dx/dt = O(1)$ away from the moon, hence

$$\alpha = \beta \qquad (4.1.188)$$

Thus
$$\alpha = \beta = 1 \qquad (4.1.189)$$

and the original equations can be rewritten in terms of the inner variable $\xi = (1 - x)\mu^{-1}$ as

$$\mu \frac{d^2\xi}{dt^2} = (1 - \mu) \frac{1 - \mu\xi}{[(1 - \mu\xi)^2 + y^2]^{3/2}} - \mu^2 \frac{\xi}{(\mu^2\xi^2 + y^2)^{3/2}} \qquad (4.1.190)$$

$$\frac{d^2y}{dt^2} = -(1 - \mu) \frac{y}{[(1 - \mu\xi)^2 + y^2]^{3/2}} - \mu \frac{y}{(\mu^2\xi^2 + y^2)^{3/2}} \qquad (4.1.191)$$

We next discuss the one-dimensional case and refer the reader to Lagerstrom and Kevorkian (1963b) for the two-dimensional case and to Lagerstrom and Kevorkian (1963a) and Breakwell and Perko (1966) for the case of rotating mass centers.

If we set $y = 0$ and integrate (2.4.7) subject to (2.4.11) with $h = \rho = 0$, we have

$$\frac{1}{2}\left(\frac{dx}{dt}\right)^2 = \frac{1 - \mu}{x} + \frac{\mu}{1 - x}, \qquad t(0) = 0 \qquad (4.1.192)$$

The outer expansion (obtained using the outer limit $\mu \to 0$ with x fixed) is (Exercise 2.12)

$$\sqrt{2}\, t^o = \tfrac{2}{3} x^{3/2} + \mu \left(\tfrac{2}{3} x^{3/2} + \sqrt{x} - \tfrac{1}{2} \ln \frac{1 + \sqrt{x}}{1 - \sqrt{x}} \right) + O(\mu^2) \qquad (4.1.193)$$

which breaks down as $x \to 1$. To describe the solution near $x = 1$, we let $\xi = (1 - x)\mu^{-1}$ in (4.1.192) and obtain

$$\tfrac{1}{2}\mu^2 \left(\frac{d\xi}{dt}\right)^2 = \frac{1 - \mu}{1 - \mu\xi} + \frac{1}{\xi} \qquad (4.1.194)$$

Now we seek an inner expansion, under the inner limit process $\mu \to 0$ with ξ fixed, of the form

$$t^i = \tau_0 + \mu T_1(\xi) + O(\mu^2) \qquad (4.1.195)$$

and equate coefficients of like powers of μ to obtain

$$\frac{1}{2}\left(\frac{d\xi}{dT_1}\right)^2 = 1 + \frac{1}{\xi} \qquad (4.1.196)$$

Its general solution is

$$\sqrt{2}\,T_1 = -\sqrt{\xi(\xi+1)} + \sinh^{-1}\sqrt{\xi} + \tau_1 \qquad (4.1.197)$$

where τ_1 is a constant, which together with τ_0 must be determined from matching because this expansion is not valid near $x = 0$; hence we cannot use the initial condition $t(x = 0) = 0$.

To perform the matching, we take $m = n = 2$ in (4.1.66) and obtain:

Two-term inner expansion of (two-term outer expansion)

$$= \tfrac{2}{3} + \mu[-\xi + \tfrac{5}{3} - \ln 2 + \tfrac{1}{2}\ln\mu + \tfrac{1}{2}\ln\xi] \quad (4.1.198)$$

Two-term outer expansion of (two-term inner expansion)

$$= \sqrt{2}\,\tau_0 - 1 + x - \mu\left[\tfrac{1}{2} - \ln 2 - \tau_1 - \frac{1}{2}\ln\frac{1-x}{\mu}\right] \quad (4.1.199)$$

Equating (4.1.198) and (4.1.199) according to the matching principle, we have

$$\sqrt{2}\tau_0 = \tfrac{2}{3}, \qquad \tau_1 = \tfrac{13}{6} - 2\ln 2 + \tfrac{1}{2}\ln\mu \qquad (4.1.200)$$

Hence

$$\sqrt{2}t^i = \tfrac{2}{3} + \mu[\tfrac{13}{6} - 2\ln 2 + \tfrac{1}{2}\ln\mu - \sqrt{\xi(\xi+1)} + \sinh^{-1}\sqrt{\xi}] + O(\mu^2)$$
$$(4.1.201)$$

which contains a $\mu \ln \mu$ term in addition to the $O(\mu)$ term.

A composite expansion uniformly valid over $[0, 1]$ can be formed according to

$$t^c = t^o + t^i - (t^o)^i$$

Hence

$$\sqrt{2}t^c = \tfrac{2}{3}x^{3/2} + \mu[\tfrac{1}{2} - \ln 2 + \tfrac{1}{2}\ln\mu + \tfrac{2}{3}x^{3/2} + \sqrt{x} - \ln(1+\sqrt{x})$$
$$+ \xi - \sqrt{\xi(\xi+1)} + \sinh^{-1}\sqrt{\xi}] + O(\mu^2) \quad (4.1.202)$$

4.1.8. SMALL REYNOLDS NUMBER FLOW PAST A SPHERE

As a last example, we consider the problem of small Reynolds number flow past a sphere discussed in Section 2.1.4. In addition to being governed by a partial differential equation, this example shows that one sometimes needs a contracting rather than a stretching transformation. The stream function is given by the fourth-order partial differential equation (2.1.37) subject to the boundary conditions (2.1.39) and (2.1.40).

Stokes' Expansion. In Section 2.1.4 we obtained the straightforward expansion (2.1.59) which was called Stokes' expansion by Lagerstrom and Cole (1955). It was obtained under what is called Stokes' limit process: $R \to 0$ with r fixed. As pointed out in Section 2.1.4, Stokes' expansion satisfies the conditions (2.1.39) on the surface of the sphere, but it does not satisfy the infinity condition (2.1.40). Thus Stokes' expansion breaks down as $r \to \infty$ (Whitehead's paradox).

Oseen's Expansion. To investigate the source of this nonuniformity, Oseen (1910) examined the relative magnitude of the (convective) terms neglected by Stokes to those retained (viscous). The right-hand side of (2.1.45) shows that

$$\text{Neglected terms} = O\left(\frac{R}{r^2}\right) \qquad (4.1.203a)$$

while the cross-product term (retained) in (2.1.42) is

$$\frac{\partial^2}{\partial r^2}\left[\frac{\sin\theta}{r^2}\frac{\partial}{\partial\theta}\left(\frac{1}{\sin\theta}\frac{\partial}{\partial\theta}\right)\right]\psi_0 = O\left(\frac{1}{r^3}\right) \qquad (4.1.203b)$$

Hence

$$\frac{\text{Neglected}}{\text{Retained}} = O(Rr) \quad \text{as} \quad r \to \infty \qquad (4.1.203c)$$

and Stokes' expansion breaks down when r increases to $O(R^{-1})$.

One can also arrive at this conclusion by noting that the source of nonuniformity is the term $-(3/16)Rr^2 \sin^2\theta \cos\theta$ in the particular integral that does not behave properly as $r \to \infty$. Consequently, Stokes' expansion is valid as long as this term is small compared to the term $-(3/4)r \sin^2\theta$ in ψ_0, and Stokes' expansion breaks down when these two terms are of the same order; that is, when $rR = O(1)$.

The above arguments led Oseen (1910) to derive an approximation to the flow field valid everywhere. It is the first two terms of what Lagerstrom and Cole (1955) called Oseen's expansion, derived using Oseen's limit process $R \to 0$ with $\rho = Rr$ fixed. It should be noted that ρ is a contracting rather than a stretching transformation. With this new variable (2.1.37) becomes

$$D^4\psi = \frac{R^2}{\rho^2 \sin\theta}\left(\psi_\theta\frac{\partial}{\partial\rho} - \psi_\rho\frac{\partial}{\partial\theta} + 2\cot\theta\,\psi_\rho - 2\frac{\psi_\theta}{\rho}\right)D^2\psi \quad (4.1.204)$$

where

$$D^2 = \frac{\partial^2}{\partial\rho^2} + \frac{\sin\theta}{\rho^2}\frac{\partial}{\partial\theta}\left(\frac{1}{\sin\theta}\frac{\partial}{\partial\theta}\right) \qquad (4.1.205)$$

Since the general solutions to partial differential equations are not known in general, it is more convenient and expedient to determine the inner and outer expansions term by term, employing the matching principle as a guide to the forms of these expansions. Since Stokes' solution (2.1.51) is uniformly valid, we can determine the first term in Oseen's expansion by taking Oseen's limit of (2.1.51). To determine the form of the second term as well, we use the matching principle

one-term Stokes (two-term Oseen) = two-term Oseen (one-term Stokes)

$$= \frac{1}{2}\frac{1}{R^2}\rho^2\sin^2\theta - \frac{3}{4}\frac{1}{R}\rho\sin^2\theta \quad (4.1.206)$$

Thus Oseen's expansion (denoted by ψ^o) must have the form

$$\psi^o = \frac{1}{2}\frac{1}{R^2}\rho^2\sin^2\theta + \frac{1}{R}\Psi_1(\rho,\theta) + \Psi_2(\rho,\theta) + \ldots \qquad (4.1.207)$$

and the first term is a consequence of the validity of Stokes' solution for all r. Substituting this expansion into (4.1.204) and equating the coefficients of R on both sides, we obtain

$$\left(D^2 - \cos\theta\frac{\partial}{\partial\rho} + \frac{\sin\theta}{\rho}\frac{\partial}{\partial\theta}\right)D^2\Psi_1 = 0 \qquad (4.1.208)$$

This equation is called Oseen's equation which he derived using physical arguments.

To solve Oseen's equation we follow Goldstein (1929) by letting

$$D^2\Psi_1 = \phi e^{(1/2)\rho\cos\theta} \qquad (4.1.209)$$

and obtaining

$$(D^2 - \tfrac{1}{4})\phi = 0 \qquad (4.1.210)$$

Rather than obtaining the general solution to (4.1.208) by solving (4.1.209) and (4.1.210) and then using the matching principle (4.1.206) to select the terms that match Stokes' solution, Proudman and Pearson (1957) matched $\mathscr{D}^2\psi^s$ and $\mathscr{D}^2\psi^o$ first. Since

$$\mathscr{D}^2\psi^s = \frac{3}{2}\frac{1}{r}\sin^2\theta + O(R)$$

$$\mathscr{D}^2\psi^o = R^2D^2\psi^o = RD^2\Psi_1(\rho,\theta) + O(R^2) \qquad (4.1.211)$$

and the matching condition demands that

> one-term Stokes (one-term $\mathscr{D}^2\psi^o$) = one-term Oseen (one-term $\mathscr{D}^2\psi^s$),
>
> $$(4.1.212)$$

we have

$$\text{one-term Stokes } (D^2\Psi'_1) = \frac{3}{2}\frac{1}{\rho}\sin^2\theta \qquad (4.1.213)$$

To satisfy this condition we seek a solution of (4.1.210) of the form

$$\phi = \sin^2\theta \, f(\rho) \qquad (4.1.214)$$

so that

$$f'' - \left(\frac{2}{\rho^2} + \frac{1}{4}\right)f = 0 \qquad (4.1.215)$$

The solution of f that does not grow exponentially with ρ is

$$f = A\left(1 + \frac{2}{\rho}\right)e^{-(1/2)\rho} \qquad (4.1.216)$$

so that

$$D^2\Psi'_1 = A\left(1 + \frac{2}{\rho}\right)\sin^2\theta \, e^{-(1/2)\rho(1-\cos\theta)} \qquad (4.1.217)$$

Hence

$$\text{one-term Stokes } (D^2\Psi'_1) = \frac{2A}{R}\frac{1}{r}\sin^2\theta \qquad (4.1.218)$$

Then (4.1.213) gives $A = 3/4$, and $D^2\Psi'_1$ becomes

$$D^2\Psi'_1 = \frac{3}{4}\left(1 + \frac{2}{\rho}\right)\sin^2\theta \, e^{-(1/2)\rho(1-\cos\theta)} \qquad (4.1.219)$$

The particular integral of (4.1.219) is

$$\Psi'_{1p} = \tfrac{3}{2}(1 + \cos\theta)e^{-(1/2)\rho(1-\cos\theta)} \qquad (4.1.220)$$

so that

$$\Psi'_1 = \Psi'_{1c} + \tfrac{3}{2}(1 + \cos\theta)e^{-(1/2)\rho(1-\cos\theta)} \qquad (4.1.221)$$

Then from (4.1.206)

$$\text{one-term Stokes } (\Psi'_{1c}) = -\tfrac{3}{2}(1 + \cos\theta) \qquad (4.1.222)$$

Hence $\Psi'_{1c} = -3(1 + \cos\theta)/2$, and Oseen's expansion becomes

$$\psi^o = \frac{1}{2R^2}\rho^2\sin^2\theta - \frac{3}{2R}(1 + \cos\theta)[1 - e^{-(1/2)\rho(1-\cos\theta)}] + O(1) \qquad (4.1.223)$$

Second Term in Stokes' Expansion. Equations (2.1.51), (2.1.52), and (2.1.58) show that the second-order Stokes' expansion is

$$\psi^s = \frac{1}{4}\left(2r^2 - 3r + \frac{1}{r}\right) \sin^2 \theta$$

$$+ R\left[-\frac{3}{32}\left(2r^2 - 3r + 1 - \frac{1}{r} + \frac{1}{r^2}\right) \sin^2 \theta \cos \theta + \psi_{1c}\right] + O(R^2)$$

$$(4.1.224)$$

where ψ_{1c} is the complementary solution of (2.1.52). To determine ψ_{1c} we match two terms of ψ^s with two terms of ψ^o according to

Two-term Stokes' expansion: $\qquad \psi \sim \psi_0 + R(\psi_{1p} + \psi_{1c})$ $\qquad$ (4.1.225a)

Rewritten in Oseen's variables: $\qquad = \frac{1}{4}\left(2\frac{\rho^2}{R^2} - \frac{3\rho}{R} + \frac{R}{\rho}\right) \sin^2 \theta$

$$+ R\left[-\frac{3}{32}\left(2\frac{\rho^2}{R^2} - \frac{3\rho}{R} + 1 - \frac{R}{\rho} + \frac{R^2}{\rho^2}\right)\right.$$

$$\times \sin^2 \theta \cos \theta + \left.\psi_{1c}\left(\frac{\rho}{R}, \theta\right)\right]$$

$$(4.1.225b)$$

Two-term Oseen's expansion: $\qquad = \frac{1}{2R^2} \rho^2 \sin^2 \theta$

$$+ \frac{1}{R}\left[-\tfrac{3}{4}(\rho \sin^2 \theta + \tfrac{1}{4}\rho^2 \sin^2 \theta \cos \theta)\right.$$

$$+ \left. \text{term } O(1) \text{ in } R^2\psi_{1c}\left(\frac{\rho}{R}, \theta\right)\right]$$

$$(4.1.225c)$$

Two-term Oseen's expansion: $\qquad \psi \sim \frac{1}{2R^2} \rho^2 \sin^2 \theta - \frac{3}{2R} (1 + \cos \theta)$

$$\times [1 - e^{-(1/2)\rho(1 - \cos \theta)}] \qquad (4.1.226a)$$

Rewritten in Stokes' variables: $\qquad = \tfrac{1}{2}r^2 \sin^2 \theta - \frac{3}{2R} (1 + \cos \theta)$

$$\times [1 - e^{-(1/2)rR(1 - \cos \theta)}] \qquad (4.1.226b)$$

Two-term Stokes' expansion: $\qquad = (\tfrac{1}{2}r^2 - \tfrac{3}{4}r) \sin^2 \theta$

$$+ \tfrac{3}{16}r^2R \sin^2 \theta (1 - \cos \theta) \qquad (4.1.226c)$$

Equating (4.1.225c) and (4.1.226c) according to the matching principle, we have

$$\text{one-term Oseen}\,(\psi_{1c}) = \tfrac{3}{16}r^2 \sin^2 \theta \qquad (4.1.227)$$

This suggests that

$$\psi_{1c} = f(r) \sin^2 \theta \qquad (4.1.228)$$

hence from (2.1.50)

$$f(r) = c_4 r^4 + c_2 r^2 + c_1 r + c_{-1} r^{-1} \qquad (4.1.229)$$

The matching condition (4.1.227) demands that $c_4 = 0$ and $c_2 = 3/16$, while the boundary conditions $\psi_1(1, \theta) = \psi_{1r}(1, \theta) = 0$ demand that $c_1 = -\tfrac{9}{32}$ and $c_{-1} = \tfrac{3}{32}$. Therefore

$$\psi^s = \frac{1}{4}\left(2r^2 - 3r + \frac{1}{r}\right)\sin^2 \theta + \tfrac{3}{32}R\left[\left(2r^2 - 3r + \frac{1}{r}\right)\sin^2 \theta\right.$$

$$\left. - \left(2r^2 - 3r + 1 - \frac{1}{r} + \frac{1}{r^2}\right)\sin^2 \theta \cos \theta\right] + O(R^2) \quad (4.1.230)$$

Higher Approximations. Proudman and Pearson (1957) found that the particular solution for ψ_2 contains $\ln r$ which leads to the appearance of $\ln R$ on matching. This is another example in which logarithms of the perturbation parameter appear naturally when one matches expansions containing logarithms of the independent variable.

4.2. The Method of Composite Expansions

The composite expansions obtained in Sections 4.1.1 through 4.1.7 are generalized expansions having the form

$$y(x; \epsilon) = y^o(x; \epsilon) + y^i(\zeta; \epsilon) - (y^o)^i = y^o + y^i - (y^i)^o \quad (4.2.1a)$$

where y is the dependent variable, ϵ is the small parameter, x is the outer variable, and ζ is the inner variable. This composite expansion can be viewed as the sum of two terms $F(x; \epsilon) = y^o$ and $G(\zeta; \epsilon) = y^i - (y^o)^i$; that is

$$y(x; \epsilon) = F(x; \epsilon) + G(\zeta; \epsilon) \qquad (4.2.1b)$$

Rather than determining outer and inner expansions, matching them, and then forming a composite expansion, Bromberg (1956) and Višik and Lyusternik (1957) assumed the solution to have the form (4.2.1b) which is valid everywhere; hence it satisfies all the boundary conditions. Taking the outer limit of (4.2.1b) gives

$$y^o(x; \epsilon) = F + G^o \qquad (4.2.1c)$$

which must satisfy the original differential equation expressed in terms of the outer variable. Similarly

$$y^i = F^i + G \tag{4.2.1d}$$

must satisfy the original differential equation expressed in terms of the inner variable. To determine an approximate solution, F and G are expanded in terms of ϵ, and equations and boundary conditions are derived for each level of approximation. This technique was applied by Chudov (1966) to viscous flow past a flat plate, and the version of Bromberg was rediscovered by O'Malley (1971).

Another method of composite expansions was devised earlier by Latta (1951). According to this technique, the solution is assumed to have the same form as (4.2.1b) except G is also a function of the outer variable and the inner variable ζ is generalized from $x/\delta(\epsilon)$ into $g(x)/\delta(\epsilon)$ with g to be determined from the analysis. Moreover, Latta inspected the inner expansion to determine special functions that can be used to represent $G(x, \zeta; \epsilon)$.

Next we illustrate both techniques by applying them to specific examples.

4.2.1. A SECOND-ORDER EQUATION WITH CONSTANT COEFFICIENTS

Let us again consider the problem

$$\epsilon y'' + y' + y = 0, \qquad 0 \le x \le 1 \tag{4.2.2a}$$

$$y(0) = \alpha, \qquad y(1) = \beta \tag{4.2.2b}$$

As discussed in Section 4.1.1, the straightforward expansion breaks down near $x = 0$, and an inner expansion using the stretching transformation $\zeta = x\epsilon^{-1}$ was introduced to describe y in the region of nonuniformity. The inner expansion was shown to involve the function $e^{-\zeta} = e^{-x/\epsilon}$. Since upon differentiation $e^{-x/\epsilon}$ reproduces itself, no other special functions are needed to represent the composite expansion. Therefore Latta assumed that y has a uniformly valid expansion of the form

$$y = \sum_{n=0}^{\infty} \epsilon^n f_n(x) + e^{-x/\epsilon} \sum_{n=0}^{\infty} \epsilon^n h_n(x) \tag{4.2.3}$$

Substituting (4.2.3) into (4.2.2a) and (4.2.2b) and equating the coefficients of ϵ^n and $\epsilon^n e^{-x/\epsilon}$ for all n to zero, we obtain equations for f_n and h_n. The equations for $n = 0$, 1, and 2 are

$$f_0' + f_0 = 0, \qquad h_0' - h_0 = 0 \tag{4.2.4}$$

$$f_1' + f_1 = -f_0'', \qquad h_1' - h_1 = h_0'' \tag{4.2.5}$$

$$f_2' + f_2 = -f_1'', \qquad h_2' - h_2 = h_1'' \tag{4.2.6}$$

The boundary conditions are

$$f_0(1) = \beta, \qquad f_0(0) + h_0(0) = \alpha \qquad (4.2.7)$$

$$f_n(1) = 0, \qquad f_n(0) + h_n(0) = 0 \quad \text{for} \quad n \geq 1 \qquad (4.2.8)$$

where the exponentially small terms $e^{-\epsilon^{-1}}h_n(1)$ are neglected.

The solution of (4.2.4) subject to (4.2.7) is

$$f_0 = \beta e^{1-x}, \qquad h_0 = (\alpha - \beta e)e^x \qquad (4.2.9)$$

Using (4.2.9) in (4.2.5) and solving the resulting equations subject to (4.2.8), we have

$$f_1 = \beta(1 - x)e^{1-x}, \qquad h_1 = [-\beta e + (\alpha - \beta e)x]e^x \qquad (4.2.10)$$

Substituting the first-order solution into (4.2.6) and solving the resulting equations subject to (4.2.8), we obtain

$$f_2 = \tfrac{1}{2}\beta(1 - x)(5 - x)e^{1-x}$$
$$h_2 = [-\tfrac{5}{2}\beta e + (2\alpha - 3\beta e)x + \tfrac{1}{2}(\alpha - \beta e)x^2]e^x \qquad (4.2.11)$$

With the above solutions the expansion (4.2.3) becomes

$$y = \beta[1 + \epsilon(1 - x) + \tfrac{1}{2}(1 - x)(5 - x)]e^{1-x}$$
$$+ \{\alpha - \beta e + \epsilon[-\beta e + (\alpha - \beta e)x]$$
$$+ \epsilon^2[-\tfrac{5}{2}\beta e + (2\alpha - 3\beta e)x + \tfrac{1}{2}(\alpha - \beta e)x^2]\}e^{-x/\epsilon} + O(\epsilon^3) \quad (4.2.12)$$

It can be easily verified that the outer expansion ($\lim \epsilon \to 0$ with x kept fixed) of the first two terms of this expansion is given by (4.1.49), while the inner expansion ($\epsilon \to 0$ with $\zeta = x/\epsilon$ kept fixed) is given by (4.1.76). Therefore the method of composite expansions gives a uniformly valid expansion directly without the need to determine an outer expansion and an inner expansion, to match them, and then to form a composite expansion.

Let us next determine an expansion for y by using the technique of Bromberg and Višik and Lyusternik. We assume that

$$y(x; \epsilon) = F(x; \epsilon) + G(\zeta; \epsilon)$$
$$= F_0(x) + G_0(\zeta) + \epsilon[F_1(x) + G_1(\zeta)] + \epsilon^2[F_2(x) + G_2(\zeta)] + \cdots$$
$$(4.2.13)$$

where $G(\zeta; \epsilon)$ is negligible outside the inner region (Bromberg, 1956); that is, $G(\zeta; \epsilon) \to 0$ as $\zeta \to \infty$ so that

$$y^o(x, \epsilon) = F(x; \epsilon) = F_0(x) + \epsilon F_1(x) + \epsilon^2 F_2(x) + \cdots \qquad (4.2.14)$$

Since $x = \epsilon\zeta$

$$y^i(x; \epsilon) = F_0(0) + G_0(\zeta) + \epsilon[F_0'(0)\zeta + F_1(0) + G_1(\zeta)]$$
$$+ \epsilon^2[\tfrac{1}{2}F_0''(0)\zeta^2 + F_1'(0)\zeta + F_2(0) + G_2(\zeta)] + \cdots \quad (4.2.15)$$

Since $G(\zeta; \epsilon)$ is assumed to be negligible outside the boundary layer, $F(x; \epsilon)$ satisfies the boundary condition $y(1) = \beta$. Hence

$$F_0(1) = \beta, \qquad F_n(1) = 0 \quad \text{for} \quad n \geq 1 \tag{4.2.16}$$

The boundary condition $y(0) = \alpha$ is satisfied by $F + G$ so that

$$F_0(0) + G_0(0) = \alpha, \qquad F_n(0) + G_n(0) = 0 \quad \text{for} \quad n \geq 1 \tag{4.2.17}$$

To determine the equations governing F_n, we substitute (4.2.14) into (4.2.2a) and equate coefficients of like powers of ϵ, assuming x to be fixed, and obtain

$$F_0' + F_0 = 0 \tag{4.2.18}$$

$$F_n' + F_n = -F_{n-1}'' \quad \text{for} \quad n \geq 1 \tag{4.2.19}$$

To determine the equations governing G_n, we first express (4.2.2a) in terms of the inner variable ζ as

$$\frac{d^2y}{d\zeta^2} + \frac{dy}{d\zeta} + \epsilon y = 0 \tag{4.2.20}$$

Substituting (4.2.15) into (4.2.20), keeping ζ fixed, and equating coefficients of like powers of ϵ, we obtain

$$G_0'' + G_0' = 0 \tag{4.2.21}$$

$$G_1'' + G_1' = -G_0 - F_0'(0) - F_0(0) \tag{4.2.22}$$

$$G_2'' + G_2' = -G_1 - [F_0'(0) + F_0''(0)]\zeta - F_0''(0) - F_1(0) - F_1'(0) \tag{4.2.23}$$

The solution of (4.2.18) subject to (4.2.16) is

$$F_0 = \beta e^{1-x} \tag{4.2.24}$$

Hence $G_0(0) = \alpha - \beta e$ from (4.2.16), and thus the solution of (4.2.21) that tends to zero as $\zeta \to \infty$ is

$$G_0 = (\alpha - \beta e)e^{-\zeta} \tag{4.2.25}$$

The solution of (4.2.19) and (4.2.17) for $n = 1$ is then

$$F_1 = \beta(1 - x)e^{1-x} \tag{4.2.26}$$

which when combined with (4.2.16) gives $G_1(0) = -\beta e$. Substituting for F_0 and G_0 into (4.2.22) gives

$$G_1'' + G_1' = -(\alpha - \beta e)e^{-\zeta}$$

whose solution subject to $G_1(0) = -\beta e$ that tends to zero as $\zeta \to \infty$ is

$$G_1 = [(\alpha - \beta e)\zeta - \beta e]e^{-\zeta} \tag{4.2.27}$$

Proceeding to second order, we find that the solution of (4.2.19) and (4.2.16) is

$$F_2 = \tfrac{1}{2}\beta(1 - x)(5 - x)e^{1-x} \tag{4.2.28}$$

Then $G_2(0) = -5\beta e/2$ from (4.2.17), and (4.2.23) becomes

$$G_2'' + G_2' = -[(\alpha - \beta e)\zeta - \beta e]e^{-\zeta}$$

whose solution subject to $G_2(0) = -5\beta e/2$ that tends to zero as $\zeta \to \infty$ is

$$G_2 = [\tfrac{1}{2}(\alpha - \beta e)\zeta^2 + (\alpha - 2\beta e)\zeta - \tfrac{5}{2}\beta e]e^{-\zeta} \tag{4.2.29}$$

Therefore the first two terms of the resulting uniformly valid expansion are exactly the same as (4.1.81) obtained earlier by using the method of matched asymptotic expansions.

4.2.2. A SECOND-ORDER EQUATION WITH VARIABLE COEFFICIENTS

As a second example, we consider the following problem which is a special case of the problem considered in Section 4.1.3

$$\epsilon y'' + (2x + 1)y' + 2y = 0, \qquad 0 \le x \le 1 \tag{4.2.30}$$

$$y(0) = \alpha, \qquad y(1) = \beta \tag{4.2.31}$$

Since the coefficient of y' is positive, the nonuniformity occurs near $x = 0$. To describe y in the region of nonuniformity, we need a stretching transformation $\zeta = x\epsilon^{-1}$, and the inner expansion is described in terms of the special function $e^{-\zeta} = e^{-x/\epsilon}$. Since the problem has variable coefficients, we assume that y possesses a uniformly valid expansion of the form

$$y = \sum_{n=0}^{\infty} \epsilon^n f_n(x) + e^{-g(x)/\epsilon} \sum_{n=0}^{\infty} \epsilon^n h_n(x) \tag{4.2.32}$$

where the function $g(x)$, which is determined from the analysis, $\to x$ as $x \to 0$.

Substituting (4.2.32) into (4.2.30) and (4.2.31) and equating coefficients of ϵ^n and $\epsilon^n e^{-g(x)/\epsilon}$ for all n to zero, we obtain equations to determine g, f_n and h_n. The first three equations are

$$h_0 g'[g' - (2x + 1)] = 0 \tag{4.2.33}$$

$$(2x + 1)f_0' + 2f_0 = 0 \tag{4.2.34}$$

$$(-2g' + 2x + 1)h_0' + (2 - g'')h_0 = 0 \tag{4.2.35}$$

$$f_0(1) = \beta, \qquad f_0(0) + h_0(0) = \alpha \tag{4.2.36}$$

To determine a nontrivial solution for h_0, (4.2.33) requires that

$$g' = 0 \quad \text{or} \quad g' = 2x + 1 \qquad (4.2.37)$$

The first case in (4.2.37) yields $g = a$ constant, which must be rejected because $g(x) \to x$ as $x \to 0$. Hence

$$g = x^2 + x \qquad (4.2.38)$$

The solution of (4.2.34) subject to $f_0(1) = \beta$ is

$$f_0 = \frac{3\beta}{2x + 1} \qquad (4.2.39)$$

Using (4.2.38) in solving (4.2.35) subject to (4.2.36) gives

$$h_0 = \alpha - 3\beta \qquad (4.2.40)$$

Therefore

$$y = \frac{3\beta}{2x + 1} + (\alpha - 3\beta)e^{-(x^2+x)/\epsilon} + O(\epsilon) \qquad (4.2.41)$$

Let us next apply the second version of the method of composite expansions to this problem. In this case (4.2.13) through (4.2.17) are applicable. Substituting (4.2.14) into (4.2.30), keeping x fixed, and equating coefficients of like powers of ϵ, we have

$$(2x + 1)F_0' + 2F_0 = 0 \qquad (4.2.42)$$

$$(2x + 1)F_1' + 2F_1 = -F_0'' \qquad (4.2.43)$$

To determine the equations governing G_n, we express (4.2.30) in terms of the inner variable $\zeta = x/\epsilon$; that is

$$\frac{d^2y}{d\zeta^2} + (1 + 2\epsilon\zeta)\frac{dy}{d\zeta} + 2\epsilon\, y = 0 \qquad (4.2.44)$$

Substituting (4.2.15) into this equation, keeping ζ fixed, and equating coefficients of like powers of ϵ, we obtain

$$G_0'' + G_0' = 0 \qquad (4.2.45)$$

$$G_1'' + G_1' = -2\zeta G_0' - 2G_0 - 2F_0(0) - F_0'(0) \qquad (4.2.46)$$

The solution of (4.2.42) subject to (4.2.16) is

$$F_0 = \frac{3\beta}{2x + 1} \qquad (4.2.47a)$$

which together with (4.2.17) gives $G_0(0) = \alpha - 3\beta$. Hence the solution of

(4.2.45) that tends to zero as $\zeta \to \infty$ is

$$G_0 = (\alpha - 3\beta)e^{-\zeta} \qquad (4.2.47b)$$

Substituting for F_0 into (4.2.43) and solving the resulting equation subject to (4.2.16), we obtain

$$F_1 = \frac{8\beta(1 - x)(2 + x)}{3(1 + 2x)^3} \qquad (4.2.48)$$

Then (4.2.17) gives $G_1(0) = -16\beta/3$, and (4.2.46) becomes

$$G_1'' + G_1' = 2(\alpha - 3\beta)(\zeta - 1)e^{-\zeta}$$

whose solution which tends to zero as $\zeta \to \infty$ is

$$G_1 = -[\tfrac{16}{3}\beta + (\alpha - 3\beta)\zeta^2]e^{-\zeta} \qquad (4.2.49)$$

Therefore

$$y = \frac{3\beta}{2x + 1} + \epsilon\,\frac{8\beta(1 - x)(2 + x)}{3(1 + 2x)^3} + (\alpha - 3\beta)e^{-\zeta}$$
$$- \epsilon[\tfrac{16}{3}\beta + (\alpha - 3\beta)\zeta^2]e^{-\zeta} + O(\epsilon^2) \quad (4.2.50)$$

4.2.3. AN INITIAL BOUNDARY VALUE PROBLEM FOR THE HEAT EQUATION

As a third example of the application of the method of composite expansions, we consider an initial boundary value problem for the heat equation which was introduced by Keller (1968). We assume that the temperature $u(x, t; \epsilon)$ depends on one space variable (i.e., x) which ranges from 0 to $b(\epsilon t)$ where b is a known function and ϵ is a small parameter. Thus b is a slowly varying function of t. The mathematical description of the problem is

$$u_t = u_{xx}, \qquad 0 \le x \le b(\epsilon t) \qquad (4.2.51)$$

$$u(0, t) = \phi(\epsilon t), \qquad u[b(\epsilon t), t] = 0 \qquad (4.2.52)$$

$$u(x, 0) = \psi(x), \qquad 0 \le x \le b(0) \qquad (4.2.53)$$

Changing from the variable t to $\tau = \epsilon t$, (4.2.51) and (4.2.52) become

$$\epsilon u_\tau = u_{xx}, \qquad 0 \le x \le b(\tau) \qquad (4.2.54)$$

$$u(0, \tau) = \phi(\tau), \qquad u[b(\tau), \tau] = 0 \qquad (4.2.55)$$

Since ϵ multiplies u_τ, the straightforward perturbation solution for small ϵ, keeping τ fixed, cannot satisfy in general the initial condition (4.2.53), and a nonuniformity exists at and near $\tau = 0$. To describe the behavior of u

near $t = 0$, a stretching transformation $t = \tau/\epsilon$ is needed. As verified below, the special function $\exp[-g(\tau)/\epsilon]$ with $g(\tau) \to \tau$ as $\tau \to 0$ describes the behavior of u in this region. Therefore we assume that u has a uniformly valid asymptotic expansion of the form

$$u = \sum_{n=0}^{\infty} \epsilon^n f_n(x, \tau) + e^{-g(\tau)/\epsilon} \sum_{n=0}^{\infty} \epsilon^n h_n(x, \tau) \tag{4.2.56}$$

Substituting this expansion into (4.2.53) through (4.2.55) and equating to zero the coefficients of ϵ^n and $\epsilon^n e^{-g(\tau)/\epsilon}$ for all n, we obtain

$$f_{0,xx} = 0, \qquad f_0(0, \tau) = \phi(\tau), \qquad f_0[b(\tau), \tau] = 0 \tag{4.2.57}$$

$$h_{0,xx} + g' h_0 = 0, \qquad h_0(0, \tau) = 0, \qquad h_0[b(\tau), \tau] = 0 \tag{4.2.58}$$

$$f_0(x, 0) + h_0(x, 0) = \psi(x), \qquad 0 \le x \le b(0) \tag{4.2.59}$$

and for $n \ge 1$

$$f_{n,xx} = f_{n-1,\tau}, \qquad f_n(0, \tau) = f_n[b(\tau), \tau] = 0 \tag{4.2.60}$$

$$h_{n,xx} + g' h_n = h_{n-1,\tau}, \qquad h_n(0, \tau) = h_n[b(\tau), \tau] = 0 \tag{4.2.61}$$

$$f_n(x, 0) + h_n(x, 0) = 0 \tag{4.2.62}$$

where g' stands for $dg/d\tau$.

The solution of (4.2.57) is

$$f_0 = \phi(\tau) \left[1 - \frac{x}{b(\tau)} \right] \tag{4.2.63}$$

Since the boundary conditions on h_0 are homogeneous, the equation for h_0 has a nontrivial solution if and only if g' is one of the eigenvalues

$$g'_k = \left[\frac{k\pi}{b(\tau)} \right]^2, \qquad k = 1, 2, \ldots \tag{4.2.64}$$

Then the corresponding normalized eigenfunctions are

$$\chi_k = \left[\frac{2}{b(\tau)} \right]^{1/2} \sin \frac{k\pi x}{b(\tau)} \tag{4.2.65}$$

Hence

$$h_0 = a_0(\tau) \chi_k(x, \tau) \tag{4.2.66}$$

where a_0 is an unknown function which is determined from the examination of the equation for h_1.

With h_0 known (4.2.61) for $n = 1$ becomes

$$h_{1,xx} + g'_k h_1 = a'_0 \chi_k + a_0 \chi_{k,\tau}$$
$$h_1(0, \tau) = h_1[b(\tau), \tau] = 0 \tag{4.2.67}$$

We assume that h_1 can be expanded in terms of the eigenfunctions χ_k; that is

$$h_1 = \sum_{s=1}^{\infty} c_s(\tau)\chi_s(x, \tau) \tag{4.2.68}$$

Substituting (4.2.68) into (4.2.67) and using the fact that $\chi_{s,xx} = -g_s'\chi_s$, we obtain

$$\sum_{s=1}^{\infty} (g_k' - g_s')c_s\chi_s = a_0'\chi_k + a_0\chi_{k,\tau} \tag{4.2.69}$$

If we multiply this equation by χ_k and integrate from $x = 0$ to $b(\tau)$, the left-hand side vanishes because χ_k is orthogonal to χ_s for $k \neq s$, and $g_s' = g_k'$ when $k = s$. Therefore

$$\int_0^{b(\tau)} (a_0'\chi_k^2 + a_0\chi_{k,\tau}\chi_k) \, dx = 0 \tag{4.2.70}$$

which is the condition for the solvability of (4.2.67). Since

$$\int_0^{b(\tau)} \chi_k^2 \, dx = 1 \tag{4.2.71}$$

$$\frac{d}{d\tau} \int_0^{b(\tau)} \chi_k^2 \, dx = 0 = b'(\tau)\chi_k^2[b(\tau), \tau] + 2\int_0^{b(\tau)} \chi_k\chi_{k,\tau} \, dx = 0 \tag{4.2.72}$$

Since $\chi_k[b(\tau), \tau] = 0$

$$\int_0^{b(\tau)} \chi_k\chi_{k,\tau} \, dx = 0 \tag{4.2.73}$$

Hence (4.2.70) and (4.2.71) lead to

$$a_0 = \text{a constant} \tag{4.2.74}$$

Therefore the zeroth-order solution is

$$u(x, \tau; \epsilon) = \phi(\tau)\left[1 - \frac{x}{b(\tau)}\right]$$

$$+ \sum_{k=1}^{\infty} a_k \left[\frac{2}{b(\tau)}\right]^{1/2} \sin\frac{k\pi x}{b(\tau)} \exp\left[-\frac{k^2\pi^2}{\epsilon}\int_0^\tau b^{-2}(\xi) \, d\xi\right] + O(\epsilon) \tag{4.2.75}$$

where a_k is a constant given by

$$a_k = \left[\frac{2}{b(0)}\right]^{1/2} \int_0^{b(0)} \left\{\psi(x) - \phi(0)\left[1 - \frac{x}{b(0)}\right]\right\} \sin\frac{k\pi x}{b(0)} \, dx \tag{4.2.76}$$

4.2.4. LIMITATIONS OF THE METHOD OF COMPOSITE EXPANSIONS

Complications might arise if one tries to apply Latta's technique to nonlinear problems. Moreover, difficulties might arise if a large number of special functions are needed to describe the behavior of the function under consideration in the inner region. In spite of these limitations, it is the starting point for the development of the method of multiple scales which is discussed in Chapter 6.

The modified method of composite expansions of Bromberg, Višik, and Lyusternik overcomes these complications as illustrated by its application to the nonlinear problem

$$\frac{1}{2}\left(\frac{dx}{dt}\right)^2 = \frac{1-\mu}{x} + \frac{\mu}{1-x}, \qquad t(0) = 0 \tag{4.2.77}$$

describing the one-dimensional earth-moon-spaceship problem treated in Section 4.1.7 using the method of matched asymptotic expansions.

We assume a composite expansion of the form

$$t(x;\mu) = F_0(x) + G_0(\xi) + \mu[F_1(x) + G_1(\xi)] + \cdots \tag{4.2.78}$$

where $\xi = (1 - x)/\mu$ is the inner variable found in Section 4.1.7 and $G_n \to 0$ as $\xi \to \infty$. Then the initial condition $t(0) = 0$ gives

$$F_0(0) = F_1(0) = 0 \tag{4.2.79}$$

From (4.2.78)

$$t^o = F_0(x) + \mu F_1(x) + \cdots \tag{4.2.80}$$

which when substituted into (4.2.77) gives

$$2F_0'^2 = x$$
$$\frac{F_1'}{F_0'^3} = \frac{1}{x} + \frac{1}{x-1} \tag{4.2.81}$$

The solutions of these equations subject to (4.2.79) are

$$\sqrt{2}\,F_0 = \tfrac{2}{3}x^{3/2}$$
$$\sqrt{2}\,F_1 = \tfrac{2}{3}x^{3/2} + \sqrt{x} - \tfrac{1}{2}\ln\frac{1+\sqrt{x}}{1-\sqrt{x}} \tag{4.2.82}$$

From (4.2.78) and (4.2.82)

$$\sqrt{2}\,t^i = \tfrac{2}{3} + \sqrt{2}\,G_0(\xi) + \mu\left[-\xi + \tfrac{5}{3} + \tfrac{1}{2}\ln\frac{\mu\xi}{4} + \sqrt{2}\,G_1(\xi)\right] + \ldots \tag{4.2.83}$$

To determine G_0 and G_1, we express (4.2.77) in terms of the inner variable ξ; that is

$$\tfrac{1}{2}\mu^2\left(\frac{d\xi}{dt}\right)^2 = \frac{1-\mu}{1-\mu\xi} + \frac{1}{\xi} \tag{4.2.84}$$

Substituting (4.2.83) into this equation and equating the coefficients of like powers of μ, we obtain

$$G_0' = 0 \tag{4.2.85}$$

$$[\sqrt{2}\,G_1 + \tfrac{1}{2}\ln\xi - \xi]' = -\left(1 + \frac{1}{\xi}\right)^{-1/2} \tag{4.2.86}$$

The solution of (4.2.85) that tends to zero as $\xi \to \infty$ is $G_0 = 0$, while the solution of (4.2.86) that tends to zero as $\xi \to \infty$ is

$$\sqrt{2}G_1 = \xi - \sqrt{\xi(\xi+1)} + \sinh^{-1}\sqrt{\xi} - \tfrac{1}{2}\ln\xi + \tfrac{1}{2} - \ln 2 \tag{4.2.87}$$

Substituting for F_0 and F_1 from (4.2.82) and using the values of G_0 and G_1 in (4.2.78), we obtain exactly the expansion (4.1.202) that we obtained using the method of matched asymptotic expansions.

Exercises

4.1. Consider the problem

$$\epsilon y'' + y' = 2x$$
$$y(0) = \alpha, \qquad y(1) = \beta$$

(a) Determine a three-term outer expansion.
(b) Determine a three-term inner expansion.
(c) Match these two expansions and then form a composite expansion.
(d) Determine a three-term uniformly valid expansion using the method of composite expansions (MCE) and compare the result with (c).

4.2. Determine second-order (three-term) expansions for

$$\epsilon y'' - y' = 2x$$
$$y(0) = \alpha, \qquad y(1) = \beta$$

using (a) the method of matched asymptotic expansions (MMAE) and (b) the MCE.

4.3. Determine second-order uniformly valid expansions for the problems

$$\epsilon y'' \pm (2x+1)y' + 2y = 0$$
$$y(0) = \alpha, \qquad y(1) = \beta$$

using (a) the MMAE (b) Latta's method, and (c) the Bromberg–Višik–Lyusternik method.

4.4. Determine first-order uniformly valid expressions for

$$\epsilon y'' - a(x)y' + b(x)y = 0, \qquad a(x) > 0$$
$$y(0) = \alpha, \qquad y(1) = \beta$$

using (a) the MMAE and (b) the MCE.

4.5. Consider the problem

$$\epsilon^2 y''' - y' + y = 0$$
$$y(0) = \alpha, \qquad y(1) = \beta, \qquad y'(1) = \gamma$$

(a) Show that boundary layers exist at both ends which are characterized by the stretching transformations

$$\eta = x/\epsilon \quad \text{and} \quad \zeta = (1 - x)/\epsilon$$

(b) Determine a second-order uniformly valid solution using the MMAE.
(c) Determine a second-order expansion using the MCE by letting

$$y = F(x; \epsilon) + G(\eta; \epsilon) + H(\zeta; \epsilon)$$

where $G \to 0$ as $\eta \to \infty$ and $H \to 0$ as $\zeta \to \infty$.

4.6. Determine first-order (two-term) uniformly valid expansions for (2.2.28) through (2.2.30) using both versions of the MCE.

4.7. Show that the MMAE cannot be used to obtain uniformly valid expansions for

$$\epsilon^2 y'' + y = f(x)$$
$$y(0) = \alpha, \qquad y(1) = \beta$$

Can you conclude from this example that the MMAE is inapplicable to oscillation problems?

4.8. Consider the problem given by (2.2.28) for $0 \le r \le b$ subject to the conditions

$$u(b) = b\alpha, \qquad \frac{du}{dr}(b) = \alpha$$

Determine two-term uniformly valid expansions using (a) the MMAE and (b) the MCE.

4.9. The vibration of a beam with clamped ends is governed by

$$\epsilon^2 \frac{d^4 u}{dx^4} - \frac{d^2 u}{dx^2} = \lambda^2 u$$
$$u(0) = u(1) = u'(0) = u'(1) = 0$$

Determine a first-order expansion for small ϵ for u and λ.

4.10. For one-dimensional nondissipative steady flow, the heat transfer is governed by (Hanks, 1971)

$$\epsilon \frac{d^2T}{dx^2} + x\frac{dT}{dx} - xT = 0$$

$$T(0) = T_0, \qquad T(l) = T_l$$

Determine first-order expansions using (a) the MMAE and (b) the two versions of the MCE.

4.11. Determine a first-order uniformly valid expansion for the equation

$$(x + \epsilon y)y' + (2 + x)y = 0, \qquad y(1) = Ae^{-1}$$

using the MMAE. Can you conclude that the method of strained coordinates (MSC) is more suited for such problems?

4.12. Determine one-term expansions for the solutions of the problems

$$\epsilon y'' \pm (2x + 1)y' + y^2 = 0$$

$$y(0) = \alpha, \qquad y(1) = \beta$$

using the MMAE and the MCE.

4.13. Determine one-term expansions for

$$y'' + a(x)y' + y^2 = 0$$

$$y(0) = \alpha, \qquad y(1) = \beta$$

using the MMAE and the MCE if $a(x)$ is (a) negative and (b) positive in $0 \leq x \leq 1$.

4.14. Determine a first-order expansion for the problem in the previous example if $a(x)$ has a simple zero at μ where $0 \leq \mu \leq 1$.

4.15. Determine first-order uniformly valid expansions for

$$\epsilon y'' \pm yy' - y = 0$$

$$y(0) = \alpha, \qquad y(1) = \beta$$

using the MMAE and the MCE.

4.16. The laminar flow through a channel with porous walls can be reduced for the suction case to (Proudman, 1960; Terrill and Shrestha, 1965)

$$\epsilon f''' - ff'' + f'^2 = c(\epsilon)$$

$$f(0) = 1 - \alpha = a, \qquad f'(0) = 0$$

$$f(1) = 1, \qquad f'(1) = 0$$

For a positive a show that (a) the first-order outer and inner expansions are

$$f^o = a \cosh xb + \epsilon f_1(x) + \cdots$$

$$f^i = 1 + \epsilon B(1 - \eta - e^{-\eta}) + \cdots, \qquad \eta = (1 - x)/\epsilon$$

$$\beta = ab + \epsilon \beta_1 + \cdots, \qquad c = -\beta^2$$

and determine b, B, β_1, and f_1; (b) form a composite expansion; and (c) determine a first-order expansion using the MCE.

4.17. Consider the problem of the previous exercise for the case $a < 0$.

(a) Show that a uniformly valid expansion is given by

$$f^o = 1 - \alpha + \alpha x + \epsilon f_1(x) + \cdots$$
$$f^i = 1 + \alpha(1 - \eta - e^{-\eta}) + \cdots, \qquad \eta = (1 - x)/\epsilon$$
$$f^I = a + \frac{\epsilon\alpha}{\alpha - 1}(1 - \zeta - e^{-\zeta}) + \cdots, \qquad \zeta = (\alpha - 1)x/\epsilon$$
$$\beta = \alpha + \epsilon\beta_1 + \cdots, \qquad c = \beta^2$$

and determine f_1 and β_1.

(b) Form a composite expansion.

(c) Determine a first-order expansion using the MCE.

4.18. Consider the problem

$$\epsilon u_{xx} + u_{yy} + a(x)u_x = 0$$
$$u(0, y) = F_1(y), \qquad u(1, y) = F_2(y)$$
$$u(x, 0) = G_1(x), \qquad u(x, 1) = G_2(x)$$

(a) Show that a boundary layer exists at $x = 0$ if $a(x) > 0$ and at $x = 1$ if $a(x) < 0$. The first is characterized by $\xi = x/\epsilon$, while the second is characterized by $\zeta = (1 - x)/\epsilon$.

(b) Determine the equations defining the first terms in the outer and inner expansions and match these expansions.

(c) Use Latta's method to show that

$$u = A(x, y) + B(x, y)e^{-Q(x)/\epsilon} + O(\epsilon)$$

where

$$Q(x) = \int_0^x a(x)\,dx \quad \text{if} \quad a(x) > 0$$

and

$$Q(x) = \int_1^x a(x)\,dx \quad \text{if} \quad a(x) < 0$$

Determine the equations defining A and B.

4.19. Consider the problem

$$\epsilon(u_{xx} + u_{yy}) + a(x, y)u_x + b(x, y)u = 0$$
$$u(x, 0) = F_1(x), \qquad u(x, 1) = F_2(x)$$
$$u(0, y) = G_1(y), \qquad u(1, y) = G_2(y)$$

(a) Determine the equations defining the first terms in the outer and inner expansions and then match the expansions.

(b) Use the MCE to obtain a first-order uniform expansion.

4.20. Consider the problem

$$\epsilon^2 \nabla^4 u + a(x, y)u_{xx} + b(x, y)u_x + c(x, y)u_y = 0$$
$$u(x, 0) = F_1(x), \qquad u(x, 1) = F_2(x)$$
$$u(0, y) = G_1(y), \qquad u(1, y) = G_2(y)$$

Determine the equations defining the first terms in the outer and inner expansions and then match them.

4.21. Use the MCE to obtain a first-order uniform expansion for small R for (2.1.37) through (2.1.40) which describe flow past a sphere.

CHAPTER 5

Variation of Parameters and Methods of Averaging

5.1. Variation of Parameters

This technique was originally developed to solve inhomogeneous linear equations when the general solutions to the corresponding homogeneous equations are known. As an example, consider the general linear second-order inhomogeneous equation

$$y'' + p(x)y' + q(x)y = R(x) \tag{5.1.1}$$

Let $y_1(x)$ and $y_2(x)$ be two linearly independent solutions of the corresponding homogeneous equation. Then we assume that the solution of (5.1.1) is

$$y = A_1(x)y_1(x) + A_2(x)y_2(x) \tag{5.1.2}$$

where the functions A_1 and A_2 need to be determined. Note that in the homogeneous problem A_1 and A_2 are constants, while in the inhomogeneous case they are allowed to vary, hence the name "variation of parameters."
  Differentiating (5.1.2) with respect to x gives

$$y' = A_1 y_1' + A_2 y_2' + A_1' y_1 + A_2' y_2 \tag{5.1.3}$$

Since we have three unknown functions A_1, A_2, and y, while we have only the two equations (5.1.1) and (5.1.2), we are free to impose one more condition on A_1, A_2, and y. Let us demand that

$$A_1' y_1 + A_2' y_2 = 0 \tag{5.1.4}$$

Then (5.1.3) becomes

$$y' = A_1 y_1' + A_2 y_2' \tag{5.1.5}$$

Differentiating (5.1.5) with respect to x gives

$$y'' = A_1 y_1'' + A_2 y_2'' + A_1' y_1' + A_2' y_2' \tag{5.1.6}$$

159

Substituting for y, y', and y'' into (5.1.1) and using the fact that y_1 and y_2 are solutions of the corresponding homogeneous equation, we obtain

$$A_1'y_1' + A_2'y_2' = R \qquad (5.1.7)$$

Solving (5.1.4) and (5.1.7) for A_1' and A_2' yields

$$A_1' = \frac{-R(x)y_2(x)}{W(x)} \qquad (5.1.8)$$

$$A_2' = \frac{R(x)y_1(x)}{W(x)} \qquad (5.1.9)$$

where $W(x)$ is called the Wronskian and is given by

$$W(x) = y_1(x)y_2'(x) - y_1'(x)y_2(x) \qquad (5.1.10)$$

The general solution of (5.1.1) is then

$$y = c_1y_1(x) + c_2y_2(x) + y_p(x) \qquad (5.1.11)$$

where c_1 and c_2 are constants, and the particular solution y_p is given by

$$y_p(x) = \int_{x_0}^{x} \frac{y_1(t)y_2(x) - y_2(t)y_1(x)}{W(t)} R(t)\,dt \qquad (5.1.12)$$

This technique has been extended for the determination of solutions of problems in which the inhomogeneity is a function of both the dependent and independent variables. The dependence on the dependent variable may be nonlinear. We next discuss two examples; the first is linear, while the second is nonlinear.

5.1.1. TIME-DEPENDENT SOLUTIONS OF THE SCHRÖDINGER EQUATION

Let us consider the Schrödinger equation

$$H_0\psi + \frac{h}{2\pi i}\frac{\partial\psi}{\partial t} = -H_1\psi \qquad (5.1.13)$$

subject to homogeneous boundary conditions, where H_0 and H_1 are time-independent and time-dependent linear operators. We assume that

$$H_0\psi + \frac{h}{2\pi i}\frac{\partial\psi}{\partial t} = 0 \qquad (5.1.14)$$

subject to the same homogeneous boundary conditions, has the solution

$$\psi = \sum_{n=1}^{\infty} a_n u_n(x)e^{-i\omega_n t}, \qquad \omega_n = \frac{2\pi}{h}E_n \qquad (5.1.15)$$

where a_n is a constant, and u_n and E_n are, respectively, the eigenfunctions and the corresponding eigenvalues of

$$H_0 u = Eu \tag{5.1.16}$$

with the same homogeneous conditions. The eigenfunctions u_n are assumed to be orthonormal over the domain D.

Following Dirac (1926), we assume that the solutions of the perturbed problem are still given by (5.1.15) but with time-varying a_n. Substituting (5.1.15) into (5.1.13) gives

$$\sum_{n=1}^{\infty} a_n [H_0 u_n(x) - E_n u_n(x)] e^{-i\omega_n t} + \frac{h}{2\pi i} \sum_{n=1}^{\infty} \frac{da_n}{dt} u_n(x) e^{-i\omega_n t}$$
$$= -\sum_{n=1}^{\infty} H_1 [a_n u_n(x) e^{-i\omega_n t}] \tag{5.1.17}$$

The first term on the left-hand side of this equation vanishes according to (5.1.16). Then (5.1.17) becomes

$$\sum_{n=1}^{\infty} \frac{da_n}{dt} u_n(x) e^{-i\omega_n t} = -\frac{2\pi i}{h} \sum_{n=1}^{\infty} H_1 [a_n u_n(x) e^{-i\omega_n t}] \tag{5.1.18}$$

Multiplying (5.1.18) by $\bar{u}_m(x)$, integrating over the domain D, and using the orthonormality of u_n, we obtain

$$\frac{da_m}{dt} = -\frac{2\pi i}{h} \sum_{n=1}^{\infty} e^{i\omega_m t} H_{1mn} \tag{5.1.19}$$

where

$$H_{1mn} = \int_D \bar{u}_m(x) H_1 [a_n u_n(x) e^{-i\omega_n t}] \, dx \tag{5.1.20}$$

If H_1 does not contain any derivative with respect to t, then (5.1.19) becomes

$$\frac{da_m}{dt} = -\frac{2\pi i}{h} \sum_{n=1}^{\infty} a_n e^{i\omega_{mn} t} \tilde{H}_{1mn} \tag{5.1.21}$$

where

$$\omega_{mn} = \frac{2\pi}{h} (E_m - E_n), \qquad \tilde{H}_{1mn} = \int_D \bar{u}_m(x) H_1 [u_n(x)] \, dx \tag{5.1.22}$$

Equation (5.1.21) is equivalent to the full problem given by (5.1.13). If H_1 is a small perturbation, then we can expand a_m as

$$a_m = a_{m0} + a_{m1} + a_{m2} + \cdots \tag{5.1.23}$$

where a_{m0} is a constant and equal to $a_m(t = 0)$, and $a_{mn} \ll a_{m(n-1)}$. Then the first approximation to a_m is given by

$$\frac{da_{m1}}{dt} = -\frac{2\pi i}{h} \sum_{n=1}^{\infty} a_{n0} e^{i\omega_{mn}t} \tilde{H}_{1mn} \qquad (5.1.24)$$

If in addition $a_{n0} = \delta_{nk}$, then (5.1.24) becomes

$$\frac{da_{m1}}{dt} = -\frac{2\pi i}{h} e^{i\omega_{mk}t} \tilde{H}_{1mk} \qquad (5.1.25)$$

As an example, let

$$H_1 \psi = \psi f(x) \sin \omega t \qquad (5.1.26)$$

Then

$$\tilde{H}_{1mk} = f_{mk} \sin \omega t = -\tfrac{1}{2} i f_{mk}(e^{i\omega t} - e^{-i\omega t}) \qquad (5.1.27)$$

where

$$f_{mk} = \int_D \bar{u}_m(x) f(x) u_k(x) \, dx$$

Substituting in (5.1.25) and solving for a_{m1}, we obtain

$$a_{m1} = i \frac{\pi f_{mk}}{h} \left[\frac{e^{i(\omega_{mk}+\omega)t} - 1}{\omega_{mk} + \omega} - \frac{e^{i(\omega_{mk}-\omega)t} - 1}{\omega_{mk} - \omega} \right], \qquad m \neq k \qquad (5.1.28)$$

5.1.2. A NONLINEAR STABILITY EXAMPLE

The method of variation of parameters in conjunction with an expansion in terms of eigenfunctions has been developed and applied extensively to nonlinear stability problems by Stuart (1958, 1960a, b, 1961), Watson (1960), Eckhaus (1965), and Reynolds and Potter (1967). This technique has been unified and presented in a coherent manner by Eckhaus (1965).

To describe this technique let us consider the following example after Eckhaus (1965)

$$L(\phi) - \frac{\partial \phi}{\partial t} = F(\phi) \qquad (5.1.29)$$

where L is a linear operator and $F(\phi)$ is a nonlinear operator. Let us assume that L depends on one space variable, say x, with $0 \leq x \leq 1$, and let ϕ satisfy the linear homogeneous boundary conditions

$$B_1(\phi) = 0 \quad \text{at} \quad x = 0 \quad \text{and} \quad B_2(\phi) = 0 \quad \text{at} \quad x = 1 \qquad (5.1.30)$$

It is obvious that the linearized problem

$$L(\phi) - \frac{\partial \phi}{\partial t} = 0 \qquad (5.1.31)$$

subject to the boundary conditions (5.1.30), admits a solution of the form

$$\phi = u(x)e^{-\lambda t} \tag{5.1.32}$$

with

$$L(u) + \lambda u = 0 \tag{5.1.33}$$

$$B_1(u) = 0 \quad \text{at} \quad x = 0 \quad \text{and} \quad B_2(u) = 0 \quad \text{at} \quad x = 1 \tag{5.1.34}$$

We assume that the eigenvalue problem (5.1.33) and (5.1.34) possesses a denumerable set of eigenvalues λ_n (real or complex), corresponding to the eigenfunctions u_n, for which we can solve. The eigenvalues are assumed to be different from each other and ordered such that Real $(\lambda_n) >$ Real (λ_{n-1}). Let us assume that L is a self-adjoint operator so that the eigenfunctions u_n are orthogonal, and we assume that they have been normalized so that

$$\int_0^1 u_n(x)\bar{u}_m(x)\,dx = \delta_{mn} \tag{5.1.35}$$

Then the general solution of the linearized problem can be written as

$$\phi = \sum_{n=1}^{\infty} a_n u_n(x)e^{-\lambda_n t} \tag{5.1.36}$$

where a_n are constants which can be determined from the initial conditions.

For the nonlinear problem we assume that the solution is still expressible in the form (5.1.36) but with time-varying a_n, and we write it as

$$\phi = \sum_{n=1}^{\infty} A_n(t)u_n(x) \tag{5.1.37}$$

where $A_n = a_n \exp(-\lambda_n t)$. Substituting (5.1.37) into (5.1.29) leads to

$$\sum_{n=1}^{\infty} A_n(t)L[u_n(x)] - \sum_{n=1}^{\infty} \frac{dA_n}{dt}(t)u_n(x) = F\left[\sum_{n=1}^{\infty} A_n(t)u_n(x)\right] \tag{5.1.38}$$

Since $L(u_n) = -\lambda_n u_n$, (5.1.38) can be rewritten as

$$-\sum_{n=1}^{\infty}\left[\frac{dA_n(t)}{dt} + \lambda_n A_n(t)\right]u_n(x) = F\left[\sum_{n=1}^{\infty} A_n(t)u_n(x)\right] \tag{5.1.39}$$

Multiplying (5.1.39) by $\bar{u}_m(x)$, integrating from $x = 0$ to $x = 1$, and using the orthonormality condition (5.1.35), we obtain

$$\frac{dA_m}{dt} + \lambda_m A_m = -\int_0^1 F\left[\sum_{n=1}^{\infty} A_n(t)u_n(x)\right]\bar{u}_m(x)\,dx \tag{5.1.40}$$

for $m = 1, 2, \ldots$.

If L is not self-adjoint, the eigenfunctions u_n are not mutually orthogonal. However, we can define the adjoint operator M by

$$\psi_1 L(\psi_2) - \psi_2 M(\psi_1) = \frac{d}{dx}[P(\psi_1, \psi_2)] \qquad (5.1.41)$$

where ψ_1 and ψ_2 are two functions of x and P is a bilinear form. With this definition the adjoint problem is given by

$$M\tilde{u} + \lambda\tilde{u} = 0 \qquad (5.1.42)$$

and the boundary conditions are chosen such that $P(u, \tilde{u})$ vanishes at both $x = 0$ and $x = 1$. Under these conditions u_n and $\tilde{u}_n$ are orthogonal, and they can be normalized so that

$$\int_0^1 u_n(x)\tilde{u}_m(x)\,dx = \delta_{mn} \qquad (5.1.43)$$

The analysis in this case would be the same as in the self-adjoint case if we replace $\bar{u}_m(x)$ by $\tilde{u}_m(x)$. In particular, (5.1.40) becomes

$$\frac{dA_m}{dt} + \lambda_m A_m = -\int_0^1 F\left[\sum_{n=1}^{\infty} A_n(t)u_n(x)\right]\tilde{u}_m(x)\,dx \qquad (5.1.44)$$

5.2. The Method of Averaging

5.2.1. VAN DER POL'S TECHNIQUE

Van der Pol (1926) devised a technique, which is described in this section, to investigate the periodic solutions of the equation

$$\frac{d^2u}{dt^2} + \omega_0^2 u = \epsilon(1 - u^2)\frac{du}{dt} + \epsilon k\lambda \cos \lambda t \qquad (5.2.1)$$

which is named for him. In (5.2.1) ϵ is assumed to be small, and λ (the frequency of the excitation) is assumed to differ from ω_0 (the natural frequency) by a small quantity which is of the order of ϵ. Under these assumptions the solution of (5.2.1) is assumed to have the form

$$u(t) = a_1(t)\cos \lambda t + a_2(t)\sin \lambda t \qquad (5.2.2)$$

where $a_1(t)$ and $a_2(t)$ are assumed to be slowly varying functions of time; that is, $da_i/dt = O(\epsilon)$ and $d^2a_i/dt^2 = O(\epsilon^2)$.

Differentiating (5.2.2) twice gives

$$\ddot{u} = -\lambda^2 a_1 \cos \lambda t - \lambda^2 a_2 \sin \lambda t - 2\dot{a}_1\lambda \sin \lambda t + 2\dot{a}_2\lambda \cos \lambda t$$

$$+ \ddot{a}_1 \cos \lambda t + \ddot{a}_2 \sin \lambda t \qquad (5.2.3)$$

where the overdots denote differentiation with respect to t. Substituting (5.2.2) and (5.2.3) into (5.2.1), neglecting terms of order higher than ϵ, keeping in mind that $\dot{a}_i = O(\epsilon)$ while $\ddot{a}_i = O(\epsilon^2)$, and equating the coefficients of $\cos \lambda t$ and $\sin \lambda t$ on both sides, we obtain

$$2\dot{a}_1 + \frac{\lambda^2 - \omega_0^2}{\lambda} a_2 - \epsilon a_1(1 - \rho) = 0 \tag{5.2.4}$$

$$2\dot{a}_2 - \frac{\lambda^2 - \omega_0^2}{\lambda} a_1 - \epsilon a_2(1 - \rho) = \epsilon k \tag{5.2.5}$$

where

$$\rho = \frac{a^2}{4} = \frac{a_1^2 + a_2^2}{4} \tag{5.2.6}$$

To analyze the periodic solutions of (5.2.1), we note that they correspond to the stationary solutions of (5.2.4) and (5.2.5); that is, they correspond to the solutions of

$$2\sigma a_{20} - a_{10}(1 - \rho_0) = 0 \tag{5.2.7}$$

$$-2\sigma a_{10} - a_{20}(1 - \rho_0) = k \tag{5.2.8}$$

where σ is the detuning factor, and it is given by

$$\sigma = \frac{\lambda - \omega_0}{\epsilon} \tag{5.2.9}$$

Terms of $O(\epsilon^2)$ in (5.2.4) and (5.2.5) have been neglected. By adding the squares of (5.2.7) and (5.2.8) and using (5.2.6), we obtain the frequency response equation

$$\rho_0[4\sigma^2 + (1 - \rho_0)^2] = \frac{k^2}{4} \tag{5.2.10}$$

5.2.2. THE KRYLOV–BOGOLIUBOV TECHNIQUE

We discuss this technique in connection with the general weakly nonlinear second-order equation

$$\frac{d^2u}{dt^2} + \omega_0^2 u = \epsilon f\left(u, \frac{du}{dt}\right) \tag{5.2.11}$$

When $\epsilon = 0$ the solution of (5.2.11) can be written as

$$u = a \cos(\omega_0 t + \theta) \tag{5.2.12}$$

where a and θ are constants. To determine an approximate solution to (5.2.11) for ϵ small but different from zero, Krylov and Bogoliubov (1947) assumed that the solution is still given by (5.2.12) but with time-varying a and θ, and subject to the condition

$$\frac{du}{dt} = -a\omega_0 \sin \phi, \qquad \phi = \omega_0 t + \theta \tag{5.2.13}$$

Thus this technique is similar to van der Pol's technique which was discussed in the previous section. The only difference is in the form of the first term.
Differentiating (5.2.12) with respect to t gives

$$\frac{du}{dt} = -a\omega_0 \sin \phi + \frac{da}{dt} \cos \phi - a\frac{d\theta}{dt} \sin \phi$$

Hence

$$\frac{da}{dt} \cos \phi - a\frac{d\theta}{dt} \sin \phi = 0 \tag{5.2.14}$$

on account of (5.2.13). Differentiating (5.2.13) with respect to t gives

$$\frac{d^2u}{dt^2} = -a\omega_0{}^2 \cos \phi - \omega_0 \frac{da}{dt} \sin \phi - a\omega_0 \frac{d\theta}{dt} \cos \phi$$

Substituting this expression into (5.2.11) and using (5.2.12), we obtain

$$\omega_0 \frac{da}{dt} \sin \phi + a\omega_0 \frac{d\theta}{dt} \cos \phi = -\epsilon f[a \cos \phi, -a\omega_0 \sin \phi] \tag{5.2.15}$$

Solving (5.2.14) and (5.2.15) for da/dt and $d\theta/dt$ yields

$$\frac{da}{dt} = -\frac{\epsilon}{\omega_0} \sin \phi f[a \cos \phi, -a\omega_0 \sin \phi] \tag{5.2.16}$$

$$\frac{d\theta}{dt} = -\frac{\epsilon}{a\omega_0} \cos \phi f[a \cos \phi, -a\omega_0 \sin \phi] \tag{5.2.17}$$

Thus the original second-order differential equation (5.2.11) for u has been replaced by the two first-order differential equations (5.2.16) and (5.2.17) for the amplitude a and the phase θ.

To solve (5.2.16) and (5.2.17), we note that the right-hand sides of these equations are periodic with respect to the variable ϕ, hence $da/dt = O(\epsilon)$ and $d\theta/dt = O(\epsilon)$. Thus a and θ are slowly varying functions of time because ϵ is small; hence they change very little during the time $T = 2\pi/\omega_0$ (the period of the terms on the right-hand sides of these equations). Averaging (5.2.16) and (5.2.17) over the interval $[t, t + T]$, during which a and θ can be taken to be constants on the right-hand side of these equations, we obtain

$$\frac{da}{dt} = -\frac{\epsilon}{2\omega_0} f_1(a) \tag{5.2.18}$$

$$\frac{d\theta}{dt} = -\frac{\epsilon}{2a\omega_0} g_1(a) \tag{5.2.19}$$

where

$$f_1(a) = \frac{2}{T} \int_0^T \sin \phi \, f[a \cos \phi, -a\omega_0 \sin \phi] \, dt$$

$$= \frac{1}{\pi} \int_0^{2\pi} \sin \phi \, f[a \cos \phi, -a\omega_0 \sin \phi] \, d\phi \tag{5.2.20}$$

$$g_1(a) = \frac{1}{\pi} \int_0^{2\pi} \cos \phi \, f[a \cos \phi, -a\omega_0 \sin \phi] \, d\phi \tag{5.2.21}$$

Note that f_1 and g_1 are simply two coefficients of the Fourier series expansion of f.

As an example, let us consider Duffing's equation (2.1.1) in which

$$f(u, \dot{u}) = -u^3 \tag{5.2.22}$$

Hence

$$f_1(a) = 0, \qquad g_1(a) = -\tfrac{3}{4}a^3 \tag{5.2.23}$$

Consequently, $a = $ a constant from (5.2.18), and

$$\theta = \tfrac{3}{8}\epsilon \frac{a^2}{\omega_0} t + \theta_0 \tag{5.2.24}$$

from (5.2.19). Therefore, to first approximation

$$u = a \cos \omega_0 \left[1 + \tfrac{3}{8}\epsilon \frac{a^2}{\omega_0^2} \right] t + O(\epsilon) \tag{5.2.25}$$

As a second example, we consider the van der Pol oscillator in which

$$f(u, \dot{u}) = (1 - u^2) \frac{du}{dt} \tag{5.2.26}$$

In this case

$$f_1 = -\omega_0 a(1 - \tfrac{1}{4}a^2), \qquad g_1 = 0 \tag{5.2.27}$$

Hence $\theta = \theta_0 = $ a constant from (5.2.19), whereas

$$\frac{da}{dt} = \frac{\epsilon a}{2}(1 - \tfrac{1}{4}a^2) \tag{5.2.28}$$

Integrating (5.2.28) yields

$$a^2 = \frac{4}{1 + \left(\dfrac{4}{a_0^{\,2}} - 1\right)e^{-\epsilon t}} \qquad (5.2.29)$$

The same basic idea of this technique underlies the method of Le Verrier (1856).

5.2.3. THE GENERALIZED METHOD OF AVERAGING

In this technique we consider (5.2.12) and (5.2.13) as a transformation from u and du/dt into a and ϕ so that

$$\frac{da}{dt} = - \frac{\epsilon}{\omega_0} \sin \phi \, f[a \cos \phi, \, -a\omega_0 \sin \phi]$$

$$\frac{d\phi}{dt} = \omega_0 - \frac{\epsilon}{a\omega_0} \cos \phi \, f[a \cos \phi, \, -a\omega_0 \sin \phi] \qquad (5.2.30)$$

where the variable ϕ is called a rapidly rotating phase. Rather than integrate these equations as in the previous section, we define a near-identity transformation (see Bogoliubov and Mitropolski, 1961, p. 412)

$$a = \bar{a} + \epsilon a_1(\bar{a}, \bar{\phi}) + \epsilon^2 a_2(\bar{a}, \bar{\phi}) + \cdots$$

$$\phi = \bar{\phi} + \epsilon \phi_1(\bar{a}, \bar{\phi}) + \epsilon^2 \phi_2(\bar{a}, \bar{\phi}) + \cdots \qquad (5.2.31)$$

from (a, ϕ) to $(\bar{a}, \bar{\phi})$, which is 2π periodic in $\bar{\phi}$ such that the transform of the system (5.2.30) has the form

$$\frac{d\bar{a}}{dt} = \epsilon A_1(\bar{a}) + \epsilon^2 A_2(\bar{a}) + \cdots$$

$$\frac{d\bar{\phi}}{dt} = \omega_0 + \epsilon B_1(\bar{a}) + \epsilon^2 B_2(\bar{a}) + \cdots \qquad (5.2.32)$$

with A_i and B_i independent of $\bar{\phi}$. In this procedure a and ϕ need not be restricted to scalar functions (Mettler 1959; Sethna 1963; Morrison, 1966b). Higher-order effects were obtained by Volosov (1961, 1962), Musen (1965) and Zabreiko and Ledovskaja (1966). The inverse transformation of (5.2.31) was proposed by Kruskal (1962) and an order-by-order algorithm for this latter procedure was obtained by Stern (1970b). Stern (1971b) used this technique to analyze slowly varying perturbed systems.

Substituting (5.2.31) and (5.2.32) into (5.2.30), expanding, and equating coefficients of like powers of ϵ, we obtain equations of the form

$$\omega_0 \frac{\partial a_n}{\partial \bar{\phi}} + A_n = F_n(\bar{a}, \bar{\phi})$$

$$\omega_0 \frac{\partial \phi_n}{\partial \bar{\phi}} + B_n = G_n(\bar{a}, \bar{\phi})$$

(5.2.33)

where the right-hand sides are known functions of the lower-order terms in (5.2.31) and (5.2.32). In general, F_n and G_n contain short-period terms (denoted by superscript s) and long-period terms (denoted by superscript l). We choose A_n and B_n to be equal to the long-period terms; that is

$$A_n = F_n{}^l, \qquad B_n = G_n{}^l \qquad (5.2.34)$$

Then

$$\omega_0 \frac{\partial a_n}{\partial \bar{\phi}} = F_n{}^s, \qquad \omega_0 \frac{\partial \phi_n}{\partial \bar{\phi}} = G_n{}^s \qquad (5.2.35)$$

which can be solved successively for a_n and ϕ_n.

As an example, we consider the van der Pol oscillator in which

$$f(u, \dot{u}) = (1 - u^2)\dot{u}, \qquad \omega_0 = 1$$

In this case (5.2.30) become

$$\frac{da}{dt} = \tfrac{1}{8}\epsilon[a(4 - a^2) - 4a \cos 2\phi + a^3 \cos 4\phi]$$

$$\frac{d\phi}{dt} = 1 + \tfrac{1}{8}\epsilon[2(2 - a^2) \sin 2\phi - a^2 \sin 4\phi]$$

(5.2.36)

Substituting (5.2.31) and (5.2.32) into (5.2.36) and equating coefficients of like powers of ϵ, we have

Order ϵ

$$\frac{\partial a_1}{\partial \bar{\phi}} + A_1 = \tfrac{1}{8}\bar{a}(4 - \bar{a}^2) - \tfrac{1}{2}\bar{a} \cos 2\bar{\phi} + \tfrac{1}{8}\bar{a}^3 \cos 4\bar{\phi}$$

$$\frac{\partial \phi_1}{\partial \bar{\phi}} + B_1 = \tfrac{1}{4}(2 - \bar{a}^2) \sin 2\bar{\phi} - \tfrac{1}{8}\bar{a}^2 \sin 4\bar{\phi}$$

(5.2.37)

Order ϵ^2

$$\frac{\partial a_2}{\partial \bar{\phi}} + A_2 = -\frac{\partial a_1}{\partial \bar{a}} A_1 - \frac{\partial a_1}{\partial \bar{\phi}} B_1$$
$$+ \tfrac{1}{8} a_1 [4 - 3\bar{a}^2 - 4\cos 2\bar{\phi} + 3\bar{a}^2 \cos 4\bar{\phi}]$$
$$+ \tfrac{1}{2} \bar{a} \phi_1 [2 \sin 2\bar{\phi} - \bar{a}^2 \sin 4\bar{\phi}] \qquad (5.2.38)$$

$$\frac{\partial \phi_2}{\partial \bar{\phi}} + B_2 = -\frac{\partial \phi_1}{\partial \bar{a}} A_1 - \frac{\partial \phi_1}{\partial \bar{\phi}} B_1$$
$$- \tfrac{1}{4} \bar{a} a_1 (2 \sin 2\bar{\phi} + \sin 4\bar{\phi})$$
$$+ \tfrac{1}{2} \phi_1 [(2 - \bar{a}^2) \cos 2\bar{\phi} - \bar{a}^2 \cos 4\bar{\phi}] \qquad (5.2.39)$$

Equating A_1 and B_1 to the long-period terms on the right-hand sides of (5.2.37), we have

$$A_1 = \tfrac{1}{8} \bar{a}(4 - \bar{a}^2), \qquad B_1 = 0 \qquad (5.2.40)$$

Consequently, (5.2.37) become

$$\frac{\partial a_1}{\partial \bar{\phi}} = -\tfrac{1}{2} \bar{a} \cos 2\bar{\phi} + \tfrac{1}{8} \bar{a}^3 \cos 4\bar{\phi}$$
$$\frac{\partial \phi_1}{\partial \bar{\phi}} = \tfrac{1}{4}(2 - \bar{a}^2) \sin 2\bar{\phi} - \tfrac{1}{8} \bar{a}^2 \sin 4\bar{\phi} \qquad (5.2.41)$$

whose solution is

$$a_1 = -\tfrac{1}{4} \bar{a} \sin 2\bar{\phi} + \tfrac{1}{32} \bar{a}^3 \sin 4\bar{\phi}$$
$$\phi_1 = -\tfrac{1}{8}(2 - \bar{a}^2) \cos 2\bar{\phi} + \tfrac{1}{32} \bar{a}^2 \cos 4\bar{\phi} \qquad (5.2.42)$$

With (5.2.40) and (5.2.42), (5.2.38) and (5.2.39) become

$$\frac{\partial a_2}{\partial \bar{\phi}} + A_2 = \text{short-period terms}$$
$$\frac{\partial \phi_2}{\partial \bar{\phi}} + B_2 = -\tfrac{1}{8} + \tfrac{3}{16} \bar{a}^2 - \tfrac{11}{256} \bar{a}^4 + \text{short-period terms} \qquad (5.2.43)$$

Equating A_2 and B_2 to the long-period terms on the right-hand sides of (5.2.43), we obtain

$$A_2 = 0, \qquad B_2 = -\tfrac{1}{8} + \tfrac{3}{16} \bar{a}^2 - \tfrac{11}{256} \bar{a}^4 \qquad (5.2.44)$$

Therefore, to second order

where

$$u = a \cos \phi \tag{5.2.45}$$

$$a = \bar{a} - \tfrac{1}{4}\epsilon\bar{a}[\sin 2\bar{\phi} - \tfrac{1}{8}\bar{a}^2 \sin 4\bar{\phi}] + O(\epsilon^2)$$

$$\phi = \bar{\phi} - \tfrac{1}{8}\epsilon[(2 - \bar{a}^2)\cos 2\bar{\phi} - \tfrac{1}{4}\bar{a}^2\cos 4\bar{\phi}] + O(\epsilon^2) \tag{5.2.46}$$

$$\frac{d\bar{a}}{dt} = \tfrac{1}{8}\epsilon\bar{a}(4 - \bar{a}^2) + O(\epsilon^3)$$

$$\frac{d\bar{\phi}}{dt} = 1 - \tfrac{1}{8}\epsilon^2[1 - \tfrac{3}{2}\bar{a}^2 + \tfrac{11}{32}\bar{a}^4] + O(\epsilon^3) \tag{5.2.47}$$

This solution is in full agreement with that obtained in Section 5.7.4 by using Kamel's algorithm.

For canonical systems the transformation (5.2.31) and (5.2.32) can be effected in a more elegant manner by using von Zeipel's procedure (Section 5.6) or Lie series and transforms (Section 5.7.5). The latter is a simple and efficient algorithm which is built from the recursive application of a few elementary operations, thereby making it well suited for implementation on a computer. For noncanonical systems an efficient recursive algorithm was formulated by using Lie transforms (Section 5.7).

5.3. Struble's Technique

Struble (1962) developed a technique for treating weakly nonlinear oscillatory systems such as those governed by

$$\ddot{u} + \omega_0^2 u = \epsilon f(u, \dot{u}, t) \tag{5.3.1}$$

He expressed the asymptotic solution of this equation for small ϵ in the form

$$u = a \cos(\omega_0 t - \theta) + \sum_{n=1}^{N} \epsilon^n u_n(t) + O(\epsilon^{N+1}) \tag{5.3.2}$$

where a and θ are slowly varying functions of time. If we set each $u_n = 0$, (5.3.2) reduces to the form of the solution used by Krylov and Bogoliubov to obtain a first approximation to u (see Section 5.2.2). Rather than carry out the details for a general function f, we use the specific function corresponding to the Duffing equation.

Thus we consider

$$\ddot{u} + \omega_0^2 u = -\epsilon u^3 \tag{5.3.3}$$

Substituting (5.3.2) into (5.3.3) gives

$$\left[2a\omega_0 \frac{d\theta}{dt} + \frac{d^2a}{dt^2} - a\left(\frac{d\theta}{dt}\right)^2 \right] \cos(\omega_0 t - \theta)$$

$$+ \left[-2\omega_0 \frac{da}{dt} + a\frac{d^2\theta}{dt^2} + 2\frac{da}{dt}\frac{d\theta}{dt} \right] \sin(\omega_0 t - \theta)$$

$$+ \epsilon\left(\frac{d^2u_1}{dt^2} + \omega_0{}^2 u_1\right) + \epsilon^2\left(\frac{d^2u_2}{dt^2} + \omega_0{}^2 u_2\right) + \cdots$$

$$= -\epsilon a^3 \cos^3(\omega_0 t - \theta) - 3\epsilon^2 u_1 a^2 \cos^2(\omega_0 t - \theta) + \cdots \quad (5.3.4)$$

If we consider terms through $O(\epsilon)$ and equate the coefficients of $\cos(\omega_0 t - \theta)$ and $\sin(\omega_0 t - \theta)$ on both sides, we obtain the following so-called variational equations

$$2a\omega_0 \frac{d\theta}{dt} + \frac{d^2a}{dt^2} - a\left(\frac{d\theta}{dt}\right)^2 = -\tfrac{3}{4}\epsilon a^3 \quad (5.3.5)$$

$$-2\omega_0 \frac{da}{dt} + a\frac{d^2\theta}{dt^2} + 2\frac{da}{dt}\frac{d\theta}{dt} = 0 \quad (5.3.6)$$

This leaves the following so-called perturbational equation

$$\frac{d^2u_1}{dt^2} + \omega_0{}^2 u_1 = -\tfrac{1}{4}a^3 \cos 3(\omega_0 t - \theta) \quad (5.3.7)$$

To first order in ϵ, (5.3.5) and (5.3.6) reduce to

$$\frac{da}{dt} = 0 \quad \text{and} \quad \frac{d\theta}{dt} = -\frac{3}{8\omega_0}\epsilon a^2 \quad (5.3.8)$$

Hence

$$a = a_0, \qquad \theta = -\frac{3}{8\omega_0}\epsilon a_0{}^2 t + \theta_0 \quad (5.3.9)$$

where a_0 and θ_0 are constants. Then the solution of (5.3.7) to first order is obtained by considering θ and a constants. The result is

$$u_1 = \frac{1}{32\omega_0{}^2} a^3 \cos 3(\omega_0 t - \theta) \quad (5.3.10)$$

Hence the first-order solution becomes

$$u = a \cos(\omega_0 t - \theta) + \frac{1}{32\omega_0{}^2}\epsilon a^3 \cos 3(\omega_0 t - \theta) \quad (5.3.11)$$

where a and θ are given by (5.3.9).

With u_1 known the term

$$-3\epsilon^2 u_1 a^2 \cos^2 (\omega_0 t - \theta)$$

$$= -\frac{3}{128\omega_0^2} \epsilon^2 a^5 [\cos (\omega_0 t - \theta) + 2 \cos 3(\omega_0 t - \theta) + \cos 5(\omega_0 t - \theta)]$$

$$(5.3.12)$$

Moreover, we have to calculate the terms of $O(\epsilon)$ in $(d^2 u_1/dt^2) + \omega_0^2 u_1$; that is, we must include the term

$$\frac{9}{16\omega_0} a^3 \frac{d\theta}{dt} \cos 3(\omega_0 t - \theta) \qquad (5.3.13)$$

Now considering terms through $O(\epsilon^2)$ leads to the variational equations

$$2a\omega_0 \frac{d\theta}{dt} + \frac{d^2 a}{dt^2} - a\left(\frac{d\theta}{dt}\right)^2 = -\tfrac{3}{4}\epsilon a^3 - \frac{3}{128\omega_0^2} \epsilon^2 a^5 \qquad (5.3.14)$$

$$-2\omega_0 \frac{da}{dt} + a \frac{d^2\theta}{dt^2} + 2 \frac{da}{dt} \frac{d\theta}{dt} = 0 \qquad (5.3.15)$$

and the perturbational equation

$$\frac{d^2 u_2}{dt^2} + \omega_0^2 u_2 = -\frac{3}{128\omega_0^2} a^5 [2 \cos 3(\omega_0 t - \theta) + \cos 5(\omega_0 t - \theta)]$$

$$-\frac{9}{16\omega_0 \epsilon} a^3 \frac{d\theta}{dt} \cos 3(\omega_0 t - \theta) \qquad (5.3.16)$$

The solution of (5.3.14) and (5.3.15) can be obtained by iteration starting with (5.3.9) to be

$$a = a_0, \qquad \theta = -\frac{3}{8\omega_0} \epsilon a_0^2 t + \frac{15}{256\omega_0^3} \epsilon^2 a_0^4 t + \theta_0 + O(\epsilon^3) \qquad (5.3.17)$$

where a_0 and θ_0 are constants. Substituting for $d\theta/dt$ from (5.3.9) into (5.3.16) and solving the resulting equation, we obtain within an error of $O(\epsilon)$ the following

$$u_2 = -\frac{21}{1024\omega_0^4} a^5 \cos 3(\omega_0 t - \theta) + \frac{1}{1024\omega_0^4} a^5 \cos 5(\omega_0 t - \theta) \qquad (5.3.18)$$

Therefore the solution to second order is

$$u = a \cos (\omega t - \theta_0) + \frac{\epsilon a^3}{32\omega_0^2}\left(1 - \frac{21\epsilon}{32\omega_0^2} a^2\right) \cos 3(\omega t - \theta_0)$$

$$+ \frac{\epsilon^2 a^5}{1024\omega_0^4} \cos 5(\omega t - \theta_0) + O(\epsilon^3) \qquad (5.3.19)$$

where

$$\omega = \omega_0\left(1 + \frac{3\epsilon a^2}{8\omega_0^2} - \frac{15\epsilon^2 a^4}{256\omega_0^4}\right) + O(\epsilon^3) \tag{5.3.20}$$

To carry out the solution to third order, we need to calculate the terms of $O(\epsilon^2)$ in d^2u_1/dt^2, and the terms of $O(\epsilon)$ in d^2u_2/dt, then write the variational and perturbational equations. This constitutes a major limitation of this technique. The second limitation is the iteration solution of the variational equations. Systematic ways of handling such problems are the Lindstedt-Poincaré technique (Section 3.1.1), the Krylov-Bogoliubov-Mitropolski technique (Section 5.4), Lie series and transforms (Section 5.7), and the method of multiple scales discussed in Ch. 6.

5.4. The Krylov–Bogoliubov–Mitropolski Technique

In the course of their refinement of the first approximation for (5.2.11), which was discussed in Section 5.2.2, Krylov and Bogoliubov (1947) developed a technique for determining the solution to any approximation. This technique has been amplified and justified by Bogoliubov and Mitropolski (1961) and extended to nonstationary vibrations by Mitropolski (1965). They assumed an asymptotic expansion of the form

$$u = a\cos\psi + \sum_{n=1}^{N}\epsilon^n u_n(a, \psi) + O(\epsilon^{N+1}) \tag{5.4.1}$$

where each u_n is a periodic function of ψ with a period 2π, and a and ψ are assumed to vary with time according to

$$\frac{da}{dt} = \sum_{n=1}^{N}\epsilon^n A_n(a) + O(\epsilon^{N+1}) \tag{5.4.2}$$

$$\frac{d\psi}{dt} = \omega_0 + \sum_{n=1}^{N}\epsilon^n \psi_n(a) + O(\epsilon^{N+1}) \tag{5.4.3}$$

where the functions u_n, A_n, and ψ_n are chosen such that (5.4.1) through (5.4.3) satisfy the differential equation (5.2.11). In order to uniquely determine A_n and ψ_n, we require that no u_n contains $\cos\psi$. The derivatives are transformed according to

$$\frac{d}{dt} = \frac{da}{dt}\frac{\partial}{\partial a} + \frac{d\psi}{dt}\frac{\partial}{\partial\psi} \tag{5.4.4}$$

$$\frac{d^2}{dt^2} = \left(\frac{da}{dt}\right)^2\frac{\partial^2}{\partial a^2} + \frac{d^2a}{dt^2}\frac{\partial}{\partial a} + 2\frac{da}{dt}\frac{d\psi}{dt}\frac{\partial^2}{\partial a\,\partial\psi} + \left(\frac{d\psi}{dt}\right)^2\frac{\partial^2}{\partial\psi^2} + \frac{d^2\psi}{dt^2}\frac{\partial}{\partial\psi} \tag{5.4.5}$$

$$\frac{d^2a}{dt^2} = \frac{d}{dt}\left(\frac{da}{dt}\right) = \frac{da}{dt}\frac{d}{da}\left(\frac{da}{dt}\right) = \frac{da}{dt}\sum_{n=1}^{N}\epsilon^n\frac{dA_n}{da} = \epsilon^2 A_1\frac{dA_1}{da} + O(\epsilon^3) \tag{5.4.6}$$

$$\frac{d^2\psi}{dt^2} = \frac{d}{dt}\left(\frac{d\psi}{dt}\right) = \frac{da}{dt}\frac{d}{da}\left(\frac{d\psi}{dt}\right) = \frac{da}{dt}\sum_{n=1}^{N}\epsilon^n\frac{d\psi_n}{da} = \epsilon^2 A_1\frac{d\psi_1}{da} + O(\epsilon^3) \tag{5.4.7}$$

We next illustrate this technique by its application to the Duffing, van der Pol, and Klein-Gordon equations.

5 4.1. THE DUFFING EQUATION

We consider the nonlinear oscillator

$$\ddot{u} + \omega_0^2 u = -\epsilon u^3 \qquad (5.4.8)$$

which was treated previously in Sections 3.1.1, 5.2.2, and 5.3. Substituting (5.4.1) through (5.4.7) into (5.4.8) and equating coefficients of like powers of ϵ through ϵ^2, we obtain

$$\omega_0^2 \frac{\partial^2 u_1}{\partial \psi^2} + \omega_0^2 u_1 = 2\omega_0 \psi_1 a \cos \psi + 2\omega_0 A_1 \sin \psi - a^3 \cos^3 \psi \qquad (5.4.9)$$

$$\omega_0^2 \frac{\partial^2 u_2}{\partial \psi^2} + \omega_0^2 u_2 = \left[(2\omega_0 \psi_2 + \psi_1^2)a - A_1 \frac{dA_1}{da} \right] \cos \psi$$

$$+ \left[2(\omega_0 A_2 + A_1 \psi_1) + a A_1 \frac{d\psi_1}{da} \right] \sin \psi$$

$$- 3u_1 a^2 \cos^2 \psi - 2\omega_0 \psi_1 \frac{\partial^2 u_1}{\partial \psi^2} - 2\omega_0 A_1 \frac{\partial^2 u_1}{\partial a \, \partial \psi} \qquad (5.4.10)$$

In order that u_1 be periodic, the terms that produce secular terms on the right-hand side of (5.4.9) must vanish. Since $\cos^3 \psi = (3 \cos \psi + \cos 3\psi)/4$, this condition gives

$$A_1 = 0, \qquad \psi_1 = \frac{3a^2}{8\omega_0} \qquad (5.4.11)$$

Then the solution of (5.4.9) is

$$u_1 = \frac{a^3}{32\omega_0^2} \cos 3\psi \qquad (5.4.12)$$

Substituting this first-order solution into (5.4.10) gives

$$\omega_0^2 \frac{\partial^2 u_2}{\partial \psi^2} + \omega_0^2 u_2 = \left(2\omega_0 \psi_2 + \frac{15a^4}{128\omega_0^2} \right) a \cos \psi + 2\omega_0 A_2 \sin \psi$$

$$+ \frac{a^5}{128\omega_0^2} (21 \cos 3\psi - 3 \cos 5\psi) \qquad (5.4.13)$$

Elimination of secular terms yields

$$A_2 = 0, \qquad \psi_2 = -\frac{15a^4}{256\omega_0^3} \qquad (5.4.14)$$

Then the solution of (5.4.13) is

$$u_2 = \frac{-a^5}{1024\omega_0{}^4} (21 \cos 3\psi - \cos 5\psi) \qquad (5.4.15)$$

Therefore, to second order, u is given by

$$u = a \cos \psi + \frac{\epsilon a^3}{32\omega_0{}^2} \cos 3\psi$$
$$- \frac{\epsilon^2 a^5}{1024\omega_0{}^4} (21 \cos 3\psi - \cos 5\psi) + O(\epsilon^3) \quad (5.4.16)$$

where

$$\frac{da}{dt} = 0 \quad \text{or} \quad a = a_0 = \text{a constant} \qquad (5.4.17)$$

$$\frac{d\psi}{dt} = \omega_0 + \frac{3\epsilon a^2}{8\omega_0} - \frac{15\epsilon^2 a^4}{256\omega_0{}^3} + O(\epsilon^3)$$

$$\psi = \omega_0 \left[1 + \frac{3\epsilon a^2}{8\omega_0{}^2} - \frac{15\epsilon^2 a^4}{256\omega_0{}^4} \right] t + \psi_0 + O(\epsilon^3) \qquad (5.4.18)$$

where ψ_0 is a constant. This solution agrees with (5.3.19) and (5.3.20) obtained using Struble's method.

5.4.2. THE VAN DER POL OSCILLATOR
We discuss next the nonlinear oscillator

$$\ddot{u} + u = \epsilon(1 - u^2)\dot{u} \qquad (5.4.19)$$

treated in Sections 5.2.2 and 5.2.3. Substituting (5.4.1) through (5.4.7) into (5.4.19) and equating coefficients of equal powers of ϵ through ϵ^2, we obtain

$$\frac{\partial^2 u_1}{\partial \psi^2} + u_1 = 2\psi_1 a \cos \psi + 2A_1 \sin \psi$$
$$- a(1 - \tfrac{1}{4}a^2) \sin \psi + \tfrac{1}{4}a^3 \sin 3\psi \qquad (5.4.20)$$

$$\frac{\partial^2 u_2}{\partial \psi^2} + u_2 = \left[(2\psi_2 + \psi_1{}^2)a - A_1 \frac{dA_1}{da} \right] \cos \psi$$
$$+ \left[2(A_2 + A_1\psi_1) + aA_1 \frac{d\psi_1}{da} \right] \sin \psi$$
$$- 2\psi_1 \frac{\partial^2 u_1}{\partial \psi^2} - 2A_1 \frac{\partial^2 u_1}{\partial a\,\partial \psi} + (1 - a^2 \cos^2 \psi)$$
$$\times \left(A_1 \cos \psi - a\psi_1 \sin \psi + \frac{\partial u_1}{\partial \psi} \right) + a^2 u_1 \sin 2\psi \quad (5.4.21)$$

Elimination of secular terms from the right-hand side of (5.4.20) gives

$$\psi_1 = 0, \qquad A_1 = \tfrac{1}{2}a(1 - \tfrac{1}{4}a^2) \qquad (5.4.22)$$

Hence

$$u_1 = -\frac{a^3}{32} \sin 3\psi \qquad (5.4.23)$$

With this solution (5.4.21) becomes

$$\frac{\partial^2 u_2}{\partial \psi^2} + u_2 = \left[2a\psi_2 - A_1 \frac{dA_1}{da} + (1 - \tfrac{3}{4}a^2)A_1 + \frac{a^5}{128} \right] \cos \psi$$

$$+ 2A_2 \sin \psi + \frac{a^3(a^2 + 8)}{128} \cos 3\psi + \frac{5a^5}{128} \cos 5\psi \quad (5.4.24)$$

For there to be no secular terms in u_2

$$A_2 = 0, \qquad \psi_2 = \frac{A_1}{2a}\left(\frac{dA_1}{da} - 1 + \tfrac{3}{4}a^2\right) - \frac{a^4}{256} \qquad (5.4.25)$$

Hence

$$u_2 = -\frac{5a^5}{3072} \cos 5\psi - \frac{a^3(a^2 + 8)}{1024} \cos 3\psi \qquad (5.4.26)$$

Therefore the solution to second order is given by

$$u = a \cos \psi - \frac{\epsilon a^3}{32} \sin 3\psi$$

$$- \frac{\epsilon^2 a^3}{1024} [\tfrac{5}{3}a^2 \cos 5\psi + (a^2 + 8) \cos 3\psi] + O(\epsilon^3) \quad (5.4.27)$$

where

$$\frac{da}{dt} = \frac{\epsilon a}{2}(1 - \tfrac{1}{4}a^2), \qquad a^2 = \frac{4}{1 + \left(\dfrac{4}{a_0^2} - 1\right)e^{-\epsilon t}} \qquad (5.4.28)$$

$$\frac{d\psi}{dt} = 1 + \epsilon^2 \left[\frac{A_1}{2a}\left(\frac{dA_1}{da} - 1 + \tfrac{3}{4}a^2\right) - \frac{a^4}{256} \right] \qquad (5.4.29)$$

where a_0 is a constant. Using (5.4.22) and (5.4.28), we can write (5.4.29) as

$$\frac{d\psi}{dt} = 1 - \frac{\epsilon^2}{16} - \frac{\epsilon}{8a}(1 - \tfrac{7}{4}a^2)\frac{da}{dt}$$

Hence

$$\psi = t - \frac{\epsilon^2}{16}t - \frac{\epsilon}{8} \ln a + \frac{7\epsilon}{64} a^2 + \psi_0 \qquad (5.4.30)$$

where ψ_0 is a constant.

5.4.3. THE KLEIN–GORDON EQUATION

As a third example, we consider nonlinear waves governed by

$$u_{tt} - c^2 u_{xx} + \lambda^2 u = \epsilon f(u, u_t, u_x) \tag{5.4.31}$$

after Montgomery and Tidman (1964). If $\epsilon = 0$, (5.4.31) admits solutions of the form

$$u = a \cos (k_0 x - \omega_0 t + \phi) \tag{5.4.32}$$

where a and ϕ are constants, and k_0 and ω_0 satisfy the dispersion relationship

$$\omega_0^2 = c^2 k_0^2 + \lambda^2 \tag{5.4.33}$$

For small but finite ϵ, we seek an expansion of the form

$$u = a \cos \psi + \epsilon u_1(a, \psi) + \cdots \tag{5.4.34}$$

where a is a slowly varying function of both time and position according to

$$\frac{\partial a}{\partial t} = \epsilon A_1(a) + \epsilon^2 A_2(a) + \ldots \tag{5.4.35}$$

$$\frac{\partial a}{\partial x} = \epsilon B_1(a) + \epsilon^2 B_2(a) + \ldots \tag{5.4.36}$$

and ψ is a new phase variable which coincides with the phase of (5.4.32) for $\epsilon = 0$

$$\frac{\partial \psi}{\partial t} = -\omega_0 + \epsilon C_1(a) + \epsilon^2 C_2(a) + \ldots \tag{5.4.37}$$

$$\frac{\partial \psi}{\partial x} = k_0 + \epsilon D_1(a) + \epsilon^2 D_2(a) + \ldots \tag{5.4.38}$$

In this case also, no u_n contains the fundamental $\cos \psi$.

Substituting (5.4.34) through (5.4.38) into (5.4.31), using (5.4.33), and equating the coefficients of ϵ on both sides, we obtain

$$\lambda^2 \left(\frac{\partial^2 u_1}{\partial \psi^2} + u_1 \right) = -2(\omega_0 A_1 + c^2 k_0 B_1) \sin \psi$$
$$- 2a(\omega_0 C_1 + c^2 k_0 D_1) \cos \psi$$
$$+ f[a \cos \psi, a\omega_0 \sin \psi, -ak_0 \sin \psi] \tag{5.4.39}$$

We now Fourier analyze f in terms of ψ as

$$f[a \cos \psi, a\omega_0 \sin \psi, -ak_0 \sin \psi]$$
$$= g_0(a) + \sum_{n=1}^{\infty} [f_n(a) \sin n\psi + g_n(a) \cos n\psi] \tag{5.4.40}$$

Eliminating secular terms, we have

$$2\omega_0 A_1 + 2c^2 k_0 B_1 = f_1(a) \qquad (5.4.41)$$

$$2a(\omega_0 C_1 + c^2 k_0 D_1) = g_1(a) \qquad (5.4.42)$$

Then the solution of (5.4.39) is

$$u_1 = \frac{g_0(a)}{\lambda^2} + \sum_{n=2}^{\infty} \frac{f_n(a)\sin n\psi + g_n(a)\cos n\psi}{\lambda^2(1 - n^2)} \qquad (5.4.43)$$

Substituting for A_1, B_1, C_1, and D_1 from (5.4.35) through (5.4.38) into (5.4.41) and (5.4.42), we obtain

$$\frac{\partial a}{\partial t} + \omega_0'\frac{\partial a}{\partial x} = \epsilon\frac{f_1(a)}{2\omega_0} \qquad (5.4.44)$$

$$\frac{\partial \beta}{\partial t} + \omega_0'\frac{\partial \beta}{\partial x} = \epsilon\frac{g_1(a)}{2a\omega_0} \qquad (5.4.45)$$

where $\omega_0' = d\omega_0/dk_0$, the group velocity, and

$$\beta = \psi - k_0 x + \omega_0 t \qquad (5.4.46)$$

If $f_1 = 0$

$$a = h_1(x - \omega_0' t) \qquad (5.4.47)$$

$$\beta = \epsilon(x + \omega_0' t)\frac{g_1(a)}{4a\omega_0\omega_0'} + h_2(x - \omega_0' t) \qquad (5.4.48)$$

where h_1 and h_2 are determined from the initial or boundary conditions. Equations (5.4.44) and (5.4.45) can be solved easily if a and β are functions of either time or position only.

5.5. The Method of Averaging by Using Canonical Variables

Let us consider a conservative dynamic system governed by the following Lagrange equations

$$\frac{d}{dt}\left(\frac{\partial L}{\partial \dot{q}_i}\right) - \frac{\partial L}{\partial q_i} = 0, \qquad i = 1, 2, \ldots, N \qquad (5.5.1)$$

where $\mathbf{q} = \{q_1, q_2, \ldots, q_N\}$ is the generalized coordinate vector, t is the independent variable, overdots denote differentiation with respect to t, $L(\dot{\mathbf{q}}, \mathbf{q}, t) \equiv T - V$ is the Lagrangian, and T and V are the kinetic and potential energies. Let us define the generalized momentum vector $\mathbf{p} = \{p_1, p_2, \ldots, p_N\}$ by

$$p_i = \frac{\partial L}{\partial \dot{q}_i} \qquad (5.5.2)$$

and the Hamiltonian H by

$$H = \mathbf{p}^T \dot{\mathbf{q}} - L \tag{5.5.3}$$

where $\mathbf{p}^T$ denotes the transpose of $\mathbf{p}$ (if $\mathbf{p}$ is a column vector, $\mathbf{p}^T$ is a row vector). Considering H as a function of $\mathbf{p}$, $\mathbf{q}$, and t only, we can write

$$dH = \sum_{i=1}^{N} \frac{\partial H}{\partial q_i} dq_i + \sum_{i=1}^{N} \frac{\partial H}{\partial p_i} dp_i + \frac{\partial H}{\partial t} dt \tag{5.5.4}$$

Also, from (5.5.3)

$$dH = \sum_{i=1}^{N} \dot{q}_i \, dp_i + \sum_{i=1}^{N} p_i \, d\dot{q}_i - \sum_{i=1}^{N} \frac{\partial L}{\partial \dot{q}_i} d\dot{q}_i - \sum_{i=1}^{N} \frac{\partial L}{\partial q_i} dq_i - \frac{\partial L}{\partial t} dt \tag{5.5.5}$$

The second and third terms on the right-hand side of (5.5.5) cancel according to (5.5.2). Moreover, since $\dot{p}_i = \partial L / \partial q_i$ from (5.5.1) and (5.5.2), we can write (5.5.5) as

$$dH = \sum_{i=1}^{N} \dot{q}_i \, dp_i - \sum_{i=1}^{N} \dot{p}_i \, dq_i - \frac{\partial L}{\partial t} dt \tag{5.5.6}$$

Comparing (5.5.4) and (5.5.6), we obtain the following canonical equations of Hamilton

$$\dot{q}_i = \frac{\partial H}{\partial p_i} \tag{5.5.7}$$

$$\dot{p}_i = -\frac{\partial H}{\partial q_i} \tag{5.5.8}$$

$$\frac{\partial L}{\partial t} = -\frac{\partial H}{\partial t} \tag{5.5.9}$$

These equations replace Lagrange's equations.

Under a transformation from $\mathbf{q}$ and $\mathbf{p}$ to $\mathbf{Q}(\mathbf{q}, \mathbf{p}, t)$ and $\mathbf{P}(\mathbf{q}, \mathbf{p}, t)$, (5.5.7) and (5.5.8) are transformed into

$$\dot{Q}_i = f_i(\mathbf{P}, \mathbf{Q}, t) \tag{5.5.10}$$

$$\dot{P}_i = g_i(\mathbf{P}, \mathbf{Q}, t) \tag{5.5.11}$$

If there exists a function $K(\mathbf{P}, \mathbf{Q}, t)$ such that

$$f_i = \frac{\partial K}{\partial P_i} \quad \text{and} \quad g_i = -\frac{\partial K}{\partial Q_i} \tag{5.5.12}$$

then (5.5.10) and (5.5.11) assume the form

$$\dot{Q}_i = \frac{\partial K}{\partial P_i}, \qquad \dot{P}_i = -\frac{\partial K}{\partial Q_i} \tag{5.5.13}$$

and $\mathbf{Q}$ and $\mathbf{P}$ are called *canonical variables*, and the transformation from $\mathbf{q}$ and $\mathbf{p}$ to $\mathbf{Q}$ and $\mathbf{P}$ is said to be a *canonical transformation* with respect to the function K.

Canonical transformations can be generated using a so-called generating function $S(\mathbf{P},\mathbf{q},t)$ according to (e.g., Goldstein, 1965, Chapter 8; Meirovitch, 1970, Chapter 9)

$$p_i = \frac{\partial S}{\partial q_i}, \qquad Q_i = \frac{\partial S}{\partial P_i} \tag{5.5.14}$$

Once these equations are solved for $\mathbf{q} = \mathbf{q}(\mathbf{P},\mathbf{Q},t)$ and $\mathbf{p} = \mathbf{p}(\mathbf{P},\mathbf{Q},t)$, K is related to H by

$$K(\mathbf{P},\mathbf{Q},t) = H[\mathbf{p}(\mathbf{P},\mathbf{Q},t), \mathbf{q}(\mathbf{P},\mathbf{Q},t), t] + \frac{\partial S}{\partial t} \tag{5.5.15}$$

If a canonical transformation can be found such that $K \equiv 0$, then $\mathbf{P}$ is a constant vector according to the second relationship in (5.5.13), hence S is a function of $\mathbf{q}$ and t only. Since $p_i = \partial S/\partial q_i$ from the first relationship in (5.5.14), S must satisfy the following so-called *Hamilton–Jacobi equation*

$$H\left(\frac{\partial S}{\partial q_1}, \frac{\partial S}{\partial q_2}, \dots, \frac{\partial S}{\partial q_N}, q_1, q_2, \dots, q_N, t\right) + \frac{\partial S}{\partial t} = 0 \tag{5.5.16}$$

If S is a complete solution of (5.5.16), then (5.5.14) furnishes a complete set of integrals of the equations

$$\dot{q}_i = \frac{\partial H}{\partial p_i}, \qquad \dot{p}_i = -\frac{\partial H}{\partial q_i} \tag{5.5.17}$$

Complete solutions of (5.5.16) are not available for a general H. However, if $H = H_0 + \tilde{H}$ where $\tilde{H}$ is small compared to H_0 and a complete solution $S_0(P_1, \dots, P_N, q_1, \dots, q_N, t)$ is available for

$$H_0\left(\frac{\partial S_0}{\partial q_1}, \dots, \frac{\partial S_0}{\partial q_N}, q_1, \dots, q_N, t\right) + \frac{\partial S_0}{\partial t} = 0 \tag{5.5.18}$$

then an approximate solution of (5.5.17) can be obtained by using the method of averaging in conjunction with the method of variation of parameters. Thus we use a generating function

$$S = S_0(P_1, \dots, P_N, q_1, \dots, q_N, t) \tag{5.5.19}$$

where $\mathbf{P}$ is time-varying rather than being a constant. Hence $\mathbf{P}$ and $\mathbf{Q}$ are given by

$$\dot{P}_i = -\frac{\partial K}{\partial Q_i}, \qquad \dot{Q}_i = \frac{\partial K}{\partial P_i} \tag{5.5.20}$$

where

$$K = H_0 + \tilde{H} + \frac{\partial S_0}{\partial t} = \tilde{H} \tag{5.5.21}$$

as a result of (5.5.18). If the solution $\mathbf{q}_0(\mathbf{P}, \mathbf{Q}, t)$ and $\mathbf{p}_0(\mathbf{P}, \mathbf{Q}, t)$ of (5.5.17) with $H = H_0$ has a period T with respect to t, then an approximate solution to (5.5.17) is still given by $\mathbf{q}_0$ and $\mathbf{p}_0$ but $\mathbf{P}$ and $\mathbf{Q}$ are given by (5.5.20) with K replaced by its average over T; that is, by

$$\langle K \rangle = \frac{1}{T} \int_0^T K(\mathbf{P}, \mathbf{Q}, t)\, dt \tag{5.5.22}$$

In (5.5.22), $\mathbf{P}$ and $\mathbf{Q}$ are held constant.

We next illustrate this technique by its application to three specific examples.

5.5.1. THE DUFFING EQUATION
We consider again the equation

$$\ddot{q} + \omega_0^2 q + \epsilon q^3 = 0 \tag{5.5.23}$$

The Hamiltonian corresponding to this equation is

$$H = \tfrac{1}{2}(p^2 + \omega_0^2 q^2) + \tfrac{1}{4}\epsilon q^4 \tag{5.5.24}$$

The Hamilton–Jacobi equation corresponding to the case $\epsilon = 0$ is

$$\frac{1}{2}\left[\left(\frac{\partial S}{\partial q}\right)^2 + \omega_0^2 q^2\right] + \frac{\partial S}{\partial t} = 0 \tag{5.5.25}$$

This equation can be solved by separation of variables; that is, by letting

$$S = S_1(q) + \sigma(t) \tag{5.5.26}$$

Equation (5.5.25) separates into

$$\dot{\sigma} = -\alpha \quad \text{or} \quad \sigma = -\alpha t \tag{5.5.27}$$

and

$$\left(\frac{dS_1}{dq}\right)^2 + \omega_0^2 q^2 = 2\alpha \quad \text{or} \quad S_1 = \int \sqrt{2\alpha - \omega_0^2 q^2}\, dq \tag{5.5.28}$$

Hence

$$S = -\alpha t + \int \sqrt{2\alpha - \omega_0^2 q^2}\, dq \tag{5.5.29}$$

with α the new momentum. Consequently, the new coordinate β is given by

$$\beta = \frac{\partial S}{\partial \alpha} = -t + \int (2\alpha - \omega_0^2 q^2)^{-1/2}\, dq = -t + \frac{1}{\omega_0}\arcsin\frac{\omega_0 q}{\sqrt{2\alpha}} \tag{5.5.30}$$

so that

$$q = \frac{\sqrt{2\alpha}}{\omega_0} \sin \omega_0(t + \beta) \qquad (5.5.31)$$

We could have written this solution for (5.5.23) by inspection if $\epsilon = 0$. However, the canonical variables α and β were obtained naturally by solving the Hamilton–Jacobi equation (5.5.25).

Since $\tilde{H} = (1/4)\epsilon q^4 = (\epsilon \alpha^2/\omega_0^4) \sin^4 \omega_0(t + \beta)$, the variational equations (5.5.20) become

$$\dot{\alpha} = -\frac{\partial \tilde{H}}{\partial \beta}, \qquad \dot{\beta} = \frac{\partial \tilde{H}}{\partial \alpha} \qquad (5.5.32a)$$

Now

$$\tilde{H} = \frac{\epsilon \alpha^2}{\omega_0^4} [\tfrac{3}{8} - \tfrac{1}{2} \cos 2\omega_0(t + \beta) + \tfrac{1}{8} \cos 4\omega_0(t + \beta)]$$

so that

$$\langle \tilde{H} \rangle = \frac{3\epsilon \alpha^2}{8\omega_0^4} \qquad (5.5.32b)$$

Hence (5.5.32a) become

$$\alpha = \text{a constant} \quad \text{and} \quad \beta = \frac{3\epsilon \alpha}{4\omega_0^4} t + \beta_0 \qquad (5.5.33)$$

where β_0 is a constant. Therefore, to first approximation

$$q = \frac{\sqrt{2\alpha}}{\omega_0} \sin \left[\omega_0 \left(1 + \frac{3}{4} \frac{\epsilon \alpha}{\omega_0^4}\right) t + \omega_0 \beta_0 \right] \qquad (5.5.34)$$

in agreement with the expansions obtained in Sections 5.4.1 and 5.3 by using the Krylov–Bogoliubov–Mitropolski and Struble techniques if we identify $\sqrt{2\alpha}/\omega_0$ by a_0.

5.5.2. THE MATHIEU EQUATION
As a second example, we consider

$$\ddot{q} + (\omega^2 + \epsilon \cos 2t)q = 0 \qquad (5.5.35)$$

for positive ω. If we let

$$\dot{q} = p \qquad (5.5.36)$$

then

$$\dot{p} = -(\omega^2 + \epsilon \cos 2t)q \qquad (5.5.37)$$

from (5.5.35). These equations can be written in the form

$$\dot{p} = -\frac{\partial H}{\partial q}, \qquad \dot{q} = \frac{\partial H}{\partial p} \qquad (5.5.38)$$

where

$$H = \tfrac{1}{2}(p^2 + \omega^2 q^2) + \tfrac{1}{2}\epsilon q^2 \cos 2t \qquad (5.5.39)$$

As in Section 5.5.1, the solution of (5.5.38) for $\epsilon = 0$ is

$$q = \frac{\sqrt{2\alpha}}{\omega} \cos \omega(t + \beta) \qquad (5.5.40)$$

Hence $\tilde{H} = (1/2)\epsilon q^2 \cos 2t = (\epsilon\alpha/\omega^2) \cos 2t \cos^2 \omega(t + \beta)$ and the variational equations (5.5.20) become

$$\dot{\alpha} = -\frac{\partial \tilde{H}}{\partial \beta}, \qquad \dot{\beta} = \frac{\partial \tilde{H}}{\partial \alpha} \qquad (5.5.41)$$

Since

$$\tilde{H} = \frac{\epsilon\alpha}{2\omega^2} \{\cos 2t + \tfrac{1}{2} \cos 2[(\omega + 1)t + \omega\beta] + \tfrac{1}{2} \cos 2[(\omega - 1)t + \omega\beta]\}$$

$$\langle \tilde{H} \rangle = \begin{cases} 0 & \text{if } \omega \text{ is away from 1} \\ \dfrac{\epsilon\alpha}{4\omega^2} \cos 2[(\omega - 1)t + \omega\beta] & \text{if } \omega - 1 = O(\epsilon) \end{cases} \qquad (5.5.42)$$

In the former case α and β are constants to first approximation. In the second case we introduce the new canonical variables α^* and β^* using the generating function

$$S^* = \alpha^*[(\omega - 1)t + \omega\beta] \qquad (5.5.43)$$

so that

$$\alpha = \frac{\partial S^*}{\partial \beta} = \omega\alpha^* \qquad (5.5.44)$$

$$\beta^* = \frac{\partial S^*}{\partial \alpha^*} = (\omega - 1)t + \omega\beta \qquad (5.5.45)$$

Consequently, α^* and β^* are canonical variables with respect to the Hamiltonian

$$K = \langle \tilde{H} \rangle + \frac{\partial S^*}{\partial t} = \frac{\epsilon\alpha^*}{4\omega} \cos 2\beta^* + (\omega - 1)\alpha^* \qquad (5.5.46)$$

Hence

$$\dot{\alpha}^* = -\frac{\partial K}{\partial \beta^*} = \frac{\epsilon\alpha^*}{2\omega} \sin 2\beta^* \qquad (5.5.47)$$

$$\dot{\beta}^* = \frac{\partial K}{\partial \alpha^*} = \omega - 1 + \frac{\epsilon}{4\omega} \cos 2\beta^* \qquad (5.5.48)$$

Elimination of t from (5.5.47) and (5.5.48) gives

$$\frac{d\alpha^*}{\alpha^*} = -\frac{\dfrac{\epsilon}{4\omega} d(\cos 2\beta^*)}{\omega - 1 + \dfrac{\epsilon}{4\omega} \cos 2\beta^*}$$

Hence

$$\ln \alpha^* = -\ln\left[\omega - 1 + \frac{\epsilon}{4\omega} \cos 2\beta^*\right] + \text{a constant} \qquad (5.5.49)$$

Thus the motion is unstable (α^* unbounded) if

$$\frac{\epsilon}{4\omega} > |\omega - 1|$$

That is, to first approximation

$$\omega < 1 + \tfrac{1}{4}\epsilon \quad \text{or} \quad \omega > 1 - \tfrac{1}{4}\epsilon \qquad (5.5.50)$$

The curves

$$\omega = 1 \pm \tfrac{1}{4}\epsilon \quad \text{or} \quad \omega^2 = 1 \pm \tfrac{1}{2}\epsilon \qquad (5.5.51)$$

separate the stable from the unstable regions in the $\omega - \epsilon$ plane. These curves are in agreement with those obtained in Section 3.1.2 by using the Lindstedt–Poincaré method and in Section 3.1.3 by using Whittaker's method.

5.5.3.   A SWINGING SPRING

Following Kane and Kahn (1968), we consider the nonlinear oscillations of a spring swinging in a vertical plane as shown in Figure 5-1. This problem

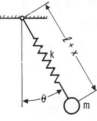

Figure 5-1

was introduced by Gorelik and Witt (1933) to illustrate internal resonances. The kinetic and potential energies of the mass m are

$$T = \tfrac{1}{2}m[\dot{x}^2 + (l + x)^2\dot{\theta}^2] \qquad (5.5.52)$$

$$V = \tfrac{1}{2}kx^2 + mg(l + x)(1 - \cos\theta) \qquad (5.5.53)$$

where x is the stretch in the spring beyond its equilibrium. Therefore

$$L = T - V = \tfrac{1}{2}m[\dot{x}^2 + (l + x)^2\dot{\theta}^2] - mg(l + x)(1 - \cos\theta) - \tfrac{1}{2}kx^2$$
$$(5.5.54)$$

Since

$$p_x = \frac{\partial L}{\partial \dot{x}} = m\dot{x}, \qquad p_\theta = \frac{\partial L}{\partial \dot{\theta}} = m(l + x)^2\dot{\theta} \qquad (5.5.55)$$

$$H = \dot{x}p_x + \dot{\theta}p_\theta - L$$

$$= \frac{1}{2}\left[\frac{p_x^2}{m} + \frac{p_\theta^2}{m(l + x)^2}\right] + mg(l + x)(1 - \cos\theta) + \tfrac{1}{2}kx^2 \quad (5.5.56)$$

For small x and θ and $x = O(\theta)$, H can be expanded into

$$H = \frac{1}{2}\left[\frac{p_x^2}{m} + \frac{p_\theta^2}{ml^2}\right] + \tfrac{1}{2}mgl\theta^2 + \tfrac{1}{2}kx^2 + \tfrac{1}{2}mgx\theta^2$$

$$- \frac{xp_\theta^2}{ml^3} - \tfrac{1}{24}mgl\theta^4 + \frac{3x^2p_\theta^2}{2ml^4} + O(\theta^5) \quad (5.5.57)$$

If we keep the quadratic terms in H, then a complete solution of the corresponding Hamilton–Jacobi equation can be obtained as follows. The Hamilton–Jacobi equation in this case is

$$\frac{1}{2}\left[\frac{1}{m}\left(\frac{\partial S}{\partial x}\right)^2 + kx^2\right] + \frac{1}{2}\left[\frac{1}{ml^2}\left(\frac{\partial S}{\partial \theta}\right)^2 + mgl\theta^2\right] + \frac{\partial S}{\partial t} = 0 \quad (5.5.58)$$

where $S = S(x, \theta, t)$. To solve this equation we let

$$S = -(\alpha_1 + \alpha_2)t + W_1(x) + W_2(\theta) \qquad (5.5.59)$$

hence

$$\frac{1}{m}\left(\frac{dW_1}{dx}\right)^2 + kx^2 = 2\alpha_1 \qquad (5.5.60)$$

$$\frac{1}{ml^2}\left(\frac{dW_2}{d\theta}\right)^2 + mgl\theta^2 = 2\alpha_2 \qquad (5.5.61)$$

Therefore

$$p_x = \frac{\partial S}{\partial x} = \sqrt{m(2\alpha_1 - kx^2)} \qquad (5.5.62)$$

$$p_\theta = \frac{\partial S}{\partial \theta} = \sqrt{ml^2(2\alpha_2 - mgl\theta^2)} \qquad (5.5.63)$$

$$S = -(\alpha_1 + \alpha_2)t + \int\sqrt{m(2\alpha_1 - kx^2)}\,dx + \int\sqrt{ml^2(2\alpha_2 - mgl\theta^2)}\,d\theta$$
$$(5.5.64)$$

Consequently

$$\beta_1 = \frac{\partial S}{\partial \alpha_1} = -t + \int \frac{m \, dx}{\sqrt{m(2\alpha_1 - kx^2)}} = -t + \sqrt{\frac{m}{k}} \arcsin x \sqrt{\frac{k}{2\alpha_1}} \qquad (5.5.65)$$

$$\beta_2 = \frac{\partial S}{\partial \alpha_2} = -t + \int \frac{ml^2 \, d\theta}{\sqrt{ml^2(2\alpha_2 - mgl\theta^2)}} = -t + \sqrt{\frac{l}{g}} \arcsin \theta \sqrt{\frac{mgl}{2\alpha_2}} \quad (5.5.66)$$

Hence

$$x = \sqrt{\frac{2\alpha_1}{k}} \sin B_1 \qquad (5.5.67)$$

$$\theta = \sqrt{\frac{2\alpha_2}{mgl}} \sin B_2 \qquad (5.5.68)$$

$$p_x = \sqrt{2m\alpha_1} \cos B_1 \qquad (5.5.69)$$

$$p_\theta = l\sqrt{2m\alpha_2} \cos B_2 \qquad (5.5.70)$$

where

$$B_i = \omega_i(t + \beta_i), \qquad \omega_1 = \sqrt{\frac{k}{m}}, \qquad \omega_2 = \sqrt{\frac{g}{l}}$$

To first approximation the variational equations correspond to

$$\tilde{H} = \tfrac{1}{2} mgx\theta^2 - \frac{xp_\theta^2}{ml^3}$$

$$= -\frac{\alpha_2\sqrt{\alpha_1}}{l\sqrt{2k}} \{\sin B_1 + \tfrac{3}{2} \sin (B_1 + 2B_2)$$

$$+ \tfrac{3}{2} \sin [(\omega_1 - 2\omega_2)t + \omega_1\beta_1 - 2\omega_2\beta_2]\} \qquad (5.5.71)$$

Thus $\tilde{H}$ is fast varying unless $\omega_1 - 2\omega_2 = \epsilon$, where ϵ is a small quantity. In the latter case the slowly varying part of $\tilde{H}$ is

$$\langle \tilde{H} \rangle = -\frac{3}{2l\sqrt{2k}} \alpha_2\sqrt{\alpha_1} \sin (\epsilon t + \omega_1\beta_1 - 2\omega_2\beta_2) \qquad (5.5.72)$$

To eliminate the explicit dependence of $\langle \tilde{H} \rangle$ on t, we make a further canonical transformation from α_1 and β_1 to $\alpha_1{}^*$ and $\beta_1{}^*$ according to

$$S^*(\alpha_1{}^*, \beta_1, t) = \frac{\epsilon\alpha_1{}^*}{2\omega_2} t + \frac{\omega_1}{2\omega_2} \alpha_1{}^*\beta_1 \qquad (5.5.73)$$

Thus

$$\alpha_1 = \frac{\partial S^*}{\partial \beta_1} = \frac{\omega_1}{2\omega_2}\alpha_1^*$$

$$\beta_1^* = \frac{\partial S^*}{\partial \alpha_1^*} = \frac{\epsilon}{2\omega_2}t + \frac{\omega_1}{2\omega_2}\beta_1$$

(5.5.74)

$$K = \langle \tilde{H} \rangle + \frac{\partial S^*}{\partial t} = \frac{\epsilon\alpha_1^*}{2\omega_2} - \frac{3}{4l}\sqrt{\frac{\omega_1}{\omega_2 k}}\alpha_2\sqrt{\alpha_1^*}\sin 2\omega_2(\beta_1^* - \beta_2) \quad (5.5.75)$$

Therefore $K = $ a constant because $\partial K/\partial t = 0$. The variational equations become

$$\dot{\alpha}_1^* = -\frac{\partial K}{\partial \beta_1^*} = 2\omega_2 C\alpha_2\sqrt{\alpha_1^*}\cos \gamma \qquad (5.5.76)$$

$$\dot{\alpha}_2 = -\frac{\partial K}{\partial \beta_2} = -2\omega_2 C\alpha_2\sqrt{\alpha_1^*}\cos \gamma \qquad (5.5.77)$$

$$\dot{\beta}_1^* = \frac{\partial K}{\partial \alpha_1^*} = \frac{\epsilon}{2\omega_2} - \tfrac{1}{2}C\alpha_2\alpha_1^{*-1/2}\sin \gamma \qquad (5.5.78)$$

$$\dot{\beta}_2 = \frac{\partial K}{\partial \alpha_2} = -C\sqrt{\alpha_1^*}\sin \gamma \qquad (5.5.79)$$

where

$$C = \frac{3}{4l}\sqrt{\frac{\omega_1}{\omega_2 k}}, \qquad \gamma = 2\omega_2(\beta_1^* - \beta_2) \qquad (5.5.80)$$

Equations similar to (5.5.76) through (5.5.80) were obtained by Mettler (1959) and Sethna (1965) using the method of averaging.

Adding (5.5.76) and (5.5.77) and integrating, we obtain

$$\alpha_1^* + \alpha_2 = E = \text{a constant} \qquad (5.5.81)$$

Hence the motion is completely bounded. Elimination of γ from (5.5.75) and (5.5.77) gives

$$\left(\frac{\dot{\alpha}_2}{2\omega_2}\right)^2 = C^2\alpha_2^2(E - \alpha_2) - \left[\frac{\epsilon(E - \alpha_2)}{2\omega_2} - K\right]^2$$

$$= C^2[F^2(\alpha_2) - G^2(\alpha_2)] \qquad (5.5.82)$$

where

$$F = \pm\alpha_2\sqrt{E - \alpha_2}, \qquad G = \frac{1}{C}\left[\frac{\epsilon(E - \alpha_2)}{2\omega_2} - K\right] \qquad (5.5.83)$$

The functions $F(\alpha_2)$ and $G(\alpha_2)$ are shown schematically in Figure 5-2. For

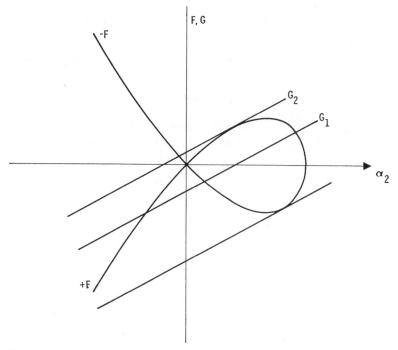

Figure 5-2

real motion, F^2 must be greater than or equal to G^2. The points where G meets F correspond to the vanishing of both $\dot{\alpha}_2$ and $\dot{\alpha}_1{}^*$. A curve such as G_1 which meets both branches of F or one branch of F at two different points corresponds to a periodic motion for the amplitudes and the phases, hence it corresponds to an aperiodic motion. The solution for the amplitudes and the phases can be written in terms of Jacobi elliptic functions. However, the points where G_2 touches the branches of F represent periodic motions where the nonlinearity adjusts the frequencies ω_1 and ω_2 to produce perfect resonance.

5.6. Von Zeipel's Procedure

To determine a first approximation to Hamiltonian systems, the method of variation of parameters in conjunction with the method of averaging was used in Section 5.5. To determine higher approximations, von Zeipel (1916) devised a technique which is described and applied in this section to the first two examples discussed in the previous section. The essence of this technique is to expand the generating function S in powers of a small parameter ϵ as

$\sum_{n=0}^{\infty} \epsilon^n S_n$, and then determine S_n recursively by solving a chain of partial differential equations.

Let the system under consideration be described by the Hamiltonian

$$H(\mathbf{p}, \mathbf{q}, t) = \sum_{n=0}^{\infty} \epsilon^n H_n(\mathbf{p}, \mathbf{q}, t), \qquad \epsilon \ll 1 \tag{5.6.1}$$

with $\mathbf{q}$ the generalized coordinate vector and $\mathbf{p}$ the conjugate momenta vector. Let $S_0 = S_0(\mathbf{P}, \mathbf{q}, t)$ be a complete solution of the Hamilton–Jacobi equation

$$H_0\left[\frac{\partial S_0}{\partial \mathbf{q}}, \mathbf{q}, t\right] + \frac{\partial S_0}{\partial t} = 0 \tag{5.6.2}$$

and let $\mathbf{p} = \mathbf{p}(\mathbf{P}, \mathbf{Q}, t)$ and $\mathbf{q} = \mathbf{q}(\mathbf{P}, \mathbf{Q}, t)$ be the solutions of

$$p_i = \frac{\partial S_0}{\partial q_i}, \qquad Q_i = \frac{\partial S_0}{\partial P_i} \tag{5.6.3}$$

If we assume $\mathbf{P}$ and $\mathbf{Q}$ to be time varying rather than being constants and use $S = S_0(\mathbf{P}, \mathbf{q}, t)$ as a generating function to transform from the canonical system $\mathbf{p}$ and $\mathbf{q}$ to the canonical system $\mathbf{P}$ and $\mathbf{Q}$, we transform the Hamiltonian from H into

$$\begin{aligned}
\tilde{H}(\mathbf{P}, \mathbf{Q}, t) &= \sum_{n=0}^{\infty} \epsilon^n H_n[\mathbf{p}(\mathbf{P}, \mathbf{Q}, t), \mathbf{q}(\mathbf{P}, \mathbf{Q}, t), t] + \frac{\partial S_0}{\partial t} \\
&= \sum_{n=1}^{\infty} \epsilon^n H_n[\mathbf{p}(\mathbf{P}, \mathbf{Q}, t), \mathbf{q}(\mathbf{P}, \mathbf{Q}, t), t] \\
&= \sum_{n=1}^{\infty} \epsilon^n \tilde{H}_n(\mathbf{P}, \mathbf{Q}, t)
\end{aligned} \tag{5.6.4}$$

Hence $\mathbf{P}$ and $\mathbf{Q}$ are governed by the variational equations

$$\dot{\mathbf{P}} = -\sum_{n=1}^{\infty} \epsilon^n \frac{\partial \tilde{H}_n}{\partial \mathbf{Q}}(\mathbf{P}, \mathbf{Q}, t) \tag{5.6.5}$$

$$\dot{\mathbf{Q}} = \sum_{n=1}^{\infty} \epsilon^n \frac{\partial \tilde{H}_n}{\partial \mathbf{P}}(\mathbf{P}, \mathbf{Q}, t) \tag{5.6.6}$$

To determine an approximate solution to (5.6.5) through (5.6.6) to any order, we introduce a near-identity transformation from the canonical system $\mathbf{P}$ and $\mathbf{Q}$ to the canonical system $\mathbf{P}^*$ and $\mathbf{Q}^*$ using the generating function

$$S = \sum_{i=1}^{N} P_i^* Q_i + \sum_{n=1}^{\infty} \epsilon^n S_n(\mathbf{P}^*, \mathbf{Q}, t) \tag{5.6.7}$$

so that

$$P_i = P_i{}^* + \sum_{n=1}^{\infty} \epsilon^n \frac{\partial S_n}{\partial Q_i}(\mathbf{P}^*, \mathbf{Q}, t) \tag{5.6.8}$$

and $\tilde{H}$ is transformed into

$$K \equiv \sum_{n=1}^{\infty} \epsilon^n K_n(\mathbf{P}^*, \mathbf{Q}, t) = \sum_{n=1}^{\infty} \epsilon^n \tilde{H}_n \left[\left(\mathbf{P}^* + \sum_{m=1}^{\infty} \epsilon^m \frac{\partial S_m}{\partial \mathbf{Q}} \right), \mathbf{Q}, t \right] + \sum_{n=1}^{\infty} \epsilon^n \frac{\partial S_n}{\partial t} \tag{5.6.9}$$

To determine K_n we expand the right-hand side of (5.6.9) for small ϵ and then equate the coefficients of equal powers of ϵ on both sides to obtain

$$K_1 = \tilde{H}_1(\mathbf{P}^*, \mathbf{Q}, t) + \frac{\partial S_1}{\partial t} \tag{5.6.10}$$

$$K_2 = \tilde{H}_2(\mathbf{P}^*, \mathbf{Q}, t) + \sum_{i=1}^{N} \frac{\partial S_1}{\partial Q_i} \frac{\partial \tilde{H}_1}{\partial P_i}(\mathbf{P}^*, \mathbf{Q}, t) + \frac{\partial S_2}{\partial t} \tag{5.6.11}$$

$$K_n = F_n + \frac{\partial S_n}{\partial t} \tag{5.6.12}$$

where $F_n = F_n(\mathbf{P}^*, \mathbf{Q}, t)$ is a known function of $\tilde{H}_1, \tilde{H}_2, \ldots, \tilde{H}_n$ and S_1, $S_2, \ldots, S_{n-1}$. The functions S_n are still undetermined and they can be chosen in any manner desired. Since F_n contains, in general, a short-period term $F_n{}^s$ and a long-period term $F_n{}^l$, we choose

$$K_n = F_n{}^l \quad \text{and} \quad \frac{\partial S_n}{\partial t} = -F_n{}^s \tag{5.6.13}$$

Thus K_n contains long-period terms only, while S_n contains short-period terms only. The functions S_n can be obtained by solving successively the chain of partial differential equations in (5.6.13).

The basic idea underlying this technique is the same as that underlying the generalized method of averaging of Section 5.2.3. Stern (1971c) showed that, for Hamiltonian systems, Kruskal's technique is equivalent to von Zeipel's technique. In both techniques we introduce near-identity transformations from the old dependent variables, which contain long- as well as short-period terms, into new dependent variables which contain long-period terms. The basic difference between the two techniques is that the transformation in the von Zeipel method must be canonical, while the transformation in the generalized method of averaging need not be canonical and the system need not be described by a Hamiltonian. Morrison (1966b) showed that up to second order the von Zeipel procedure is a particular case of the generalized method of averaging. Giacaglia (1964) carried out the expansion to any order, while Barrar (1970) investigated the convergence of the von Zeipel procedure. Musen (1965) showed that the equations governing the expansion take a concise form if written in terms of Faa de Bruno operators (1857). We next determine second-order expansions for the first two problems discussed in the previous section.

5.6.1. THE DUFFING EQUATION

As a first example, we consider the Duffing equation (5.5.23) which corresponds to the Hamiltonain (5.5.24). The solution of the problem corresponding to H_0 is given by (5.5.31) as found in Section 5.5.1. Hence

$$\tilde{H} = \frac{\epsilon \alpha^2}{\omega_0^4} \sin^4 \omega_0(t + \beta) \qquad (5.6.14)$$

$$\dot{\alpha} = -\frac{\partial \tilde{H}}{\partial \beta}, \qquad \dot{\beta} = \frac{\partial \tilde{H}}{\partial \alpha} \qquad (5.6.15)$$

To determine an approximate solution to (5.6.15), we introduce a near-identity transformation from α and β to α^* and β^* using the generating function

$$S = \alpha^* \beta + \epsilon S_1(\alpha^*, \beta, t) + \epsilon^2 S_2(\alpha^*, \beta, t) + \cdots \qquad (5.6.16)$$

Hence

$$\alpha = \alpha^* + \epsilon \frac{\partial S_1}{\partial \beta} + \epsilon^2 \frac{\partial S_2}{\partial \beta} + \cdots \qquad (5.6.17)$$

and the Hamiltonian $\tilde{H}$ is transformed into

$$K = \sum_{n=1}^{\infty} \epsilon^n K_n(\alpha^*, \beta, t)$$

$$= \frac{\epsilon}{\omega_0^4} \sin^4 \omega_0(t + \beta) \left[\alpha^* + \epsilon \frac{\partial S_1}{\partial \beta} + \epsilon^2 \frac{\partial S_2}{\partial \beta} + \cdots \right]^2$$

$$+ \epsilon \frac{\partial S_1}{\partial t} + \epsilon^2 \frac{\partial S_2}{\partial t} + \cdots \qquad (5.6.18)$$

Equating coefficients of equal powers of ϵ on both sides, we have

$$K_1 = \frac{\alpha^{*2}}{\omega_0^4} [\tfrac{3}{8} - \tfrac{1}{2} \cos 2\omega_0(t + \beta) + \tfrac{1}{8} \cos 4\omega_0(t + \beta)] + \frac{\partial S_1}{\partial t} \qquad (5.6.19)$$

$$K_2 = \frac{2\alpha^*}{\omega_0^4} \frac{\partial S_1}{\partial \beta} \sin^4 \omega_0(t + \beta) + \frac{\partial S_2}{\partial t} \qquad (5.6.20)$$

Equating K_1 to the long-period term on the right-hand side of (5.6.19), we obtain

$$K_1 = \frac{3\alpha^{*2}}{8\omega_0^4} \qquad (5.6.21)$$

Hence (5.6.19) becomes

$$\frac{\partial S_1}{\partial t} + \frac{\alpha^{*2}}{\omega_0^4} [-\tfrac{1}{2}\cos 2\omega_0(t + \beta) + \tfrac{1}{8}\cos 4\omega_0(t + \beta)] = 0 \qquad (5.6.22)$$

The solution of (5.6.22) is

$$S_1 = \frac{\alpha^{*2}}{4\omega_0^5} [\sin 2\omega_0(t + \beta) - \tfrac{1}{8}\sin 4\omega_0(t + \beta)] \qquad (5.6.23)$$

With this value of S_1, (5.6.20) becomes

$$K_2 = \frac{\alpha^{*3}}{\omega_0^8} [\cos 2\omega_0(t + \beta) - \tfrac{1}{4}\cos 4\omega_0(t + \beta)] \sin^4 \omega_0(t + \beta) + \frac{\partial S_2}{\partial t} \qquad (5.6.24)$$

Equating K_2 to the long-period term on the right-hand side of this equation, we have

$$K_2 = -\frac{17\alpha^{*3}}{64\omega_0^8} \qquad (5.6.25)$$

Hence, to second order

$$K = \epsilon \frac{3\alpha^{*2}}{8\omega_0^4} - \epsilon^2 \frac{17\alpha^{*3}}{64\omega_0^8} \qquad (5.6.26)$$

and

$$\dot{\alpha}^* = -\frac{\partial K}{\partial \beta^*} = 0 \quad \text{or} \quad \alpha^* = \text{a constant} \qquad (5.6.27)$$

$$\dot{\beta}^* = \frac{\partial K}{\partial \alpha^*} = \tfrac{3}{4}\epsilon \frac{\alpha^*}{\omega_0^4} - \tfrac{51}{64}\epsilon^2 \frac{\alpha^{*2}}{\omega_0^8}$$

or

$$\beta^* = \left(\tfrac{3}{4}\epsilon \frac{\alpha^*}{\omega_0^4} - \tfrac{51}{64}\epsilon^2 \frac{\alpha^{*2}}{\omega_0^8}\right) t + \frac{\beta_0}{\omega_0} \qquad (5.6.28)$$

where β_0 is a constant.

Having determined S_1, we obtain

$$\alpha = \alpha^* + \epsilon \frac{\partial S_1}{\partial \beta} + \cdots$$

$$= \alpha^* + \epsilon \frac{\alpha^{*2}}{2\omega_0^4} [\cos 2\omega_0(t + \beta) - \tfrac{1}{4}\cos 4\omega_0(t + \beta)] + O(\epsilon^2) \quad (5.6.29)$$

$$\beta^* = \beta + \epsilon \frac{\partial S_1}{\partial \alpha^*} + \cdots$$

$$= \beta + \epsilon \frac{\alpha^*}{2\omega_0^5} [\sin 2\omega_0(t + \beta) - \tfrac{1}{8}\sin 4\omega_0(t + \beta)] + O(\epsilon^2) \quad (5.6.30)$$

Solving (5.6.29) and (5.6.30) for α and β in terms of α^* and β^* gives

$$\alpha = \alpha^* + \frac{\epsilon \alpha^{*2}}{2\omega_0^4} [\cos 2\omega_0(t + \beta^*) - \tfrac{1}{4} \cos 4\omega_0(t + \beta^*)] + O(\epsilon^2) \quad (5.6.31)$$

$$\beta = \beta^* - \frac{\epsilon \alpha^*}{2\omega_0^5} [\sin 2\omega_0(t + \beta^*) - \tfrac{1}{8} \sin 4\omega_0(t + \beta^*)] + O(\epsilon^2) \quad (5.6.32)$$

To compare this expansion with those obtained by other methods, we substitute for α and β from (5.6.31) and (5.6.32) into (5.5.31), expand for small ϵ, keeping α^* and β^* fixed, and obtain

$$q = \frac{\sqrt{2\alpha^*}}{\omega_0}\left(1 - \tfrac{3}{8}\epsilon\, \frac{\alpha^*}{\omega_0^4}\right) \sin(\omega t + \beta_0) - \epsilon\, \frac{\sqrt{2\alpha^*}\, \alpha^*}{16\omega_0^5} \sin 3(\omega t + \beta_0) + O(\epsilon^2)$$

$$(5.6.33)$$

where

$$\omega = \omega_0(1 + \dot\beta^*) \quad (5.6.34)$$

If we let

$$a = \frac{\sqrt{2\alpha^*}}{\omega_0}\left(1 - \tfrac{3}{8}\epsilon\, \frac{\alpha^*}{\omega_0^4}\right)$$

then

$$\frac{\sqrt{2\alpha^*}}{\omega_0} = a + \tfrac{3}{16}\epsilon\, \frac{a^3}{\omega_0^2} + O(\epsilon^2) \quad (5.6.35)$$

Hence

$$q = a \sin(\omega t + \beta_0) - \tfrac{1}{32}\epsilon\, \frac{a^3}{\omega_0^2} \sin 3(\omega t + \beta_0) + O(\epsilon^2) \quad (5.6.36)$$

where

$$\omega = \omega_0\left[1 + \tfrac{3}{8}\epsilon\, \frac{a^2}{\omega_0^2} - \tfrac{15}{256}\epsilon^2\, \frac{a^4}{\omega_0^4}\right] + O(\epsilon^3) \quad (5.6.37)$$

This expansion is in agreement with those obtained in Section 5.3 by using Struble's method and in Section 5.4.1 by using the Krylov–Bogoliubov–Mitropolski method.

5.6.2. THE MATHIEU EQUATION

Next we determine a second-order expansion for Mathieu's equation (5.5.35) which corresponds to the Hamiltonian (5.5.39). The solution corresponding to H_0 can be written as (see Section 5.5.1)

$$q = \frac{\sqrt{2\alpha}}{\omega} \cos \omega(t + \beta), \qquad p = -\sqrt{2\alpha} \sin \omega(t + \beta) \quad (5.6.38a)$$

Hence α and β are canonical variables with respect to

$$\tilde{H} = \frac{\epsilon\alpha}{\omega^2}\cos^2\omega(t + \beta)\cos 2t \qquad (5.6.38b)$$

Transforming from α and β to α^* and β^* using the generating function (5.6.16), we obtain

$$K = \epsilon K_1 + \epsilon^2 K_2 + \cdots$$

$$= \frac{\epsilon}{\omega^2}\left(\alpha^* + \epsilon\frac{\partial S_1}{\partial\beta} + \cdots\right)\cos^2\omega(t + \beta)\cos 2t + \epsilon\frac{\partial S_1}{\partial t} + \epsilon^2\frac{\partial S_2}{\partial t} + \cdots$$

$$(5.6.39)$$

since α is given by (5.6.17). Equating coefficients of equal powers of ϵ in (5.6.39) yields

$$K_1 = \frac{\partial S_1}{\partial t} + \frac{\alpha^*}{2\omega^2}\{\cos 2t + \tfrac{1}{2}\cos 2[(\omega + 1)t + \omega\beta]$$

$$+ \tfrac{1}{2}\cos 2[(\omega - 1)t + \omega\beta]\} \qquad (5.6.40)$$

$$K_2 = \frac{\partial S_2}{\partial t} + \frac{1}{\omega^2}\frac{\partial S_1}{\partial\beta}\cos^2\omega(t + \beta)\cos 2t \qquad (5.6.41)$$

Two cases arise depending on whether ω is near 1 (resonance) or ω is away from 1. We consider both cases next starting with the latter case.

The Case of ω Away from 1. In this case all the terms on the right-hand side of (5.6.40) are fast varying. Hence $K_1 = 0$, and

$$S_1 = -\frac{\alpha^*}{4\omega^2}\left\{\sin 2t + \frac{\sin 2[(\omega + 1)t + \omega\beta]}{2(\omega + 1)} + \frac{\sin 2[(\omega - 1)t + \omega\beta]}{2(\omega - 1)}\right\}$$

$$(5.6.42)$$

Substituting for S_1 into (5.6.41), we obtain

$$K_2 = \frac{\partial S_2}{\partial t} - \frac{\alpha^*}{4\omega^3}\cos^2\omega(t + \beta)$$

$$\times\cos 2t\left\{\frac{\cos 2[(\omega + 1)t + \omega\beta]}{\omega + 1} + \frac{\cos 2[(\omega - 1)t + \omega\beta]}{\omega - 1}\right\}$$

$$= \frac{\partial S_2}{\partial t} - \frac{\alpha^*}{16\omega^3}\left\{\frac{\omega}{\omega^2 - 1} + \frac{\cos 2[(\omega - 2)t + \omega\beta]}{\omega - 1} + \frac{\cos 2[(\omega + 2)t + \omega\beta]}{\omega + 1}\right.$$

$$+ \frac{\cos 4[(\omega + 1)t + \omega\beta]}{2(\omega + 1)} + \frac{\cos 4[(\omega - 1)t + \omega\beta]}{2(\omega - 1)}$$

$$\left.+ \omega\frac{\cos 4\omega(t + \beta) + \cos 4t + 2\cos 2\omega(t + \beta)}{\omega^2 - 1}\right\}$$

$$(5.6.43)$$

If ω is also away from 2

$$K_2 = -\frac{1}{16}\frac{\alpha^*}{\omega^2(\omega^2 - 1)} \qquad (5.6.44)$$

because the rest of the terms on the right-hand side of (5.6.43) are fast varying and must be set equal to $-\partial S_2/\partial t$. In this case

$$K = -\frac{1}{16}\epsilon^2\frac{\alpha^*}{\omega^2(\omega^2 - 1)} + O(\epsilon^3) \qquad (5.6.45)$$

Hence

$$\dot\alpha^* = -\frac{\partial K}{\partial\beta^*} = 0 \quad \text{or} \quad \alpha^* = \text{a constant} \qquad (5.6.46)$$

$$\dot\beta^* = \frac{\partial K}{\partial\alpha^*} = -\frac{1}{16}\frac{\epsilon^2}{\omega^2(\omega^2 - 1)} + O(\epsilon^3) \qquad (5.6.47)$$

Consequently

$$\beta^* = -\frac{1}{16}\frac{\epsilon^2 t}{\omega^2(\omega^2 - 1)} + \beta_0 \qquad (5.6.48)$$

and q is given by (6.2.95) if we make the following replacement of variables

$$u \to q, \quad \omega_0 \to \omega, \quad a \to \sqrt{2\alpha^*}/\omega$$
$$\phi_0 \to \omega\beta_0 \quad \text{and} \quad \omega \to \omega[1 - \epsilon^2/16\omega^2(\omega^2 - 1)] \qquad (5.6.49)$$

Thus q is bounded and the motion in this case is stable.

If ω is near 2, however, $\cos 2[(\omega - 2)t + \omega\beta]$ is slowly varying, and it has to be included in K_2; otherwise S_2 will contain secular terms or a small divisor depending on whether ω is exactly equal to 2 or not. Equating K_2 to the long-period terms on the right-hand side of (5.6.43), we obtain

$$K_2 = -\frac{\alpha^*}{16\omega^3}\left\{\frac{\omega}{\omega^2 - 1} + \frac{\cos 2[(\omega - 2)t + \omega\beta]}{\omega - 1}\right\} \qquad (5.6.50)$$

To an error of $O(\epsilon)$, β can be replaced by β^* in (5.6.50). To analyze the motion in this case, we remove the explicit dependence of K on t by transforming from α^* and β^* to α' and β' using the generating function

$$S' = \alpha'[(\omega - 2)t + \omega\beta^*] \qquad (5.6.51)$$

Thus

$$\alpha^* = \frac{\partial S'}{\partial\beta^*} = \omega\alpha' \qquad (5.6.52)$$

$$\beta' = \frac{\partial S'}{\partial\alpha'} = (\omega - 2)t + \omega\beta^* \qquad (5.6.53)$$

and

$$K' = K + (\omega - 2)\alpha' = (\omega - 2)\alpha' - \epsilon^2 \frac{\alpha'}{16\omega^2}\left[\frac{\cos 2\beta'}{\omega - 1} + \frac{\omega}{\omega^2 - 1}\right] \quad (5.6.54)$$

Hence

$$\dot{\alpha}' = -\frac{\partial K'}{\partial \beta'} = -\epsilon^2 \frac{\alpha'}{8\omega^2(\omega - 1)} \sin 2\beta' \quad (5.6.55)$$

$$\dot{\beta}' = \frac{\partial K'}{\partial \alpha'} = \omega - 2 - \frac{\epsilon^2}{16\omega^2}\left[\frac{\cos 2\beta'}{\omega - 1} + \frac{\omega}{\omega^2 - 1}\right] \quad (5.6.56)$$

The solution of (5.6.55) and (5.6.56) can be obtained as in Section 5.5.2 to be

$$\ln \alpha' = -\ln\left[\omega - 2 - \frac{\epsilon^2}{16\omega(\omega^2 - 1)} - \frac{\epsilon^2}{16\omega^2(\omega - 1)} \cos 2\beta'\right]$$

$$+ \text{ a constant} \quad (5.6.57)$$

Therefore, for instability

$$\frac{\epsilon^2}{16\omega^2(\omega - 1)} > \left| \omega - 2 - \frac{\epsilon^2}{16\omega(\omega^2 - 1)} \right|$$

or

$$\omega < 2 + \frac{5\epsilon^2}{192} + O(\epsilon^3), \ \omega > 2 - \frac{\epsilon^2}{192} + O(\epsilon^3) \quad (5.6.58)$$

Consequently, the transition curves that separate stable from unstable motions in the $\omega^2 - \epsilon$ plane and emanating from $\omega = 2$ are given by

$$\omega^2 = 4 + \frac{5\epsilon^2}{48} + O(\epsilon^3) \quad \text{and} \quad \omega^2 = 4 - \frac{\epsilon^2}{48} + O(\epsilon^3) \quad (5.6.59)$$

in agreement with those obtained in Section 3.1.2.

The Case of ω Near 1. In this case $\cos 2[(\omega - 1)t + \omega\beta]$ is slowly varying, and therefore it should remain in K_1; otherwise S_1 is singular at $\omega = 1$ as evident from (5.6.42). Equating K_1 to the long-period terms in (5.6.40), we have

$$K_1 = \frac{\alpha^*}{4\omega^2} \cos 2[(\omega - 1)t + \omega\beta] \quad (5.6.60)$$

Hence

$$\frac{\partial S_1}{\partial t} = -\frac{\alpha^*}{2\omega^2}\{\cos 2t + \tfrac{1}{2}\cos 2[(\omega + 1)t + \omega\beta]\} \quad (5.6.61)$$

The solution of (5.6.61) is

$$S_1 = -\frac{\alpha^*}{4\omega^2}\left\{\sin 2t + \frac{\sin 2[(\omega + 1)t + \omega\beta]}{2(\omega + 1)}\right\} \qquad (5.6.62)$$

Substituting for S_1 into (5.6.41) gives

$$K_2 = \frac{\partial S_2}{\partial t} - \frac{\alpha^*}{4\omega^3(\omega + 1)}\cos 2[(\omega + 1)t + \omega\beta]\cos^2 \omega(t + \beta)\cos 2t \qquad (5.6.63)$$

Equating K_2 to the long-period terms in (5.6.63), we have

$$K_2 = -\frac{\alpha^*}{32\omega^3(\omega + 1)} \qquad (5.6.64)$$

Therefore, to second order

$$K = \frac{\epsilon\alpha^*}{4\omega^2}\cos 2[(\omega - 1)t + \omega\beta] - \frac{\epsilon^2\alpha^*}{32\omega^3(\omega + 1)} + O(\epsilon^3) \qquad (5.6.65)$$

Moreover

$$\alpha = \alpha^* + \epsilon\frac{\partial S_1}{\partial \beta} + \cdots$$

$$= \alpha^* - \frac{\epsilon\alpha^*}{4\omega(\omega + 1)}\cos 2[(\omega + 1)t + \omega\beta] + O(\epsilon^2) \qquad (5.6.66)$$

$$\beta^* = \beta + \epsilon\frac{\partial S_1}{\partial \alpha^*} + \cdots$$

$$= \beta - \frac{\epsilon}{4\omega^2}\left\{\sin 2t + \frac{\sin 2[(\omega + 1)t + \omega\beta]}{2(\omega + 1)}\right\} + O(\epsilon^2) \qquad (5.6.67)$$

Solving (5.6.66) and (5.6.67) for α and β in terms of α^* and β^* gives

$$\alpha = \alpha^* - \frac{\epsilon\alpha^*}{4\omega(\omega + 1)}\cos 2[(\omega + 1)t + \omega\beta^*] + O(\epsilon^2) \qquad (5.6.68)$$

$$\beta = \beta^* + \frac{\epsilon}{4\omega^2}\left\{\sin 2t + \frac{\sin 2[(\omega + 1)t + \omega\beta^*]}{2(\omega + 1)}\right\} + O(\epsilon^2) \qquad (5.6.69)$$

Substituting for β into (5.6.65) gives

$$K = \frac{\epsilon\alpha^*}{4\omega^2}\cos 2[(\omega - 1)t + \omega\beta^*] - \frac{\epsilon^2\alpha^*}{32\omega^3(\omega + 1)}$$

$$+ \frac{\epsilon^2\alpha^*}{2\omega}\frac{\partial S_1}{\partial \alpha^*}\sin 2[(\omega - 1)t + \omega\beta^*] + O(\epsilon^3) \qquad (5.6.70)$$

The presence of the last term on the right-hand side of (5.6.70) exhibits a shortcoming of the von Zeipel procedure in its present form, in which the fast and slowly varying terms in (5.6.40) were determined using mixed variables (new momenta but old coordinates). Had we expressed (5.6.39) in terms of the new variables α^* and β^*, this last term would have been absorbed into S_2, which might contribute to the slowly varying part of K_3. In fact, such a representation in mixed variables was recognized by Breakwell (see Schechter, 1968) to lead to invalid results for the motion of a particle near the triangular points in the restricted problem of three bodies (Breakwell and Pringle, 1966). Using this suggestion, Schechter obtained a valid expansion by expressing the Hamiltonian in terms of the new variables before averaging to determine the slowly varying part (long period). Musen (1965) developed algorithms for effecting the transformation, to any order, of variables and arbitrary functions from the old to the new variables and vice versa. Lacina (1969a, b) and Stern (1970a, 1971a) obtained expressions for general near-identity canonical transformations from the old variables to the new variables. With these transformations, they modified the Hamilton–Jacobi equation. The resulting perturbation schemes may be related to other perturbation schemes which use canonical variables by the proper choice of certain expressions entering into these transformations.

A technique of determining integrals of motion of a system governed by a Hamiltonian was developed by Whittaker (1916, 1937), Cherry (1927), Contopoulos (1963), McNamara and Whiteman (1967), and Coffey (1969). The technique is based on the fact that the equation

$$[\phi, H] = \frac{\partial \phi}{\partial \mathbf{q}} \cdot \frac{\partial H}{\partial \mathbf{p}} - \frac{\partial \phi}{\partial \mathbf{p}} \cdot \frac{\partial H}{\partial \mathbf{q}} = 0$$

is satisfied by any integral of the canonical equations of motion

$$\dot{\mathbf{q}} = \frac{\partial H}{\partial \mathbf{p}}, \dot{\mathbf{p}} = -\frac{\partial H}{\partial \mathbf{q}}$$

An effective and powerful technique of effecting the transformation of variables and arbitrary functions to new variables has been developed by Hori (1966, 1967) using the Lie series, and Deprit (1969) and Kamel (1969, 1970) using the Lie transforms. This technique is described in Section 5.7.

Had we expressed (5.6.39) in terms of the new variables, we would have found that

$$K = \frac{\epsilon \alpha^*}{4\omega^2} \cos 2[(\omega - 1)t + \omega\beta^*] - \frac{\epsilon^2 \alpha^*}{32\omega^3(\omega + 1)} + O(\epsilon^3) \quad (5.6.71)$$

We remove the explicit dependence of K on t by changing from α^* and β^*

to α' and β' by using the generating function

$$S' = \alpha'[(\omega - 1)t + \omega\beta^*] \qquad (5.6.72)$$

Hence

$$\alpha^* = \frac{\partial S'}{\partial \beta^*} = \omega\alpha' \qquad (5.6.73)$$

$$\beta' = \frac{\partial S'}{\partial \alpha'} = (\omega - 1)t + \omega\beta^* \qquad (5.6.74)$$

and

$$K' = K + \frac{\partial S'}{\partial t} = \frac{\epsilon\alpha'}{4\omega}\cos 2\beta' - \frac{\epsilon^2\alpha'}{32\omega^2(\omega + 1)} + (\omega - 1)\alpha' \qquad (5.6.75)$$

Therefore

$$\dot{\alpha}' = -\frac{\partial K'}{\partial \beta'} = \frac{\epsilon\alpha'}{2\omega}\sin 2\beta' \qquad (5.6.76)$$

$$\dot{\beta}' = \frac{\partial K'}{\partial \alpha'} = \frac{\epsilon}{4\omega}\cos 2\beta' - \frac{\epsilon^2}{32\omega^2(\omega + 1)} + \omega - 1 \qquad (5.6.77)$$

As in Section 5.5.2, the solution of (5.6.76) and (5.6.77) is

$$\ln \alpha' = -\ln\left[\frac{\epsilon}{4\omega}\cos 2\beta' - \frac{\epsilon^2}{32\omega^2(\omega + 1)} + \omega - 1\right] + \text{a constant} \qquad (5.6.78)$$

Hence the transition curves are given by

$$\frac{\epsilon}{4\omega} = \left| \omega - 1 - \frac{\epsilon^2}{32\omega^2(\omega + 1)} \right|$$

Consequently

$$\omega = 1 \pm \tfrac{1}{4}\epsilon - \tfrac{3}{64}\epsilon^2 + O(\epsilon^3) \qquad (5.6.79)$$

or

$$\omega^2 = 1 \pm \tfrac{1}{2}\epsilon - \tfrac{1}{32}\epsilon^2 + O(\epsilon^3) \qquad (5.6.80)$$

These curves are in agreement with those obtained in Section 3.1.2 by using the Lindstedt–Poincaré technique, and in Section 3.1.3 by using the Whittaker technique.

5.7. Averaging by Using the Lie Series and Transforms

In analyzing the oscillations of a weakly nonlinear system, the method of variation of parameters is usually used to transform the equations governing these oscillations into the standard form

$$\dot{\mathbf{x}} = \mathbf{f}(\mathbf{x}; \epsilon) = \sum_{m=0}^{\infty} \frac{\epsilon^m}{m!} \mathbf{f}_m(\mathbf{x}) \qquad (5.7.1)$$

where

$$\mathbf{f}_m(\mathbf{x}) = \frac{\partial^m \mathbf{f}}{\partial \epsilon^m}\bigg|_{\epsilon=0}$$

Here $\mathbf{x}$ and $\mathbf{f}$ are vectors with N components. The vector $\mathbf{x}$ may represent, for example, the amplitudes and the phases of the system, or the orbital parameters of the unperturbed two-body problem. If we denote the components of the vector $\mathbf{f}_m$ by f_{mn}, then a component x_k of the vector $\mathbf{x}$ is said to be a rapidly rotating phase if $f_{0k} \not\equiv 0$.

To analyze this standard system, we found it to be useful (see Section 5.2.3) to introduce a near-identity transformation

$$\mathbf{x} = \mathbf{X}(\mathbf{y}; \epsilon) = \mathbf{y} + \epsilon \mathbf{X}_1(\mathbf{y}) + \epsilon^2 \mathbf{X}_2(\mathbf{y}) + \cdots \qquad (5.7.2a)$$

from $\mathbf{x}$ to $\mathbf{y}$ such that the system (5.7.1) is transformed into

$$\dot{\mathbf{y}} = \mathbf{g}(\mathbf{y}; \epsilon) = \sum_{n=0}^{\infty} \frac{\epsilon^n}{n!} \mathbf{g}_n(\mathbf{y}) \qquad (5.7.2b)$$

where $\mathbf{g}_n$ contains long-period terms only. In Section 5.2.3, $\mathbf{X}_n$ and $\mathbf{g}_n$ were determined by substituting (5.7.2) into (5.7.1) and separating the short- and the long-period terms assuming that $\mathbf{X}_n$ contains short-period terms only.

5.7.1. THE LIE SERIES AND TRANSFORMS

In this section we define the transformation (5.7.2a) as the solution of the N differential equations

$$\frac{d\mathbf{x}}{d\epsilon} = \mathbf{W}(\mathbf{x}; \epsilon), \qquad \mathbf{x}(\epsilon = 0) = \mathbf{y} \qquad (5.7.3)$$

The vector $\mathbf{W}$ is called the generating vector. It seems at first that we are turning in circles because we are proposing to simplify the original system of differential equations by solving a system of N differential equations. This is not the case, because we are interested in the solution of (5.7.1) for large t, whereas we need the solution of (5.7.3) for small ϵ; which is a significant simplification.

Equation (5.7.3) generates the so-called Lie transforms (Kamel, 1970), which are invertible because they are close to the identity. If $\mathbf{W}$ does not depend on ϵ, (5.7.3) generates the so-called Lie series. For a canonical system, Hori (1966, 1967) and Deprit (1969) took

$$\mathbf{x} = \begin{bmatrix} \mathbf{q} \\ \mathbf{p} \\ t \end{bmatrix}, \qquad \mathbf{y} = \begin{bmatrix} \mathbf{Q} \\ \mathbf{P} \\ t \end{bmatrix} \qquad (5.7.4a)$$

with $\mathbf{q}$ the system's coordinates, $\mathbf{p}$ the conjugate momenta, and t the time, and defined

$$\mathbf{W} = \begin{bmatrix} S_{\mathbf{p}} \\ -S_{\mathbf{q}} \\ 0 \end{bmatrix}, \qquad S = S(\mathbf{q}, \mathbf{p}, t; \epsilon) \qquad (5.7.4\text{b})$$

with S the generating function.

Hori (1966) constructed a nonrecursive algorithm using the Lie series to determine the transform $K = \sum_{n=0}^{\infty} (\epsilon^n/n!)K_n(\mathbf{Q}, \mathbf{P}, t)$ of a Hamiltonian $H = \sum_{n=0}^{\infty} (\epsilon^n/n!)H_n(\mathbf{q}, \mathbf{p}, t)$. Deprit (1969) constructed another algorithm to generate K recursively using the Lie transforms. Kamel (1969b), Campbell and Jefferys (1970), and Mersman (1970) showed the equivalence of Hori's and Deprit's theories, while Kamel (1969a) simplified Deprit's algorithm. Hori (1970) showed that, to second order, Lie transforms are equivalent to von Zeipel's technique. Shniad (1970) proved that the von Zeipel transformation is equivalent to the Deprit transformation, while Mersman (1971) established the equivalence of the Hori, Deprit and von Zeipel transformations. It should be mentioned that a perturbation theory based on the Lie series and transforms has several advantages over von Zeipel's procedure. The generating function is not a mixed function of the old and new variables, the theory is canonically invariant, and it is possible to give a direct expansion of any function of the old variables in terms of the new variables.

Kamel (1970) introduced the transformation (5.7.3) and constructed an algorithm to transform the standard system (5.7.1) into (5.7.2b). Moreover, he constructed algorithms (1) to transform any vector function from the old to the new variables and (2) to determine (5.7.2a) and its inverse. Henrard (1970) and Kamel (1971) investigated more deeply the mathematical and operational significance of these algorithms. We next construct these generalized algorithms and then specialize them to the canonical case in Section 5.7.5.

5.7.2. GENERALIZED ALGORITHMS

Let the solution of (5.7.3) be $\mathbf{x} = \mathbf{X}(\mathbf{y}; \epsilon)$ and let its inverse be $\mathbf{y} = \mathbf{Y}(\mathbf{x}; \epsilon)$ so that

$$d\mathbf{x} = \mathbf{X_Y}\, d\mathbf{y} \quad \text{and} \quad d\mathbf{y} = \mathbf{Y_X}\, d\mathbf{x} \qquad (5.7.5)$$

where

$$\mathbf{X_Y} = \frac{\partial X_i}{\partial Y_j} \qquad \text{(Jacobian matrix)}$$

and

$$\mathbf{X_Y}\, d\mathbf{y} = \frac{\partial X_i}{\partial Y_j}\, dy_j$$

From (5.7.5) we have
$$dx = X_Y Y_X \, dx$$
so that
$$X_Y Y_X = I \qquad \text{(identity matrix)}$$
Hence
$$Y_X = (X_Y)^{-1} \qquad \text{(inverse of } X_Y) \tag{5.7.6}$$
The second relationship in (5.7.5) gives
$$\dot{y} = Y_X \dot{x}$$
which can be rewritten using (5.7.1) as
$$\dot{y} = g(y; \epsilon) = Y_x f \big|_{x=X(y;\epsilon)} \tag{5.7.7}$$

We are interested in developing the right-hand side of (5.7.7) in powers of ϵ so that
$$\dot{y} = \sum_{n=0}^{\infty} \frac{\epsilon^n}{n!} \frac{d^n g}{d\epsilon^n} \bigg|_{\epsilon=0} \tag{5.7.8}$$
From (5.7.7)
$$\frac{dg}{d\epsilon} = \frac{d}{d\epsilon} [(Y_X f)_{x=X(y;\epsilon)}]$$
$$= \left[\frac{\partial}{\partial X}(Y_X f)\frac{dX}{d\epsilon} + \frac{\partial}{\partial \epsilon}(Y_X f) \right]_{x=X(y;\epsilon)} \tag{5.7.9}$$
Now
$$\frac{\partial}{\partial \epsilon}(Y_X f) = Y_X \frac{\partial f}{\partial \epsilon} + \frac{\partial}{\partial X}\left(\frac{\partial Y}{\partial \epsilon}\right) f \tag{5.7.10}$$
Since $y = Y(x; \epsilon)$ is the inverse of $x = X(y; \epsilon)$ [the solution of $dx/d\epsilon = W(x; \epsilon)$ subject to $x(\epsilon = 0) = y$]
$$\frac{dy}{d\epsilon} = 0 = \frac{\partial Y}{\partial \epsilon} + Y_X \frac{dx}{d\epsilon} = \frac{\partial Y}{\partial \epsilon} + Y_X W$$
so that
$$\frac{\partial Y}{\partial \epsilon} = -Y_X W$$
Hence (5.7.10) can be rewritten as
$$\frac{\partial}{\partial \epsilon}(Y_X f) = Y_X \frac{\partial f}{\partial \epsilon} - \frac{\partial}{\partial X}(Y_X W) f \tag{5.7.11}$$
Using this expression and noting that $dX/d\epsilon = W$, we rewrite (5.7.9) as
$$\frac{dg}{d\epsilon} = \frac{d}{d\epsilon} [(Y_X f)_{x=X(y;\epsilon)}]$$
$$= \left[\frac{\partial}{\partial X}(Y_X f)W + Y_X \frac{\partial f}{\partial \epsilon} - \frac{\partial}{\partial X}(Y_X W)f \right]_{x=X(y;\epsilon)}$$

which can be simplified to

$$\frac{d\mathbf{g}}{d\epsilon} = \frac{d}{d\epsilon}\left[(\mathbf{Y_X f})_{x=X(y;\epsilon)}\right] = [\mathbf{Y_X} D\mathbf{f}]_{x=X(y;\epsilon)} \tag{5.7.12}$$

where

$$D\mathbf{f} = \frac{\partial \mathbf{f}}{\partial \epsilon} + \mathbf{f_X W} - \mathbf{W_X f} \tag{5.7.13}$$

By recurrence, (5.7.12) gives

$$\frac{d^n \mathbf{g}}{d\epsilon^n} = [\mathbf{Y_X} D^n \mathbf{f}]_{x=X(y;\epsilon)} \tag{5.7.14}$$

Since $\mathbf{x}(\epsilon = 0) = \mathbf{y}$

$$\mathbf{Y_X}\big|_{\epsilon=0} = I \qquad \text{(identity matrix)}$$

so that

$$\frac{d^n \mathbf{g}}{d\epsilon^n}\bigg|_{\epsilon=0} = D^n \mathbf{f}\big|_{x=y,\ \epsilon=0} \tag{5.7.15}$$

To determine $D^n \mathbf{f}$, we assume that $\mathbf{W}$ can be expanded as

$$\mathbf{W} = \sum_{n=0}^{\infty} \frac{\epsilon^n}{n!} \mathbf{W}_{n+1} \tag{5.7.16}$$

so that the transformation (5.7.3) can be generated, successively, to any order. If $\mathbf{f} = \sum_{n=0}^{\infty} (\epsilon^n/n!)\mathbf{f}_n$, (5.7.13) becomes

$$D\mathbf{f} = \sum_{n=1}^{\infty} \frac{\epsilon^{n-1}}{(n-1)!}\mathbf{f}_n + \sum_{n=0}^{\infty}\sum_{m=0}^{\infty} \frac{\epsilon^n}{n!}\frac{\epsilon^m}{m!}\left[\frac{\partial \mathbf{f}_n}{\partial \mathbf{X}}\mathbf{W}_{m+1} - \frac{\partial \mathbf{W}_{m+1}}{\partial X}\mathbf{f}_n\right]$$

Letting $n = k + 1$ in the first term, and $n = k - m$ in the second term, we can rewrite this expression as

$$D\mathbf{f} = \sum_{k=0}^{\infty} \frac{\epsilon^k}{k!}\mathbf{f}_{k+1} + \sum_{k=0}^{\infty} \frac{\epsilon^k}{k!}\sum_{m=0}^{k} C_m{}^k\left[\frac{\partial \mathbf{f}_{k-m}}{\partial \mathbf{X}}\mathbf{W}_{m+1} - \frac{\partial \mathbf{W}_{m+1}}{\partial \mathbf{X}}\mathbf{f}_{k-m}\right]$$

or

$$D\mathbf{f} = \sum_{k=0}^{\infty} \frac{\epsilon^k}{k!}\mathbf{f}_k^{(1)} \tag{5.7.17}$$

where

$$\mathbf{f}_k^{(1)} = \mathbf{f}_{k+1} + \sum_{m=0}^{k} C_m{}^k L_{m+1}\mathbf{f}_{k-m} \tag{5.7.18a}$$

$$L_m \mathbf{g} = \frac{\partial \mathbf{g}}{\partial \mathbf{X}}\mathbf{W}_m - \frac{\partial \mathbf{W}_m}{\partial \mathbf{X}}\mathbf{g} \tag{5.7.18b}$$

$$C_m{}^k = \frac{k!}{(k-m)!\,m!}$$

By recurrence, (5.7.17) and (5.7.18) with $\mathbf{f}_k = \mathbf{f}_k^{(0)}$ give

$$D^n\mathbf{f} = \sum_{k=0}^{\infty} \frac{\epsilon^k}{k!} \mathbf{f}_k^{(n)} \qquad (5.7.19)$$

where

$$\mathbf{f}_k^{(n)} = \mathbf{f}_{k+1}^{(n-1)} + \sum_{m=0}^{k} C_m{}^k L_{m+1} \mathbf{f}_{k-m}^{(n-1)} \qquad (5.7.20)$$

Hence

$$\dot{\mathbf{y}} = \sum_{n=0}^{\infty} \frac{\epsilon^n}{n!} \mathbf{f}^{(n)} \quad \text{where} \quad \mathbf{f}^{(n)} = \mathbf{f}_0^{(n)}\big|_{\mathbf{x}=\mathbf{y}} \qquad (5.7.21)$$

The recurrence relationship (5.7.20) can be best visualized using the Lie triangle introduced by Deprit (1969); it is somewhat reminiscent of the Pascal triangle and shown in Figure 5-3.

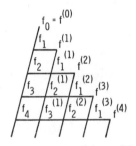

Figure 5-3. Lie triangle. $(\mathbf{f}^{(n)} = \mathbf{f}_0{}^{(n)}, \mathbf{f}_n = \mathbf{f}_n{}^{(0)})$

For example

$$\mathbf{f}^{(1)} = \mathbf{f}_1 + L_1\mathbf{f}_0$$
$$\mathbf{f}_1^{(1)} = \mathbf{f}_2 + L_1\mathbf{f}_1 + L_2\mathbf{f}_0$$
$$\mathbf{f}^{(2)} = \mathbf{f}_1^{(1)} + L_1\mathbf{f}_0^{(1)}$$
$$\mathbf{f}_2^{(1)} = \mathbf{f}_3 + L_1\mathbf{f}_2 + 2L_2\mathbf{f}_1 + L_3\mathbf{f}_0 \qquad (5.7.22)$$
$$\mathbf{f}_1^{(2)} = \mathbf{f}_2^{(1)} + L_1\mathbf{f}_1^{(1)} + L_2\mathbf{f}^{(1)}$$
$$\mathbf{f}^{(3)} = \mathbf{f}_1^{(2)} + L_1\mathbf{f}^{(2)}$$

In carrying out perturbation solutions, we often need to express a vector

$$\mathbf{F}(\mathbf{x}; \epsilon) = \sum_{n=0}^{\infty} \frac{\epsilon^n}{n!} \mathbf{F}_n(\mathbf{x}), \qquad \mathbf{F}_n(\mathbf{x}) = \frac{\partial^n \mathbf{F}}{\partial \epsilon^n}\bigg|_{\epsilon=0} \qquad (5.7.23a)$$

in terms of the new variable y in the form

$$G(y; \epsilon) = F[X(y; \epsilon); \epsilon]$$

$$= \sum_{n=0}^{\infty} \frac{\epsilon^n}{n!} F^{(n)}(y), \qquad F^{(n)}(y) = \frac{d^n F}{d\epsilon^n}\bigg|_{x=y, \ \epsilon=0} \qquad (5.7.23b)$$

where

$$\frac{dF}{d\epsilon} = \frac{\partial F}{\partial \epsilon} + \mathcal{L}F, \qquad \mathcal{L}F = \frac{\partial F}{\partial X} W \qquad (5.7.23c)$$

Using (5.7.16) and (5.7.23a), we express this [as we expressed (5.7.17)] in the form

$$\frac{dF}{d\epsilon} = \sum_{k=0}^{\infty} \frac{\epsilon^k}{k!} F_k^{(1)} \qquad (5.7.24)$$

where

$$F_k^{(1)} = F_{k+1} + \sum_{m=0}^{k} C_m{}^k \mathcal{L}_{m+1} F_{k-m} \qquad (5.7.25)$$

$$\mathcal{L}_m G = \frac{\partial G}{\partial X} W_m \qquad (5.7.26)$$

By recurrence (5.7.24) and (5.7.25) give

$$\frac{d^n F}{d\epsilon^n} = \sum_{k=0}^{\infty} \frac{\epsilon^k}{k!} F_k^{(n)} \qquad (5.7.27)$$

where

$$F_k^{(n)} = F_{k+1}^{(n-1)} + \sum_{m=0}^{k} C_m{}^k \mathcal{L}_{m+1} F_{k-m}^{(n-1)} \qquad (5.7.28)$$

with $F_k^{(0)} = F_k$. Hence

$$G(y; \epsilon) = \sum_{n=0}^{\infty} \frac{\epsilon^n}{n!} F^{(n)}, \qquad F^{(n)} = F_0^{(n)}\bigg|_{x=y} \qquad (5.7.29)$$

Equations (5.7.27) through (5.7.29) have the same form as (5.7.19) through (5.7.21) except for the different operator $\mathcal{L}$, so that (5.7.28) can also be visualized using the Lie triangle.

5.7.3.   SIMPLIFIED GENERAL ALGORITHMS

To simplify an algorithm such as (5.7.20), Kamel (1969, 1970) wrote it first as

$$f_k^{(n)} = f_{k-1}^{(n+1)} - \sum_{m=0}^{k-1} C_m^{k-1} L_{m+1} f_{k-m-1}^{(n)} \qquad (5.7.30)$$

He then eliminated successively the functions on the right-hand side to obtain eventually $\mathbf{f}_k^{(n)}$ as a linear functional of $\mathbf{f}^{(n+k)}, \mathbf{f}^{(n+k-1)}, \ldots, \mathbf{f}^{(n)}$. Thus we assume that

$$\mathbf{f}_k^{(n)} = \mathbf{f}^{(n+k)} - \sum_{j=1}^{k} C_j{}^k G_j \mathbf{f}^{(n+k-j)} \tag{5.7.31}$$

where G_j is a linear operator which is a functional of $L_j, L_{j-1}, \ldots, L_1$. Substituting (5.7.31) into (5.7.30) yields the following recursion relationship

$$G_j = L_j - \sum_{m=1}^{j-1} C_{m-1}^{j-1} L_m G_{j-m}, \qquad 1 \leq j \leq n \tag{5.7.32}$$

For example

$$\begin{aligned}
G_1 &= L_1 \\
G_2 &= L_2 - L_1 L_1 \\
G_3 &= L_3 - L_1(L_2 - L_1 L_1) - 2L_2 L_1
\end{aligned} \tag{5.7.33}$$

For $n = 0$ and $n = 1$, (5.7.31) gives

$$\mathbf{f}^{(k)} = \mathbf{f}_k + \sum_{j=1}^{k} C_j{}^k \mathbf{f}_{j,k-j} \tag{5.7.34}$$

$$\mathbf{f}_k^{(1)} = \mathbf{f}^{(k+1)} - \sum_{j=1}^{k} C_j{}^k \mathbf{f}_{j,k-j+1} \tag{5.7.35}$$

where

$$\mathbf{f}_{j,i} = G_j \mathbf{f}^{(i)} = L_j f^{(i)} - \sum_{m=1}^{j-1} C_{m-1}^{j-1} L_m \mathbf{f}_{j-m,i} \tag{5.7.36}$$

This is the simplified algorithm of Kamel in which $\mathbf{f}_{j,i} = 0$ if $\mathbf{f}^{(i)} = 0$ because the second index i, in the recursion relationship, is fixed.

A more convenient form of this algorithm was obtained by Kamel by writing (5.7.35) as

$$\mathbf{f}^{(k)} = \mathbf{f}_{k-1}^{(1)} + \sum_{j=1}^{k-1} C_j^{k-1} \mathbf{f}_{j,k-j} \tag{5.7.37}$$

However, from (5.7.18a)

$$\mathbf{f}_{k-1}^{(1)} = \mathbf{f}_k + \sum_{j=1}^{k-1} C_{j-1}^{k-1} L_j \mathbf{f}_{k-j} + L_k \mathbf{f}_0$$

so that (5.7.37) can be rewritten as

$$\mathbf{f}^{(k)} = \mathbf{f}_k + \sum_{j=1}^{k-1} [C_{j-1}^{k-1} L_j \mathbf{f}_{k-j} + C_j^{k-1} \mathbf{f}_{j,k-j}] + L_k \mathbf{f}_0 \tag{5.7.38}$$

Since $dx/d\epsilon = \mathbf{W}$ from (5.7.3)

$$\mathbf{x} = \mathbf{y} + \sum_{n=1}^{\infty} \frac{\epsilon^n}{n!} \mathbf{x}^{(n)}(\mathbf{y}) \tag{5.7.39}$$

where

$$\mathbf{x}^{(n+1)}(\mathbf{y}) = \frac{d^n \mathbf{W}}{d\epsilon^n}\bigg|_{\epsilon=0,\,\mathbf{x}=\mathbf{y}} \qquad \text{for } n \geq 1$$

Hence from (5.7.16) and (5.7.34)

$$\mathbf{x}^{(k)} = \mathbf{W}_k + \sum_{j=1}^{k-1} C_j^{k-1} \mathbf{x}_{j,k-j}, \qquad k \geq 1 \qquad (5.7.40)$$

where

$$\mathbf{x}_{j,i} = \mathscr{L}_j \mathbf{x}^{(i)} - \sum_{m=1}^{j-1} C_{m-1}^{j-1} \mathscr{L}_m \mathbf{x}_{j-m,i} \qquad (5.7.41)$$

To determine the inverse transform

$$\mathbf{y} = \mathbf{x} + \sum_{n=1}^{\infty} \frac{\epsilon^n}{n!} \mathbf{y}^{(n)}(\mathbf{x}) \qquad (5.7.42)$$

we eliminate $\mathbf{x} - \mathbf{y}$ from (5.7.39) and (5.7.42) to obtain

$$\mathbf{u} = \sum_{n=1}^{\infty} \frac{\epsilon^n}{n!} \mathbf{y}^{(n)}(\mathbf{x}) = -\sum_{n=1}^{\infty} \frac{\epsilon^n}{n!} \mathbf{x}^{(n)}(\mathbf{y}) \qquad (5.7.43a)$$

However

$$\mathbf{u} = \sum_{n=1}^{\infty} \frac{\epsilon^n}{n!} \mathbf{u}_n(\mathbf{x}) = \sum_{n=1}^{\infty} \frac{\epsilon^n}{n!} \mathbf{u}^{(n)}(\mathbf{y}) \qquad (5.7.43b)$$

so that

$$\mathbf{u}_n(\mathbf{x}) = \mathbf{y}^{(n)}(\mathbf{x}), \qquad \mathbf{u}^{(n)}(\mathbf{y}) = -\mathbf{x}^{(n)}(\mathbf{y}), \qquad n \geq 1 \qquad (5.7.43c)$$

Then (5.7.34) gives

$$\mathbf{y}^{(k)}(\mathbf{x}) = -\mathbf{x}^{(k)}(\mathbf{x}) + \sum_{j=1}^{k-1} C_j^k \mathbf{x}_{j,k-j} \quad \text{for} \quad k \geq 1 \qquad (5.7.44)$$

where $\mathbf{x}_{j,i}$ is defined by (5.7.41).

5.7.4. PROCEDURE OUTLINE

Consider a system of differential equations written in the standard form

$$\dot{\mathbf{x}} = \sum_{n=0}^{\infty} \frac{\epsilon^n}{n!} \mathbf{f}_n(\mathbf{x}) \qquad (5.7.45)$$

The essence of the algorithms of the previous section is to introduce a transformation from $\mathbf{x}$ to $\mathbf{y}$ so that (5.7.45) becomes

$$\dot{\mathbf{y}} = \sum_{n=0}^{\infty} \frac{\epsilon^n}{n!} \mathbf{g}_n(\mathbf{y}) \qquad (5.7.46)$$

where $\mathbf{g}_n$ does not contain short-period terms. To do this we construct a near-identity transformation

$$\mathbf{x} = \mathbf{y} + \sum_{n=1}^{\infty} \frac{\epsilon^n}{n!} \mathbf{x}^{(n)}(\mathbf{y}) \tag{5.7.47}$$

Under such a transformation a vector

$$\mathbf{F}(\mathbf{x};\epsilon) = \sum_{n=0}^{\infty} \frac{\epsilon^n}{n!} \mathbf{F}_n(\mathbf{x}) \tag{5.7.48}$$

becomes

$$\mathbf{F}(\mathbf{x};\epsilon) = \sum_{n=0}^{\infty} \frac{\epsilon^n}{n!} \mathbf{F}^{(n)}(\mathbf{y}) \tag{5.7.49}$$

The algorithms described in the previous section to generate these transformations can be implemented on a computer because they can be effected by the recursive application of elementary operations. We describe below the procedure to second order. It is initiated by putting

$$\begin{aligned} \mathbf{g}_0(\mathbf{y}) &= \mathbf{f}_0(\mathbf{y}) \\ \mathbf{F}^{(0)}(\mathbf{y}) &= \mathbf{F}_0(\mathbf{y}) \end{aligned} \tag{5.7.50}$$

Then we begin the first-order expansion by writing the linear partial differential equation

$$\mathbf{g}_1(\mathbf{y}) = \mathbf{f}_1(\mathbf{y}) + L_1 \mathbf{f}_0 \tag{5.7.51}$$

We choose $\mathbf{g}_1$ to equal the long-period terms in $\mathbf{f}_1$, solve the resulting equation for $\mathbf{W}_1$, and compute

$$\begin{aligned} \mathbf{x}^{(1)} &= \mathbf{W}_1 \\ \mathbf{F}_{1,0} &= \mathscr{L}_1 \mathbf{F}^{(0)} \\ \mathbf{F}^{(1)} &= \mathbf{F}_1 + \mathbf{F}_{1,0} \end{aligned} \tag{5.7.52}$$

To prepare for the second-order expansion, we compute

$$\mathbf{g}_{1,1} = L_1 \mathbf{g}_1$$

Then we set up the differential equation

$$\mathbf{g}_2 = \mathbf{f}_2 + L_1 \mathbf{f}_1 + \mathbf{g}_{1,1} + L_2 \mathbf{f}_0 \tag{5.7.53}$$

and choose $\mathbf{g}_2$ to be equal to the long-period part of the right-hand side. This completes the expansion to second order.

We illustrate this procedure by applying it to van der Pol's equation

$$\ddot{q} + q = \epsilon(1 - q^2)\dot{q} \tag{5.7.54}$$

whose solution for $\epsilon = 0$ can be written as

$$q = a \cos \phi, \qquad \phi = t + \beta \tag{5.7.55}$$

By the method of variation of parameters, (5.7.54) can be replaced by (see Section 5.2.3)

$$\dot{a} = \tfrac{1}{2}\epsilon[a(1 - \tfrac{1}{4}a^2) - aC_2 + \tfrac{1}{4}a^3C_4] \tag{5.7.56}$$

$$\dot{\phi} = 1 + \tfrac{1}{2}\epsilon[(1 - \tfrac{1}{2}a^2)S_2 - \tfrac{1}{4}a^2S_4] \tag{5.7.57}$$

where

$$C_n = \cos n\phi \quad \text{and} \quad S_n = \sin n\phi$$

Equations (5.7.56) and (5.7.57) have the same form as (5.7.1) with

$$\mathbf{x} = \begin{bmatrix} a \\ \phi \end{bmatrix} \tag{5.7.58}$$

$$\mathbf{f}_0 = \begin{bmatrix} 0 \\ 1 \end{bmatrix} \tag{5.7.59}$$

$$\mathbf{f}_1 = \begin{bmatrix} \tfrac{1}{2}a(1 - \tfrac{1}{4}a^2) - \tfrac{1}{2}aC_2 + \tfrac{1}{8}a^3C_4 \\ \tfrac{1}{2}(1 - \tfrac{1}{2}a^2)S_2 - \tfrac{1}{8}a^2S_4 \end{bmatrix} \tag{5.7.60}$$

$$\mathbf{f}_n = 0 \quad \text{for} \quad n > 1 \tag{5.7.61}$$

We transform now from $\mathbf{x} = \begin{bmatrix} a \\ \phi \end{bmatrix}$ to $\mathbf{y} = \begin{bmatrix} a^* \\ \phi^* \end{bmatrix}$. From (5.7.50) and (5.7.59) we obtain

$$\mathbf{g}_0 = \begin{bmatrix} 0 \\ 1 \end{bmatrix} \tag{5.7.62}$$

From (5.7.18b) and (5.7.59), we find that

$$L_n\mathbf{f}_0 = -\frac{\partial \mathbf{W}_n}{\partial \phi^*} \tag{5.7.63}$$

Hence (5.7.51) becomes

$$\mathbf{g}_1 = \mathbf{f}_1 - \frac{\partial \mathbf{W}_1}{\partial \phi^*} \tag{5.7.64}$$

Choosing $\mathbf{W}_1$ to remove the short-period terms in $\mathbf{f}_1$ gives

$$\mathbf{g}_1 = \begin{bmatrix} \tfrac{1}{2}a^*(1 - \tfrac{1}{4}a^{*2}) \\ 0 \end{bmatrix} \tag{5.7.65}$$

Solving the resultant equation gives

$$\mathbf{W}_1 = \begin{bmatrix} -\tfrac{1}{4}a^*S_2^* + \tfrac{1}{32}a^{*3}S_4^* \\ -\tfrac{1}{4}(1 - \tfrac{1}{2}a^{*2})C_2^* + \tfrac{1}{32}a^{*2}C_4^* \end{bmatrix} \tag{5.7.66}$$

where $S_n^* = \sin n\phi^*$ and $C_n^* = \cos n\phi^*$.

With (5.7.61) and (5.7.63), (5.7.53) becomes

$$g_2 = L_1(f_1 + g_1) - \frac{\partial W_2}{\partial \phi^*} \tag{5.7.67}$$

Choosing W_2 to remove the short-period terms, we obtain

$$g_2 = \langle L_1 f_1 \rangle + \langle L_1 g_1 \rangle \tag{5.7.68}$$

Since g_1 consists of long-period terms only, and W_1 consists of short-period terms only, $\langle L_1 g_1 \rangle = 0$. Hence

$$g_2 = \langle L_1 f_1 \rangle = \begin{bmatrix} 0 \\ -\frac{1}{4} + \frac{3}{8}a^{*2} - \frac{11}{128}a^{*4} \end{bmatrix} \tag{5.7.69}$$

Therefore

$$\dot{y} = \begin{bmatrix} \dot{a}^* \\ \dot{\phi}^* \end{bmatrix} = \begin{bmatrix} \frac{1}{2}\epsilon a^*(1 - \frac{1}{4}a^{*2}) \\ 1 - \frac{1}{8}\epsilon^2(1 - \frac{3}{2}a^{*2} + \frac{11}{32}a^{*4}) \end{bmatrix} \tag{5.7.70}$$

in agreement with the expansion obtained in Section 5.2.3 using the generalized method of averaging.

To compare the expansion obtained in this section with that obtained in Section 5.4.2 by using the Krylov–Bogoliubov–Mitropolski technique, we need to express (5.7.55) in terms of the new variables. Here

$$q = \sum_{n=0}^{\infty} \frac{\epsilon^n}{n!} F_n(\mathbf{x}) \tag{5.7.71}$$

with

$$F_0 = a \cos \phi, \qquad F_n = 0 \quad \text{for} \quad n \geq 1 \tag{5.7.72}$$

From (5.7.50)

$$F^{(0)} = a^* \cos \phi^* \tag{5.7.73}$$

Then (5.7.52) and (5.7.26) give

$$\begin{aligned} F_{1,0} &= \mathscr{L}_1(a^* \cos \phi^*) = [\cos \phi^*, -a^* \sin \phi^*]W_1 \\ &= -\frac{1}{4}a^*(1 - \frac{1}{4}a^{*2}) \sin \phi^* - \frac{1}{32}a^{*3} \sin 3\phi^* \end{aligned}$$

Hence

$$q = a^* \cos \phi^* - \frac{1}{4}\epsilon a^*[(1 - \frac{1}{4}a^{*2}) \sin \phi^* + \frac{1}{8}a^{*2} \sin 3\phi^*] + O(\epsilon^2) \tag{5.7.74}$$

which can be rewritten as

$$q = a^* \cos \psi - \frac{\epsilon a^{*3}}{32} \sin 3\psi + O(\epsilon^2) \tag{5.7.75a}$$

where

$$\frac{d\psi}{dt} = 1 - \epsilon^2 \left(\frac{1}{8} - \frac{a^{*2}}{8} + \frac{7a^{*4}}{256} \right) + O(\epsilon^3) \tag{5.7.75b}$$

in agreement with the expansion obtained in Section 5.4.2 using the Krylov–Bogoliubov–Mitropolski technique.

5.7.5.   ALGORITHMS FOR CANONICAL SYSTEMS

Hori (1966, 1967) and Deprit (1969) used the Lie series and transforms, respectively, to transform a Hamiltonian

$$H(\mathbf{p}, \mathbf{q}, t; \epsilon) = \sum_{n=0}^{\infty} \frac{\epsilon^n}{n!} H_n(\mathbf{p}, \mathbf{q}, t) \qquad (5.7.76)$$

into a new Hamiltonian

$$K(\mathbf{P}, \mathbf{Q}, t; \epsilon) = \sum_{n=0}^{\infty} \frac{\epsilon^n}{n!} K_n(\mathbf{P}, \mathbf{Q}, t) \qquad (5.7.77)$$

If we let

$$\mathbf{x} = \begin{bmatrix} \mathbf{q} \\ \mathbf{p} \\ t \end{bmatrix}, \qquad \mathbf{y} = \begin{bmatrix} \mathbf{Q} \\ \mathbf{P} \\ t \end{bmatrix} \qquad (5.7.78)$$

$$\mathbf{W} = \begin{bmatrix} S_\mathbf{p} \\ -S_\mathbf{q} \\ 0 \end{bmatrix}, \qquad \mathbf{f} = \begin{bmatrix} H_\mathbf{p} \\ -H_\mathbf{q} \\ 1 \end{bmatrix} \qquad (5.7.79)$$

then $\mathbf{g}$ can be generated from a Hamiltonian K as

$$\mathbf{g} = \begin{bmatrix} K_\mathbf{P} \\ -K_\mathbf{Q} \\ 1 \end{bmatrix} \qquad (5.7.80)$$

In this case the algorithm of Section 5.7.3 reduces to the scalar form (Kamel, 1969a)

$$K_0 = H_0(\mathbf{Q}, \mathbf{P}, t)$$

$$K_n = H_n + \sum_{j=1}^{n-1} [C_{j-1}^{n-1} L_j' H_{n-j} + C_j^{n-1} K_{j,n-j}] - \frac{\mathscr{D} S_n}{\mathscr{D} t} \qquad (5.7.81)$$

where

$$L_j' f = \frac{\partial f}{\partial \mathbf{Q}} \frac{\partial S_j}{\partial \mathbf{P}} - \frac{\partial f}{\partial \mathbf{P}} \frac{\partial S_j}{\partial \mathbf{Q}} \qquad (5.7.82a)$$

$$\frac{\mathscr{D} S_n}{\mathscr{D} t} = \frac{\partial S_n}{\partial t} - L_n' H_0 \qquad (5.7.82b)$$

and

$$K_{j,i} = L_j' K_i - \sum_{m=1}^{j-1} C_{m-1}^{j-1} L_j' K_{j-m,i} \qquad (5.7.82c)$$

Under the above transformation the old variables are given in terms of the new variables as

$$\mathbf{q} = \mathbf{Q} + \sum_{n=1}^{\infty} \frac{\epsilon^n}{n!} \mathbf{q}^{(n)}(\mathbf{Q}, \mathbf{P}, t)$$

$$\mathbf{p} = \mathbf{P} + \sum_{n=1}^{\infty} \frac{\epsilon^n}{n!} \mathbf{p}^{(n)}(\mathbf{Q}, \mathbf{P}, t)$$

(5.7.83)

where

$$\mathbf{q}^{(n)} = \frac{\partial S_n}{\partial \mathbf{P}} + \sum_{j=1}^{n-1} C_j^{n-1} \mathbf{q}_{j,n-j}$$

$$\mathbf{p}^{(n)} = -\frac{\partial S_n}{\partial \mathbf{Q}} + \sum_{j=1}^{n-1} C_j^{n-1} \mathbf{p}_{j,n-j}$$

(5.7.84a)

and

$$\mathbf{q}_{j,i} = L_j' \mathbf{q}^{(i)} - \sum_{m=1}^{j-1} C_{m-1}^{j-1} L_m' \mathbf{q}_{j-m,i}$$

$$\mathbf{p}_{j,i} = L_j' \mathbf{p}^{(i)} - \sum_{m=1}^{j-1} C_{m-1}^{j-1} L_m' \mathbf{p}_{j-m,i}$$

(5.7.84b)

To third order, the above algorithms become

$$K_0 = H_0 \tag{5.7.85}$$

$$K_1 = H_1 - \frac{\mathscr{D}S_1}{\mathscr{D}t} \tag{5.7.86}$$

$$K_2 = H_2 + L_1' H_1 + L_1' K_1 - \frac{\mathscr{D}S_2}{\mathscr{D}t} \tag{5.7.87}$$

$$K_3 = H_3 + L_1' H_2 + 2L_2' H_1 + 2L_1' K_2 + L_2' K_1 - L_1'^2 K_1 - \frac{\mathscr{D}S_3}{\mathscr{D}t} \tag{5.7.88}$$

$$\left. \begin{aligned} \mathbf{q}^{(1)} &= \frac{\partial S_1}{\partial \mathbf{P}}, \qquad \mathbf{q}^{(2)} = \frac{\partial S_2}{\partial \mathbf{P}} + L_1' \mathbf{q}^{(1)} \\ \mathbf{q}^{(3)} &= \frac{\partial S_3}{\partial \mathbf{P}} + 2L_1' \mathbf{q}^{(2)} + L_2' \mathbf{q}^{(1)} - L_1'^2 \mathbf{q}^{(1)} \end{aligned} \right\} \tag{5.7.89}$$

$$\left. \begin{aligned} \mathbf{p}^{(1)} &= -\frac{\partial S_1}{\partial \mathbf{Q}}, \qquad \mathbf{p}^{(2)} = -\frac{\partial S_2}{\partial \mathbf{Q}} + L_1' \mathbf{p}^{(1)} \\ \mathbf{p}^{(3)} &= -\frac{\partial S_3}{\partial \mathbf{Q}} + 2L_1' \mathbf{p}^{(2)} + L_2' \mathbf{p}^{(1)} - L_1'^2 \mathbf{p}^{(1)} \end{aligned} \right\} \tag{5.7.90}$$

We next illustrate this procedure by its application to the swinging spring represented by the Hamiltonian (5.5.57). Using the solution (5.5.67) through (5.5.70), we transform this Hamiltonian into

$$H = H_1 + \tfrac{1}{2}H_2 + \cdots \tag{5.7.91}$$

where

$$H_1 = \frac{\alpha_2}{l}\sqrt{\frac{2\alpha_1}{k}} \, [\sin^2 B_2 - 2\cos^2 B_2] \sin B_1 \tag{5.7.92}$$

$$H_2 = -\frac{1}{3}\frac{{\alpha_2}^2}{mgl} \sin^4 B_2 + \frac{12\alpha_1\alpha_2}{kl^2} \sin^2 B_1 \cos^2 B_2 \tag{5.7.93}$$

and $B_i = \omega_i(t + \beta_i)$.

We transform α, β, and H into α^*, β^*, and $K = K_0 + K_1 + \tfrac{1}{2}K_2 + \cdots$ using the algorithm defined by (5.7.85) through (5.7.87). Since $H_0 = 0$, $K_0 = 0$ by (5.7.85), and $\mathscr{D}S_n/\mathscr{D}t = \partial S_n/\partial t$ by (5.7.82b). With (5.7.92), (5.7.86) becomes

$$K_1 = -\frac{{\alpha_2}^*}{l}\sqrt{\frac{{\alpha_1}^*}{2k}} \left\{ \sin {B_1}^* + \tfrac{3}{2}\sin({B_1}^* + 2{B_2}^*) \right.$$

$$\left. + \tfrac{3}{2}\sin[(\omega_1 - 2\omega_2)t + \omega_1{\beta_1}^* - 2\omega_2{\beta_2}^*] \right\} - \frac{\partial S_1}{\partial t} \tag{5.7.94}$$

All the terms in K_1 have short periods unless $\omega_1 \approx 2\omega_2$. In the latter case $\sin[(\omega_1 - 2\omega_2)t + \omega_1{\beta_1}^* - 2\omega_2{\beta_2}^*]$ has a long period (slowly varying). Choosing S_1 to eliminate the short-period terms results in

$$K_1 = -\frac{3{\alpha_2}^*}{2l}\sqrt{\frac{{\alpha_1}^*}{2k}} \sin[(\omega_1 - 2\omega_2)t + \omega_1{\beta_1}^* - 2\omega_2{\beta_2}^*] \tag{5.7.95}$$

and

$$S_1 = \frac{{\alpha_2}^*}{l}\sqrt{\frac{{\alpha_1}^*}{2k}}\left[\frac{\cos {B_1}^*}{\omega_1} + \frac{3\cos({B_1}^* + 2{B_2}^*)}{2(\omega_1 + 2\omega_2)}\right] \tag{5.7.96}$$

If we choose S_2 to eliminate the short-period terms in (5.7.87), we obtain

$$K_2 = \langle H_2 \rangle + \langle L_1' H_1 \rangle + \langle L_1' K_1 \rangle \tag{5.7.97}$$

The averaged values in K_2 are given by

$$\langle L_1' K_1 \rangle = \left\langle \frac{\partial K_1}{\partial \beta_1}\frac{\partial S_1}{\partial \alpha_1} \right\rangle + \left\langle \frac{\partial K_1}{\partial \beta_2}\frac{\partial S_1}{\partial \alpha_2} \right\rangle - \left\langle \frac{\partial K_1}{\partial \alpha_1}\frac{\partial S_1}{\partial \beta_1} \right\rangle - \left\langle \frac{\partial K_1}{\partial \alpha_2}\frac{\partial S_1}{\partial \beta_2} \right\rangle = 0 \tag{5.7.98}$$

$$\langle H_2 \rangle = -\frac{{\alpha_2^*}^2}{8mgl} + \frac{3{\alpha_1}^*{\alpha_2}^*}{kl^2} \approx -\frac{{\alpha_2^*}^2}{2kl^2} + \frac{3{\alpha_1}^*{\alpha_2}^*}{kl^2} \tag{5.7.99}$$

$$\langle L_1' H_1 \rangle = \left\langle \frac{\partial H_1}{\partial \beta_1} \frac{\partial S_1}{\partial \alpha_1} \right\rangle + \left\langle \frac{\partial H_1}{\partial \beta_2} \frac{\partial S_1}{\partial \alpha_2} \right\rangle - \left\langle \frac{\partial H_1}{\partial \alpha_1} \frac{\partial S_1}{\partial \beta_1} \right\rangle - \left\langle \frac{\partial H_1}{\partial \alpha_2} \frac{\partial S_1}{\partial \beta_2} \right\rangle$$

$$= -\left[1 + \frac{9\omega_1}{4(\omega_1 + 2\omega_2)} \right] \frac{\alpha_2^{*^2}}{4kl^2} - \frac{9\alpha_1^* \alpha_2^*}{4kl^2} \frac{\omega_2}{\omega_1 + 2\omega_2}$$

$$\approx -\frac{17\alpha_2^{*^2}}{32kl^2} - \frac{9\alpha_1^* \alpha_2^*}{16kl^2} \tag{5.7.100}$$

In (5.7.99) and (5.7.100) use has been made of the fact that $\omega_1 \approx 2\omega_2$ (i.e., $kl \approx 4mg$ from the definitions of ω_1 and ω_2). Hence

$$K_2 = -\frac{33\alpha_2^{*^2}}{32kl^2} + \frac{39\alpha_1^* \alpha_2^*}{16kl^2} \tag{5.7.101}$$

and

$$K = K_0 + K_1 + \tfrac{1}{2} K_2 + \cdots$$

$$= -\frac{3\alpha_2^*}{2l} \sqrt{\frac{\alpha_1^*}{2k}} \sin \left[(\omega_1 - 2\omega_2)t + \omega_1 \beta_1^* - 2\omega_2 \beta_2^* \right]$$

$$- \frac{33\alpha_2^{*^2}}{64kl^2} + \frac{39\alpha_1^* \alpha_2^*}{32kl^2} + \cdots \tag{5.7.102}$$

To remove the explicit dependence of K on t, we transform from α^* and β^* to α' and β' using the generating function

$$S' = \alpha_1' \left[\frac{(\omega_1 - 2\omega_2)t}{\omega_1} + \beta_1^* \right] + \alpha_2' \beta_2^* \tag{5.7.103}$$

so that

$$\alpha_1^* = \frac{\partial S'}{\partial \beta_1^*} = \alpha_1' \tag{5.7.104}$$

$$\alpha_2^* = \frac{\partial S'}{\partial \beta_2^*} = \alpha_2' \tag{5.7.105}$$

$$\beta_1' = \frac{\partial S'}{\partial \alpha_1'} = \frac{(\omega_1 - 2\omega_2)t}{\omega_1} + \beta_1^* \tag{5.7.106}$$

$$\beta_2' = \frac{\partial S'}{\partial \alpha_2'} = \beta_2^* \tag{5.7.107}$$

and

$$K' = K + \frac{\partial S'}{\partial t} = -\frac{3\alpha_2' \sqrt{\alpha_1'}}{2l\sqrt{2k}} \sin (\omega_1 \beta_1' - 2\omega_2 \beta_2')$$

$$- \frac{33\alpha_2'^2}{64kl^2} + \frac{39\alpha_1' \alpha_2'}{32kl^2} + \frac{\omega_1 - 2\omega_2}{\omega_1} \alpha_1' \tag{5.7.108}$$

Since $\dot{\alpha}'_i = -\partial K'/\partial \beta'_i$ and $\dot{\beta}'_i = \partial K'/\partial \alpha'_i$

$$\dot{\alpha}'_1 = \frac{3\omega_1 \alpha'_2 \sqrt{\alpha'_1}}{2l\sqrt{2k}} \cos \gamma \qquad (5.7.109)$$

$$\dot{\alpha}'_2 = -\frac{3\omega_2 \alpha'_2 \sqrt{\alpha'_1}}{l\sqrt{2k}} \cos \gamma \qquad (5.7.110)$$

$$\dot{\beta}'_1 = -\frac{3\alpha'_2}{4l\sqrt{2k}\alpha'_1} \sin \gamma + \frac{39\alpha'_2}{32kl^2} + \frac{\omega_1 - 2\omega_2}{\omega_1} \qquad (5.7.111)$$

$$\dot{\beta}'_2 = -\frac{3\sqrt{\alpha'_1}}{2l\sqrt{2k}} \sin \gamma - \frac{33\alpha'_2}{32kl^2} + \frac{39\alpha'_1}{32kl^2} \qquad (5.7.112)$$

where

$$\gamma = \omega_1 \beta'_1 - 2\omega_2 \beta'_2 \qquad (5.7.113)$$

5.8. Averaging by Using Lagrangians

Instead of using canonical variables, which require the use of Hamiltonians, Sturrock (1958, 1962) developed a technique that does not require canonical variables. It consists of averaging the Lagrangian and then writing down the corresponding Euler–Lagrange equations. Whitham (1965a, 1967a, b, 1970) developed a similar technique for waves in which the frequency and wave number as well as the amplitude are slowly varying functions of space and time. Bisshop (1969) supplied a more rigorous justification of this technique. Although this technique is not as elegant as those using canonical variables, it has the advantage of being directly applicable to partial as well as ordinary differential equations. Kawakami (1970) and Kawakami and Yagishita (1971) used canonical variables in conjunction with Hamiltonians to solve the nonlinear Vlasov equation.

This technique was applied to a variety of problems of wave propagation in fluids and plasmas. Lighthill (1965, 1967) applied Whitham's theory to moderate waves in deep water where pseudofrequencies are absent, while Karpman and Krushkal' (1969) used Whitham's theory to study the decay of a plane wave into separate wave packets. Howe (1967) studied open-channel steady flow past a solid surface of finite wave group shape. Bretherton (1968) treated linear wave propagation in slowly varying wave guides, while Bretherton and Garrett (1968) investigated slowly varying waves in inhomogeneous media. Garrett (1968), Drazin (1969), and Rarity (1969) analyzed nonlinear internal gravity waves; the effects of shear and slight atmospheric stratification were determined by Garrett and Drazin, respectively. Simmons (1969) studied the interaction of capillary and gravity waves, while Grimshaw

(1970) discussed solitary waves in water of variable depth. Crapper (1970) investigated the generation of capillary waves by gravity waves. Dougherty (1970), Galloway and Crawford (1970), and Galloway and Kim (1971) treated nonlinear waves in plasmas. Dewar (1970) investigated the interaction between hydromagnetic waves and an inhomogeneous medium, while Tang and Sivasubramanian (1971) studied the nonlinear instability of modulated waves in a magnetoplasma. Lowell (1970) analyzed wave propagation in lattices with an anharmonic potential. We describe this technique and its application to three examples.

5.8.1. A MODEL FOR DISPERSIVE WAVES

As a first example, we analyze slowly varying wave train solutions for Bretherton's (1964) model equation

$$\phi_{tt} + \phi_{xxxx} + \phi_{xx} + \phi = \epsilon\phi^3 \tag{5.8.1}$$

with the nonlinear term $\epsilon\phi^3$ rather than $\epsilon\phi^2$. If we neglect the nonlinear term $\epsilon\phi^3$, (5.8.1) admits the uniform traveling wave solution

$$\phi = a \cos \theta, \qquad \theta = kx - \omega t \tag{5.8.2}$$

where ω and k satisfy the dispersion relationship

$$\omega^2 = k^4 - k^2 + 1 \tag{5.8.3}$$

To determine a slowly varying wave train solution by using the variational approach, we first write the Lagrangian corresponding to (5.8.1); that is

$$L = \tfrac{1}{2}\phi_t^2 - \tfrac{1}{2}\phi_{xx}^2 + \tfrac{1}{2}\phi_x^2 - \tfrac{1}{2}\phi^2 + \tfrac{1}{4}\epsilon\phi^4 \tag{5.8.4}$$

It can be easily verified that (5.8.1) is the Euler–Lagrange equation corresponding to this Lagrangian. We assume an expansion of the form

$$\phi = a \cos \theta + \epsilon \sum_{n=2}^{\infty} A_n \cos n\theta + O(\epsilon^2) \tag{5.8.5}$$

where

$$k = \theta_x, \qquad \omega = -\theta_t \tag{5.8.6}$$

and a, ω, k, and A_i are slowly varying functions of x and t. If θ is twice continuously differentiable, (5.8.6) gives the compatability relationship

$$k_t + \omega_x = 0 \tag{5.8.7}$$

Since secular terms appear first at $O(\epsilon)$ in the straightforward expansion, we assume that a_x, a_t, ω_x, ω_t, k_x, and k_t are $O(\epsilon)$. Thus

$$\phi_t = a\omega \sin\theta + a_t \cos\theta + \epsilon\omega \sum_{n=2}^{\infty} nA_n \sin n\theta + O(\epsilon^2)$$

$$\phi_x = -ak \sin\theta + a_x \cos\theta - \epsilon k \sum_{n=2}^{\infty} nA_n \sin n\theta + O(\epsilon^2)$$

$$\phi_{xx} = -ak^2 \cos\theta - 2a_x k \sin\theta - \epsilon k^2 \sum_{n=2}^{\infty} n^2 A_n \cos n\theta + O(\epsilon^2)$$

Substituting these expressions into (5.8.4), we find that the resulting Lagrangian depends implicitly on x and t through θ, a, ω, k, and A_i. Its variation with x and t through θ is much faster than that through the other parameters, because when θ varies over $[0, 2\pi]$, the period of the θ terms in L, the other parameters hardly change. As in the other versions of the method of averaging, we average L over θ from 0 to 2π, keeping a, ω, k, and A_i constant. To do this we first average each term in L separately. Thus

$$\overline{\phi_t^2} = \frac{1}{2\pi} \int_0^{2\pi} \phi_t^2 \, d\theta = \tfrac{1}{2}a^2\omega^2 + O(\epsilon^2)$$

$$\overline{\phi_x^2} = \tfrac{1}{2}a^2 k^2 + O(\epsilon^2)$$

$$\overline{\phi_{xx}^2} = \tfrac{1}{2}a^2 k^4 + O(\epsilon^2)$$

$$\overline{\phi^4} = \tfrac{3}{8}a^4 + O(\epsilon)$$

Hence

$$\mathscr{L} = \overline{L} = \tfrac{1}{4}(\omega^2 - k^4 + k^2 - 1)a^2 + \frac{3\epsilon}{32}a^4 + O(\epsilon^2) \qquad (5.8.8)$$

The averaged Lagrangian $\mathscr{L}$ is an explicit function of a and an implicit function of θ through ω and k. Now we write the Euler–Lagrange equations with respect to the variables a and θ. The Euler–Lagrange equation with respect to a is $\partial\mathscr{L}/\partial a = 0$; it gives the dispersion relationship

$$\omega^2 = k^4 - k^2 + 1 + \tfrac{3}{4}\epsilon a^2 + O(\epsilon^2) \qquad (5.8.9)$$

Note that this dispersion relationship can be obtained by using the principle of harmonic balance; that is, we let $\phi = a\cos\theta$ in (5.8.1) and equate the coefficients of $\cos\theta$ on both sides. Since $\omega = -\theta_t$ and $k = \theta_x$, the Euler–Lagrange equation with respect to θ is

$$\frac{\partial}{\partial t}\left(\frac{\partial\mathscr{L}}{\partial\theta_t}\right) + \frac{\partial}{\partial x}\left(\frac{\partial\mathscr{L}}{\partial\theta_x}\right) = \frac{\partial\mathscr{L}}{\partial\theta} = 0 \qquad (5.8.10)$$

or

$$-\frac{\partial}{\partial t}\left(\frac{\partial\mathscr{L}}{\partial\omega}\right) + \frac{\partial}{\partial x}\left(\frac{\partial\mathscr{L}}{\partial k}\right) = 0 \qquad (5.8.11)$$

However,

$$\frac{\partial \mathscr{L}}{\partial \omega} = \tfrac{1}{2}\omega a^2 + O(\epsilon^2), \qquad \frac{\partial \mathscr{L}}{\partial k} = -\tfrac{1}{2}(2k^3 - k)a^2 + O(\epsilon^2)$$

Hence

$$\frac{\partial}{\partial t}(\omega a^2) + \frac{\partial}{\partial x}[(2k^3 - k)a^2] = 0 \qquad (5.8.12)$$

To simplify (5.8.12), we differentiate (5.8.9) with respect to k to obtain

$$\omega\omega' = 2k^3 - k + O(\epsilon^2)$$

where $\omega' = d\omega/dk$ is the group velocity. Hence we rewrite (5.8.12) as

$$\frac{\partial}{\partial t}(\omega a^2) + \frac{\partial}{\partial x}(\omega\omega' a^2) = 0$$

or

$$\omega\left[\frac{\partial a^2}{\partial t} + \frac{\partial}{\partial x}(\omega' a^2)\right] + a^2[\omega_t + \omega'\omega_x] = 0 \qquad (5.8.13)$$

Since $\omega = \omega(k)$, $\omega_t = \omega' k_t$, and the second term in (5.8.13) vanishes according to (5.8.7). Hence (5.8.13) simplifies into

$$\frac{\partial a^2}{\partial t} + \frac{\partial}{\partial x}(\omega' a^2) = 0 \qquad (5.8.14)$$

Moreover, since $\omega = \omega(k)$, (5.8.7) can be rewritten as

$$\frac{\partial k}{\partial t} + \omega'\frac{\partial k}{\partial x} = 0 \qquad (5.8.15)$$

Therefore the spatial and temporal variations of the amplitude a, frequency ω, and wave number k are given by (5.8.9), (5.8.14), and (5.8.15).

5.8.2. A MODEL FOR WAVE–WAVE INTERACTION

Had we carried out the expansion of the previous section to second order, we would have obtained

$$\phi = a\cos\theta + \frac{\epsilon a^3}{32(9k^4 - 1)}\cos 3\theta + O(\epsilon^2) \qquad (5.8.16)$$

Although valid for a wide range of values of k, this expansion breaks down near $k^2 = 1/3$. This case is referred to as the third-harmonic resonance case in which both $\cos\theta$ and $\cos 3\theta$ satisfy the same dispersion relationship; that is, both the fundamental and its third harmonic have the same phase speed ω/k.

To determine an expansion valid near $k^2 = 1/3$ for (5.8.1), we assume the expansion to be

$$\phi = a_1 \cos \theta_1 + a_3 \cos \theta_3 + \epsilon \sum_{n \neq 1,3}^{\infty} A_n \cos (\theta_n + \nu_n) + O(\epsilon^2) \quad (5.8.17)$$

where

$$\theta_n = k_n x - \omega_n t + \beta_n, \qquad \omega_n^2 = k_n^4 - k_n^2 + 1 \quad (5.8.18)$$

with

$$k_3 \approx 3k_1 \quad \text{and} \quad \omega_3 \approx 3\omega_1 \quad (5.8.19)$$

Note that the first-order term contains the fundamental $\cos \theta_1$ and its third harmonic θ_3. Since we are interested in the case $k_1^2 \approx 1/3$, we take ω_i and k_i to be constants, whereas we take β_i, ν_i, a_i, and A_i to be slowly varying functions of x and t. We next substitute this expansion into (5.8.4) and average the resulting Lagrangian over θ_i, keeping β_i, ν_i, a_i, and A_i constants.

In performing the averaging we note that, although θ_i are fast varying, $\theta_3 - 3\theta_1$ is slowly varying. Thus

$$\overline{\phi^4} = \tfrac{3}{8}(a_1^4 + 4a_1^2 a_3^2 + a_3^4) + \tfrac{1}{2}a_1^3 a_3 \cos (\theta_3 - 3\theta_1) + O(\epsilon)$$

$$\overline{\phi^2} = \tfrac{1}{2}(a_1^2 + a_3^2) + O(\epsilon^2)$$

$$\overline{\phi_t^2} = \frac{1}{2} \sum_{i=1,3} (\omega_i^2 - 2\omega_i \beta_{it})a_i^2 + O(\epsilon^2)$$

$$\overline{\phi_x^2} = \frac{1}{2} \sum_{i=1,3} (k_i^2 + 2k_i \beta_{ix})a_i^2 + O(\epsilon^2)$$

$$\overline{\phi_{xx}^2} = \frac{1}{2} \sum_{i=1,3} (k_i^4 + 4k_i^3 \beta_{ix})a_i^2 + O(\epsilon^2)$$

With these expressions $\mathscr{L}$ becomes

$$\mathscr{L} = -\frac{1}{2} \sum_{i=1,3} \omega_i(\beta_{it} + \omega_i' \beta_{ix})a_i^2$$
$$+ \frac{3\epsilon}{32} (a_1^4 + 4a_1^2 a_3^2 + a_3^4) + \frac{\epsilon}{8} a_1^3 a_3 \cos \delta \quad (5.8.20)$$

where

$$\delta = \theta_3 - 3\theta_1 = (k_3 - 3k_1)x - (\omega_3 - 3\omega_1)t + \beta_3 - 3\beta_1 \quad (5.8.21)$$

In arriving at (5.8.20) we used the dispersion relationship (5.8.18) and the definition of the group velocity $\omega_i' = (2k_i^3 - k_i)/\omega_i$.

With ω_i and k_i constants, a_i and β_i are governed by the Euler–Lagrange equations

$$\frac{\partial \mathscr{L}}{\partial a_i} = 0 \quad (5.8.22)$$

and

$$\frac{\partial}{\partial t}\left(\frac{\partial \mathscr{L}}{\partial \beta_{it}}\right) + \frac{\partial}{\partial x}\left(\frac{\partial \mathscr{L}}{\partial \beta_{ix}}\right) = \frac{\partial \mathscr{L}}{\partial \beta_i} \tag{5.8.23}$$

Substituting for $\mathscr{L}$ from (5.8.20) into these equations and using (5.8.21), we obtain

$$\beta_{1t} + \omega_1'\beta_{1x} = \frac{3\epsilon}{8\omega_1}[a_1^2 + 2a_3^2 + a_1 a_3 \cos \delta] \tag{5.8.24}$$

$$\beta_{3t} + \omega_3'\beta_{3x} = \frac{\epsilon}{8\omega_3}[6a_1^2 + 3a_3^2 + a_1^3 a_3^{-1} \cos \delta] \tag{5.8.25}$$

$$\frac{\partial a_1}{\partial t} + \omega_1'\frac{\partial a_1}{\partial x} = -\frac{3\epsilon}{8\omega_1} a_1^2 a_3 \sin \delta \tag{5.8.26}$$

$$\frac{\partial a_3}{\partial t} + \omega_3'\frac{\partial a_3}{\partial x} = \frac{\epsilon}{8\omega_3} a_1^3 \sin \delta \tag{5.8.27}$$

These equations are in agreement with those obtained in Section 6.2.9 by using the method of multiple scales.

5.8.3. THE NONLINEAR KLEIN–GORDON EQUATION

As a last example, we consider after Whitham (1965a) the nonlinear Klein–Gordon equation

$$u_{tt} - u_{xx} + V'(u) = 0 \tag{5.8.28}$$

where $V(u)$ is any nonlinear potential function which yields oscillatory solutions. Scott (1970) treated the special case $V(u) = -\cos u$, which represents the propagation of a magnetic flux on a long Josephson tunnel junction. The Lagrangian corresponding to this equation is

$$L = \tfrac{1}{2}u_t^2 - \tfrac{1}{2}u_x^2 - V(u) \tag{5.8.29}$$

For a uniform wave train solution $u(\theta)$, (5.8.28) becomes

$$(\omega^2 - k^2)u_{\theta\theta} + V'(u) = 0 \tag{5.8.30}$$

where

$$\omega = -\theta_t \quad \text{and} \quad k = \theta_x \tag{5.8.31}$$

A first integral of (5.8.30) is

$$\tfrac{1}{2}(\omega^2 - k^2)u_\theta^2 + V(u) = E \tag{5.8.32}$$

Integrating this equation gives

$$\theta = \sqrt{\frac{\omega^2 - k^2}{2}} \int \frac{du}{\sqrt{E - V(u)}} \tag{5.8.33}$$

We assume that u is periodic with a period that can be normalized to unity so that

$$\sqrt{\frac{\omega^2 - k^2}{2}} \oint \frac{du}{\sqrt{E - V(u)}} = 1 \qquad (5.8.34)$$

To determine approximate equations for slowly varying E, ω, and k, we substitute $u(\theta)$ for u in (5.8.29) and average the resulting Lagrangian over the interval $[0, 1]$. Thus

$$L = \tfrac{1}{2}(\omega^2 - k^2)u_\theta{}^2 - V(u) = (\omega^2 - k^2)u_\theta{}^2 - E \qquad (5.8.35)$$

as a result of (5.8.32), hence

$$\mathscr{L} = \int_0^1 L[u(\theta)] \, d\theta = (\omega^2 - k^2) \int_0^1 u_\theta{}^2 \, d\theta - E \int_0^1 d\theta$$

$$= (\omega^2 - k^2) \oint u_\theta \, du - E$$

$$= \sqrt{2(\omega^2 - k^2)} \oint \sqrt{E - V(u)} \, du - E \qquad (5.8.36)$$

The variation of $\mathscr{L}$ with respect to E gives the dispersion relation (5.8.34). Since $\omega = -\theta_t$ and $k = \theta_x$, the Euler–Lagrange equation corresponding to the variable θ is

$$-\frac{\partial}{\partial t}\left(\frac{\partial \mathscr{L}}{\partial \omega}\right) + \frac{\partial}{\partial x}\left(\frac{\partial \mathscr{L}}{\partial k}\right) = 0 \qquad (5.8.37)$$

Substituting for $\mathscr{L}$ into (5.8.37), we obtain

$$\frac{\partial}{\partial t}(\omega W) + \frac{\partial}{\partial x}(kW) = 0 \qquad (5.8.38)$$

where

$$W = \sqrt{\frac{2}{\omega^2 - k^2}} \oint \sqrt{E - V(u)} \, du \qquad (5.8.39)$$

The problem formulation is completed by augmenting (5.8.34) and (5.8.38) by the compatability relation

$$\frac{\partial k}{\partial t} + \frac{\partial \omega}{\partial x} = 0 \qquad (5.8.40)$$

Exercises

5.1. Use the Stuart–Watson–Eckhaus technique to determine approximate solutions to

(a) $\ddot{u} + \lambda u = \epsilon u^3$

$u(0) = u(\pi) = 0$

(b) $u_{tt} - u_{xx} + u = \epsilon u^3$

$u(x, 0) = a \cos x, \qquad u_t(x, 0) = 0$

5.2. Use Struble's technique to determine uniform second-order expansions for (Struble, 1962)

(a) $\ddot{u} + u = \epsilon(1 - u^2)\dot{u}$

(b) $\ddot{u} + (\delta + \epsilon \cos 2t)u = 0$

5.3. Use the Krylov–Bogoliubov technique to determine approximate solutions for

(a) $\ddot{u} + \omega_0^2 u = -\epsilon \dot{u}|\dot{u}|$

(b) $\ddot{u} + \omega_0^2 u = \epsilon(1 - u^2)\dot{u} + \epsilon u^3$

(c) $\ddot{u} + (\delta + \epsilon \cos 2t)u = 0$

5.4. Consider Mathieu's equation

$$\ddot{u} + (\delta + \epsilon \cos 2t)u = 0$$

Determine uniform second-order expansions using (a) the Krylov–Bogoliubov–Mitropolski technique, (b) the generalized method of averaging, and (c) the Lie transforms.

5.5. Consider the equation

$$\ddot{q} + \omega_0^2 q = \epsilon q^3 + K \cos \omega t$$

(a) Show that it corresponds to the Hamiltonian

$$H = \tfrac{1}{2}(p^2 + \omega_0^2 q^2) - \tfrac{1}{4}\epsilon q^4 - Kq \cos \omega t$$

(b) Determine a first-order expansion when

$K = O(1)$ and ω is away from $3\omega_0$, ω_0, $\omega_0/3$.

(c) $K = O(1)$ and ω is near $3\omega_0$.

(d) $K = O(1)$ and ω is near $\omega_0/3$.

(e) $K = O(\epsilon)$ and ω is near ω_0.

5.6. Consider the equation

$$\ddot{q} + \omega_0^2 q = \epsilon(1 - q^2)\dot{q} + K \cos \omega t$$

Use the Krylov–Bogoliubov method to determine first-order expansions for the cases enumerated in Exercise 5.5.

5.7. Use the generalized method of averaging, the Krylov–Bogoliubov–Mitropolski method, and Kamel's method to determine second-order expansions for

$$\ddot{u} + u = \epsilon(1 - u^2)\dot{u} + \epsilon u^3$$

Compare the results of the three methods. Which of these techniques would you recommend for such problems?

5.8. Consider the problem

$$\ddot{u} + 2\mu\dot{u} + \nu^2 u = -\epsilon f(u, \dot{u})$$

As $\epsilon \to 0$

$$u = a_0 e^{-\mu t} \cos(\omega_0 t + \phi_0), \qquad \omega_0 = \sqrt{\nu^2 - \mu^2}$$

(a) For $\epsilon \neq 0$ determine a uniform expansion following Mendelson (1970) by letting $u = u(a, \psi)$, where

$$u = u_0 + \epsilon u_1 + \cdots$$

$$\frac{da}{dt} = -\mu a + \epsilon A_1(a) + \cdots$$

$$\frac{d\psi}{dt} = \omega_0 + \epsilon B_1(a) + \cdots$$

(b) Determine an expansion using the Krylov–Bogoliubov–Mitropolski technique.

(c) Which of these expansions is more accurate?

5.9. Consider the problem

$$\ddot{u} + \delta u + \epsilon b_1 u^m + \epsilon b_2 u^{n-1} - \epsilon b_3 u^{n-1} \cos \lambda t = 0$$

where δ, ϵ, b_i, and λ are constants and m is an odd while n is an even integer with $m > n$.

(a) For small ϵ find a solution of the form

$$u = a(t) \cos \theta, \qquad \theta = \omega t - \phi(t), \qquad \omega = \lambda/n$$

and use the method of averaging to determine equations for a and ϕ (Tso and Caughey, 1965).

(b) Determine a Hamiltonian corresponding to the above equation and then determine a first-order expansion using canonical variables for δ near ω^2.

(c) Compare the results of the two techniques.

5.10. The problem of a spherical pendulum (i.e., a particle moving under the action of gravity on the surface of a smooth fixed sphere) is represented by the Hamiltonian (Johansen and Kane, 1969)

$$H = \frac{p_1^2 + p_2^2}{2m} - mg\sqrt{l^2 - q_1^2 - q_2^2} - \frac{(p_1 q_1 + p_2 q_2)^2}{2ml^2}$$

where q_i and p_i are the coordinates and momenta of the particle, m is its mass, g is the gravitational acceleration, and l is the radius of the sphere.

(a) Determine a first-order solution for small amplitudes using the method of averaging with canonical variables.

(b) Determine a second-order expansion using Lie transforms.

(c) Determine a second-order expansion using the generalized method of averaging.

(d) Compare the three resulting expansions.

5.11. Consider the problem of a swinging spring with damping

$$\ddot{x} + \delta_1 \dot{x} + \frac{k}{m} x + g(1 - \cos\theta) - (l + x)\dot{\theta}^2 = 0$$

$$\ddot{\theta} + \delta_2\dot{\theta} + \frac{g}{l + x}\sin\theta + \frac{2}{l + x}\dot{x}\dot{\theta} = 0$$

If $\omega_1^2 = k/m$ and $\omega_2^2 = g/l$, determine second-order uniform expansions using the generalized method of averaging, and the Lie transforms for the cases (a) $\omega_1 \approx 2\omega_2$ and (b) $\omega_1 \approx 3\omega_2$.

5.12. Consider (3.1.63) through (3.1.65). (a) Show that the corresponding Hamiltonian is

$$H = \tfrac{1}{2}(p_1^2 + p_2^2) + q_2 p_1 - q_1 p_2 + \tfrac{1}{2}(q_1^2 + q_2^2) - \tfrac{1}{2}(h_1 q_2^2 + h_2 q_1^2)(1 + e\cos f)^{-1}$$

(b) Use the method of averaging with canonical variables to determine a first-order expansion near the transition curves starting at (μ_0, e) where $\mu_0 = (1 - 2\sqrt{2}/3)/2$.

(c) Use the Lie transforms to determine a second-order expansion near these transition curves.

5.13. Consider the motion of a particle described by the Hamiltonian

$$H = \tfrac{1}{2}(p_1^2 + p_2^2) + \tfrac{1}{2}(q_2 p_1 - q_1 p_2) + \tfrac{9}{8}q_1^2 + (\delta + \tfrac{1}{8})q_2^2 + q_1^3 + 2q_1 q_2^2$$

where δ is a constant parameter. (a) Show that the circular frequencies of the linearized problem are 1 and 2 when $\delta = 1$, (b) use the method of averaging with canonical variables to determine a first-order expansion for small amplitudes when $\delta \approx 1$, (c) determine a first-order expansion using the method of averaging, and (d) which of these techniques would you recommend for such problems?

5.14. Consider the problem (Sethna, 1965)

$$\ddot{x} + \omega_1^2 x = \epsilon(3b_1 x^2 + 2b_2 xy + b_3 y^2) - \epsilon\delta_1\dot{x} + K_1\cos\lambda t$$

$$\ddot{y} + \omega_2^2 y = \epsilon(b_2 x^2 + 2b_3 xy + 3b_4 y^2) - \epsilon\delta_2\dot{y} + K_2\sin\lambda t$$

If internal resonance means $\omega_1 \approx 2\omega_2$ or $\omega_2 \approx 2\omega_1$, resonant excitation means $\lambda \approx \omega_1$ or ω_2, soft excitation is denoted by $K_i = \epsilon k_i$ with $k_i = O(1)$, and hard excitation is denoted by $K_i = O(1)$, use the method of averaging to determine first-order expansions for the following cases:

(a) Hard nonresonant excitation in the absence of internal resonance.

(b) hard nonresonant excitation in the presence of internal resonance.

(c) soft resonant excitation in the absence of internal resonance.

(d) soft resonant excitation in the presence of internal resonance.

5.15. Traveling waves in a cold plasma are governed by

$$\frac{\partial \rho}{\partial t} + \frac{\partial}{\partial x} (\rho u) = 0$$

$$\frac{\partial u}{\partial t} + u \frac{\partial u}{\partial x} + E = 0$$

$$\frac{\partial E}{\partial x} = \omega_p{}^2 (1 - \rho)$$

Let $\rho = 1 + O(\epsilon)$, $u = O(\epsilon)$, and $E = O(\epsilon)$. Use the method of averaging to determine the temporal as well as the spatial variation of the amplitude and phase of a monochromatic traveling wave.

5.16. Consider the problem

$$\phi_{tt} - \phi_{xx} + \phi = 0$$

(a) Show that

$$\phi = a \cos \theta, \qquad \theta = kx - \omega t$$

is a solution of this equation if $\omega^2 = k^2 + 1$.

(b) Show that the above equation can be written in the conservation forms (Whitham, 1965b)

$$\frac{\partial}{\partial t} [\tfrac{1}{2}(\phi_t{}^2 + \phi_x{}^2 + \phi^2)] + \frac{\partial}{\partial x} (-\phi_x \phi_t) = 0$$

$$\frac{\partial}{\partial t} (-\phi_x \phi_t) + \frac{\partial}{\partial x} [\tfrac{1}{2}(\phi_t{}^2 + \phi_x{}^2 - \phi^2)] = 0$$

(c) Let $\phi = a \cos \theta$ with $a = a(x, t)$, $k = \theta_x$, and $\omega = -\theta_t$ in the conservation equations. Assume that a, ω, and k are slowly varying functions of x and t. Hence average these equations over $\theta = 0$ to 2π, keeping a, ω, and k constants, and obtain

$$\frac{\partial k}{\partial t} + \omega' \frac{\partial k}{\partial x} = 0$$

$$\frac{\partial a^2}{\partial t} + \frac{\partial}{\partial x} (\omega' a^2) = 0$$

5.17. Consider the equation

$$u_{tt} - c^2 u_{xx} + u = \epsilon u^3$$

(a) Write down the Lagrangian corresponding to this equation, (b) determine a first-order expansion for traveling waves with constant wave number and frequency but both spatially and temporally varying amplitude and phase.

5.18. The problem of nonlinear transverse oscillations in a homogeneous free-free beam with a nonlinear moment-curvature relationship is described by the Lagrangian

$$L = \tfrac{1}{2}\rho \left(\frac{\partial w}{\partial t}\right)^2 - \tfrac{1}{2}\beta \left[\left(\frac{\partial^2 w}{\partial x^2}\right)^2 + \tfrac{1}{2}\epsilon \left(\frac{\partial^2 w}{\partial x^2}\right)^4\right]$$

where ρ, β, and ϵ are constants. Determine a first-order expansion for traveling waves with slowly varying amplitudes and phases using (a) the variational approach, and (b) writing down the governing equation and then using the method of averaging.

5.19. Consider Bretherton's (1964) model equation

$$\phi_{tt} + \phi_{xxxx} + \phi_{xx} + \phi = \phi^2$$

(a) Show that the linear problem has the dispersion relationship

$$\omega^2 = k^4 - k^2 + 1$$

(b) Determine the wave number corresponding to nth-harmonic resonance.

(c) Use the method of averaging to determine a first-order expansion near the second-harmonic resonance condition (let the amplitudes and phases be functions of x and t).

(d) Write down the corresponding Lagrangian, and then use the variational approach to determine a first-order expansion near the second-harmonic resonance condition.

CHAPTER 6

The Method of Multiple Scales

6.1. Description of the Method

There are three variants of the method of multiple scales. We describe them by discussing the linear damped oscillator

$$\ddot{x} + x = -2\epsilon\dot{x} \qquad (6.1.1)$$

We chose this example because its exact solution is available for comparison with the approximate solution obtained, and because we will be able to display the different variants of the method more clearly without involving ourselves in algebra.

To start, let us determine a straightforward asymptotic expansion for small ϵ. Thus we assume that

$$x = x_0 + \epsilon x_1 + \epsilon^2 x_2 + \cdots \qquad (6.1.2)$$

Substituting (6.1.2) into (6.1.1) and equating coefficients of equal powers of ϵ to zero lead to

$$\ddot{x}_0 + x_0 = 0 \qquad (6.1.3)$$

$$\ddot{x}_1 + x_1 = -2\dot{x}_0 \qquad (6.1.4)$$

$$\ddot{x}_2 + x_2 = -2\dot{x}_1 \qquad (6.1.5)$$

The general solution of (6.1.3) is

$$x_0 = a \cos(t + \phi) \qquad (6.1.6)$$

where a and ϕ are arbitrary constants. Substituting for x_0 into (6.1.4) and solving the resulting equation, we obtain

$$x_1 = -at \cos(t + \phi) \qquad (6.1.7)$$

Substituting for x_1 in (6.1.5) and solving for x_2, we obtain

$$x_2 = \tfrac{1}{2}at^2 \cos(t + \phi) + \tfrac{1}{2}at \sin(t + \phi) \qquad (6.1.8)$$

228

Therefore

$$x = a \cos (t + \phi) - \epsilon a t \cos (t + \phi)$$
$$+ \tfrac{1}{2}\epsilon^2 a[t^2 \cos (t + \phi) + t \sin (t + \phi)] + O(\epsilon^3) \quad (6.1.9)$$

It is obvious that (6.1.9) is a poor approximation to x when t is as large as ϵ^{-1}, because then the second (ϵx_1) and the third ($\epsilon^2 x_2$) terms are not small compared to x_0 and ϵx_1, respectively (x_1 and x_2 contain secular terms), as was assumed when we carried out the above expansion. Thus the straight-forward expansion is not valid when t increases to $O(\epsilon^{-1})$, and the source of the difficulty is the infinite domain as discussed in Section 2.1.

The failure of the above straightforward expansion can be seen by investigating the exact solution of (6.1.1), which is given by

$$x = ae^{-\epsilon t} \cos [\sqrt{1 - \epsilon^2}\, t + \phi] \quad (6.1.10)$$

Equation (6.1.9) can be obtained by expanding (6.1.10) for small ϵ with t kept fixed. Thus the exponent and cosine factors are represented by

$$\exp (-\epsilon t) = 1 - \epsilon t + \tfrac{1}{2}\epsilon^2 t^2 + \cdots \quad (6.1.11)$$

$$\cos (\sqrt{1 - \epsilon^2}\, t + \phi) = \cos (t + \phi) + \tfrac{1}{2}\epsilon^2 t \sin (t + \phi) + \cdots \quad (6.1.12)$$

It is clear that $\exp (-\epsilon t)$ can be approximated by a finite number of terms only if the combination ϵt is small. Since ϵ is small, this means that $t = O(1)$. When t is as large as ϵ^{-1}, ϵt is not small and the truncated expansion breaks down. The above truncated series is satisfactory up to a certain value of t after which $\exp (-\epsilon t)$ and the truncated series differ from each other by a quantity that exceeds the prescribed limit of accuracy. Adding more terms to the truncated series increases the value of t to a new value t' for which this truncated series is satisfactory. However, for $t > t'$, the difference between $\exp (-\epsilon t)$ and the new truncated series again exceeds the prescribed limit of accuracy. All terms of the series are needed to give a satisfactory expansion for $\exp (-\epsilon t)$ for all t. Thus to determine an expansion valid for times as large as ϵ^{-1}, the combination ϵt should be considered a single variable $T_1 = O(1)$. Then any truncated expansion for $\exp (-\epsilon t)$ valid for times as large as ϵ^{-1} is of the form

$$\exp (-\epsilon t) = \exp (-T_1) \quad (6.1.13)$$

Similarly, the truncated expansion (6.1.12) is not satisfactory when t is as large as $O(\epsilon^{-2})$. To obtain a truncated asymptotic expansion for $\cos [\sqrt{1 - \epsilon^2}\, t + \phi]$ valid for $t = O(\epsilon^{-2})$, $\epsilon^2 t$ should be considered a single variable $T_2 = O(1)$. With this condition

$$\cos\left[\sqrt{1 - \epsilon^2}\, t + \phi\right] = \cos\left[t - \tfrac{1}{2}T_2 + \phi - \tfrac{1}{8}\epsilon^4 t + \cdots\right]$$
$$= \cos\left(t - \tfrac{1}{2}T_2 + \phi\right) + \tfrac{1}{8}\epsilon^4 t \sin\left(t - \tfrac{1}{2}T_2 + \phi\right) + \cdots$$

$$(6.1.14)$$

Expansion (6.1.14) is valid when $t = O(\epsilon^{-2})$ because the correction term (second term) is $O(\epsilon^2)$ or less for all times up to $O(\epsilon^{-2})$. However, this expansion breaks down when $t = O(\epsilon^{-4})$ because the second term ceases to be small compared to the first. To obtain an expansion valid for times as large as $O(\epsilon^{-4})$, another variable, $T_4 = \epsilon^4 t = O(1)$, should be introduced.

The above discussion suggests that $x(t; \epsilon)$ depends explicitly on t, ϵt, $\epsilon^2 t, \ldots$, as well as ϵ itself. This can also be seen from the exact solution. Thus, in order to determine a truncated expansion valid for all t up to $O(\epsilon^{-M})$, where M is a positive integer, we must determine the dependence of x on the $M + 1$ different time scales $T_0, T_1, \ldots, T_M$, where

$$T_m = \epsilon^m t \qquad (6.1.15)$$

The time scale T_1 is slower than T_0, while the time scale T_2 is slower than T_1. In general, T_n is slower than T_{n-1}. Thus we assume that

$$x(t; \epsilon) = \tilde{x}(T_0, T_1, \ldots, T_M; \epsilon)$$
$$= \sum_{m=0}^{M-1} \epsilon^m x_m(T_0, T_1, \ldots, T_M) + O(\epsilon T_M) \qquad (6.1.16)$$

The error in (6.1.16) is stated $O(\epsilon T_M)$ to remind the reader that this expansion is valid for times up to $O(\epsilon^{-M})$. Beyond these times, we must use other time scales to keep the expansion uniformly valid. Equations (6.1.15) and (6.1.16) show that the problem has been transformed from an ordinary differential equation to a partial differential equation. If the original problem is a partial differential equation, then the introduction of different time scales increases the number of independent variables. By using the chain rule, the time derivative is transformed according to

$$\frac{d}{dt} = \frac{\partial}{\partial T_0} + \epsilon \frac{\partial}{\partial T_1} + \epsilon^2 \frac{\partial}{\partial T_2} + \cdots \qquad (6.1.17)$$

Equations (6.1.15) through (6.1.17) formulate one version of the method of multiple scales; namely, the many-variable version. This technique has been developed by Sturrock (1957, 1963), Frieman (1963), Nayfeh (1965c, d, 1968), and Sandri (1965, 1967). Equations (6.1.16) and (6.1.17) show that a uniformly valid expansion is obtained by expanding the derivatives as well as the dependent variables in powers of the small parameter. Hence Sturrock and Nayfeh called this technique the derivative-expansion method.

Substitution of (6.1.16) and (6.1.17) into (6.1.1) and equating coefficients of like powers of ϵ, we obtain equations for determining $x_0, x_1, \ldots, x_M$. The solutions of these equations contain arbitrary functions of the time scales $T_1, T_2, \ldots, T_M$. In order to determine these functions, additional conditions need to be imposed. If (6.1.16) is to be valid for times as large as ϵ^{-M}, $\epsilon^m x_m$ should be a small correction to $\epsilon^{m-1} x_{m-1}$, which in turn should be a small correction to $\epsilon^{m-2} x_{m-2}$. Thus we require that

$$\frac{x_m}{x_{m-1}} < \infty \quad \text{for all} \quad T_0, T_1, \ldots, T_M$$

This condition does not mean that each x_m is bounded. In fact, each x_m may be unbounded. However, this condition requires, as in Lighthill's technique (Section 3.2), that higher approximations be no more singular than the first term. This condition is equivalent to the elimination of secular terms.

The second version of the method of multiple scales was introduced by Cole and Kevorkian (1963) and applied by Kevorkian (1966a) and Cole (1968) to several examples. Morrison (1966a) showed that this procedure is equivalent to the method of averaging to second order, while Perko (1969) established their equivalence to nth order. Kevorkian (1966b) showed the equivalence of this procedure and von Zeipel's method to first order. If we investigate the exact solution (6.1.10), we find that t appears in either of the combinations ϵt or $\sqrt{1 - \epsilon^2}\, t$. Hence to determine an expansion valid for large times, one introduces the two time scales

$$\xi = \epsilon t \quad \text{and} \quad \eta = \sqrt{1 - \epsilon^2}\, t = (1 - \tfrac{1}{2}\epsilon^2 - \tfrac{1}{8}\epsilon^4 + \cdots)t \quad (6.1.18)$$

Therefore Cole and Kevorkian (1963) assumed that

$$x(t; \epsilon) = \tilde{x}(\xi, \eta; \epsilon)$$
$$= \sum_{m=0}^{M-1} \epsilon^m x_m(\xi; \eta) + O(\epsilon^M) \quad (6.1.19)$$

where

$$\xi = \epsilon t, \qquad \eta = (1 + \epsilon^2 \omega_2 + \epsilon^3 \omega_3 + \cdots + \epsilon^M \omega_M)t \quad (6.1.20)$$

with constant ω_n. In this case ξ is slower than η, and the time derivative is transformed according to

$$\frac{d}{dt} = \epsilon \frac{\partial}{\partial \xi} + (1 + \epsilon^2 \omega_2 + \epsilon^3 \omega_3 + \cdots + \epsilon^M \omega_M) \frac{\partial}{\partial \eta} \quad (6.1.21)$$

These two versions can be generalized considerably. Thus the many-variable version can be generalized (Nayfeh, 1967b) by using an asymptotic sequence $\delta_n(\epsilon)$ rather than powers of ϵ. Thus

$$T_n = \delta_n(\epsilon)t \tag{6.1.22}$$

$$\frac{d}{dt} = \sum_{n=0}^{M} \delta_n(\epsilon) \frac{\partial}{\partial T_n} \tag{6.1.23}$$

Equations (6.1.22) and (6.1.23) can be generalized further by letting

$$T_n = \delta_n(\epsilon)g_n[\mu_n(\epsilon)t] \tag{6.1.24}$$

$$\frac{d}{dt} = \sum_{n=0}^{M} \delta_n(\epsilon)\mu_n(\epsilon)g_n'[\mu_n(\epsilon)t] \frac{\partial}{\partial T_n} \tag{6.1.25}$$

where $\mu_n(\epsilon)$ is another asymptotic sequence. Thus (6.1.24) allows for linear as well as nonlinear time scales.

Similarly the two-variable expansion procedure can also be generalized. Thus (6.1.20) and (6.1.21) can be generalized to

$$\xi = \mu(\epsilon)t, \qquad \eta = \sum_{n=0}^{M} \delta_n(\epsilon)g_n[\mu(\epsilon)t] \tag{6.1.26}$$

$$\frac{d}{dt} = \mu(\epsilon)\frac{\partial}{\partial \xi} + \left(\sum_{n=0}^{M} \delta_n(\epsilon)\mu(\epsilon)g_n'[\mu(\epsilon)t]\right)\frac{\partial}{\partial \eta} \tag{6.1.27}$$

This general form was developed by several investigators including Kuzmak (1959), Cochran (1962), Mahony (1962), and Nayfeh (1964, 1965b). Klimas, Ramnath, and Sandri (1970) investigated the role of gauge transformations for uniformization of asymptotic expansions.

The method of multiple scales is so popular that it is being rediscovered just about every 6 months. It has been applied to a wide variety of problems in physics, engineering, and applied mathematics.

Cole and Kevorkian (1963), Nayfeh (1965c, 1967b, 1968), Kevorkian (1966a), Davis and Alfriend (1967), Schwertassek (1969), and Musa (1967), Rasmussen (1970), and Reiss (1971) analyzed weakly linear and nonlinear vibrations governed by second- or third-order ordinary differential equations. Kuzmak (1959) studied nonlinear oscillations in second-order differential equations with slowly varying coefficients. Cochran (1962), Nayfeh (1964, 1965b), and Fowkes (1968I) used the generalized version to analyze turning point problems for second-order linear differential equations. Cochran (1962), Nayfeh (1964, 1965b), and Ramnath and Sandri (1969) used the generalized method to study linear equations with variable coefficients, while Cheng and Wu (1970) analyzed the effect of the scales on the problem of an aging spring. Noerdlinger and Petrosian (1971) discussed a linear in-homogeneous equation with slowly varying coefficients which describes the effect of cosmological expansion on self-gravitating ensembles of particles, while Kevorkian (1971) investigated the problem of passage through resonance for a one-dimensional oscillator with a slowly varying frequency. Cochran (1962), O'Malley (1968a, b), and Searl (1971) applied the generalized method to boundary value problems for certain nonlinear second-order differential equations, while Cochran (1962) and Ackerberg and O'Malley (1970) applied this method to second-order equations that exhibit turning points and boundary layers. Tam (1968) used the generalized version to solve the Orr–Sommerfeld equation.

In orbital mechanics, Nayfeh (1965a) used the generalized version to analyze the earth-moon-spaceship problem. Ting and Brofman (1964) and Nayfeh (1966) analyzed the problem of takeoff of a satellite from a circular orbit with a small thrust, Shi and Eckstein (1966) investigated takeoff from an elliptical orbit with a small thrust, Kevorkian (1966a) and Brofman (1967) studied the motion of a satellite subjected to a small thrust or drag, and Eckstein and Shi (1967) analyzed the motion of a satellite with variable mass and low thrust. Eckstein, Shi, and Kevorkian (1966a) determined the motion of a satellite around the primary in the restricted problem of three bodies, while Alfriend and Rand (1969) determined the stability of the triangular points in the elliptic restricted problem of three bodies. Eckstein, Shi, and Kevorkian (1966c) evaluated higher-order terms in the motion of a satellite using the energy integral and evaluated the effects of eccentricity and inclination (1966b). Shi and Eckstein (1968) analyzed the motion of an artificial satellite having a period commensurable with the rotation period of the primary. Alfriend (1970) and Nayfeh (1971b) studied the two-to-one resonances, while Nayfeh and Kamel (1970b) and Alfriend (1971b) studied the three-to-one resonances near the equilateral libration points. Alfriend (1971a) analyzed two-to-one resonances in two-degree-of-freedom Hamiltonian systems.

In flight mechanics, Ashley (1967) discussed the role of different time scales in flight mechanics, while Nayfeh and Saric (1971b) analyzed nonlinear resonances in the motion of a missile with slight asymmetries. Nayfeh (1969a) used the generalized version to study the motion of a rolling missile with variable roll rate and dynamic pressure but linear aerodynamics, while Nayfeh and Saric (1972a) studied the motion of a missile with nonlinear aerodynamics and variable roll rate and dynamic pressure. Ramnath (1970b) studied the transition dynamics of VTOL aircraft.

In solid mechanics, Amazigo, Budiansky and Carrier (1970) analyzed the nonlinear buckling of imperfect columns, while Reiss and Matkowsky (1971) investigated the nonlinear dynamic buckling of a compressed elastic column. Mortell (1968) analyzed the problem of a traveling wave on a cylindrical shell and the propagation of waves on a spherical shell (1969). Kelly (1965) and Morino (1969) studied nonlinear panel flutter, while Spriggs, Messiter, and Anderson (1969) discussed membrane flutter.

In partial differential equations, Cochran (1962), Nayfeh (1965b), and Comstock (1971) treated elliptic equations. Fowkes (1968, Part II) obtained uniformly valid expansions for caustic problems. Neubert (1970) obtained solutions for the Helmholtz equation for turbulent water. Wingate and Davis (1970) discussed the propagation of waves in an inhomogeneous rod. Keller and Kogelman (1970) treated a nonlinear initial boundary value problem for a partial differential equation.

Luke (1966) studied the Klein–Gordon equation and general variational equations of second order, while Emery (1970) treated the case of several dependent variables and several rapidly rotating phases. Ablowitz and Benney (1970) investigated the evolution of multiphase modes for the Klein–Gordon equation. Nayfeh and Hassan (1971) and Nayfeh and Saric (1972b) discussed nonlinear dispersive waves on the interface of two fluids and in a hot electron plasma. Parker (1969) analyzed the effects of relaxation and diffusive damping on dispersive waves.

In wave interactions, Benney and Saffman (1966), Benney (1967), Davidson (1967), Benney and Newell (1967), Hoult (1968), Newell (1968), and Benney and Newell (1969) investigated the nonlinear interaction of random waves in a dispersive medium. Davidson (1969) studied the time evolution of wave correlations in a uniformly turbulent ensemble of weakly nonlinear dispersive systems.

In water waves, Carrier (1966) analyzed gravity waves in water of variable depth, while Hoogstraten (1968) and Freeman and Johnson (1970) studied shallow water waves in shear flows. Jacobs (1967) solved the tidal equations. Murray (1968) treated free surface oscillations in a tank resulting from drainage. Chu and Mei (1970) studied slowly varying Stokes' waves. McGoldrick (1970) and Nayfeh (1970b) treated the second-harmonic resonance case, while Nayfeh (1970d, 1971a) investigated the third-harmonic resonance case in the interaction of capillary and gravity waves.

In atmospheric science, Newell (1969) treated the resonant interaction of Rossby wave packets, while Stone (1969) analyzed the problem of baroclinic waves. Shabbar (1971) discussed the side-band resonance mechanism in the atmosphere supporting Rossby waves, while Lindzen (1971) studied the propagation of equatorial Tanai and Kelvin waves through shear.

In plasma physics, Ball (1964), Taussig (1969), and Tam (1969, 1970) analyzed the propagation of nonlinear waves in a cold plasma, while Nayfeh (1965d) and Das (1971) investigated nonlinear oscillations in a hot electron plasma. Davidson (1968) treated nonlinear oscillations in a Vlasov–Maxwell plasma. Peyret (1966) analyzed plasma waves in an accelerator, while Butler and Gribben (1968) discussed nonlinear waves in a nonuniform plasma.

Maroli and Pozzoli (1969) studied the penetration of high-frequency electromagnetic waves into a slightly ionized plasma. Abraham–Shrauner (1970a, b) investigated the suppression of runaway of electrons in a Lorentz gas. Chen and Lewak (1970), Chen (1971), and Prasad (1971) studied parametric excitation in a plasma, while Lewak (1971) discussed the interaction of electrostatic waves in a plasma. Dobrowolny and Rogister (1971) and Rogister (1971) analyzed the propagation of hydromagnetic waves in a high-beta plasma.

In hydrodynamic and plasma stability, Frieman and Rutherford (1964) developed a kinetic theory for weakly unstable plasmas, while Albright (1970) analyzed the stabilization of transverse plasma instability. Kelly (1967) investigated the stability of an inviscid shear layer. Benney and Roskes (1969) analyzed the instability of gravity waves. Kiang (1969) and Nayfeh (1969b) studied Rayleigh–Taylor instability, while Newell and Whitehead (1969) analyzed postcritical Rayleigh–Bénard convection. Nayfeh (1970c) investigated the nonlinear stability of a liquid jet. Nayfeh and Saric (1971a) studied nonlinear Kelvin–Helmholtz instability, while Puri (1971) analyzed the effects of viscosity and membrane on the oscillation of two superposed fluids. Stewartson and Stuart (1971) treated the nonlinear stability of plane Poiseuille flow. Mitchell (1971) applied this technique to combustion instability.

In fluid mechanics, Germain (1967) and Lick (1970) reviewed recent developments in aerodynamics and nonlinear wave propagation in fluids including the methods of matched asymptotic expansions, strained coordinates, and multiple scales. Benney (1965) analyzed the flow field produced by finite-amplitude oscillation of a disk about a steady state of rotation, while Barcilon (1970) studied the linear viscous theory of steady rotating fluid flows. Rubbert and Landahl (1967) discussed the transonic airfoil problem. Peyret (1970) treated the problem of steady flow of a conducting perfectly compressible fluid in a channel. Chong and Sirovich (1971) studied the problem of steady supersonic dissipative gas dynamics. Cheng, Kirsch, and Lee (1971) analyzed the behavior of a strong shock wave initiated by a point explosion and driven continuously outward by an inner contact surface.

In general physics, Caughey and Payne (1967) used a combination of the method of multiple scales and the method of matched asymptotic expansions to solve the Fokker–Planck equation arising from the response of self-excited oscillators to random excitations. Brau (1967) obtained a stochastic theory for the dissociation and recombination of diatomic molecules. Ramnath (1970a) obtained an approximation for the Thomas–Fermi model in atomic physics and treated a class of nonlinear differential equations arising in astrophysics (1971). Meyer (1971) investigated Rayleigh scattering of a laser beam from a massive relativistic two-level atom, while Nienhuis (1970) studied Brownian motion with a rotational degree of freedom.

In statistical mechanics, Maroli (1966) solved Boltzmann's equation to obtain a kinetic theory of high-frequency resonance gas discharge breakdown, while Caldirola, De Barbieri, and Maroli (1966) solved Boltzmann's equation for the electronic distribution function. De Barbieri and Maroli (1967) solved the Liouville equation to analyze the dynamics of weakly ionized gases, while Goldberg and Sandri (1967) and Ramanathan and Sandri (1969) derived sets of hierarchical equations.

In the remainder of this section, we describe the three versions of the method of multiple scales and their application to the simple linear damped oscillator given by (6.1.1). In the following sections we apply these techniques to different problems in mathematical physics.

6.1.1.   MANY-VARIABLE VERSION (THE DERIVATIVE-EXPANSION PROCEDURE)

Substituting (6.1.16) and (6.1.17) into (6.1.1) and equating coefficients of like powers of ϵ, we obtain the following equations for x_0, x_1, and x_2

$$\frac{\partial^2 x_0}{\partial T_0^2} + x_0 = 0 \tag{6.1.28}$$

$$\frac{\partial^2 x_1}{\partial T_0^2} + x_1 = -2\frac{\partial x_0}{\partial T_0} - 2\frac{\partial^2 x_0}{\partial T_0 \partial T_1} \tag{6.1.29}$$

$$\frac{\partial^2 x_2}{\partial T_0^2} + x_2 = -2\frac{\partial x_1}{\partial T_0} - 2\frac{\partial^2 x_1}{\partial T_0 \partial T_1} - \frac{\partial^2 x_0}{\partial T_1^2} - 2\frac{\partial^2 x_0}{\partial T_0 \partial T_2} - 2\frac{\partial x_0}{\partial T_1} \tag{6.1.30}$$

The general solution of (6.1.28) is

$$x_0 = A_0(T_1, T_2)e^{iT_0} + \bar{A}_0(T_1, T_2)e^{-iT_0} \tag{6.1.31}$$

where $\bar{A}_0$ is the complex conjugate of A_0. This solution is simply equivalent to (6.1.6) where a and ϕ are taken to be functions of the slow time scales T_1 and T_2 rather than being constants. Substituting for x_0 from (6.1.31) into (6.1.29), we obtain

$$\frac{\partial^2 x_1}{\partial T_0^2} + x_1 = -2i\left(A_0 + \frac{\partial A_0}{\partial T_1}\right)e^{iT_0} + 2i\left(\bar{A}_0 + \frac{\partial \bar{A}_0}{\partial T_1}\right)e^{-iT_0} \tag{6.1.32a}$$

The general solution of (6.1.32a) is

$$x_1 = A_1(T_1, T_2)e^{iT_0} + \bar{A}_1(T_1, T_2)e^{-iT_0}$$
$$- \left(A_0 + \frac{\partial A_0}{\partial T_1}\right)T_0 e^{iT_0} - \left(\bar{A}_0 + \frac{\partial \bar{A}_0}{\partial T_1}\right)T_0 e^{-iT_0} \tag{6.1.32b}$$

Comparing (6.1.32b) with (6.1.31) shows that ϵx_1 is a small correction to x only when $\epsilon T_0 = \epsilon t$ is small. In order to obtain an expansion valid for times as large as $O(\epsilon^{-1})$, the secular terms, $T_0 \exp(\pm i T_0)$, in (6.1.32b) must vanish; that is

$$A_0 + \frac{\partial A_0}{\partial T_1} = 0 \tag{6.1.33}$$

or

$$A_0 = a_0(T_2)e^{-T_1} \tag{6.1.34}$$

Then (6.1.32b) becomes

$$x_1 = A_1(T_1, T_2)e^{iT_0} + \bar{A}_1(T_1, T_2)e^{-iT_0} \tag{6.1.35}$$

Using x_0 and x_1 in (6.1.30), we obtain

$$\frac{\partial^2 x_2}{\partial T_0^2} + x_2 = -Q(T_1, T_2)e^{iT_0} - \bar{Q}(T_1, T_2)e^{-iT_0} \tag{6.1.36}$$

where

$$Q(T_1, T_2) = 2iA_1 + 2i\frac{\partial A_1}{\partial T_1} - a_0 e^{-T_1} + 2i\frac{\partial a_0}{\partial T_2}e^{-T_1} \tag{6.1.37}$$

The terms on the right-hand side of (6.1.36) produce secular terms because the particular solution is

$$x_2 = \tfrac{1}{2}iQ(T_1, T_2)T_0 e^{iT_0} - \tfrac{1}{2}i\bar{Q}(T_1, T_2)T_0 e^{-iT_0} \tag{6.1.38}$$

These secular terms make $\epsilon^2 x_2$ the same order as ϵx_1 when t is as large as $O(\epsilon^{-1})$. In order to eliminate these secular terms, Q must vanish; that is

$$\frac{\partial A_1}{\partial T_1} + A_1 = \tfrac{1}{2}i\left(-a_0 + 2i\frac{\partial a_0}{\partial T_2}\right)e^{-T_1} \tag{6.1.39}$$

In general, one does not need to solve for x_2 in order to arrive at (6.1.39). One needs only to inspect (6.1.36) and eliminate terms that produce secular terms. The general solution of (6.1.39) is

$$A_1 = \left[a_1(T_2) + \tfrac{1}{2}i\left(-a_0 + 2i\frac{\partial a_0}{\partial T_2}\right)T_1\right]e^{-T_1} \tag{6.1.40}$$

Substituting for A_1 into (6.1.35), we obtain

$$x_1 = \left[a_1(T_2) + \tfrac{1}{2}i\left(-a_0 + 2i\frac{\partial a_0}{\partial T_2}\right)T_1\right]e^{-T_1}e^{iT_0} + CC \tag{6.1.41}$$

where CC stands for the complex conjugate of the preceding expression. However

$$x_0 = [a_0 e^{iT_0} + \bar{a}_0 e^{-iT_0}]e^{-T_1} \tag{6.1.42}$$

Therefore, as $T_1 \to \infty$, although x_0 and $x_1 \to 0$, ϵx_1 becomes $O(x_0)$ as t increases to $O(\epsilon^{-2})$. Thus the expansion $x_0 + \epsilon x_1$ breaks down for t as large as $O(\epsilon^{-2})$ unless the coefficients of T_1 in the brackets in (6.1.41) vanish; that is, unless

$$-a_0 + 2i\frac{\partial a_0}{\partial T_2} = 0 \qquad (6.1.43)$$

or

$$a_0 = a_{00}e^{-iT_2/2} \qquad (6.1.44)$$

where a_{00} is a constant. Then (6.1.40) becomes

$$A_1 = a_1(T_2)e^{-T_1} \qquad (6.1.45)$$

Therefore

$$x = e^{-T_1}\{a_{00}e^{i(T_0 - T_2/2)} + \bar{a}_{00}e^{-i(T_0 - T_2/2)}$$
$$+ \epsilon[a_1(T_2)e^{iT_0} + \bar{a}_1(T_2)e^{-iT_0}]\} + O(\epsilon^2) \qquad (6.1.46)$$

The function $a_1(T_2)$ can be determined by carrying out the expansion to third order

$$a_1(T_2) = a_{11}e^{-iT_2/2} \qquad (6.1.47)$$

where a_{11} is a constant. If we assume that the initial conditions are such that $x(0) = a \cos \phi$ and $\dot{x}(0) = -a(\sin \phi \sqrt{1 - \epsilon^2} + \epsilon \cos \phi)$ and replace T_n by $\epsilon^n t$, we obtain

$$x = ae^{-\epsilon t} \cos(t - \tfrac{1}{2}\epsilon^2 t + \phi) + R \qquad (6.1.48)$$

where R is the remainder. From (6.1.10) and (6.1.48), we find that

$$R = ae^{-\epsilon t}[\cos(t\sqrt{1 - \epsilon^2} + \phi) - \cos(t - \tfrac{1}{2}\epsilon^2 t + \phi)]$$
$$= -2ae^{-\epsilon t}\sin[\tfrac{1}{2}(\sqrt{1 - \epsilon^2} + 1 - \tfrac{1}{2}\epsilon^2)t + \phi]\sin[\tfrac{1}{2}(\sqrt{1 - \epsilon^2} - 1 + \tfrac{1}{2}\epsilon^2)t]$$
$$= -2ae^{-\epsilon t}\sin[\tfrac{1}{2}(\sqrt{1 - \epsilon^2} + 1 - \tfrac{1}{2}\epsilon^2)t + \phi]\sin[(-\tfrac{1}{16}\epsilon^4 + \cdots)t]$$
$$= O(\epsilon^4 t) \qquad (6.1.49)$$

For linear equations such as (6.1.1), we may introduce the different time scales without expanding x. Thus using (6.1.17) in (6.1.1), we obtain

$$\left[\frac{\partial^2}{\partial T_0^2} + 2\epsilon\frac{\partial^2}{\partial T_0\,\partial T_1} + \epsilon^2\left(\frac{\partial^2}{\partial T_1^2} + 2\frac{\partial^2}{\partial T_0\,\partial T_2}\right) + \cdots\right]x + x$$
$$= -2\epsilon\left(\frac{\partial}{\partial T_0} + \epsilon\frac{\partial}{\partial T_1} + \epsilon^2\frac{\partial}{\partial T_2} + \cdots\right)x \qquad (6.1.50)$$

Equating the coefficients of like powers of ϵ to zero yields

$$\frac{\partial^2 x}{\partial T_0^2} + x = 0 \tag{6.1.51}$$

$$2\frac{\partial^2 x}{\partial T_0\,\partial T_1} = -2\frac{\partial x}{\partial T_0} \tag{6.1.52}$$

$$\frac{\partial^2 x}{\partial T_1^2} + 2\frac{\partial^2 x}{\partial T_0\,\partial T_2} = -2\frac{\partial x}{\partial T_1} \tag{6.1.53}$$

The general solution of (6.1.51) is

$$x = A(T_1, T_2)e^{iT_0} + \bar{A}(T_1, T_2)e^{-iT_0} \tag{6.1.54}$$

Substituting into (6.1.52), we obtain

$$\left(\frac{\partial A}{\partial T_1} + A\right)e^{iT_0} + \left(\frac{\partial \bar{A}}{\partial T_1} + \bar{A}\right)e^{-iT_0} = 0 \tag{6.1.55}$$

Since (6.1.55) is valid for all T_0, the coefficients of $\exp(iT_0)$ and $\exp(-iT_0)$ must vanish; that is

$$\frac{\partial A}{\partial T_1} + A = 0 \tag{6.1.56}$$

or

$$A = a(T_2)e^{-T_1} \tag{6.1.57}$$

Substituting (6.1.54) into (6.1.53) yields

$$\left(\frac{\partial^2 A}{\partial T_1^2} + 2i\frac{\partial A}{\partial T_2} + 2\frac{\partial A}{\partial T_1}\right)e^{iT_0} + CC = 0 \tag{6.1.58}$$

Thus

$$\frac{\partial^2 A}{\partial T_1^2} + 2\frac{\partial A}{\partial T_1} + 2i\frac{\partial A}{\partial T_2} = 0 \tag{6.1.59}$$

Substituting for A from (6.1.57) into (6.1.59) gives

$$2i\frac{\partial a}{\partial T_2} - a = 0 \tag{6.1.60}$$

Hence

$$a = a_0 e^{-iT_2/2} \tag{6.1.61}$$

where a_0 is a constant.

Therefore (6.1.54) becomes

$$x = a_0 e^{-T_1}e^{i(T_0 - T_2/2)} + CC \tag{6.1.62}$$

Expressing (6.1.62) in terms of t yields

$$x = ae^{-\epsilon t}\cos(t - \tfrac{1}{2}\epsilon^2 t + \phi) \tag{6.1.63}$$

where $a_0 = (1/2)a \exp(i\phi)$. This result is in full agreement with (6.1.48).

6.1.2. THE TWO-VARIABLE EXPANSION PROCEDURE

Changing the independent variable from t to ξ and η as defined by (6.1.21), we transform (6.1.1) into

$$(1 + \epsilon^2\omega_2 + \cdots)^2 \frac{\partial^2 x}{\partial \eta^2} + 2\epsilon(1 + \epsilon^2\omega_2 + \cdots)\frac{\partial^2 x}{\partial \xi\,\partial \eta} + \epsilon^2 \frac{\partial^2 x}{\partial \xi^2} + x$$
$$= -2\epsilon(1 + \epsilon^2\omega_2 + \cdots)\frac{\partial x}{\partial \eta} - 2\epsilon^2 \frac{\partial x}{\partial \xi} \tag{6.1.64}$$

We assume that

$$x = x_0(\xi, \eta) + \epsilon x_1(\xi, \eta) + \epsilon^2 x_2(\xi, \eta) + \cdots \tag{6.1.65}$$

Substituting (6.1.65) into (6.1.64) and equating coefficients of equal powers of ϵ on both sides, we obtain

$$\frac{\partial^2 x_0}{\partial \eta^2} + x_0 = 0 \tag{6.1.66}$$

$$\frac{\partial^2 x_1}{\partial \eta^2} + x_1 + 2\frac{\partial^2 x_0}{\partial \xi\,\partial \eta} = -2\frac{\partial x_0}{\partial \eta} \tag{6.1.67}$$

$$\frac{\partial^2 x_2}{\partial \eta^2} + x_2 + 2\omega_2\frac{\partial^2 x_0}{\partial \eta^2} + 2\frac{\partial^2 x_1}{\partial \xi\,\partial \eta} + \frac{\partial^2 x_0}{\partial \xi^2} = -2\frac{\partial x_1}{\partial \eta} - 2\frac{\partial x_0}{\partial \xi} \tag{6.1.68}$$

The general solution of (6.1.66) is

$$x_0 = A_0(\xi)e^{i\eta} + \bar{A}_0(\xi)e^{-i\eta} \tag{6.1.69}$$

With this solution, (6.1.67) becomes

$$\frac{\partial^2 x_1}{\partial \eta^2} + x_1 = -2i\left(\frac{dA_0}{d\xi} + A_0\right)e^{i\eta} + CC \tag{6.1.70}$$

Eliminating the terms that produce secular terms in (6.1.70) gives

$$\frac{dA_0}{d\xi} + A_0 = 0 \tag{6.1.71}$$

Hence

$$x_1 = A_1(\xi)e^{i\eta} + \bar{A}_1(\xi)e^{-i\eta} \tag{6.1.72}$$

The solution of (6.1.71) is
$$A_0 = a_0 e^{-\xi} \tag{6.1.73}$$
where a_0 is a constant.

Substituting the above solutions for x_0 and x_1 into (6.1.68) gives

$$\frac{\partial^2 x_2}{\partial \eta^2} + x_2 = \left[-2i\left(\frac{dA_1}{d\xi} + A_1\right) + (2\omega_2 + 1)a_0 e^{-\xi} \right]e^{i\eta} + CC \tag{6.1.74}$$

Eliminating the terms that produce secular terms in (6.1.74) yields

$$\frac{dA_1}{d\xi} + A_1 = -\tfrac{1}{2}i(2\omega_2 + 1)a_0 e^{-\xi} \tag{6.1.75}$$

whose solution is
$$A_1 = a_1 e^{-\xi} - \tfrac{1}{2}i(2\omega_2 + 1)a_0 \xi e^{-\xi} \tag{6.1.76}$$

Substituting for A_1 into (6.1.72) and comparing the result with (6.1.69) show that x_1/x_0 is unbounded as $\xi \to \infty$ unless

$$\omega_2 = -\tfrac{1}{2} \tag{6.1.77}$$

Therefore in terms of t (6.1.65) becomes

$$x = ae^{-\epsilon t}\cos(t - \tfrac{1}{2}\epsilon^2 t + \phi) + O(\epsilon^2) \tag{6.1.78}$$

where $a_0 + \epsilon a_1 = (1/2)a \exp(i\phi)$. This expression is in full agreement with that obtained using the many-variable version (derivative-expansion method).

6.1.3. GENERALIZED METHOD—NONLINEAR SCALES

We first introduce a new variable $\tau = \epsilon t$ to transform (6.1.1) into

$$\epsilon^2\left(\frac{d^2 x}{d\tau^2} + 2\frac{dx}{d\tau}\right) + x = 0 \tag{6.1.79}$$

In order to determine a uniformly valid expansion, we let

$$\xi = \tau, \qquad \eta = \frac{g_{-1}(\tau)}{\epsilon} + g_0(\tau) + \epsilon g_1(\tau) + \cdots, \qquad g_i(0) = 0 \tag{6.1.80}$$

where g_i is determined in the course of analysis. The derivatives with respect to τ are then transformed according to

$$\frac{d}{d\tau} = \frac{\partial}{\partial \xi} + \left[\frac{g'_{-1}(\xi)}{\epsilon} + g'_0(\xi) + \epsilon g'_1(\xi) + \cdots\right]\frac{\partial}{\partial \eta} \tag{6.1.81}$$

$$\frac{d^2}{d\tau^2} = \frac{\partial^2}{\partial \xi^2} + 2\left[\frac{g'_{-1}(\xi)}{\epsilon} + g'_0(\xi) + \epsilon g'_1(\xi) + \cdots\right]\frac{\partial^2}{\partial \xi \, \partial \eta}$$

$$+ \left[\frac{g''_{-1}(\xi)}{\epsilon} + g''_0(\xi) + \epsilon g''_1(\xi) + \cdots\right]\frac{\partial}{\partial \eta}$$

$$+ \left[\frac{g'_{-1}(\xi)}{\epsilon} + g'_0(\xi) + \epsilon g'_1(\xi) + \cdots\right]^2\frac{\partial^2}{\partial \eta^2} \tag{6.1.82}$$

We assume that x possesses a uniformly valid expansion of the form

$$x = x_0(\xi, \eta) + \epsilon x_1(\xi, \eta) + \epsilon^2 x_2(\xi, \eta) + \cdots \qquad (6.1.83)$$

Substituting (6.1.81) through (6.1.83) into (6.1.79) and equating coefficients of like powers of ϵ, we obtain

$$g_{-1}'^2 \frac{\partial^2 x_0}{\partial \eta^2} + x_0 = 0 \qquad (6.1.84)$$

$$g_{-1}'^2 \frac{\partial^2 x_1}{\partial \eta^2} + x_1 + 2g_{-1}' g_0' \frac{\partial^2 x_0}{\partial \eta^2} + g_{-1}'' \frac{\partial x_0}{\partial \eta} + 2g_{-1}' \frac{\partial^2 x_0}{\partial \xi \, \partial \eta} + 2g_{-1}' \frac{\partial x_0}{\partial \eta} = 0$$

$$(6.1.85)$$

The general solution of (6.1.84) is

$$x_0 = A_0(\xi) \exp\left(i \frac{\eta}{g_{-1}'}\right) + \bar{A}_0(\xi) \exp\left(-i \frac{\eta}{g_{-1}'}\right) \qquad (6.1.86)$$

Substituting for x_0 into (6.1.85) gives

$$g_{-1}'^2 \frac{\partial^2 x_1}{\partial \eta^2} + x_1 = -\left[\left(-\frac{2g_0'}{g_{-1}'} + i \frac{g_{-1}''}{g_{-1}'} + 2i\right) A_0 \right.$$

$$\left. + 2ig_{-1}'\left(\frac{A_0}{g_{-1}'}\right)' + 2\frac{g_{-1}''}{g_{-1}'^2} A_0 \eta\right] \exp\left(i \frac{\eta}{g_{-1}'}\right) + CC \quad (6.1.87)$$

The terms on the right-hand side of (6.1.87) produce secular terms. In order to eliminate secular terms

$$\left(-\frac{2g_0'}{g_{-1}'} + i \frac{g_{-1}''}{g_{-1}'} + 2i\right) A_0 + 2ig_{-1}'\left(\frac{A_0}{g_{-1}'}\right)' + 2\frac{g_{-1}''}{g_{-1}'^2} \eta A_0 = 0 \quad (6.1.88)$$

Since (6.1.88) must be valid for all η, and $A_0 \neq 0$ for a nontrivial solution, we require that

$$g_{-1}'' = 0 \quad \text{or} \quad g_{-1} = c\xi \quad \text{since} \quad \eta(0) = 0 \qquad (6.1.89)$$

where c is an arbitrary constant which can be taken to be unity without loss of generality. Then (6.1.88) becomes

$$A_0' + (1 + ig_0')A_0 = 0 \qquad (6.1.90)$$

whose solution is

$$A_0 = a_0 e^{-\xi - ig_0(\xi)} \qquad (6.1.91)$$

where a_0 is a constant. With A_0 and g_{-1} known

$$x_0 = a_0 e^{-\tau} e^{i(\tau/\epsilon)} + \bar{a}_0 e^{-\tau} e^{-i(\tau/\epsilon)} \qquad (6.1.92)$$

Equation (6.1.92) shows that g_0 cancels out, hence the solution is independent of the value of g_0. Therefore we might as well set it equal to zero without loss of generality. Hence A_0 becomes

$$A_0 = a_0 e^{-\xi} \tag{6.1.93}$$

As a result of (6.1.88), the solution for x_1 is

$$x_1 = A_1(\xi)e^{i\eta} + \bar{A}_1(\xi)e^{-i\eta} \tag{6.1.94}$$

With $g_{-1} = \xi$ and $g_0 = 0$, the equation for x_2 can be determined by substituting (6.1.81) through (6.1.83) into (6.1.79) and equating the coefficient of ϵ^2 to zero. Thus

$$\frac{\partial^2 x_2}{\partial \eta^2} + x_2 + 2\frac{\partial^2 x_1}{\partial \xi\, \partial \eta} + 2\frac{\partial x_1}{\partial \eta} + 2g_1'\frac{\partial^2 x_0}{\partial \eta^2} + \frac{\partial^2 x_0}{\partial \xi^2} + 2\frac{\partial x_0}{\partial \xi} = 0 \tag{6.1.95}$$

Substituting for x_0 and x_1 into (6.1.95) yields

$$\frac{\partial^2 x_2}{\partial \eta^2} + x_2 = -[2i(A_1' + A_1) - (2g_1' + 1)a_0 e^{-\xi}]e^{i\eta} + CC \tag{6.1.96}$$

Eliminating the terms that produce secular terms in (6.1.96), we obtain

$$A_1' + A_1 = -\tfrac{1}{2}i(2g_1' + 1)a_0 e^{-\xi} \tag{6.1.97}$$

The solution of (6.1.97) is

$$A_1 = a_1 e^{-\xi} - \tfrac{1}{2}ia_0(2g_1 + \xi)e^{-\xi} \tag{6.1.98}$$

where a_1 is a constant. Equation (6.1.98) shows that x_1/x_0 is unbounded as $\xi \to \infty$ unless

$$g_1 = -\tfrac{1}{2}\xi \tag{6.1.99}$$

In terms of $t = \tau/\epsilon$, the expansion becomes

$$x = ae^{-\epsilon t}\cos{(t - \tfrac{1}{8}\epsilon^2 t + \phi)} + O(\epsilon^2) \tag{6.1.100}$$

where $a_0 + \epsilon a_1 = (1/2)a \exp{(i\phi)}$. This expansion is again in agreement with those obtained using the derivative-expansion and the two-variable expansion versions of the method of multiple scales.

6.2. Applications of the Derivative-Expansion Method

6.2.1. THE DUFFING EQUATION
The second example to which we apply the derivative-expansion method is the Duffing equation

$$\frac{d^2u}{dt^2} + \omega_0^2 u + \epsilon u^3 = 0 \tag{6.2.1}$$

We assume that

$$u = \sum_{n=0}^{2} \epsilon^n u_n(T_0, T_1, T_2) + O(\epsilon^3) \qquad (6.2.2)$$

Then

$$\frac{d}{dt} = D_0 + \epsilon D_1 + \epsilon^2 D_2 + \cdots, \qquad D_n = \frac{\partial}{\partial T_n} \qquad (6.2.3)$$

Substituting (6.2.2) and (6.2.3) into (6.2.1) and equating coefficients of each power of ϵ to zero, we have

$$D_0^2 u_0 + \omega_0^2 u_0 = 0 \qquad (6.2.4)$$

$$D_0^2 u_1 + \omega_0^2 u_1 = -2D_0 D_1 u_0 - u_0^3 \qquad (6.2.5)$$

$$D_0^2 u_2 + \omega_0^2 u_2 = -2D_0 D_1 u_1 - 2D_0 D_2 u_0 - D_1^2 u_0 - 3u_0^2 u_1 \qquad (6.2.6)$$

The solution of (6.2.4) is

$$u_0 = A(T_1, T_2)e^{i\omega_0 T_0} + \bar{A}(T_1, T_2)e^{-i\omega_0 T_0} \qquad (6.2.7)$$

Equation (6.2.5) then becomes

$$D_0^2 u_1 + \omega_0^2 u_1 = -[2i\omega_0 D_1 A + 3A^2 \bar{A}]e^{i\omega_0 T_0} - A^3 e^{3i\omega_0 T_0} + CC \qquad (6.2.8)$$

In order that u_1/u_0 be bounded for all T_0, terms that produce secular terms must be eliminated. Hence

$$2i\omega_0 D_1 A + 3A^2 \bar{A} = 0 \qquad (6.2.9)$$

and the solution for u_1 becomes

$$u_1 = B(T_1, T_2)e^{i\omega_0 T_0} + \frac{A^3}{8\omega_0^2} e^{3i\omega_0 T_0} + CC \qquad (6.2.10)$$

To solve (6.2.9), we let $A = (1/2)ae^{i\phi}$ with real a and ϕ, separate real and imaginary parts, and obtain

$$\frac{\partial a}{\partial T_1} = 0, \qquad -\omega_0 \frac{\partial \phi}{\partial T_1} + \tfrac{3}{8}a^2 = 0 \qquad (6.2.11)$$

Hence

$$a = a(T_2), \qquad \phi = \frac{3}{8\omega_0} a^2 T_1 + \phi_0(T_2) \qquad (6.2.12)$$

Substituting for u_0 and u_1 into (6.2.6) yields

$$D_0^2 u_2 + \omega_0^2 u_2 = -\frac{3}{8\omega_0^2} A^5 e^{5i\omega_0 T_0} + \left[\frac{21}{8\omega_0^2} A^4 \bar{A} - 3BA^2\right]e^{3i\omega_0 T_0}$$

$$- Q(T_1, T_2)e^{i\omega_0 T_0} + CC \qquad (6.2.13)$$

where

$$Q = 2i\omega_0 D_1 B + 3A^2 \bar{B} + 6A\bar{A}B + 2i\omega_0 D_2 A - \frac{15A^3 \bar{A}^2}{8\omega_0^2} \qquad (6.2.14)$$

Secular terms are eliminated if

$$B = 0 \tag{6.2.15}$$

and

$$2i\omega_0 D_2 A = \frac{15 A^3 \bar{A}^2}{8\omega_0{}^2} \tag{6.2.16}$$

With $Q = 0$, the solution of u_2, disregarding the homogeneous solution, is

$$u_2 = \frac{A^5}{64\omega_0{}^4} e^{5i\omega_0 T_0} - \frac{21 A^4 \bar{A}}{64\omega_0{}^4} e^{3i\omega_0 T_0} + CC \tag{6.2.17}$$

Letting $A = (1/2)ae^{i\phi}$ in (6.2.16) and separating real and imaginary parts, we obtain

$$\frac{\partial a}{\partial T_2} = 0, \qquad -\omega_0 \frac{\partial \phi}{\partial T_2} = \frac{15}{256\omega_0{}^2} a^4 \tag{6.2.18}$$

Equations (6.2.12) and (6.2.18) lead to $a = $ a constant, hence

$$\phi_0 = -\frac{15}{256\omega_0{}^3} a^4 T_2 + \chi \tag{6.2.19a}$$

where χ is a constant. Therefore

$$\phi = \frac{3}{8\omega_0} a^2 T_1 - \frac{15}{256\omega_0{}^3} a^4 T_2 + \chi \tag{6.2.19b}$$

Substituting for u_0, u_1, and u_2 into (6.2.2), keeping in mind that $A = (1/2)a \exp(i\phi)$ and expressing the result in terms of t, we obtain

$$u = a \cos(\omega t + \chi) + \frac{\epsilon a^3}{32\omega_0{}^2}\left(1 - \epsilon \frac{21 a^2}{32\omega_0{}^2}\right) \cos 3(\omega t + \chi)$$

$$+ \frac{\epsilon^2 a^5}{1024\omega_0{}^4} \cos 5(\omega t + \chi) + O(\epsilon^3) \tag{6.2.20a}$$

where

$$\omega = \omega_0 + \frac{3a^2}{8\omega_0} \epsilon - \frac{15 a^4}{256\omega_0{}^3} \epsilon^2 + O(\epsilon^3) \tag{6.2.20b}$$

In the last two terms of (6.2.20a), ω_0 is replaced by ω with an error $O(\epsilon^3)$.

6.2.2. THE VAN DER POL OSCILLATOR

As a second example, we consider the van der Pol oscillator

$$\frac{d^2 u}{dt^2} + u = \epsilon(1 - u^2)\frac{du}{dt} \tag{6.2.21}$$

Substituting (6.2.2) and (6.2.3) into (6.2.21) and equating the coefficients of like powers of ϵ, we obtain

$$D_0^2 u_0 + u_0 = 0 \tag{6.2.22}$$

$$D_0^2 u_1 + u_1 = -2D_0 D_1 u_0 + (1 - u_0^2) D_0 u_0 \tag{6.2.23}$$

$$D_0^2 u_2 + u_2 = -2D_0 D_1 u_1 - D_1^2 u_0 - 2D_0 D_2 u_0 + (1 - u_0^2) D_0 u_1$$
$$+ (1 - u_0^2) D_1 u_0 - 2u_0 u_1 D_0 u_0 \tag{6.2.24}$$

The solution of (6.2.22) is

$$u_0 = A(T_1, T_2)e^{iT_0} + \bar{A}(T_1, T_2)e^{-iT_0} \tag{6.2.25}$$

Substituting for u_0 into (6.2.23) gives

$$D_0^2 u_1 + u_1 = -i(2D_1 A - A + A^2 \bar{A})e^{iT_0} - iA^3 e^{3iT_0} + CC \tag{6.2.26}$$

To eliminate terms that produce secular terms, we require the vanishing of the coefficients of $\exp(\pm iT_0)$; that is

$$2D_1 A = A - A^2 \bar{A} \tag{6.2.27}$$

Then the solution of (6.2.26) is

$$u_1 = B(T_1, T_2)e^{iT_0} + \tfrac{1}{8}iA^3 e^{3iT_0} + CC \tag{6.2.28}$$

In order to solve (6.2.27), we let

$$A = \tfrac{1}{2}a(T_1, T_2) \exp i\phi(T_1, T_2) \tag{6.2.29}$$

Separating real and imaginary parts in (6.2.27), we obtain

$$\frac{\partial \phi}{\partial T_1} = 0, \qquad \frac{\partial a}{\partial T_1} = \tfrac{1}{2}(1 - \tfrac{1}{4}a^2)a \tag{6.2.30}$$

Hence

$$\phi = \phi(T_2), \qquad a^2 = \frac{4}{1 + c(T_2)e^{-T_1}} \tag{6.2.31}$$

If we are interested in the first approximation to u, then we consider B, ϕ, and c as constants. Moreover, if $u(0) = a_0$ and $du(0)/dt = 0$, then

$$u = a \cos t + O(\epsilon) \tag{6.2.32}$$

where

$$a^2 = \frac{4}{1 + \left(\dfrac{4}{a_0^2} - 1\right)e^{-\epsilon t}} \tag{6.2.33}$$

which is in agreement with the expansion obtained in Section 5.4.2 using the Krylov–Bogoliubov–Mitropolski technique.

To determine the second approximation, we need to determine the functions B, ϕ, and c. Thus we substitute for u_0 and u_1 into (6.2.24) and obtain

$$D_0{}^2 u_2 + u_2 = Q(T_1, T_2)e^{iT_0} + \bar{Q}(T_1, T_2)e^{-iT_0} + \text{NST} \quad (6.2.34a)$$

where

$$Q = -2iD_1 B + i(1 - 2A\bar{A})B - iA^2\bar{B} - 2iD_2 A - D_1{}^2 A$$
$$+ (1 - 2A\bar{A})D_1 A - A^2 D_1\bar{A} + \tfrac{1}{8}A^3\bar{A}^2 \quad (6.2.34b)$$

Secular terms will be eliminated if $Q = 0$. To solve (6.2.34b) with $Q = 0$, we let $B = (1/2)ib \exp i\phi$ with b real and ϕ defined in (6.2.29). We substitute for A and B into (6.2.34b) with $Q = 0$, separate real and imaginary parts, and obtain

$$\frac{\partial a}{\partial T_2} = 0, \quad \text{or} \quad a = a(T_1) \quad (6.2.35a)$$

$$2\frac{\partial b}{\partial T_1} - (1 - \tfrac{1}{4}a^2)b = -2a\frac{d\phi}{dT_2} + \frac{d^2 a}{dT_1{}^2} - (1 - \tfrac{3}{4}a^2)\frac{da}{dT_1} - \frac{1}{128}a^5 \quad (6.2.35b)$$

With the help of (6.2.30), (6.2.35b) can be expressed in the form

$$2\frac{\partial b}{\partial T_1} - \frac{2}{a}\frac{da}{dT_1}b = -2a\left(\frac{d\phi}{dT_2} + \frac{1}{16}\right) + (\tfrac{7}{16}a^2 - \tfrac{1}{4})\frac{da}{dT_1} \quad (6.2.36a)$$

Thus

$$d\left(\frac{b}{a}\right) = -\left(\frac{d\phi}{dT_2} + \frac{1}{16}\right)dT_1 + \left(\tfrac{7}{32}a - \frac{1}{8a}\right)da \quad (6.2.36b)$$

Integrating, we obtain

$$b = -a\left(\frac{d\phi}{dT_2} + \frac{1}{16}\right)T_1 + \tfrac{7}{64}a^3 - \tfrac{1}{8}a \ln a + ab_0(T_2) \quad (6.2.36c)$$

In order that u_1/u_0 be bounded for all T_1, the coefficient of T_1 in the above expression for b must vanish. This condition gives

$$\phi = -\tfrac{1}{16}T_2 + \phi_0 \quad (6.2.37)$$

where ϕ_0 is a constant. The expansion of u to second approximation is then

$$u = a \cos\left[(1 - \tfrac{1}{16}\epsilon^2)t + \phi_0\right]$$
$$- \epsilon\{(\tfrac{7}{64}a^3 - \tfrac{1}{8}a \ln a + ab_0)\sin\left[(1 - \tfrac{1}{16}\epsilon^2)t + \phi_0\right]$$
$$+ \tfrac{1}{32}a^3 \sin 3[(1 - \tfrac{1}{16}\epsilon^2)t + \phi_0]\} + O(\epsilon^2) \quad (6.2.38)$$

where a is defined by (6.2.33) and b_0 is considered a constant to within the order of error indicated. To an error of $O(\epsilon^2)$, this expression can be written as

$$u = a \cos (t - \theta) - \tfrac{1}{32}\epsilon a^3 \sin 3(t - \theta) + O(\epsilon^2) \qquad (6.2.39a)$$

where

$$\theta = \tfrac{1}{16}\epsilon^2 t + \tfrac{1}{8}\epsilon \ln a - \tfrac{7}{64}\epsilon a^2 + \theta_0 \qquad (6.2.39b)$$

and $\theta_0 = -\phi_0 - \epsilon b_0 = $ a constant. This last form of the solution is in full agreement with that obtained in Section 5.4.2 using the Krylov–Bogoliubov–Mitropolski method.

6.2.3. FORCED OSCILLATIONS OF THE VAN DER POL EQUATION

We consider next the response of the van der Pol oscillator, discussed in the previous section, to a periodic external force; that is, the oscillations of the equation

$$\frac{d^2 u}{dt^2} + \omega_0{}^2 u = \epsilon(1 - u^2)\frac{du}{dt} + K \cos \lambda t \qquad (6.2.40)$$

where K and λ are real constants. Four cases arise depending on whether the excitation (external force) is "soft" [i.e., $K = O(\epsilon)$] or "hard" [i.e., $K = O(1)$], and whether the excitation is resonant [i.e., $\lambda - \omega_0 = O(\epsilon)$] or nonresonant [i.e., $\lambda - \omega_0 = O(1)$].

Soft Nonresonant Excitation.   In this case $K = \epsilon k$, where $k = O(1)$, and we express $\cos \lambda t$ in the form $\cos \lambda T_0$. To determine a first approximation to u, we let

$$u = u_0(T_0, T_1) + \epsilon u_1(T_0, T_1) + O(\epsilon^2) \qquad (6.2.41)$$

with $T_0 = t$ and $T_1 = \epsilon t$. Substituting (6.2.41) into (6.2.40) and equating the coefficients of ϵ^0 and ϵ on both sides, we obtain

$$D_0{}^2 u_0 + \omega_0{}^2 u_0 = 0 \qquad (6.2.42)$$

$$D_0{}^2 u_1 + \omega_0{}^2 u_1 = -2D_0 D_1 u_0 + (1 - u_0{}^2)D_0 u_0 + k \cos \lambda T_0 \quad (6.2.43)$$

The solution of (6.2.42) is

$$u_0 = A(T_1)e^{i\omega_0 T_0} + \bar{A}(T_1)e^{-i\omega_0 T_0} \qquad (6.2.44)$$

Substituting for u_0 into (6.2.43) gives

$$D_0{}^2 u_1 + \omega_0{}^2 u_1 = i\omega_0(-2A' + A - A^2\bar{A})e^{i\omega_0 T_0}$$
$$+ \tfrac{1}{2}ke^{i\lambda T_0} - i\omega_0 A^3 e^{3i\omega_0 T_0} + CC \quad (6.2.45)$$

For there to be no secular terms, we require that

$$2A' = A - A^2\bar{A} \qquad (6.2.46)$$

where the prime denotes differentiation with respect to T_1. Then the solution of (6.2.45) is

$$u_1 = B(T_1)e^{i\omega_0 T_0} + \frac{1}{2}\frac{k}{\omega_0^2 - \lambda^2} e^{i\lambda T_0} + \frac{iA^3}{8\omega_0} e^{3i\omega_0 T_0} + CC \quad (6.2.47)$$

Letting $A = (1/2)a \exp i\phi$ in (6.2.46), separating real and imaginary parts, and solving the resulting equations, we find that ϕ is a constant, while a is given by (6.2.33).

Therefore, to first approximation

$$u = a \cos \omega_0 t + O(\epsilon) \quad (6.2.48)$$

where a is given by (6.2.33).

Equations (6.2.33) and (6.2.48) show that, to first approximation, neither the phase nor the amplitude is affected by the presence of a soft nonresonant excitation. Moreover, the natural response (i.e., the case with $k = 0$) dominates the forced response, as expected, since the forcing function is soft. However, as the forcing frequency λ approaches the natural frequency ω_0, the forced response becomes more significant and approaches infinity as can be seen from (6.2.47), and the above expansion is no longer valid.

Hard Nonresonant Excitation. In this case $K = O(1)$ and (6.2.42) and (6.2.43) are modified to

$$D_0^2 u_0 + \omega_0^2 u_0 = K \cos \lambda T_0 \quad (6.2.49)$$

$$D_0^2 u_1 + \omega_0^2 u_1 = -2D_0 D_1 u_0 + (1 - u_0^2)D_0 u_0 \quad (6.2.50)$$

The solution of (6.2.49) is

$$u_0 = A(T_1)e^{i\omega_0 T_0} + \bar{A}(T_1)e^{-i\omega_0 T_0} + \frac{K}{\omega_0^2 - \lambda^2} \cos \lambda T_0 \quad (6.2.51)$$

Substituting for u_0 into (6.2.50) gives

$$D_0^2 u_1 + \omega_0^2 u_1 = i\omega_0[-2A' + A\eta - A^2\bar{A}]e^{i\omega_0 T_0} + CC + NST \quad (6.2.52)$$

where $\eta = 1 - K^2/2(\omega_0^2 - \lambda^2)^2$. In order to eliminate the secular terms, we require that

$$2A' = A\eta - A^2\bar{A} \quad (6.2.53)$$

To solve (6.2.53), we let $A = (1/2)a \exp i\phi$, separate real and imaginary parts, and obtain $\phi = $ a constant and

$$\frac{da}{dT_1} = \tfrac{1}{2}a(\eta - \tfrac{1}{4}a^2) \tag{6.2.54}$$

The solution of (6.2.54) can be obtained by separation of variables to be

$$\ln a^2 - \ln (\eta - \tfrac{1}{4}a^2) = \eta T_1 + \text{a constant}$$

If $u(0) = a_0 + [K/(\omega_0^2 - \lambda^2)]$ and $du(0)/dt = 0$, the first approximation to u is given by

$$u = a \cos \omega_0 t + \frac{K}{\omega_0^2 - \lambda^2} \cos \lambda t + O(\epsilon) \tag{6.2.55}$$

where

$$a^2 = \frac{4\eta}{1 + \left(\dfrac{4\eta}{a_0^2} - 1\right) e^{-\epsilon \eta t}} \tag{6.2.56}$$

The steady-state solution (i.e., $t \to \infty$) depends on whether η is positive or negative [i.e., K^2 is less than or greater than $2(\omega_0^2 - \lambda^2)^2$]. For negative η, $\exp(-\epsilon \eta t) \to \infty$ as $t \to \infty$, hence $a \to 0$ as $t \to \infty$, and the steady-state solution is

$$u_s = \frac{K}{\omega_0^2 - \lambda^2} \cos \lambda t + O(\epsilon) \tag{6.2.57}$$

However, for positive η, $\exp(-\epsilon \eta t) \to 0$ as $t \to \infty$, and $a \to 2\sqrt{\eta}$ as $t \to \infty$. Consequently, the steady-state solution is

$$u_s = 2\sqrt{\eta} \cos \omega_0 t + \frac{K}{\omega_0^2 - \lambda^2} \cos \lambda t + O(\epsilon) \tag{6.2.58}$$

Therefore, if η is negative, the natural response fades away and the steady-state solution consists of the forced response only. However, if η is positive, the steady-state solution is a combination of the natural and forced responses, with the amplitude of the natural response modified by the presence of the hard excitation.

Soft Resonant Excitation. In this case $K = \epsilon k$ with $k = O(1)$, and $\lambda - \omega_0 = \sigma \epsilon$ with the detuning $\sigma = O(1)$. In order to determine a valid asymptotic expansion in this case, we express the excitation in terms of T_0 and T_1 according to

$$K \cos \lambda t = \epsilon k \cos (\omega_0 t + \sigma \epsilon t) = \epsilon k \cos (\omega_0 T_0 + \sigma T_1) \tag{6.2.59}$$

With this expression for the excitation, the equations for u_0 and u_1 of (6.2.41) are

$$D_0{}^2 u_0 + \omega_0{}^2 u_0 = 0 \tag{6.2.60}$$

$$D_0{}^2 u_1 + \omega_0{}^2 u_1 = -2D_0 D_1 u_0 + (1 - u_0{}^2)D_0 u_0 + k \cos(\omega_0 T_0 + \sigma T_1) \tag{6.2.61}$$

The general solution of (6.2.60) is

$$u_0 = A(T_1)e^{i\omega_0 T_0} + \bar{A}(T_1)e^{-i\omega_0 T_0} \tag{6.2.62}$$

Hence (6.2.61) becomes

$$D_0{}^2 u_1 + \omega_0{}^2 u_1 = [i\omega_0(-2A' + A - A^2\bar{A}) + \tfrac{1}{2}ke^{i\sigma T_1}]e^{i\omega_0 T_0}$$
$$- i\omega_0 A^3 e^{3i\omega_0 T_0} + CC \tag{6.2.63}$$

The terms proportional to $\exp(\pm i\omega_0 T_0)$ in (6.2.63) produce secular terms with respect to the time scale T_0 because the terms in the brackets are functions of T_1 only. In order that u_1/u_0 be bounded for all T_0

$$2A' = A - A^2\bar{A} - \frac{1}{2\omega_0} ike^{i\sigma T_1} \tag{6.2.64}$$

To solve (6.2.64), we let $A = (1/2)a \exp i\phi$, separate real and imaginary parts, and obtain

$$\frac{da}{dT_1} = \tfrac{1}{2}a(1 - \tfrac{1}{4}a^2) + \frac{k}{2\omega_0} \sin(\sigma T_1 - \phi) \tag{6.2.65}$$

$$\frac{d\phi}{dT_1} = -\frac{k}{2\omega_0 a} \cos(\sigma T_1 - \phi) \tag{6.2.66}$$

To eliminate the explicit time dependence of the right-hand sides of (6.2.65) and (6.2.66), we let

$$\psi = \sigma T_1 - \phi \quad \text{or} \quad \frac{d\psi}{dT_1} = \sigma - \frac{d\phi}{dT_1} \tag{6.2.67}$$

Hence (6.2.65) and (6.2.66) become

$$\frac{da}{dT_1} = \tfrac{1}{2}a(1 - \tfrac{1}{4}a^2) + \frac{k}{2\omega_0} \sin\psi \tag{6.2.68}$$

$$\frac{d\psi}{dT_1} = \sigma + \frac{k}{2\omega_0 a} \cos\psi \tag{6.2.69}$$

Periodic solutions of the externally excited oscillator (6.2.40) correspond to the stationary solutions of (6.2.68) and (6.2.69); that is, $da/dT_1 = d\psi/dT_1 = 0$, or

$$\tfrac{1}{2}\tilde{a}(1 - \tfrac{1}{4}\tilde{a}^2) + \frac{k}{2\omega_0} \sin\tilde{\psi} = 0 \tag{6.2.70}$$

$$\sigma + \frac{k}{2\omega_0\tilde{a}} \cos\tilde{\psi} = 0 \tag{6.2.71}$$

where the tilde refers to the stationary solution. Elimination of $\tilde{\psi}$ from these equations leads to the following frequency response equation

$$\rho(1 - \rho)^2 + 4\sigma^2\rho = \frac{k^2}{4\omega_0^2} = F^2, \qquad \rho = \frac{\tilde{a}^2}{4} \qquad (6.2.72)$$

For a given excitation amplitude ϵk and frequency $\lambda = \omega_0 + \epsilon\sigma$, (6.2.72) furnishes ρ, hence the amplitude of harmonic oscillations. To first approximation the harmonic oscillation is given by

$$u = \tilde{a}\cos(\omega_0 t + \tilde{\phi}) + O(\epsilon) \qquad (6.2.73)$$

while the frequency of oscillation is

$$\omega = \frac{d}{dt}(\omega_0 t + \tilde{\phi}) = \omega_0 + \frac{d\tilde{\phi}}{dt} = \omega_0 + \epsilon\sigma = \lambda \qquad (6.2.74)$$

Therefore, as λ approaches ω_0, the natural response is entrained by the forced response. The result is a synchronization of the output at the excitation frequency.

The stability of these harmonic oscillations can be obtained by letting

$$a = \tilde{a} + \Delta a, \qquad \psi = \tilde{\psi} + \Delta\psi \qquad (6.2.75)$$

Developing the right-hand sides of (6.2.68) and (6.2.69) in powers of Δa and $\Delta\psi$ and keeping only linear terms, we have

$$\frac{d(\Delta a)}{dT_1} = \tfrac{1}{2}(1 - \tfrac{3}{4}\tilde{a}^2)\Delta a + \frac{k}{2\omega_0}\cos\tilde{\psi}\,\Delta\psi \qquad (6.2.76)$$

$$\frac{d(\Delta\psi)}{dT_1} = -\frac{k}{2\omega_0\tilde{a}^2}\cos\tilde{\psi}\,\Delta a - \frac{k}{2\omega_0\tilde{a}}\sin\tilde{\psi}\,\Delta\psi \qquad (6.2.77)$$

If we let $\Delta a \propto \exp mT_1$ and $\Delta\psi \propto \exp mT_1$, then m must satisfy the equation

$$m^2 - \Omega m + \Delta = 0 \qquad (6.2.78)$$

where

$$\Omega = 1 - 2\rho, \qquad \Delta = \tfrac{1}{4}(1 - 4\rho + 3\rho^2) + \sigma^2 \qquad (6.2.79)$$

where use has been made of (6.2.70) and (6.2.71). The discriminant of (6.2.78) is

$$D = \rho^2 - 4\sigma^2 \qquad (6.2.80)$$

The loci $\Omega = \Delta = D = 0$ are called separatrices and shown in Figure 6-1. The locus $\Delta = 0$ is an ellipse whose center is $\rho = 2/3$, $\sigma = 0$, while the locus $D = 0$ is the two straight lines $\rho = \pm 2\sigma$. The interior points of the ellipse correspond to saddle points, hence the corresponding harmonic oscillations are unstable. The points exterior to the ellipse are nodes if

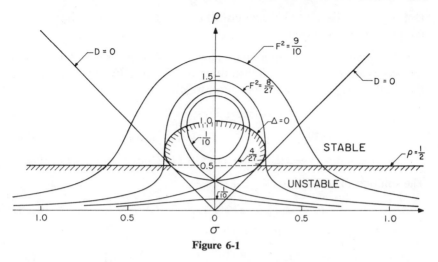

Figure 6-1

$D \geq 0$ and foci if $D < 0$. The harmonic oscillations corresponding to these points are stable or unstable according to whether ρ is greater or less than $1/2$.

Hard Resonant Excitation. The analysis for this case can be obtained as a special case of the previous case, with $k = K/\epsilon$, where the amplitude of excitation $K = O(1)$. Hence k is very large because ϵ is small. Thus, for σ near zero, (6.2.72) shows that there exists only one amplitude ρ for harmonic oscillation and it is stable. As k increases without bound the amplitude increases without bound also.

6.2.4. PARAMETRIC RESONANCE—MATHIEU EQUATION

Let us return to the Mathieu equation discussed in Section 3.1.2, namely

$$\ddot{u} + (\delta + \epsilon \cos 2t)u = 0 \qquad (6.2.81)$$

According to the Floquet theory of linear differential equations with periodic coefficients, the δ-ϵ plane is divided into regions of stability and instability which are separated by transition curves along which u is periodic with a period of either π or 2π. In Section 3.1.2, we determined approximations to the transition curves using the Lindstedt–Poincaré method. In this section we find not only the transition curves but also the solutions, hence the degree of stability or instability, as we found in Section 3.1.3 using Whittaker's method. To this end we let

$$\delta = \omega_0{}^2 \qquad \text{with positive } \omega_0 \qquad (6.2.82)$$

and assume that

$$u = u_0(T_0, T_1, T_2) + \epsilon u_1(T_0, T_1, T_2) + \epsilon^2 u_2(T_0, T_1, T_2) + \cdots \quad (6.2.83)$$

Different cases have to be distinguished depending on whether ω_0 is near or far away from an integer n.

Solution for ω_0 Far Away from an Integer. We express cos $2t$ in terms of the time scale T_0 as cos $2T_0$. Substituting (6.2.83) into (6.2.81) and equating the coefficients of ϵ^0, ϵ, and ϵ^2 to zero, we obtain

$$D_0^2 u_0 + \omega_0^2 u_0 = 0 \tag{6.2.84}$$

$$D_0^2 u_1 + \omega_0^2 u_1 = -2D_0 D_1 u_0 - u_0 \cos 2T_0 \tag{6.2.85}$$

$$D_0^2 u_2 + \omega_0^2 u_2 = -2D_0 D_1 u_1 - (D_1^2 + 2D_0 D_2)u_0 - u_1 \cos 2T_0 \tag{6.2.86}$$

The solution of (6.2.84) is

$$u_0 = A(T_1, T_2)e^{i\omega_0 T_0} + \bar{A}(T_1, T_2)e^{-i\omega_0 T_0} \tag{6.2.87}$$

Substituting for u_0 into (6.2.85) yields

$$D_0^2 u_1 + \omega_0^2 u_1 = -2i\omega_0 D_1 A e^{i\omega_0 T_0} - \tfrac{1}{2}A e^{i(\omega_0+2)T_0} - \tfrac{1}{2}A e^{i(\omega_0-2)T_0} + CC \tag{6.2.88}$$

Since ω_0 is far away from 1, secular terms will be eliminated if $D_1 A = 0$ or $A = A(T_2)$. Then the solution of u_1 is

$$u_1 = \frac{1}{8(\omega_0 + 1)} A e^{i(\omega_0+2)T_0} - \frac{1}{8(\omega_0 - 1)} A e^{i(\omega_0-2)T_0} + CC \tag{6.2.89}$$

Substituting for u_0 and u_1 into (6.2.86) yields

$$D_0^2 u_2 + \omega_0^2 u_2 = -2\left[i\omega_0 D_2 A - \frac{A}{16(\omega_0^2 - 1)}\right]e^{i\omega_0 T_0}$$

$$- \frac{1}{16(\omega_0 + 1)} A e^{i(\omega_0+4)T_0} + \frac{1}{16(\omega_0 - 1)} A e^{i(\omega_0-4)T_0} + CC \tag{6.2.90}$$

Since ω_0 is far away from 1 or 2, secular terms will be eliminated if

$$D_2 A = - \frac{i}{16\omega_0(\omega_0^2 - 1)} A \tag{6.2.91}$$

If we let $A = (1/2)a \exp i\phi$ and separate real and imaginary parts, we obtain

$$\frac{da}{dT_2} = 0, \qquad \frac{d\phi}{dT_2} = - \frac{1}{16\omega_0(\omega_0^2 - 1)} \tag{6.2.92}$$

Therefore

$$a = \text{a constant,} \quad \text{and} \quad \phi = - \frac{1}{16\omega_0(\omega_0^2 - 1)} T_2 + \phi_0 \tag{6.2.93}$$

where ϕ_0 is a constant. With (6.2.91) the solution of (6.2.90) is

$$u_2 = \frac{1}{128(\omega_0 + 1)(\omega_0 + 2)} Ae^{i(\omega_0+4)T_0}$$

$$+ \frac{1}{128(\omega_0 - 1)(\omega_0 - 2)} Ae^{i(\omega_0-4)T_0} + CC \quad (6.2.94)$$

Summarizing, to $O(\epsilon^2)$, the solution for u is

$$u = a \cos (\omega t + \phi_0)$$

$$+ \frac{\epsilon a}{8}\left\{\frac{1}{\omega_0 + 1} \cos [(\omega + 2)t + \phi_0] - \frac{1}{\omega_0 - 1} \cos [(\omega - 2)t + \phi_0]\right\}$$

$$+ \frac{\epsilon^2 a}{128}\left\{\frac{1}{(\omega_0 + 1)(\omega_0 + 2)} \cos [(\omega + 4)t + \phi_0]\right.$$

$$\left. + \frac{1}{(\omega_0 - 1)(\omega_0 - 2)} \cos [(\omega - 4)t + \phi_0]\right\} + O(\epsilon^3) \quad (6.2.95)$$

where

$$\omega = \omega_0 - \frac{\epsilon^2}{16\omega_0(\omega_0^2 - 1)} + O(\epsilon^3) \quad (6.2.96)$$

We emphasize again that this expansion is valid only when ω_0 is away from 1 and 2. As $\omega_0 \to 1$ or 2, $u \to \infty$. An expansion valid near $\omega_0 = 1$ is obtained next.

Solution for ω_0 Near 1.   In this case we let

$$\delta = 1 + \epsilon\delta_1 + \epsilon^2\delta_2 + \cdots \quad (6.2.97)$$

with δ_1 and $\delta_2 = O(1)$. Equation (6.2.97) modifies (6.2.84) through (6.2.86) into

$$D_0^2 u_0 + u_0 = 0 \quad (6.2.98)$$

$$D_0^2 u_1 + u_1 = -2D_0D_1u_0 - \delta_1 u_0 - u_0 \cos 2T_0 \quad (6.2.99)$$

$$D_0^2 u_2 + u_2 = -2D_0D_1u_1 - (D_1^2 + 2D_0D_2)u_0 - \delta_1 u_1 - \delta_2 u_0 - u_1 \cos 2T_0$$

$$(6.2.100)$$

The solution of (6.2.98) is

$$u_0 = A(T_1, T_2)e^{iT_0} + \bar{A}(T_1, T_2)e^{-iT_0} \quad (6.2.101)$$

Substituting for u_0 into (6.2.99) gives

$$D_0^2 u_1 + u_1 = (-2iD_1A - \delta_1 A - \tfrac{1}{2}\bar{A})e^{iT_0} - \tfrac{1}{2}Ae^{3iT_0} + CC \quad (6.2.102)$$

Secular terms with respect to the time scale T_0 will be eliminated if

$$D_1A = \tfrac{1}{2}i(\delta_1 A + \tfrac{1}{2}\bar{A}) \quad (6.2.103)$$

Then the solution of (6.2.102) is

$$u_1 = \tfrac{1}{16}(Ae^{3iT_0} + \bar{A}e^{-3iT_0}) \tag{6.2.104}$$

To solve (6.2.103), we assume that

$$A = A_r + iA_i \tag{6.2.105}$$

where A_r and A_i are real, and separate real and imaginary parts to obtain

$$\frac{\partial A_r}{\partial T_1} = \tfrac{1}{2}(\tfrac{1}{2} - \delta_1)A_i \tag{6.2.106}$$

$$\frac{\partial A_i}{\partial T_1} = \tfrac{1}{2}(\tfrac{1}{2} + \delta_1)A_r \tag{6.2.107}$$

The solution of these equations is

$$A_r = a_1(T_2)e^{\gamma_1 T_1} + a_2(T_2)e^{-\gamma_1 T_1} \tag{6.2.108}$$

$$A_i = \frac{2\gamma_1}{\tfrac{1}{2} - \delta_1} [a_1(T_2)e^{\gamma_1 T_1} - a_2(T_2)e^{-\gamma_1 T_1}] \tag{6.2.109}$$

where

$$\gamma_1^2 = \tfrac{1}{4}(\tfrac{1}{4} - \delta_1^2) \tag{6.2.110}$$

Here a_1 and a_2 are real valued functions of the time scale T_2; however, to first approximation, a_1 and a_2 are constant.

Equations (6.2.108) through (6.2.110) show that A grows exponentially with T_1 (i.e., with t) if γ_1 is real or $|\delta_1| \leq 1/2$, and A oscillates with T_1 if γ_1 is imaginary or $|\delta_1| \geq 1/2$ (in this case the solution is written in terms of $\cos \gamma_1 T_1$ and $\sin \gamma_1 T_1$ to keep A_r and A_i real). Hence the boundaries (transition curves) that separate the stable from the unstable domains emanating from $\delta = 1$, $\epsilon = 0$ are given to a first approximation by $\delta_1 = \pm 1/2$ or

$$\delta = 1 \pm \tfrac{1}{2}\epsilon + O(\epsilon^2) \tag{6.2.111}$$

To determine a second approximation to u and the transition curves, we substitute for u_0 and u_1 into (6.2.100) and obtain

$$D_0^2 u_2 + u_2 = -[2iD_2A + D_1^2A + (\delta_2 + \tfrac{1}{32})A]e^{iT_0} + CC + NST \tag{6.2.112}$$

The condition that must be satisfied for there to be no secular terms is

$$2iD_2A + D_1^2A + (\delta_2 + \tfrac{1}{32})A = 0 \tag{6.2.113}$$

Since $A = A_r + iA_i$, (6.2.113) gives the following equations for A_r and A_i upon separation of real and imaginary parts

$$2\frac{\partial A_r}{\partial T_2} + \alpha A_i = 0 \tag{6.2.114}$$

$$-2\frac{\partial A_i}{\partial T_2} + \alpha A_r = 0 \tag{6.2.115}$$

where

$$\alpha = \gamma_1^2 + \delta_2 + \tfrac{1}{32} \qquad (6.2.116)$$

Replacing A_r and A_i by their expressions from (6.2.108) and (6.2.109) and equating the coefficients of $\exp(\pm\gamma_1 T_1)$ to zero because they are functions of T_2, we obtain

$$2\frac{da_1}{dT_2} + \frac{2\gamma_1}{\tfrac{1}{2} - \delta_1}\alpha a_1 = 0, \qquad -\frac{4\gamma_1}{\tfrac{1}{2} - \delta_1}\frac{da_1}{dT_2} + \alpha a_1 = 0 \quad (6.2.117)$$

$$2\frac{da_2}{dT_2} - \frac{2\gamma_1}{\tfrac{1}{2} - \delta_1}\alpha a_2 = 0, \qquad \frac{4\gamma_1}{\tfrac{1}{2} - \delta_1}\frac{da_2}{dT_2} + \alpha a_2 = 0 \quad (6.2.118)$$

These equations lead to

$$\frac{da_1}{dT_2} = \frac{da_2}{dT_2} = 0 \quad \text{or} \quad a_1 = \text{a constant} \quad \text{and} \quad a_2 = \text{a constant} \quad (6.2.119)$$

and

$$\alpha = 0 \quad \text{or} \quad \delta_2 = -\gamma_1^2 - \tfrac{1}{32} \qquad (6.2.120)$$

Therefore, to second approximation, the solution is given by (3.1.57) through (3.1.62) which were obtained by Whittaker's method.

6.2.5. THE VAN DER POL OSCILLATOR WITH DELAYED AMPLITUDE LIMITING

The next example is a third-order problem as opposed to the previous examples which are of the second order. It is given in dimensionless quantities by

$$\frac{d^2v}{dt^2} + \omega_0^2 v = 2\mu\frac{d}{dt}[(1 - Z)v] + 2\frac{de}{dt} \qquad (6.2.121)$$

$$\tau\frac{dZ}{dt} + Z = v^2 \qquad (6.2.122)$$

where v is voltage, t is time, e is excitation, ω_0 is natural frequency, τ is delay time, Z is output of the low-pass filter, and μ is a measure of the servo-loop gain. This oscillator was first studied by Golay (1964) and then by Scott (1966) and Nayfeh (1967b, 1968). We consider here free oscillations (i.e., $e \equiv 0$) and refer the reader to Nayfeh (1968) for the forced oscillation case.

To determine a first approximation to the above equations, we assume that

$$v = v_0(T_0, T_1) + \mu v_1(T_0, T_1) + \cdots \qquad (6.2.123)$$

$$Z = Z_0(T_0, T_1) + \mu Z_1(T_0, T_1) + \cdots \qquad (6.2.124)$$

with

$$T_0 = t, \qquad T_1 = \mu t \qquad (6.2.125)$$

Substituting (6.2.123) through (6.2.125) into (6.2.121) and (6.2.122) and equating coefficients of like powers of μ, we obtain

$$D_0{}^2v_0 + \omega_0{}^2v_0 = 0 \tag{6.2.126}$$

$$\tau D_0Z_0 + Z_0 = v_0{}^2 \tag{6.2.127}$$

$$D_0{}^2v_1 + \omega_0{}^2v_1 = 2D_0[(1 - Z_0)v_0 - D_1v_0] \tag{6.2.128}$$

$$\tau D_0Z_1 + Z_1 = -\tau D_1Z_0 + 2v_0v_1 \tag{6.2.129}$$

The solution of (6.2.126) is

$$v_0 = A(T_1)e^{i\omega_0T_0} + \bar{A}(T_1)e^{-i\omega_0T_0} \tag{6.2.130}$$

Substituting for v_0 into (6.2.127) gives

$$\tau D_0Z_0 + Z_0 = A\bar{A} + A^2e^{2i\omega_0T_0} + CC \tag{6.2.131}$$

Its solution is

$$Z_0 = B(T_1)e^{-T_0/\tau} + 2A\bar{A} + \frac{A^2e^{2i\omega_0T_0}}{1 + 2i\omega_0\tau} + \frac{\bar{A}^2e^{-2i\omega_0T_0}}{1 - 2i\omega_0\tau} \tag{6.2.132}$$

With v_0 and Z_0 known (6.2.128) becomes

$$D_0{}^2v_1 + \omega_0{}^2v_1 = 2i\omega_0Q(T_1)e^{i\omega_0T_0} - 2AB\left(i\omega_0 - \frac{1}{\tau}\right)e^{[i\omega_0-(1/\tau)]T_0}$$

$$- 6i\omega_0 \frac{A^3e^{3i\omega_0T_0}}{1 + 2i\omega_0\tau} + CC \tag{6.2.133}$$

where

$$Q = A - 2A^2\bar{A} - \frac{A^2\bar{A}}{1 + 2i\omega_0\tau} - D_1A \tag{6.2.134}$$

Secular terms will be eliminated if $Q = 0$. Then the solution for v_1 is

$$v_1 = 2AB\tau \frac{1 - i\omega_0\tau}{1 - 2i\omega_0\tau} e^{[i\omega_0-(1/\tau)]T_0} + \frac{3i}{4\omega_0} \frac{A^3e^{3i\omega_0T_0}}{1 + 2i\omega_0\tau} + CC \tag{6.2.135}$$

To solve the equation $Q = 0$, we let $A = (1/2)ae^{i\phi}$ with real a and ϕ, separate real and imaginary parts in (6.2.134), and obtain

$$\frac{da}{dT_1} = a(1 - \tfrac{1}{4}\alpha_ra^2) \tag{6.2.136}$$

$$\frac{d\phi}{dT_1} = -\tfrac{1}{4}\alpha_ia^2 \tag{6.2.137}$$

where

$$\alpha_r = \frac{3 + 8\omega_0{}^2\tau^2}{1 + 4\omega_0{}^2\tau^2}, \qquad \alpha_i = -\frac{2\omega_0\tau}{1 + 4\omega_0{}^2\tau^2} \tag{6.2.138}$$

The solutions of (6.2.136) and (6.2.137) are

$$a^2 = \frac{4}{\alpha_r + \left(\dfrac{4}{a_0^2} - \alpha_r\right)e^{-2\mu t}} \tag{6.2.139}$$

$$\phi = \frac{\tau\omega_0}{3 + 8\omega_0^2\tau^2} \ln\left[\frac{4}{a_0^2} + \alpha_r(e^{2\mu t} - 1)\right] + \phi_0 \tag{6.2.140}$$

where a_0 is the initial amplitude and ϕ_0 is a constant.

To determine B, we substitute for v_0, Z_0, and v_1 into (6.2.129); that is

$$\tau D_0 Z_1 + Z_1 = \left[-\tau D_1 B + 8\tau\frac{1 + 2\omega_0^2\tau^2}{1 + 4\omega_0^2\tau^2}A\bar{A}B\right]e^{-T_0/\tau} + \text{NST} \tag{6.2.141}$$

In order that Z_1/Z_0 be bounded for all T_0, the coefficient of $\exp(-T_0/\tau)$ must vanish; hence

$$D_1 B = 2\frac{1 + 2\omega_0^2\tau^2}{1 + 4\omega_0^2\tau^2}a^2 B \tag{6.2.142}$$

where we made use of $A = (1/2)a \exp i\phi$. Substituting for a^2 from (6.2.139) into (6.2.142) and solving the resulting equation, we obtain

$$B = b\left[\frac{4}{a_0^2} + \alpha_r(e^{2\mu t} - 1)\right]^\zeta \tag{6.2.143}$$

where

$$\zeta = \frac{4(1 + 2\omega_0^2\tau^2)}{3 + 8\omega_0^2\tau^2}$$

Therefore, to first approximation

$$v = a\cos(\omega_0 t + \phi) + O(\mu) \tag{6.2.144}$$

$$Z = Be^{-t/\tau} + \frac{a^2}{2\sqrt{1 + 4\omega_0^2\tau^2}}\cos[2\omega_0 t + 2\phi - \tan^{-1}2\omega_0\tau] + \tfrac{1}{2}a^2 + O(\mu) \tag{6.2.145}$$

with a, ϕ, and B given, respectively, by (6.2.139), (6.2.140), and (6.2.143).

6.2.6. THE STABILITY OF THE TRIANGULAR POINTS IN THE ELLIPTIC RESTRICTED PROBLEM OF THREE BODIES

The next two examples are fourth-order problems—one is linear and the second is nonlinear. Consider first the stability of the triangular points in the restricted problem of three bodies treated in Sections 3.1.4 and 3.1.5 using the Lindstedt–Poincaré and Whittaker techniques. This problem was treated first using the method of multiple scales by Alfriend and Rand (1969). The problem reduces mathematically to the stability of solutions of (3.1.63) through (3.1.65). In this section we use the method of multiple scales to determine the transition curves intersecting the μ axis at $\mu_0 = (1 - 2\sqrt{2}/3)/2$ and determine the behavior of x and y near these transition curves.

We set $\cos f = \cos T_0$ and assume that

$$x = x_0(T_0, T_1) + ex_1(T_0, T_1) + \cdots \qquad (6.2.146)$$

$$y = y_0(T_0, T_1) + ey_1(T_0, T_1) + \cdots \qquad (6.2.147)$$

$$\mu = \mu_0 + e\mu_1 + \cdots \qquad (6.2.148)$$

where

$$T_0 = f \quad \text{and} \quad T_1 = ef \qquad (6.2.149)$$

Hence

$$\frac{d}{df} = D_0 + eD_1 + \cdots, \qquad D_n = \frac{\partial}{\partial T_n} \qquad (6.2.150)$$

Substituting (6.2.146) through (6.2.150) into (3.1.63) through (3.1.65) and equating coefficients of e^0 and e to zero, we obtain

Order e^0

$$D_0{}^2 x_0 - 2D_0 y_0 - b_0 x_0 = 0 \qquad (6.2.151)$$

$$D_0{}^2 y_0 + 2D_0 x_0 - a_0 y_0 = 0 \qquad (6.2.152)$$

Order e

$$D_0{}^2 x_1 - 2D_0 y_1 - b_0 x_1 = -2D_0 D_1 x_0 + 2D_1 y_0 + b_1 x_0 - b_0 x_0 \cos T_0 \quad (6.2.153)$$

$$D_0{}^2 y_1 + 2D_0 x_1 - a_0 y_1 = -2D_0 D_1 y_0 - 2D_1 x_0 - b_1 y_0 - a_0 y_0 \cos T_0 \quad (6.2.154)$$

where a_i and b_i are given by (3.1.71) and (3.1.72).

The solution of (6.2.151) and (6.2.152) is

$$x_0 = A(T_1) \cos \tfrac{1}{2} T_0 + B(T_1) \sin \tfrac{1}{2} T_0 \qquad (6.2.155)$$

$$y_0 = \alpha B(T_1) \cos \tfrac{1}{2} T_0 - \alpha A(T_1) \sin \tfrac{1}{2} T_0 \qquad (6.2.156)$$

where

$$\alpha = (a_0 + \tfrac{1}{4})^{-1} = b_0 + \tfrac{1}{4} = \tfrac{1}{4}(7 - \sqrt{33}) \qquad (6.2.157)$$

The zeroth-order solution determines the right-hand sides of (6.2.153) and (6.2.154). Thus they become

$$D_0{}^2 x_1 - 2D_0 y_1 - b_0 x_1 = P_1 \cos \tfrac{1}{2} T_0 + Q_1 \sin \tfrac{1}{2} T_0 + \text{NST} \quad (6.2.158)$$

$$D_0{}^2 y_1 + 2D_0 x_1 - a_0 y_1 = P_2 \cos \tfrac{1}{2} T_0 + Q_2 \sin \tfrac{1}{2} T_0 + \text{NST} \quad (6.2.159)$$

where

$$P_1 = (2\alpha - 1)B' + (b_1 - \tfrac{1}{2}b_0)A \qquad (6.2.160)$$

$$P_2 = (\alpha - 2)A' - \alpha(b_1 + \tfrac{1}{2}a_0)B \qquad (6.2.161)$$

$$Q_1 = -(2\alpha - 1)A' + (b_1 + \tfrac{1}{2}b_0)B \qquad (6.2.162)$$

$$Q_2 = (\alpha - 2)B' + \alpha(b_1 - \tfrac{1}{2}a_0)A \qquad (6.2.163)$$

To determine a first approximation, we need not solve for x_1 and y_1 but only insure that x_1/x_0 and y_1/y_0 are bounded for all T_0. This is the reason why we spelled out the terms that produce secular terms. To eliminate the secular terms, we can find the particular secular solution and then determine the condition for its vanishing. The resulting particular solution is of the form

$$x = 0, \qquad y = R_1 \cos \tfrac{1}{2}T_0 + S_1 \sin \tfrac{1}{2}T_0 \qquad (6.2.164)$$

or

$$y = 0, \qquad x = R_2 \cos \tfrac{1}{2}T_0 + S_2 \sin \tfrac{1}{2}T_0 \qquad (6.2.165)$$

Thus we can determine the conditions for the elimination of the secular terms by assuming a particular solution of the form (6.2.164) or (6.2.165). The results of using either form are the same. Substituting (6.2.164) into (6.2.158) and (6.2.159) and equating the coefficients of $\cos(T_0/2)$ and $\sin(T_0/2)$ on both sides, we obtain

$$R_1 = Q_1, \qquad S_1 = -P_1 \qquad (6.2.166)$$

$$R_1 = -\alpha P_2, \qquad S_1 = -\alpha Q_2 \qquad (6.2.167)$$

Elimination of R_1 and S_1 from (6.2.166) and (6.2.167) leads to the required conditions; that is

$$P_1 = \alpha Q_2, \qquad Q_1 = -\alpha P_2 \qquad (6.2.168)$$

Substituting for P_1, P_2, Q_1, and Q_2 from (6.2.160) through (6.2.163) into (6.2.168) leads to the following two equations for A and B

$$(1 - 4\alpha + \alpha^2)A' + [(1 - \alpha^2)b_1 + \tfrac{1}{2}(b_0 - \alpha^2 a_0)]B = 0 \qquad (6.2.169)$$

$$(1 - 4\alpha + \alpha^2)B' - [(1 - \alpha^2)b_1 - \tfrac{1}{2}(b_0 - \alpha^2 a_0)]A = 0 \qquad (6.2.170)$$

To solve these equations, we let

$$A = a \exp \gamma_1 T_1 \qquad B = b \exp \gamma_1 T_1 \qquad (6.2.171)$$

and obtain

$$(1 - 4\alpha + \alpha^2)\gamma_1 a + [(1 - \alpha^2)b_1 + \tfrac{1}{2}(b_0 - \alpha^2 a_0)]b = 0 \qquad (6.2.172)$$

$$-[(1 - \alpha^2)b_1 - \tfrac{1}{2}(b_0 - \alpha^2 a_0)]a + (1 - 4\alpha + \alpha^2)b\gamma_1 = 0 \qquad (6.2.173)$$

These equations are the same as (3.1.105) and (3.1.106) obtained using Whittaker's method. Hence γ_1 and b/a are given by (3.1.107) and (3.1.108), while x and y are given by (3.1.109). This expansion was continued to second order by Alfriend and Rand (1969).

6.2.7. A SWINGING SPRING

We consider next the nonlinear swinging spring discussed in Sections 5.5.3 and 5.7.5 and described by the Lagrangian (5.5.54). The equations of motion corresponding to (5.5.54) are

$$\ddot{x} + \frac{k}{m} x + g(1 - \cos \theta) - (l + x)\dot{\theta}^2 = 0 \qquad (6.2.174)$$

$$\ddot{\theta} + \frac{g}{l + x} \sin \theta + \frac{2}{l + x} \dot{x}\dot{\theta} = 0 \qquad (6.2.175)$$

We seek an asymptotic solution of these equations for small but finite x and θ of the form

$$x(t) = \epsilon x_1(T_0, T_1) + \epsilon^2 x_2(T_0, T_1) + \cdots \qquad (6.2.176)$$

$$\theta(t) = \epsilon \theta_1(T_0, T_1) + \epsilon^2 \theta_2(T_0, T_1) + \cdots \qquad (6.2.177)$$

where $T_n = \epsilon^n t$ and ϵ is of the order of the amplitudes of oscillation.

Substituting (6.2.176) and (6.2.177) into (6.2.174) and (6.2.175) and equating coefficients of like powers of ϵ, we obtain

Order ϵ

$$D_0^2 x_1 + \omega_1^2 x_1 = 0, \qquad \omega_1^2 = \frac{k}{m} \qquad (6.2.178)$$

$$D_0^2 \theta_1 + \omega_2^2 \theta_1 = 0, \qquad \omega_2^2 = \frac{g}{l} \qquad (6.2.179)$$

Order ϵ^2

$$D_0^2 x_2 + \omega_1^2 x_2 = -2D_0 D_1 x_1 - \tfrac{1}{2} g \theta_1^2 + l(D_0 \theta_1)^2 \qquad (6.2.180)$$

$$D_0^2 \theta_2 + \omega_2^2 \theta_2 = -2D_0 D_1 \theta_1 + \frac{\omega_2^2}{l} x_1 \theta_1 - \frac{2}{l} (D_0 x_1)(D_0 \theta_1) \qquad (6.2.181)$$

The solution of the first-order equations is

$$x_1 = A(T_1)e^{i\omega_1 T_0} + \bar{A}(T_1)e^{-i\omega_1 T_0} \qquad (6.2.182)$$

$$\theta_1 = B(T_1)e^{i\omega_2 T_0} + \bar{B}(T_1)e^{-i\omega_2 T_0} \qquad (6.2.183)$$

Then (6.2.180) and (6.2.181) become

$$D_0^2 x_2 + \omega_1^2 x_2 = -2i\omega_1 D_1 A e^{i\omega_1 T_0} - \tfrac{3}{2}gB^2 e^{2i\omega_2 T_0} + \tfrac{1}{2}gB\bar{B} + CC \quad (6.2.184)$$

$$D_0^2 \theta_2 + \omega_2^2 \theta_2 = -2i\omega_2 D_1 B e^{i\omega_2 T_0} + \frac{\omega_2(\omega_2 + 2\omega_1)}{l} A B e^{i(\omega_1 + \omega_2)T_0}$$

$$+ \frac{\omega_2(\omega_2 - 2\omega_1)}{l} A\bar{B} e^{i(\omega_1 - \omega_2)T_0} + CC \quad (6.2.185)$$

If A and B are constants, the particular solutions of (6.2.184) and (6.2.185) are

$$x_2 = \frac{1}{2}\frac{g}{\omega_1^2} B\bar{B} - \frac{3}{2}\frac{g}{\omega_1^2 - 4\omega_2^2} B^2 e^{2i\omega_2 T_0} + CC \quad (6.2.186)$$

$$\theta_2 = - \frac{\omega_2(\omega_2 + 2\omega_1)}{l\omega_1(\omega_1 + 2\omega_2)} A B e^{i(\omega_1 + \omega_2)T_0}$$

$$- \frac{\omega_2(\omega_2 - 2\omega_1)}{l\omega_1(\omega_1 - 2\omega_2)} A\bar{B} e^{i(\omega_1 - \omega_2)T_0} + CC \quad (6.2.187)$$

which tend to ∞ as $\omega_1 \to 2\omega_2$. Consequently, the expansions (6.2.176) and (6.2.177) break down when $\omega_1 \approx 2\omega_2$.

To obtain an expansion valid when $\omega_1 \approx 2\omega_2$, we let

$$\omega_1 - 2\omega_2 = \epsilon\sigma, \qquad \sigma = O(1) \quad (6.2.188)$$

and let A and B be functions of T_1 rather than being constants. Moreover, using (6.2.188) we express $\exp(2i\omega_2 T_0)$ and $\exp[i(\omega_1 - \omega_2)T_0]$ in (6.2.184) and (6.2.185) as

$$\exp(2i\omega_2 T_0) = \exp(i\omega_1 T_0 - i\sigma T_1)$$
$$\exp[i(\omega_1 - \omega_2)T_0] = \exp(i\omega_2 T_0 + i\sigma T_1)$$

to obtain

$$D_0^2 x_2 + \omega_1^2 x_2 = -(2i\omega_1 D_1 A + \tfrac{3}{2}gB^2 e^{-i\sigma T_1})e^{i\omega_1 T_0} + CC + \text{NST} \quad (6.2.189)$$

$$D_0^2 \theta_2 + \omega_2^2 \theta_2 = -\left[2i\omega_2 D_1 B - \frac{\omega_2(\omega_2 - 2\omega_1)}{l} A\bar{B} e^{i\sigma T_1}\right]e^{i\omega_2 T_0} + CC + \text{NST}$$

$$(6.2.190)$$

Eliminating secular terms, we have

$$2i\omega_1 D_1 A = -\tfrac{3}{2}gB^2 \exp(-i\sigma T_1)$$

$$2i\omega_2 D_1 B = \frac{\omega_2(\omega_2 - 2\omega_1)}{l} A\bar{B} \exp(i\sigma T_1) \qquad (6.2.191)$$

Letting $A = -(1/2)ia_1 \exp(i\omega_1\beta_1)$ and $B = -(1/2)ia_2 \exp(i\omega_2\beta_2)$, with real a_i and β_i, and separating real and imaginary parts, we obtain

$$\dot{a}_1 = \frac{3\epsilon g}{8\omega_1} a_2^2 \cos\gamma \tag{6.2.192}$$

$$\dot{a}_2 = -\frac{3\epsilon\omega_2}{4l} a_1 a_2 \cos\gamma \tag{6.2.193}$$

$$a_1\dot{\beta}_1 = -\frac{3\epsilon g}{8\omega_1^2} a_2^2 \sin\gamma \tag{6.2.194}$$

$$\dot{\beta}_2 = -\frac{3\epsilon}{4l} a_1 \sin\gamma \tag{6.2.195}$$

where

$$\gamma = \omega_1\beta_1 - 2\omega_2\beta_2 + (\omega_1 - 2\omega_2)t \tag{6.2.196}$$

If we let $a_1^2 = \omega_1\alpha_1^*/\omega_2 k$ and $a_2^2 = 2\alpha_2/mgl$, (6.2.192) through (6.2.196) go over into (5.5.76) through (5.5.80) which were obtained using the method of averaging in conjunction with canonical variables.

6.2.8. A MODEL FOR WEAK NONLINEAR INSTABILITY

We next consider the model problem

$$u_{tt} - u_{xx} - u = u^3$$
$$u(x, 0) = \epsilon \cos kx, \qquad u_t(x, 0) = 0 \tag{6.2.197}$$

for weak nonlinear instability of standing waves, which was discussed in Sections 2.1.2, 3.4.2, and 3.5.1. Away from $k = 1$, a uniformly valid solution for standing waves is (Section 3.4.2)

$$u = \epsilon \cos \sigma t \cos kx + O(\epsilon^3) \tag{6.2.198}$$

where

$$\sigma = \sqrt{k^2 - 1}\left[1 - \frac{9\epsilon^2}{32(k^2 - 1)}\right] + \cdots$$

It is clear that this expansion breaks down when $k - 1 = O(\epsilon^2)$.

To determine an expansion valid near $k = 1$, we introduce the new variable $\xi = kx$ in (6.2.197) to obtain

$$u_{tt} - k^2 u_{\xi\xi} - u = u^3$$
$$u(\xi, 0) = \epsilon \cos \xi, \qquad u_t(\xi, 0) = 0 \tag{6.2.199}$$

Moreover, we let

$$k = 1 + \epsilon^2 k_2 \quad \text{with} \quad k_2 = O(1) \tag{6.2.200}$$

and assume that

$$u(\xi, t; \epsilon) = u(\xi, T_0, T_1, T_2; \epsilon) = \epsilon u_1 + \epsilon^2 u_2 + \epsilon^3 u_3 + \cdots \qquad (6.2.201)$$

where $T_n = \epsilon^n t$.

Substituting (6.2.200) and (6.2.201) into (6.2.199) and equating coefficients of equal powers of ϵ, we obtain

Order ϵ

$$\frac{\partial^2 u_1}{\partial T_0^2} - \frac{\partial^2 u_1}{\partial \xi^2} - u_1 = 0$$

$$u_1 = \cos \xi, \qquad \frac{\partial u_1}{\partial T_0} = 0 \quad \text{at} \quad T_n = 0 \qquad (6.2.202)$$

Order ϵ^2

$$\frac{\partial^2 u_2}{\partial T_0^2} - \frac{\partial^2 u_2}{\partial \xi^2} - u_2 = -2 \frac{\partial^2 u_1}{\partial T_0 \partial T_1}$$

$$u_2 = 0, \qquad \frac{\partial u_2}{\partial T_0} = -\frac{\partial u_1}{\partial T_1} \quad \text{at} \quad T_n = 0 \qquad (6.2.203)$$

Order ϵ^3

$$\frac{\partial^2 u_3}{\partial T_0^2} - \frac{\partial^2 u_3}{\partial \xi^2} - u_3 = u_1{}^3 + 2k_2 \frac{\partial^2 u_1}{\partial \xi^2} - \frac{\partial^2 u_1}{\partial T_1{}^2} - 2 \frac{\partial^2 u_2}{\partial T_0 \partial T_1} - 2 \frac{\partial^2 u_1}{\partial T_0 \partial T_2}$$

$$(6.2.204)$$

$$u_3 = 0, \qquad \frac{\partial u_3}{\partial T_0} = -\frac{\partial u_1}{\partial T_2} - \frac{\partial u_2}{\partial T_1} \quad \text{at} \quad T_n = 0$$

The solution of the first-order problem is

$$u_1 = a(T_1, T_2) \cos \xi, \qquad a(0, 0) = 1 \qquad (6.2.205)$$

Then (6.2.203) becomes

$$\frac{\partial^2 u_2}{\partial T_0^2} - \frac{\partial^2 u_2}{\partial \xi^2} - u_2 = 0$$

$$u_2 = 0, \qquad \frac{\partial u_2}{\partial T_0} = -\left(\frac{\partial a}{\partial T_1}\right) \cos \xi \quad \text{at} \quad T_n = 0 \qquad (6.2.206)$$

The solution of (6.2.206) will contain a term proportional to T_0 making u_2/u_1 unbounded as $T_0 \to \infty$ unless $\partial a/\partial T_1 = 0$ at $T_1 = T_2 = 0$. Then

$$u_2 = b(T_1, T_2) \cos \xi, \qquad b(0, 0) = 0 \qquad (6.2.207)$$

With the first- and second-order solutions known, the equation for u_3

becomes

$$\frac{\partial^2 u_3}{\partial T_0^2} - \frac{\partial^2 u_3}{\partial \xi^2} - u_3 = \left(\tfrac{3}{4}a^3 - 2k_2 a - \frac{\partial^2 a}{\partial T_1^2} \right) \cos \xi + \tfrac{1}{4}a^3 \cos 3\xi \quad (6.2.208)$$

Secular terms will be eliminated if

$$\frac{\partial^2 a}{\partial T_1^2} + (2k_2 - \tfrac{3}{4}a^2)a = 0 \qquad (6.2.209)$$

The initial conditions for a were obtained above as

$$a = 1 \quad \text{and} \quad \frac{\partial a}{\partial T_1} = 0 \quad \text{at} \quad T_n = 0 \qquad (6.2.210)$$

To determine $b(T_1, T_2)$ and the dependence of a on T_2, we need to carry out the expansion to higher order. If we limit the analysis to $O(\epsilon^3)$, we can regard a as a function of T_1 within an error of $O(\epsilon^2 t)$.

A first integral for (6.2.209) and (6.2.210) is

$$\left(\frac{\partial a}{\partial T_1} \right)^2 = \tfrac{3}{8}(a^2 - 1)(a^2 - \beta), \qquad \beta = \frac{16k_2}{3} - 1 \qquad (6.2.211)$$

Since $a(T_1)$ is real, the right-hand side of (6.2.211) must be positive, hence a^2 must be outside the interval whose ends are 1 and β. Since $a(0) = 1$, a^2 increases without bound if $\beta < 1$ and oscillates between 0 and 1 if $\beta > 1$. Therefore $\beta = 1$ or $k_2 = 3/8$ separates the stable from the unstable regions. Hence the condition of neutral stability is

$$k = 1 + \tfrac{3}{8}\epsilon^2 \qquad (6.2.212)$$

in agreement with (3.5.6). The solution for a is a Jacobian elliptic function.

6.2.9.   A MODEL FOR WAVE–WAVE INTERACTION
We again consider Bretherton's (1964) model equation

$$\phi_{tt} + \phi_{xxxx} + \phi_{xx} + \phi = \epsilon\phi^3 \qquad (6.2.213)$$

which was treated in Sections 5.8.1 and 5.8.2 using the variational approach. The linear problem admits the uniform traveling wave solution

$$\phi = a \cos(kx - \omega t + \beta) \qquad (6.2.214)$$

where a, k, ω, and β are constants and ω and k satisfy the dispersion relationship

$$\omega^2 = k^4 - k^2 + 1 \qquad (6.2.215)$$

Harmonic resonance may occur whenever (ω, k) and $(n\omega, nk)$ for some integer $n \geq 2$ satisfy (6.2.215). This occurs at all $k^2 = 1/n$ for $n \geq 2$. At these wave numbers the fundamental and its nth harmonic have the same phase speed.

Since the nonlinearity is cubic in our equation, the fundamental corresponding to $k^2 = 1/3$ interacts to $O(\epsilon)$ only with its third harmonic ($k^2 = 3$). If the nonlinearity is $\epsilon\phi^m$ for some integer m, then the fundamental ($k^2 = 1/m$) interacts to $O(\epsilon)$ with its mth harmonic ($k^2 = m$). If we consider interactions to orders higher than ϵ, harmonic resonances other than the third can occur even for a cubic nonlinearity.

To determine a first-order expansion valid near $k^2 = 1/3$, we let

$$\phi = \phi_0(T_0, T_1, X_0, X_1) + \epsilon\phi_1(T_0, T_1, X_0, X_1) + \cdots \quad (6.2.216)$$

where

$$T_n = \epsilon^n t, \qquad X_n = \epsilon^n x$$

Substituting this expansion into (6.2.213) and equating coefficients of like powers of ϵ, we obtain

$$L(\phi_0) = \frac{\partial^2 \phi_0}{\partial T_0^2} + \frac{\partial^4 \phi_0}{\partial X_0^4} + \frac{\partial^2 \phi_0}{\partial X_0^2} + \phi_0 = 0 \quad (6.2.217)$$

$$L(\phi_1) = \phi_0^3 - 2\frac{\partial^2 \phi_0}{\partial T_0 \partial T_1} - 4\frac{\partial^4 \phi_0}{\partial X_0^3 \partial X_1} - 2\frac{\partial^2 \phi_0}{\partial X_0 \partial X_1} \quad (6.2.218)$$

The solution of (6.2.217) is taken to be

$$\phi_0 = A_1(T_1, X_1)e^{i\theta_1} + A_3(T_1, X_1)e^{i\theta_3} + CC \quad (6.2.219)$$

where

$$\theta_n = k_n X_0 - \omega_n T_0$$

$$\omega_n^2 = k_n^4 - k_n^2 + 1 \quad (6.2.220)$$

$$\omega_3 \approx 3\omega_1, \qquad k_3 \approx 3k_1$$

Note that ϕ_0 is assumed to contain the two interacting harmonics. Had we assumed it to contain $\exp(i\theta_1)$, we would have found it to be invalid (see Sections 5.8.2 and 6.4.8).

Substituting for ϕ_0 into (6.2.218), we obtain

$$L(\phi_1) = \sum_{n=1,3} 2i\omega_n\left(\frac{\partial A_n}{\partial T_1} + \omega_n'\frac{\partial A_n}{\partial X_1}\right)e^{i\theta_n} + 3(A_1\bar{A}_1 + 2A_3\bar{A}_3)A_1e^{i\theta_1}$$

$$+ 3(2A_1\bar{A}_1 + A_3\bar{A}_3)A_3e^{i\theta_3} + A_1^3e^{3i\theta_1} + A_3^3e^{3i\theta_3}$$

$$+ 3A_1^2A_3e^{i(\theta_3+2\theta_1)} + 3\bar{A}_1^2A_3e^{i(\theta_3-2\theta_1)}$$

$$+ 3A_1A_3^2e^{i(2\theta_3+\theta_1)} + 3\bar{A}_1A_3^2e^{i(2\theta_3-\theta_1)} + CC \quad (6.2.221)$$

where $\omega_n = d\omega_n/dk$: group velocity.

Because of the interaction between the two modes, terms that produce secular terms other than the usual exp $(i\theta_n)$ occur in (6.2.221). To recognize these terms we consider the perfect resonance case in which $\theta_3 = 3\theta_1$ so that exp $(i\theta_1)$ and exp $(3i\theta_1)$ produce secular terms. We immediately see that exp $(3i\theta_1)$ and exp $[i(\theta_3 - 2\theta_1)]$ produce secular terms. For the near-resonance case, we indicate their secular behavior by expressing them in terms of exp $(i\theta_1)$ and exp $(i\theta_3)$. To do this we observe that

$$\theta_3 - 3\theta_1 \equiv \Gamma = (k_3 - 3k_1)X_0 - (\omega_3 - 3\omega_1)T_0$$

Although X_0 and T_0 are $O(1)$, Γ becomes slowly varying as $k_3 \to 3k_1$ and $\omega_3 \to 3\omega_1$, hence we express this slow variation by rewriting Γ as

$$\Gamma = \frac{k_3 - 3k_1}{\epsilon} X_1 - \frac{\omega_3 - 3\omega_1}{\epsilon} T_1 \tag{6.2.222}$$

With this function we express exp $(3i\theta_1)$ and exp $[i(\theta_3 - 2\theta_1)]$ as

$$\exp(3i\theta_1) = \exp[i(\theta_3 - \Gamma)], \quad \exp[i(\theta_3 - 2\theta_1)] = \exp[i(\theta_1 + \Gamma)].$$

Eliminating the terms that produce secular terms on the right-hand side of (6.2.221), we have

$$2i\omega_1\left(\frac{\partial A_1}{\partial T_1} + \omega_1'\frac{\partial A_1}{\partial X_1}\right) = -3(A_1\bar{A}_1 + 2A_3\bar{A}_3)A_1 - 3\bar{A}_1^2 A_3 e^{i\Gamma} \tag{6.2.223}$$

$$2i\omega_3\left(\frac{\partial A_3}{\partial T_1} + \omega_3'\frac{\partial A_3}{\partial X_1}\right) = -3(2A_1\bar{A}_1 + A_3\bar{A}_3)A_3 - A_1^3 e^{-i\Gamma} \tag{6.2.224}$$

Letting $A_n = (1/2)a_n \exp(i\beta_n)$ with real a_n and β_n in (6.2.223) and (6.2.224) and separating real and imaginary parts, we have

$$\frac{\partial a_1}{\partial T_1} + \omega_1'\frac{\partial a_1}{\partial X_1} = -\frac{3}{8\omega_1}a_1^2 a_3 \sin\delta \tag{6.2.225}$$

$$\frac{\partial \beta_1}{\partial T_1} + \omega_1'\frac{\partial \beta_1}{\partial X_1} = \frac{3}{8\omega_1}(a_1^2 + 2a_3^2 + a_1 a_3 \cos\delta) \tag{6.2.226}$$

$$\frac{\partial a_3}{\partial T_1} + \omega_3'\frac{\partial a_3}{\partial X_1} = \frac{1}{8\omega_3}a_1^3 \sin\delta \tag{6.2.227}$$

$$\frac{\partial \beta_3}{\partial T_1} + \omega_3'\frac{\partial \beta_3}{\partial X_1} = \frac{1}{8\omega_3}(6a_1^2 + 3a_3^2 + a_1^3 a_3^{-1} \cos\delta) \tag{6.2.228}$$

where

$$\delta = \Gamma + \beta_3 - 3\beta_1 \tag{6.2.229}$$

Equations (6.2.225) through (6.2.228) are in full agreement with (5.8.24) through (5.8.27) obtained using the variational approach.

6.2.10. LIMITATIONS OF THE DERIVATIVE-EXPANSION METHOD

This method applies to wave-type problems only. It does not apply to unstable cases except when the instability is weak, such as the nonlinear stability problem discussed in Section 6.2.8. For $k > 1$, u is bounded and the expansion (6.2.198) is valid for times as large as ϵ^{-2} if k is away from 1. This expansion is valid only for small times for $k < 1$ and away from 1. Near $k = 1$, the instability is weak and a valid solution is given by (6.2.211) for times as large as ϵ^{-1}.

In the case of hyperbolic equations, this method applies to dispersive waves only when the initial conditions can be represented by the superposition of a finite number of sinusoidal functions. For a linearly nondispersive wave problem such as

$$\frac{\partial^2 u}{\partial t^2} - \frac{\partial^2 u}{\partial x^2} = \epsilon u^2 \qquad (6.2.230)$$

$$u(x, 0) = f(x), \qquad \frac{\partial u}{\partial t}(x, 0) = 0 \qquad (6.2.231)$$

this method does not provide a solution even if $f(x)$ is a sinusoidal function such as $\cos x$. To see this we let

$$u = u_0(x, T_0, T_1) + \epsilon u_1(x, T_0, T_1) + \cdots \qquad (6.2.232)$$

where

$$T_0 = t, \qquad T_1 = \epsilon t$$

Substituting (6.2.232) into (6.2.230) and (6.2.231) and equating like powers of ϵ, we obtain

Order ϵ^0

$$\frac{\partial^2 u_0}{\partial T_0^2} - \frac{\partial^2 u_0}{\partial x^2} = 0 \qquad (6.2.233)$$

$$u_0(x, 0, 0) = \cos x, \qquad \frac{\partial u_0}{\partial T_0}(x, 0, 0) = 0 \qquad (6.2.234)$$

Order ϵ

$$\frac{\partial^2 u_1}{\partial T_0^2} - \frac{\partial^2 u_0}{\partial x^2} = -2\frac{\partial^2 u_0}{\partial T_0 \partial T_1} + u_0^2 \qquad (6.2.235)$$

$$u(x, 0, 0) = 0, \qquad \frac{\partial u_1}{\partial T_0}(x, 0, 0) = -\frac{\partial u_0}{\partial T_1}(x, 0, 0) \qquad (6.2.236)$$

The solution of (6.2.233) and (6.2.234) is

$$u_0 = A(T_1)e^{i(x-T_0)} + \bar{A}(T_1)e^{-i(x-T_0)} \qquad (6.2.237)$$

where

$$A(0) = \tfrac{1}{2}$$

Substituting this zeroth-order solution into (6.2.235) gives

$$\frac{\partial^2 u_1}{\partial T_0^2} - \frac{\partial^2 u_1}{\partial x^2} = 2i\frac{\partial A}{\partial T_1}e^{i(x-T_0)} - 2i\frac{\partial \bar{A}}{\partial T_1}e^{-i(x-T_0)}$$
$$+ 2A\bar{A} + A^2 e^{2i(x-T_0)} + \bar{A}^2 e^{-2i(x-T_0)} \quad (6.2.238)$$

The right-hand side of this equation contains terms that produce secular terms. They are the terms proportional to $\exp[\pm 2i(x - T_0)]$ in addition to $\exp[\pm i(x - T_0)]$. In order that u_1/u_0 be bounded for all T_0, all these terms must be eliminated. However, there is no way in which this can be done. In the previous examples such terms were proportional to $\exp[\pm i(x - T_0)]$, hence A was chosen to eliminate them. In this case, if a nontrivial solution is desired, A can be chosen in such a way as to eliminate the terms proportional to $\exp[\pm i(x - T_0)]$. The resulting expansion contains secular terms, hence it is not valid for large times.

Expansions valid for large times and general initial conditions for nondispersive waves were obtained in Sections 3.2.4 and 3.2.5 using the method of strained coordinates.

6.3.   The Two-Variable Expansion Procedure

6.3.1.   THE DUFFING EQUATION
Let us consider again the equation

$$\frac{d^2u}{dt^2} + \omega_0^2 u + \epsilon u^3 = 0 \qquad (6.3.1)$$

We assume that

$$u = u_0(\xi, \eta) + \epsilon u_1(\xi, \eta) + \epsilon^2 u_2(\xi, \eta) + \cdots \qquad (6.3.2)$$

where

$$\xi = \epsilon t, \qquad \eta = (1 + \epsilon^2\omega_2 + \epsilon^3\omega_3 + \cdots)t \qquad (6.3.3)$$

Substituting (6.3.2) and (6.3.3) into (6.3.1) and equating like powers of ϵ, we obtain

$$\frac{\partial^2 u_0}{\partial \eta^2} + \omega_0^2 u_0 = 0 \qquad (6.3.4)$$

$$\frac{\partial^2 u_1}{\partial \eta^2} + \omega_0^2 u_1 = -2\frac{\partial^2 u_0}{\partial \xi\, \partial \eta} - u_0^3 \qquad (6.3.5)$$

$$\frac{\partial^2 u_2}{\partial \eta^2} + \omega_0^2 u_2 = -2\frac{\partial^2 u_1}{\partial \xi\, \partial \eta} - \frac{\partial^2 u_0}{\partial \xi^2} - 2\omega_2\frac{\partial^2 u_0}{\partial \eta^2} - 3u_0^2 u_1 \qquad (6.3.6)$$

The general solution of (6.3.4) is

$$u_0 = A_0(\xi) \cos \omega_0 \eta + B_0(\xi) \sin \omega_0 \eta \qquad (6.3.7)$$

Then (6.3.5) becomes

$$\frac{\partial^2 u_1}{\partial \eta^2} + \omega_0{}^2 u_1 = -[2\omega_0 B_0' + \tfrac{3}{4}(A_0{}^3 + A_0 B_0{}^2)] \cos \omega_0 \eta$$

$$+ [2\omega_0 A_0' - \tfrac{3}{4}(B_0{}^3 + A_0{}^2 B_0)] \sin \omega_0 \eta$$

$$- \tfrac{1}{4}(A_0{}^3 - 3A_0 B_0{}^2) \cos 3\omega_0 \eta + \tfrac{1}{4}(B_0{}^3 - 3A_0{}^2 B_0) \sin 3\omega_0 \eta \quad (6.3.8)$$

Secular terms will be eliminated if

$$2\omega_0 B_0' + \tfrac{3}{4}(A_0{}^3 + A_0 B_0{}^2) = 0 \qquad (6.3.9)$$

$$2\omega_0 A_0' - \tfrac{3}{4}(B_0{}^3 + A_0{}^2 B_0) = 0 \qquad (6.3.10)$$

Adding B_0 times (6.3.9) to A_0 times (6.3.10) gives

$$\frac{d}{d\xi}(A_0{}^2 + B_0{}^2) = 0 \quad \text{or} \quad A_0{}^2 + B_0{}^2 = a^2 = \text{a constant} \quad (6.3.11)$$

Using (6.3.11), we can express (6.3.9) and (6.3.10) in the form

$$B_0' + \omega_1 A_0 = 0, \qquad A_0' - \omega_1 B_0 = 0 \qquad (6.3.12)$$

where

$$\omega_1 = \frac{3}{8\omega_0} a^2.$$

Hence

$$A_0 = a \cos(\omega_1 \xi + \phi), \qquad B_0 = -a \sin(\omega_1 \xi + \phi) \qquad (6.3.13)$$

where ϕ is a constant. With the secular terms eliminated, the solution of u_1 becomes

$$u_1 = \tilde{A}_1(\xi) \cos \omega_0 \eta + \tilde{B}_1(\xi) \sin \omega_0 \eta + \frac{1}{32\omega_0{}^2}(A_0{}^3 - 3A_0 B_0{}^2) \cos 3\omega_0 \eta$$

$$- \frac{1}{32\omega_0{}^2}(B_0{}^3 - 3A_0{}^2 B_0) \sin 3\omega_0 \eta \quad (6.3.14)$$

Substituting for A_0 and B_0 into u_0 and u_1, we get

$$u_0 = a \cos \theta \qquad (6.3.15)$$

$$u_1 = A_1(\xi) \cos \theta + B_1(\xi) \sin \theta + \frac{a^3}{32\omega_0{}^2} \cos 3\theta \qquad (6.3.16)$$

where

$$\theta = \omega_0 \eta + \omega_1 \xi + \phi \qquad (6.3.17)$$

Replacing u_0 and u_1 in (6.3.6) by their expressions from (6.3.15) through (6.3.17), we obtain

$$\frac{\partial^2 u_2}{\partial \eta^2} + \omega_0^2 u_2 = -\left(2\omega_0 B_1{'} + \tfrac{9}{4}a^2 A_1 - 2\omega_0\omega_1 A_1 - a\omega_1^2 - 2a\omega_2\omega_0^2\right.$$

$$\left. + \frac{3}{128\omega_0^2}a^5\right) \cos\theta - (-2\omega_0 A_1' - 2\omega_0\omega_1 B_1 + \tfrac{3}{4}a^2 B_1) \sin\theta + \text{NST} \quad (6.3.18)$$

Secular terms are eliminated if $A_1 = B_1 = 0$ and

$$2\omega_0^2\omega_2 = -\omega_1^2 + \frac{3}{128\omega_0^2}a^4$$

or

$$\omega_2 = -\frac{15}{256\omega_0^4}a^4 \quad (6.3.19)$$

Therefore, to second approximation

$$u = a\cos(\omega t + \phi) + \frac{\epsilon a^3}{32\omega_0^2}\cos 3(\omega t + \phi) + O(\epsilon^2) \quad (6.3.20)$$

where

$$\omega = \frac{d}{dt}(\omega_0\eta + \omega_1\xi) = \frac{d}{dt}[(\omega_0 + \epsilon\omega_1 + \epsilon^2\omega_0\omega_2 + \cdots)t]$$

or

$$\omega = \omega_0 + \frac{3\epsilon}{8\omega_0}a^2 - \frac{15\epsilon^2}{256\omega_0^3}a^4 + O(\epsilon^3) \quad (6.3.21)$$

This expansion is in full agreement with those obtained in Section 3.1.1 by using the Lindstedt–Poincaré methods, in Section 5.4.1 by using methods of averaging, and in Section 6.2.1 by using the derivative-expansion method.

6.3.2. THE VAN DER POL OSCILLATOR

The second example we consider is the van der Pol oscillator

$$\frac{d^2 u}{dt^2} + u = \epsilon(1 - u^2)\frac{du}{dt} \quad (6.3.22)$$

discussed in Sections 5.4.2, 5.7.4, and 6.2.2. We assume that u possesses the following uniformly valid expansion (Cole and Kevorkian, 1963; Kevorkian, 1966a)

$$u = u_0(\xi, \eta) + \epsilon u_1(\xi, \eta) + \epsilon^2 u_2(\xi, \eta) + \cdots \quad (6.3.23)$$

where ξ and η are defined in (6.3.3). Substituting (6.3.3) and (6.3.23) into

(6.3.22), and equating like powers of ϵ, we obtain

$$\frac{\partial^2 u_0}{\partial \eta^2} + u_0 = 0 \tag{6.3.24}$$

$$\frac{\partial^2 u_1}{\partial \eta^2} + u_1 = -2\frac{\partial^2 u_0}{\partial \xi \partial \eta} + (1 - u_0{}^2)\frac{\partial u_0}{\partial \eta} \tag{6.3.25}$$

$$\frac{\partial^2 u_2}{\partial \eta^2} + u_2 = -2\frac{\partial^2 u_1}{\partial \xi \partial \eta} - \frac{\partial^2 u_0}{\partial \xi^2} - 2\omega_2\frac{\partial^2 u_0}{\partial \eta^2}$$

$$+ (1 - u_0{}^2)\left(\frac{\partial u_1}{\partial \eta} + \frac{\partial u_0}{\partial \xi}\right) - 2u_0 u_1\frac{\partial u_0}{\partial \eta} \tag{6.3.26}$$

The general solution of (6.3.24) is

$$u_0 = A_0(\xi)\cos\eta + B_0(\xi)\sin\eta \tag{6.3.27}$$

Hence (6.3.25) becomes

$$\frac{\partial^2 u_1}{\partial \eta^2} + u_1 = \left[-2B_0' + \left(1 - \frac{A_0{}^2 + B_0{}^2}{4}\right)B_0\right]\cos\eta$$

$$+ \left[2A_0' - \left(1 - \frac{A_0{}^2 + B_0{}^2}{4}\right)A_0\right]\sin\eta$$

$$+ \tfrac{1}{4}(A_0{}^3 - 3A_0 B_0{}^2)\sin 3\eta + \tfrac{1}{4}(B_0{}^3 - 3A_0{}^2 B_0)\cos 3\eta \tag{6.2.28}$$

Elimination of secular terms necessitates satisfying the following conditions

$$-2B_0' + \left(1 - \frac{A_0{}^2 + B_0{}^2}{4}\right)B_0 = 0 \tag{6.3.29}$$

$$2A_0' - \left(1 - \frac{A_0{}^2 + B_0{}^2}{4}\right)A_0 = 0 \tag{6.3.30}$$

Subtracting B_0 times (6.3.29) from A_0 times (6.3.30) gives

$$\rho' - \rho(1 - \tfrac{1}{4}\rho) = 0 \tag{6.3.31}$$

where ρ is the square of the amplitude of the zeroth-order solution; that is

$$\rho = a^2 = A_0{}^2 + B_0{}^2 \tag{6.3.32}$$

By separation of variables we integrate (6.3.31) to obtain

$$a^2 = \frac{4}{1 + \left(\frac{4}{a_0{}^2} - 1\right)e^{-\xi}} \tag{6.3.33}$$

where a_0 is the initial amplitude. Expressing A_0 and B_0 in terms of the phase ϕ and the amplitude a, we obtain

$$A_0 = a \cos \phi, \qquad B_0 = -a \sin \phi \qquad (6.3.34)$$

Substituting into either (6.3.29) or (6.3.30) and using (6.3.31), we find that

$$\phi' = 0 \quad \text{or} \quad \phi = \phi_0 = \text{a constant} \qquad (6.3.35)$$

Hence u_0 can be expressed as

$$u_0 = a \cos (\eta + \phi_0) \qquad (6.3.36)$$

With (6.3.29) and (6.3.30) satisfied, the solution of (6.3.28) is

$$u_1 = A_1(\xi) \cos (\eta + \phi_0) + B_1(\xi) \sin (\eta + \phi_0) - \frac{a^3}{32} \sin 3(\eta + \phi_0) \quad (6.3.37)$$

Substituting for u_0 and u_1 into (6.3.26) gives

$$\frac{\partial^2 u_2}{\partial \eta^2} + u_2 = \left[-2B_1' + (1 - \tfrac{1}{4}a^2)B_1 - a'' + 2\omega_2 a + (1 - \tfrac{3}{4}a^2)a' + \frac{a^5}{128} \right]$$

$$\times \cos (\eta + \phi_0) + [2A_1' - (1 - \tfrac{3}{4}a^2)A_1] \sin (\eta + \phi_0) + \text{NST}$$

$$(6.3.38)$$

To eliminate secular terms we require that

$$2B_1' - (1 - \tfrac{1}{4}a^2)B_1 = 2\omega_2 a - a'' + (1 - \tfrac{3}{4}a^2)a' + \frac{a^5}{128} \qquad (6.3.39)$$

$$2A_1' - (1 - \tfrac{3}{4}a^2)A_1 = 0 \qquad (6.3.40)$$

Using (6.3.31) and (6.3.32), we express the above two equations as

$$2B_1' - \frac{2a'}{a} B_1 = 2a(\omega_2 + \tfrac{1}{16}) - (\tfrac{7}{16}a^2 - \tfrac{1}{4})a' \qquad (6.3.41)$$

$$A_1' - \left(\frac{3a'}{a} - 1 \right) A_1 = 0 \qquad (6.3.42)$$

The solutions of these equations are

$$B_1 = a(\omega_2 + \tfrac{1}{16})\xi - b_1 a + \tfrac{1}{8}a \ln a - \tfrac{7}{64}a^3 \qquad (6.3.43)$$

$$A_1 = a_1 a^3 e^{-\xi} \qquad (6.3.44)$$

where a_1 and b_1 are constants. Since as $t \to \infty$, $\xi \to \infty$ and $a \to 2$, u_1/u_0 is unbounded as $\xi \to \infty$ unless

$$\omega_2 = -\tfrac{1}{16} \qquad (6.3.45)$$

Therefore, to second approximation

$$u = (a + \epsilon a^3 a_1 e^{-\epsilon t}) \cos [(1 - \tfrac{1}{16}\epsilon^2)t + \phi_0]$$

$$- \epsilon \bigg\{ (\tfrac{7}{64}a^3 - \tfrac{1}{8}a \ln a + b_1 a) \sin [(1 - \tfrac{1}{16}\epsilon^2)t + \phi_0]$$

$$+ \frac{a^3}{32} \sin 3[(1 - \tfrac{1}{16}\epsilon^2)t + \phi_0] \bigg\} + O(\epsilon^2) \quad (6.3.46)$$

where

$$a = \frac{2}{\sqrt{1 + \left(\dfrac{4}{\tilde{a}_0{}^2} - 1\right) e^{-\epsilon t}}} \quad (6.3.47)$$

This expansion is in full agreement with (6.2.38) obtained using the derivative-expansion method if we identify a_0 with $\tilde{a}_0 + \epsilon \tilde{a}_0{}^3 a_1$.

6.3.3. THE STABILITY OF THE TRIANGULAR POINTS IN THE ELLIPTIC RESTRICTED PROBLEM OF THREE BODIES

Let us consider again the parametric resonance problem treated in Section 6.2.6 using the derivative-expansion method. The problem is described mathematically by (3.1.63) through (3.1.65). To determine a uniformly valid expansion near the transition curves using the two-variable expansion procedure, we need to use different time scales from those given by (6.3.3). The appropriate time scales are

$$\xi = (e + \omega_2 e^2 + \cdots)t, \qquad \eta = t \quad (6.3.48)$$

We assume that x and y possess expansions of the form

$$x = x_0(\xi, \eta) + e x_1(\xi, \eta) + e^2 x_2(\xi, \eta) + \cdots \quad (6.3.49)$$

$$y = y_0(\xi, \eta) + e y_1(\xi, \eta) + e^2 y_2(\xi, \eta) + \cdots \quad (6.3.50)$$

The algebraic details of the solution are not presented here. The details for the first-order solution are the same as in Section 6.2.6, with $\xi = T_1$ and $\eta = T_0$. The reader is referred to Alfriend and Rand (1969) for details of the second-order solution. Their results are in full agreement with those obtained in Section 3.1.5 using Whittaker's method.

6.3.4. LIMITATIONS OF THIS TECHNIQUE

The above examples demonstrate that by choosing the two variables appropriately the results of the two-variable expansion procedure agree with those of the derivative-expansion method. In some cases more than two variables are needed to obtain uniformly valid expansions such as in the case of satellite motion around the smaller primary in the restricted problem of three bodies (Eckstein, Shi, and Kevorkian, 1966a) and the motion of an artificial satellite having a period commensurable with the rotational period of its primary (Shi and Eckstein, 1968).

In the case of hyperbolic equations, this technique as well as the derivative-expansion method applies only to dispersive wave problems and it does not provide solutions for nondispersive wave problems such as that discussed in Section 6.2.10.

6.4 Generalized Method

6.4.1. A SECOND-ORDER EQUATION WITH VARIABLE COEFFICIENTS

Let us consider the following special second-order problem (Nayfeh, 1964, 1965b)

$$\epsilon \frac{d^2y}{dx^2} + (2x + 1)\frac{dy}{dx} + 2y = 0 \qquad (6.4.1)$$

$$y(0) = \alpha, \qquad y(1) = \beta \qquad (6.4.2)$$

discussed in Sections 4.1.3 and 4.2.2 using the methods of matched and composite asymptotic expansions. As discussed in Section 4.1.3, the straightforward expansion possesses a nonuniformity at $x = 0$. The size of the region of nonuniformity is $x = O(\epsilon)$. To treat this problem using the method of matched asymptotic expansions, an inner expansion valid when $x = O(\epsilon)$ was introduced using the inner variable $\eta = x/\epsilon$. This inner expansion was matched to the outer expansion and then a composite expansion was formed to give a uniformly valid expansion.

To obtain a uniformly valid expansion using the generalized version of the method of multiple scales, we introduce the scales

$$\xi = x \qquad (6.4.3)$$

$$\eta = \frac{g_0(x)}{\epsilon} + g_1(x) + \epsilon g_2(x) \qquad (6.4.4)$$

where g_n is determined from the analysis. We require that $g_0(0) = g_i(0) = 0$ so that $g_0(x) \to x$ as $x \to 0$, hence η approaches the inner variable x/ϵ. Then the derivatives with respect to x are transformed according to

$$\frac{d}{dx} = \frac{d\eta}{dx}\frac{\partial}{\partial \eta} + \frac{\partial}{\partial \xi} \qquad (6.4.5)$$

$$\frac{d^2}{dx^2} = \left[\frac{d\eta}{dx}\right]^2 \frac{\partial^2}{\partial \eta^2} + \frac{d^2\eta}{dx^2}\frac{\partial}{\partial \eta} + 2\frac{d\eta}{dx}\frac{\partial^2}{\partial \eta \partial \xi} + \frac{\partial^2}{\partial \xi^2} \qquad (6.4.6)$$

These variables transform (6.4.1) into

$$
\epsilon \left(\frac{g_0'}{\epsilon} + g_1' + \epsilon g_2' + \cdots \right)^2 \frac{\partial^2 y}{\partial \eta^2} + \epsilon \left(\frac{g_0''}{\epsilon} + g_1'' + \epsilon g_2'' + \cdots \right) \frac{\partial y}{\partial \eta}
$$

$$
+ 2\epsilon \left(\frac{g_0'}{\epsilon} + g_1' + \epsilon g_2' + \cdots \right) \frac{\partial^2 y}{\partial \xi \, \partial \eta} + \epsilon \frac{\partial^2 y}{\partial \xi^2}
$$

$$
+ (2\xi + 1) \left[\left(\frac{g_0'}{\epsilon} + g_1' + \epsilon g_2' + \cdots \right) \frac{\partial y}{\partial \eta} + \frac{\partial y}{\partial \xi} \right] + 2y = 0 \quad (6.4.7)
$$

where primes denote differentiation with respect to the argument. Note that
we expressed the x variables appearing in (6.4.1) in terms of ξ; namely, we
expressed $2x + 1$ as $2\xi + 1$. Moreover, we expressed g_n and its derivatives
in terms of ξ. Now we assume that there exists a uniformly valid asymptotic
representation of the solution of (6.4.7) in the form

$$
y = \sum_{n=0}^{N-1} \epsilon^n y_n(\xi, \eta) + O(\epsilon^N) \quad (6.4.8)
$$

where

$$
\frac{y_n}{y_{n-1}} < \infty \quad (6.4.9)
$$

for all $\xi = x$ and $\eta = \eta(x; \epsilon)$ where x is in the domain of interest. This
last condition is the mathematical expression of the fact that the expansion
(6.4.8) is regular in the whole domain of interest.

Substituting (6.4.8) into (6.4.7) and equating like powers of ϵ, we obtain
the following equations for y_0, y_1, and y_2

$$
\left[g_0' \frac{\partial^2 y_0}{\partial \eta^2} + (2\xi + 1) \frac{\partial y_0}{\partial \eta} \right] g_0' = 0 \quad (6.4.10)
$$

$$
\left[g_0' \frac{\partial^2 y_1}{\partial \eta^2} + (2\xi + 1) \frac{\partial y_1}{\partial \eta} \right] g_0' + 2g_0' \frac{\partial^2 y_0}{\partial \xi \, \partial \eta} + [g_0'' + (2\xi + 1)g_1'] \frac{\partial y_0}{\partial \eta}
$$

$$
+ (2\xi + 1) \frac{\partial y_0}{\partial \xi} + 2y_0 + 2g_0' g_1' \frac{\partial^2 y_0}{\partial \eta^2} = 0 \quad (6.4.11)
$$

$$
\left[g_0' \frac{\partial^2 y_2}{\partial \eta^2} + (2\xi + 1) \frac{\partial y_2}{\partial \eta} \right] g_0' + 2g_0' g_1' \frac{\partial^2 y_1}{\partial \eta^2} + (g_1'^2 + 2g_0' g_2') \frac{\partial^2 y_0}{\partial \eta^2}
$$

$$
+ [g_0'' + (2\xi + 1)g_1'] \frac{\partial y_1}{\partial \eta} + [g_1'' + (2\xi + 1)g_2'] \frac{\partial y_0}{\partial \eta}
$$

$$
+ 2g_0' \frac{\partial^2 y_1}{\partial \xi \, \partial \eta} + 2g_1' \frac{\partial^2 y_0}{\partial \xi \, \partial \eta} + \frac{\partial^2 y_0}{\partial \xi^2} + (2\xi + 1) \frac{\partial y_1}{\partial \xi} + 2y_1 = 0 \quad (6.4.12)
$$

Since $g_0' \not\equiv 0$ because $g_0(x) \to x$ as $x \to 0$, the solution of (6.4.10) is

$$y_0 = A_0(\xi) + B_0(\xi)e^{-\gamma(\xi)\eta} \tag{6.4.13}$$

where

$$\gamma(\xi) = \frac{2\xi + 1}{g_0'} \tag{6.4.14}$$

Then (6.4.11) becomes

$$\left(\frac{\partial^2 y_1}{\partial \eta^2} + \gamma \frac{\partial y_1}{\partial \eta}\right)g_0'^2 = -[(2\xi + 1)A_0' + 2A_0]$$

$$- \{g_0'\gamma\gamma' B_0\eta + [-2g_0'(B_0\gamma)' + (2\xi + 1)B_0' + (2 - \gamma g_0'' + g_0'g_1'\gamma^2)B_0]\}e^{-\gamma\eta} \tag{6.4.15}$$

The solution of this equation is

$$y_1 = A_1(\xi) + B_1(\xi)e^{-\gamma\eta} - \frac{(2\xi + 1)A_0' + 2A_0}{g_0'^2\gamma}\eta$$

$$+ \frac{1}{g_0'^2}\Big\{\tfrac{1}{2}B_0 g_0'\gamma'\eta^2 - \frac{1}{\gamma}[2g_0'(B_0\gamma)' - (2\xi + 1)B_0'$$

$$- (2 - \gamma g_0'' + g_0'g_1'\gamma^2)B_0 - B_0 g_0'\gamma']\eta\Big\}e^{-\gamma\eta} \tag{6.4.16}$$

In order that y_1/y_0 be bounded for all η, the coefficients of η, $\eta e^{-\gamma\eta}$, and $\eta^2 e^{-\gamma\eta}$ must vanish; that is

$$(2\xi + 1)A_0' + 2A_0 = 0 \tag{6.4.17}$$

$$B_0\gamma' = 0 \tag{6.4.18}$$

$$2g_0'(B_0\gamma)' - (2\xi + 1)B_0' - (2 - \gamma g_0'' + g_0'g_1'\gamma^2)B_0 - B_0 g_0'\gamma' = 0 \tag{6.4.19}$$

The general solution of (6.4.17) is

$$A_0 = \frac{a_0}{2\xi + 1} \tag{6.4.20}$$

where a_0 is a constant. If y, hence y_0, is to satisfy two boundary conditions, $B_0 \not\equiv 0$, thus (6.4.18) yields

$$\gamma' = 0 \tag{6.4.21}$$

Therefore γ is a constant, which is taken to be unity without loss of generality. Then (6.4.14) gives

$$g_0 = \xi^2 + \xi \tag{6.4.22}$$

since $g_0(0) = 0$ to reflect the fact that the nonuniformity is at $\xi = 0$. Equation (6.4.19) becomes

$$B_0' - g_1'B_0 = 0 \tag{6.4.23}$$

whose solution is

$$B_0 = b_0 e^{g_1(\xi)} \tag{6.4.24}$$

where b_0 is another constant of integration. Therefore, to first order

$$y = \frac{a_0}{2\xi + 1} + b_0 e^{g_1(\xi)} e^{-[g_0(\xi)/\epsilon] - g_1(\xi)} + O(\epsilon)$$

$$= \frac{a_0}{2\xi + 1} + b_0 e^{-[g_0(\xi)/\epsilon]} + O(\epsilon) \tag{6.4.25}$$

Since $g_1(\xi)$ disappears from the expansion irrespective of its value, we can set $g_1 = 0$ without loss of generality.

The above analysis shows that γ in (6.4.13) must be a constant which is taken to be unity without loss of generality. If γ is not a constant, then since it multiplies η the derivatives with respect to ξ always create terms proportional to powers of η which make y_1/y_0 unbounded as $\eta \to \infty$. Therefore, whenever such a situation arises, γ is set equal to unity from the start. Moreover, we did not need to solve (6.4.15) to determine the conditions for y_1/y_0 to be bounded for all η. We could have investigated (6.4.15) and required the vanishing of all terms leading to particular solutions that make y_1/y_0 unbounded. Such terms include all those that are proportional to the solutions of the homogeneous equation. Since $e^{-\gamma\eta}$ and 1 are solutions of the homogeneous equation, we require the satisfaction of the conditions (6.4.17) and (6.4.19).

To determine the second approximation, we let $g_1 = 0$, $\gamma = 1$, substitute for y_0, y_1, and g_0 into (6.4.12), and obtain

$$\left(\frac{\partial^2 y_2}{\partial \eta^2} + y_2\right) g_0'^2 = -[(2\xi + 1)A_1' + 2A_1 + A_0''] + g_0'[B_1' - B_0 g_2']e^{-\eta} \tag{6.4.26}$$

In order that y_2/y_0 be bounded for all η

$$(2\xi + 1)A_1' + 2A_1 + A_0'' = 0 \tag{6.4.27}$$

$$B_1' - B_0 g_2' = 0 \tag{6.4.28}$$

Using (6.4.20) in solving (6.4.27), we obtain

$$A_1 = \frac{a_1}{2\xi + 1} + \frac{2a_0}{(2\xi + 1)^3} \tag{6.4.29}$$

where a_1 is a constant of integration. The condition (6.4.28) is satisfied if

$$B_1' = 0 \quad \text{and} \quad g_2' = 0 \tag{6.4.30}$$

Thus

$$B_1 = b_1 \quad \text{and} \quad g_2 = \text{a constant} \tag{6.4.31}$$

where b_1 is a constant of integration and $g_2 \equiv 0$ since $g_2(0) = 0$.

Therefore, to second approximation, y is given by

$$y = \frac{a_0}{1 + 2x} + b_0 e^{-[(x^2+x)/\epsilon]}$$

$$+ \epsilon \left[\frac{a_1}{1 + 2x} + \frac{2a_0}{(1 + 2x)^3} + b_1 e^{-[(x^2+x)/\epsilon]} \right] + O(\epsilon^2) \quad (6.4.32)$$

Imposing the boundary conditions $y(0) = \alpha$ and $y(1) = \beta$, we obtain $a_0 = 3\beta$, $b_0 = \alpha - 3\beta$, $a_1 = -2\beta/3$, and $b_1 = -16\beta/3$. Hence (6.4.32) becomes

$$y = \frac{3\beta}{1 + 2x} + (\alpha - 3\beta)e^{-[(x^2+x)/\epsilon]}$$

$$- \epsilon \left[\frac{2\beta}{3(1 + 2x)} - \frac{6\beta}{(1 + 2x)^3} + \frac{16}{3}\beta e^{-[(x^2+x)/\epsilon]} \right] + O(\epsilon^2) \quad (6.4.33)$$

If we expand $e^{-(x^2/\epsilon)}$ for small x^2/ϵ, (6.4.33) agrees with (4.2.50) obtained using the method of composite expansions. Thus the method of multiple scales gives a single uniformly valid expansion in contrast with the method of matched asymptotic expansions which gives two expansions that must be matched.

6.4.2.   A GENERAL SECOND-ORDER EQUATION WITH VARIABLE COEFFICIENTS

As a second example, we consider (Cochran, 1962; Nayfeh, 1964, 1965b)

$$\epsilon y'' + a(x)y' + b(x)y = c(x) \quad (6.4.34)$$

$$y(0) = \alpha \quad \text{and} \quad y(1) = \beta \quad (6.4.35)$$

where $a(x) > 0$ in $[0, 1]$. The case in which $a(x)$ vanishes in the interior of $[0, 1]$ is called a turning point problem. Turning point problems are discussed briefly in Section 6.4.4 and in detail in Sections 7.3.1 through 7.3.9. If $c = 0$, this example will be the same as that treated in Section 4.1.3 using the method of matched asymptotic expansions.

Since $a(x) > 0$, the nonuniformity is at $x = 0$. In Section 4.1.3, we introduced an inner variable x/ϵ to determine an expansion valid in the region $x = O(\epsilon)$, which was matched to an outer expansion. To determine a uniformly valid first approximation using the method of multiple scales, we assume that there exists an asymptotic representation for y of the form

$$y = y_0(\xi, \eta) + \epsilon y_1(\xi, \eta) + \cdots \quad (6.4.36)$$

where

$$\xi = x, \quad \eta = \frac{g(x)}{\epsilon} \quad \text{with} \quad g(x) \to x \quad \text{as} \quad x \to 0 \quad (6.4.37)$$

Substituting (6.4.36) and (6.4.37) into (6.4.34) and equating the coefficients of ϵ^0 and ϵ^{-1} to zero, we obtain

$$\left[g' \frac{\partial^2 y_0}{\partial \eta^2} + a(\xi) \frac{\partial y_0}{\partial \eta} \right] g' = 0 \tag{6.4.38}$$

$$\left[g' \frac{\partial^2 y_1}{\partial \eta^2} + a(\xi) \frac{\partial y_1}{\partial \eta} \right] g' + 2g' \frac{\partial^2 y_0}{\partial \eta \, \partial \xi} + g'' \frac{\partial y_0}{\partial \eta} + a(\xi) \frac{\partial y_0}{\partial \xi} + b(\xi) y_0 = c(\xi) \tag{6.4.39}$$

where we have expressed $a(x)$, $b(x)$, $c(x)$, and $g(x)$ in terms of ξ.

Since $g \not\equiv 0$, the general solution of (6.4.38) is

$$y_0 = A(\xi) + B(\xi) e^{-\gamma(\xi)\eta} \tag{6.4.40}$$

where

$$\gamma = \frac{a(\xi)}{g'} \tag{6.4.41}$$

As discussed in the previous section, γ must be a constant; otherwise the derivatives with respect to ξ would produce terms proportional to $\gamma'\eta$ in (6.4.39), hence make y_1/y_0 unbounded as $\eta \to \infty$. For a uniformly valid expansion, we require that $\gamma = 1$ without loss of generality. Hence

$$g = \int_0^x a(t) \, dt \quad \text{since} \quad g(x) \to x \quad \text{as} \quad x \to 0 \tag{6.4.42}$$

Substituting for y_0 into (6.4.39) gives

$$\left(\frac{\partial^2 y_1}{\partial \eta^2} + \frac{\partial y_1}{\partial \eta} \right) g'^2 = -[aA' + bA - c] + [g'B' + (g'' - b)B]e^{-\eta} \tag{6.4.43}$$

In order that y_1/y_0 be bounded for all η, we require that

$$aA' + bA = c \tag{6.4.44}$$

$$g'B' + (g'' - b)B = 0 \tag{6.4.45}$$

The solutions of these equations are

$$A = e^{-\int_1^x [b(t)/a(t)] \, dt} \left[a_0 + \int_1^x \frac{c(\tau)}{a(\tau)} e^{\int_1^\tau [b(t)/a(t)] \, dt} d\tau \right] \tag{6.4.46}$$

$$B = \frac{b_0}{a(x)} e^{\int_0^x [b(t)/a(t)] \, dt} \tag{6.4.47}$$

where a_0 and b_0 are constants of integration.

To first approximation, y is given by

$$y = e^{-\int_1^x [b(t)/a(t)]\,dt}\left[a_0 + \int_1^x \frac{c(\tau)}{a(\tau)}\, e^{\int_1^\tau [b(t)/a(t)]\,dt}\,d\tau\right]$$

$$+\frac{b_0}{a(x)}\, e^{\int_0^x [b(t)/a(t)]\,dt}e^{-\epsilon^{-1}\int_0^x a(t)\,dt} + O(\epsilon) \quad (6.4.48)$$

The limits of the integrals in (6.4.46) and (6.4.47) were chosen so that a_0 and b_0 could be expressed in a simple manner in terms of the boundary conditions (6.4.35). Thus $a_0 = \beta$ and

$$b_0 = a(0)\left\{\alpha - e^{-\int_1^0 [b(t)/a(t)]\,dt}\left[\beta + \int_1^0 \frac{c(\tau)}{a(\tau)}\, e^{\int_1^\tau [b(t)/a(t)]\,dt}\,d\tau\right]\right\} \quad (6.4.49)$$

Expansion (6.4.48) is a composite expansion which agrees with the inner and outer expansions obtained in Section 4.1.3 in the inner and outer regions, respectively. If we specialize (6.4.48) for the case

$$a(x) = 1 + 2x, \qquad b(x) = 2, \qquad c(x) = 0 \quad (6.4.50)$$

discussed in the previous section, we obtain

$$y = \frac{3\beta}{1 + 2x} + (\alpha - 3\beta)e^{-[(x^2+x)/\epsilon]} + O(\epsilon) \quad (6.4.51)$$

where we have used $a_0 = \beta$ and (6.4.49). This expansion is in full agreement with the first term in the expansion obtained in the previous section.

6.4.3. A LINEAR OSCILLATOR WITH A SLOWLY VARYING RESTORING FORCE

The two cases discussed above can be treated by using either the method of multiple scales or the method of matched asymptotic expansions. Let us consider next an example that cannot be treated by the latter method; namely

$$y'' + b(\epsilon x)y = 0 \quad (6.4.52)$$

where $b(\epsilon x) \neq 0$ and ϵ is a small parameter. To obtain an expansion uniformly valid for large x, we assume that there exists an asymptotic representation for y of the form

$$y = y_0(\xi, \eta) + \epsilon y_1(\xi, \eta) + \cdots \quad (6.4.53)$$

where

$$\xi = \epsilon x, \qquad \eta = \frac{g(\xi)}{\epsilon} + \cdots \quad (6.4.54)$$

This form of η was chosen in order that the frequency of oscillation $\omega = d\eta/dx = g'(\xi) = O(1)$. Substituting (6.4.53) and (6.4.54) into (6.4.52) and equating the coefficients of ϵ^0 and ϵ to zero, we obtain

$$g'^2 \frac{\partial^2 y_0}{\partial \eta^2} + b(\xi) y_0 = 0 \tag{6.4.55}$$

$$g'^2 \frac{\partial^2 y_1}{\partial \eta^2} + b(\xi) y_1 + g'' \frac{\partial y_0}{\partial \eta} + 2g' \frac{\partial^2 y_0}{\partial \xi \, \partial \eta} = 0 \tag{6.4.56}$$

The general solution of (6.4.55) is

$$y_0 = A(\xi) e^{i\gamma\eta} + B(\xi) e^{-i\gamma\eta} \tag{6.4.57}$$

where

$$\gamma^2 = \frac{b(\xi)}{g'^2(\xi)} \tag{6.4.58}$$

As argued in the previous two sections, we set $\gamma = 1$ to obtain an expansion in which y_1/y_0 is bounded for all η. Hence

$$g = \int_0^\xi \sqrt{b(t)} \, dt \tag{6.4.59}$$

Substituting for y_0 into (6.4.56), and remembering that $\gamma = 1$, we obtain

$$\left(\frac{\partial^2 y_1}{\partial \eta^2} + y_1 \right) g'^2 = -i(g''A + 2g'A')e^{i\eta} + i(g''B + 2g'B')e^{-i\eta} \tag{6.4.60}$$

In order that y_1/y_0 be bounded for all η, we require the vanishing of the coefficients of $\exp(\pm i\eta)$ on the right-hand side of (6.4.60); that is

$$g''A + 2g'A' = 0 \tag{6.4.61}$$

$$g''B + 2g'B' = 0 \tag{6.4.62}$$

The solutions of these equations are

$$A = \frac{\tilde{a}_0}{\sqrt{g'}}, \qquad B = \frac{\tilde{b}_0}{\sqrt{g'}} \tag{6.4.63}$$

where $\tilde{a}_0$ and $\tilde{b}_0$ are constants of integration.

If $b(\epsilon x) > 0$, y is given by

$$y = \frac{1}{\sqrt[4]{b(\epsilon x)}} \left[a_0 \cos \left(\epsilon^{-1} \int_0^{\epsilon x} \sqrt{b(t)} \, dt \right) + b_0 \sin \left(\epsilon^{-1} \int_0^{\epsilon x} \sqrt{b(t)} \, dt \right) \right] + O(\epsilon) \tag{6.4.64}$$

where a_0 and b_0 are constants. If $b(\epsilon x) < 0$, y is given by

$$y = \frac{1}{\sqrt[4]{|b(\epsilon x)|}} \left[a_0 \exp \left(\epsilon^{-1} \int_0^{\epsilon x} \sqrt{-b(t)} \, dt \right) + b_0 \exp \left(-\epsilon^{-1} \int_0^{\epsilon x} \sqrt{-b(t)} \, dt \right) \right] \tag{6.4.65}$$

Expansions (6.4.64) and (6.4.65) are called the WKB approximation to the solution of (6.4.52) (see Section 7.1.3).

These expansions are clearly not valid near a point where $b(\epsilon x)$ vanishes. In fact, they tend to infinity as x approaches a zero of $b(\epsilon x)$. The zeros of $b(\epsilon x)$ are called turning points and are discussed in detail in Sections 7.3.1 through 7.3.9. An example of a turning point problem is treated in the next section using the method of multiple scales.

If we change variables in (6.4.52) from x to ξ, we find that

$$\frac{d^2y}{d\xi^2} + \lambda^2 b(\xi)y = 0, \qquad \lambda = \frac{1}{\epsilon} \tag{6.4.66}$$

which is a problem containing a large parameter λ. Thus the above-obtained approximation is applicable to this problem as well.

6.4.4.　AN EXAMPLE WITH A TURNING POINT
Let us consider the problem

$$y'' + \lambda^2(1 - x)f(x)y = 0 \tag{6.4.67}$$

where λ is a large positive number and $f(x)$ is regular and positive. The WKB approximation tends to infinity as $x \to 1$ as can be seen from (6.4.64) and (6.4.65) if we let $b(\epsilon t) = (1 - x)f(x)$ and $\epsilon = \lambda^{-1}$. To determine an expansion valid everywhere using the method of multiple scales, we first determine the size of the nonuniformity. Thus we let $\zeta = (1 - x)\lambda^\nu$ with $\nu > 0$ in (6.4.67) and obtain

$$\frac{d^2y}{d\zeta^2} + \lambda^{2-3\nu}f(1 - \zeta\lambda^{-\nu})\zeta y = 0 \tag{6.4.68}$$

As $\lambda \to \infty$, the following different limits exist depending on the value of ν

$$\left.\begin{array}{ll} \dfrac{d^2y}{d\zeta^2} = 0 & \text{if } \nu > \tfrac{2}{3} \\[2mm] y = 0 & \text{if } \nu < \tfrac{2}{3} \\[2mm] \dfrac{d^2y}{d\zeta^2} + f(1)\zeta y = 0 & \text{if } \nu = \tfrac{2}{3} \end{array}\right\} \tag{6.4.69}$$

The last limit is the appropriate one because its solution has an exponential behavior for $\zeta < 0$ (i.e., $x > 1$) and an oscillatory behavior for $\zeta > 0$ (i.e., $x < 1$). Thus it can be used to connect (6.4.64) and (6.4.65) as the turning point is crossed.

Therefore we assume that there exists an asymptotic representation of the solution of (6.4.67) of the form (Cochran, 1962; Nayfeh, 1964, 1965b; Fowkes, 1968, Part I)

$$y = y_0(\xi, \eta) + \lambda^{-2/3}y_1(\xi, \eta) + \cdots \qquad (6.4.70)$$

where

$$\xi = x, \qquad \eta = \lambda^{2/3}g(x) + \cdots \qquad (6.4.71)$$

with

$$g(x) = (1 - x)h(x), \qquad h(x) > 0 \qquad (6.4.72)$$

The functions of the independent variable x that appear in (6.4.67) are expressed in terms of ξ, except $1 - x$ is replaced by $\eta\lambda^{-2/3}/h(x)$ because it reflects the nonuniformity. Therefore (6.4.67) becomes

$$\lambda^{4/3}g'^2\frac{\partial^2 y}{\partial \eta^2} + 2\lambda^{2/3}g'\frac{\partial^2 y}{\partial \xi\, \partial\eta} + \lambda^{2/3}g''\frac{\partial y}{\partial\eta} + \frac{\partial^2 y}{\partial\xi^2} + \lambda^{4/3}\frac{f(\xi)}{h(\xi)}\eta y = 0 \quad (6.4.73)$$

Substituting (6.4.70) into (6.4.73) and equating the coefficients of $\lambda^{4/3}$ and $\lambda^{2/3}$ to zero, we obtain

$$g'^2\frac{\partial^2 y_0}{\partial\eta^2} + \frac{f(\xi)}{h(\xi)}\eta y_0 = 0 \qquad (6.4.74)$$

$$g'^2\frac{\partial^2 y_1}{\partial\eta^2} + \frac{f(\xi)}{h(\xi)}\eta y_1 + 2g'\frac{\partial^2 y_0}{\partial\xi\, \partial\eta} + g''\frac{\partial y_0}{\partial\eta} = 0 \qquad (6.4.75)$$

The general solution of (6.4.74) is

$$y_0 = A(\xi)\eta^{1/2}J_{1/3}[\gamma(\xi)\eta^{3/2}] + B(\xi)\eta^{1/2}J_{-1/3}[\gamma(\xi)\eta^{3/2}] \qquad (6.4.76)$$

where $J_{\pm 1/3}$ are Bessel's functions of order $\pm 1/3$ and

$$\gamma = \pm\frac{2}{3}\left[\frac{f(\xi)}{h(\xi)g'^2(\xi)}\right]^{1/2} \qquad (6.4.77)$$

In order that y_1/y_0 be bounded for all η, $\gamma = 1$ as discussed in Sections 6.4.1 and 6.4.2. Therefore

$$g'(\xi)h^{1/2}(\xi) = -\tfrac{2}{3}[f(\xi)]^{1/2} \qquad (6.4.78)$$

where the negative sign in (6.4.77) was taken so that $h(x) > 0$. Multiplying both sides of (6.4.78) by $(1 - \xi)^{1/2}$, we obtain

$$g^{1/2}g' = -\tfrac{2}{3}[(1 - \xi)f(\xi)]^{1/2}$$

Since $g(1) = 0$

$$g^{3/2} = -\int_1^x [(1 - t)f(t)]^{1/2}\, dt \qquad (6.4.79)$$

With y_0 known and $\gamma = 1$, (6.4.75) becomes

$$\left(\frac{\partial^2 y_1}{\partial \eta^2} + \tfrac{9}{4}\eta y_1\right) g'^2$$

$$= -\frac{\partial}{\partial \eta}\left[(2g'A' + g''A)\eta^{1/2}J_{1/3}(\eta^{3/2}) + (2g'B' + g''B)\eta^{1/2}J_{-1/3}(\eta^{3/2})\right] \quad (6.4.80)$$

In order that y_1/y_0 be bounded for all η, the right-hand side of (6.4.80) must vanish; that is

$$\left.\begin{aligned} 2g'A' + g''A &= 0 \\ 2g'B' + g''B &= 0 \end{aligned}\right\} \quad (6.4.81)$$

Hence

$$A = \frac{a}{\sqrt{g'}}, \qquad B = \frac{b}{\sqrt{g'}} \quad (6.4.82)$$

where a and b are constants of integration.

Thus, to first approximation

$$y = \frac{g^{3/4}(x)}{[(1-x)f(x)]^{1/4}}\left[a_0 J_{1/3}(\lambda g^{3/2}) + b_0 J_{-1/3}(\lambda g^{3/2})\right] + \ldots \quad (6.4.83a)$$

where a_0 and b_0 are constants. As $x \to 1$

$$g(x) \to [\tfrac{4}{9}f(1)]^{1/3}(1-x)$$

hence

$$y \to (1-x)^{1/2}\{\tilde{a}_0 J_{1/3}[\tfrac{2}{3}\lambda\sqrt{f(1)}(1-x)^{3/2}] \\ + \tilde{b}_0 J_{-1/3}[\tfrac{2}{3}\lambda\sqrt{f(1)}(1-x)^{3/2}]\} + \ldots \quad (6.4.83b)$$

where $\tilde{a}_0$ and $\tilde{b}_0$ are constants. Since

$$J_\nu(t) = t^\nu + o(t^\nu) \quad \text{as} \quad t \to 0$$

we conclude that (6.4.83b) is bounded as $x \to 1$.

6.4.5. THE DUFFING EQUATION WITH SLOWLY VARYING COEFFICIENTS

Let us next consider the equation

$$\frac{d^2 u}{dt^2} + \alpha(\xi)u + \beta(\xi)u^3 = 0 \quad (6.4.84)$$

where

$$\xi = \epsilon t, \qquad \epsilon \ll 1$$

The asymptotic solutions of this equation were studied by Kuzmak (1959) using the method of multiple scales.

If α and β are constants, the solution of (6.4.84) can be expressed in terms of Jacobi's elliptic functions; that is, in terms of

$$u = A\,sn(Kt, v), \; A\,cn(Kt, v), \; A\,dn(Kt, v) \qquad (6.4.85)$$

Here v is the modulus and $K(v)$ is the complete elliptic integral. The differential equations satisfied by these functions are

$$\left[\frac{dsn}{d\tau}\right]^2 = (1 - sn^2)(1 - v^2 sn^2) \qquad (6.4.86a)$$

$$\left[\frac{dcn}{d\tau}\right]^2 = (1 - cn^2)(1 - v^2 + v^2 cn^2) \qquad (6.4.86b)$$

$$\left[\frac{ddn}{d\tau}\right]^2 = (1 - dn^2)(v^2 - 1 + dn^2) \qquad (6.4.86c)$$

where $\tau = Kt$. Differentiating both sides of (6.4.86) yields

$$\frac{d^2 sn}{d\tau^2} + (1 + v^2)sn - 2v^2 sn^3 = 0 \qquad (6.4.87a)$$

$$\frac{d^2 cn}{d\tau^2} + (1 - 2v^2)cn + 2v^2 cn^3 = 0 \qquad (6.4.87b)$$

$$\frac{d^2 dn}{d\tau^2} + (v^2 - 2)\,dn + 2dn^3 = 0 \qquad (6.4.87c)$$

Since these elliptic functions are tabulated for the case $0 < v < 1$, we express the solution in terms of one of these tabulated functions.

If α and β are slowly varying functions rather than constants, we expect the solution to depend on the slow time scale $\xi = \epsilon t$ as well as on the fast time scale t. Moreover, to first approximation, the solution can be expressed as in (6.4.85) but with $A = A(\xi)$, $K = K(\xi)$, and $v = v(\xi)$. Thus in the case of slowly varying coefficients, we let

$$u = u_0(\xi, \eta) + \epsilon u_1(\xi, \eta) + \cdots \qquad (6.4.88)$$
where

$$\eta = \frac{g(\xi)}{\epsilon} + \cdots \quad \text{or} \quad \frac{d\eta}{dt} = g'(\xi) + \cdots$$

This form of solution differs from that of Kuzmak in that he assumed $\eta = g'(\xi)t$. Substituting (6.4.88) into (6.4.84) and equating coefficients of like powers of ϵ, we obtain

$$g'^2 \frac{\partial^2 u_0}{\partial \eta^2} + \alpha(\xi)u_0 + \beta(\xi)u_0^3 = 0 \tag{6.4.89}$$

$$g'^2 \frac{\partial^2 u_1}{\partial \eta^2} + \alpha(\xi)u_1 + 3\beta(\xi)u_0^2 u_1 = -2g' \frac{\partial^2 u_0}{\partial \xi \partial \eta} - g'' \frac{\partial u_0}{\partial \eta} \tag{6.4.90}$$

We write the solution of (6.4.89) in terms of one of the elliptic functions in (6.4.85), say sn; that is

$$u_0 = A(\xi)sn[\eta, \nu(\xi)] \tag{6.4.91}$$

Hence u_0/A should satisfy (6.4.87a) with $\eta = \tau$; that is

$$\frac{\partial^2 u_0}{\partial \eta^2} + [1 + \nu^2(\xi)]u_0 - \frac{2\nu^2(\xi)}{A^2(\xi)} u_0^3 = 0 \tag{6.4.92}$$

In order that (6.4.89) and (6.4.92) be identical

$$[1 + \nu^2(\xi)]g'^2(\xi) = \alpha(\xi) \tag{6.4.93}$$

$$2\nu^2(\xi)g'^2(\xi) = -\beta(\xi)A^2(\xi) \tag{6.4.94}$$

These are two relationships among $A(\xi)$, $\nu(\xi)$, and $g(\xi)$. A third relationship is determined from the condition that u_1/u_0 be bounded for all η in order that (6.4.88) be a uniformly valid asymptotic expansion.

Differentiating (6.4.89) with respect to η leads to the homogeneous part of (6.4.90). Hence $\partial u_0/\partial \eta$ is a solution of the homogeneous part of (6.4.90). In order that u_1/u_0 be bounded for all η, the inhomogeneous part in (6.4.90) must be orthogonal to the solution of the homogeneous part; that is

$$\int_{\eta_1}^{\eta_1 + T} \left[2g' \frac{\partial^2 u_0}{\partial \xi \partial \eta} + g'' \frac{\partial u_0}{\partial \eta} \right] \frac{\partial u_0}{\partial \eta} d\eta = 0 \tag{6.4.95}$$

where $sn(\eta_1, \nu) = 0$ and T is the period of $sn(\eta, \nu)$ with respect to η. This condition is a generalization of the elimination of terms that produce secular terms. Equation (6.4.95) can be rewritten in the form

$$\frac{d}{d\xi}\left[g'(\xi) \int_{\eta_1}^{\eta_1 + T} \left(\frac{\partial u_0}{\partial \eta} \right)^2 d\eta \right] = 0$$

or

$$g'(\xi) \int_{\eta_1}^{\eta_1 + T} \left(\frac{\partial u_0}{\partial \eta} \right)^2 d\eta = \text{a constant} \tag{6.4.96}$$

Since $u_0 = A\, sn(n, \nu)$, η_1 can be taken to be zero and $T = 4K$ where K is the following complete elliptic integral of the second kind

$$K = \int_0^1 \frac{dx}{[(1 - x^2)(1 - \nu^2 x^2)]^{1/2}} \tag{6.4.97}$$

Substituting for u_0 from (6.4.91) into (6.4.96), we have

$$g'(\xi)A^2(\xi)L[v^2(\xi)] = c \qquad (6.4.98)$$

where c is a constant, and

$$L = \int_0^K \left(\frac{\partial \zeta}{\partial \eta}\right)^2 d\eta = \int_0^K \frac{\partial \zeta}{\partial \eta} \, d\zeta \qquad (6.4.99)$$

with $\zeta = sn(\eta, v)$. Using (6.4.86a), we express L as

$$L = \int_0^1 \sqrt{(1 - \zeta^2)(1 - v^2\zeta^2)} \, d\zeta$$

or

$$L = \frac{(1 + v^2)E(v) - (1 - v^2)K(v)}{3v^2} \qquad (6.4.100)$$

where $E(v)$ is the following complete elliptic integral of the first kind

$$E(v) = \int_0^1 \sqrt{\frac{1 - v^2x^2}{1 - x^2}} \, dx \qquad (6.4.101)$$

The conditions (6.4.93), (6.4.94), and (6.4.98) constitute three relationships for the determination of $A(\xi)$, $v(\xi)$, and $g'(\xi)$. Solving for g' from (6.4.93) gives

$$g'(\xi) = \sqrt{\frac{\alpha(\xi)}{1 + v^2(\xi)}} \qquad (6.4.102)$$

Eliminating g' from (6.4.93) and (6.4.94) and solving for A, we obtain

$$A(\xi) = \sqrt{-\frac{2\alpha(\xi)v^2(\xi)}{\beta(\xi)[1 + v^2(\xi)]}} \qquad (6.4.103)$$

Squaring (6.4.98) and substituting for g' and A from (6.4.102) and (6.4.103), we obtain

$$\frac{4v^4(\xi)L^2[v(\xi)]}{[1 + v^2(\xi)]^3} = \rho(\xi) = \frac{c^2\beta^2(\xi)}{\alpha^3(\xi)} \qquad (6.4.104)$$

With the aid of these last-mentioned relationships, we can compute $v(\xi)$ from (6.4.104), then g' and A from (6.4.102) and (6.4.103). The graph for the solution of (6.4.104) was given by Kuzmak and is shown in Figure 6-2.

Three different cases arise depending on the signs of $\alpha(\xi)$ and $\beta(\xi)$:

(1) $\alpha(\xi) > 0$, $\beta(\xi) < 0$. In this case $\rho > 0$ and (6.4.94) shows that $\gamma = v^2(\xi) > 0$. Hence the curve that determines γ lies in the first quadrant. The solution for γ exists if $0 < \rho < 2/9$. At ξ_c such that $\rho(\xi_c) = 2/9$, the

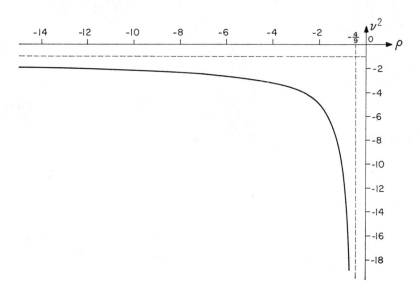

(a)

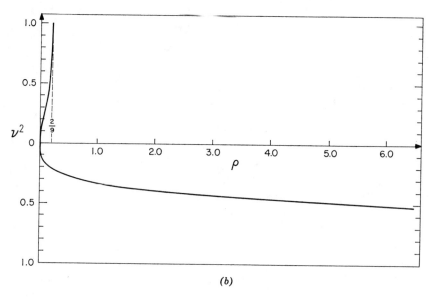

(b)

Figure 6-2

290

asymptotic solution ceases to be oscillatory. If $\rho > 2/9$, or if $\alpha(\xi) < 0$ and $\beta(\xi) < 0$, (6.4.89) has no periodic solutions.

(2) $\alpha(\xi) > 0$, $\beta(\xi) > 0$. In this case $\rho > 0$ and (6.4.94) shows that $\gamma < 0$. Hence the curve that determines γ lies in the fourth quadrant. The solution for γ exists for $0 < \rho < \infty$.

(3) $\alpha(\xi) < 0$, $\beta(\xi) > 0$. In this case $\rho < 0$ and (6.4.94) shows that $\gamma < 0$. Hence the curve that determines γ lies in the third quadrant. The solution for γ exists if $-\infty < \rho < -4/9$.

Since the elliptic functions and integrals are usually tabulated for real ν such that $0 < \nu < 1$, alternative asymptotic solutions in terms of $cn(\eta, \nu)$ and $dn(\eta, \nu)$ are preferred in cases (2) and (3).

6.4.6.   REENTRY DYNAMICS

The motion of a reentry rolling body with variable spin under the influence of nonlinear aerodynamic forces and slight center of gravity offset and aerodynamic asymmetries is governed by (Nayfeh and Saric, 1972a)

$$\ddot{\xi} - i\frac{I_x}{I}p\dot{\xi} + \omega_0^2\xi = \epsilon Ke^{i(\phi+\phi_0)} + \gamma|\xi|^2\xi + \epsilon^2\mu_1\dot{\xi} + \mu_2|\xi|^2\dot{\xi}$$

$$+ i\epsilon^2\chi_1\xi + i\chi_2|\xi|^2\xi \quad (6.4.105)$$

$$\dot{\phi} = p \quad (6.4.106)$$

$$\dot{p} = \epsilon^2\nu_0 + \epsilon\nu_1\tilde{\alpha} + \epsilon^2\nu_2 p, \quad \tilde{\alpha} = \text{Imaginary}\{\xi e^{-i\phi}\} \quad (6.4.107)$$

where $\xi = \beta + i\alpha$, $|\xi|$ is the sine of the total angle of attack, p is the roll rate, ϵK is the amplitude of the excitation due to aerodynamic asymmetry, and ϵ is a small but finite quantity of the order of the sine of the initial total angle of attack. Here ω_0, K, γ, μ_i, χ_i, and ν_i are slowly varying functions of time, and I and I_x are constants.

In the absence of damping and nonlinear terms (i.e., $\gamma = \mu_i = \chi_i = 0$), the solution of (6.4.105) for constant p, K, and ω_0 is

$$\xi = A_1 e^{i\omega_1 t} + A_2 e^{i\omega_2 t} + \frac{\epsilon K}{(\omega_1 - p)(p - \omega_2)} e^{i(\phi+\phi_0)} \quad (6.4.108)$$

where A_1 and A_2 are complex constants and

$$\omega_{1,2} = \frac{pI_x}{2I} \pm \sqrt{\left(\frac{pI_x}{2I}\right)^2 + \omega_0^2} \quad (6.4.109)$$

The frequencies ω_1 and ω_2 are called the nutation and precession frequencies.

For statically stable bodies (i.e., $\omega_0{}^2 > 0$) and positive p, ω_1 is positive while ω_2 is negative. Two cases arise depending on whether p is near ω_1 or not. The first case is called roll resonance, and the forced response tends to infinity as $p \to \omega_1$. Before p approaches ω_1, the damping as well as the nonlinear aerodynamic forces significantly modify the response. The roll resonance case is discussed in this section for the case $K = \epsilon^2 k$, and we refer the reader to Nayfeh and Saric (1972a) for the nonresonant case.

To determine an approximate solution to (6.4.105) through (6.4.107) when $p \approx \omega_1$ using the generalized version of the method of multiple scales, we make use of the fact that actual flight test data and six-degree-of-freedom numerical calculations show that there are at least four time scales: a slow time scale $T_2 = \epsilon^2 t$ characterizing K, ω_0, γ, ν_i, χ_i, and μ_i; and three fast scales characterizing the nutation, precession, and forced components of the angle of attack. Thus we assume expansions of the form

$$\xi(t; \epsilon) = \epsilon \xi_1(\eta_1, \eta_2, T_2) + \epsilon^3 \xi_3(\eta_1, \eta_2, \phi, T_2) + \cdots \quad (6.4.110)$$

$$p(t; \epsilon) = p_0(T_2) + \epsilon^2 p_2(\eta_1, \eta_2, \phi, T_2) + \cdots \quad (6.4.111)$$

where

$$\frac{d\eta_i}{dt} = \omega_i, \qquad \omega_{1,2} = \frac{p_0 I_x}{2I} \pm \sqrt{\left(\frac{p_0 I_x}{2I}\right)^2 + \omega_0{}^2} \quad (6.4.112)$$

where $\omega_1(T_2)$ is the nutation frequency and $\omega_2(T_2)$ is the precession frequency. In terms of these variables, the time derivatives are transformed according to

$$\frac{d}{dt} = \omega_1 \frac{\partial}{\partial \eta_1} + \omega_2 \frac{\partial}{\partial \eta_2} + p \frac{\partial}{\partial \phi} + \epsilon^2 \frac{\partial}{\partial T_2} \quad (6.4.113)$$

$$\frac{d^2}{dt^2} = \omega_1{}^2 \frac{\partial^2}{\partial \eta_1{}^2} + \omega_2{}^2 \frac{\partial^2}{\partial \eta_2{}^2} + p^2 \frac{\partial^2}{\partial \phi^2} + 2\omega_1\omega_2 \frac{\partial^2}{\partial \eta_1 \partial \eta_2} + 2\omega_1 p \frac{\partial^2}{\partial \eta_1 \partial \phi}$$

$$+ 2\omega_2 p \frac{\partial^2}{\partial \eta_2 \partial \phi} + \omega_1 \frac{\partial p}{\partial \eta_1} \frac{\partial}{\partial \phi} + \omega_2 \frac{\partial p}{\partial \eta_2} \frac{\partial}{\partial \phi} + p \frac{\partial p}{\partial \phi} \frac{\partial}{\partial \phi}$$

$$+ 2\epsilon^2\omega_1 \frac{\partial^2}{\partial \eta_1 \partial T_2} + 2\epsilon^2\omega_2 \frac{\partial^2}{\partial \eta_2 \partial T_2} + 2\epsilon^2 p \frac{\partial^2}{\partial \phi \partial T_2} + \epsilon^2\omega_1' \frac{\partial}{\partial \eta_1}$$

$$+ \epsilon^2\omega_2' \frac{\partial}{\partial \eta_2} + \epsilon^2 \frac{\partial p}{\partial T_2} \frac{\partial}{\partial \phi} + \epsilon^4 \frac{\partial^2}{\partial T_2{}^2} \quad (6.4.114)$$

Substituting (6.4.110) through (6.4.114) into (6.4.105) and (6.4.107) and equating coefficients of like powers of ϵ, we have

$$L(\xi_1) = 0 \tag{6.4.115}$$

$$L(\xi_3) = ke^{i(\phi+\phi_0)} - 2\omega_1 \frac{\partial^2 \xi_1}{\partial \eta_1 \, \partial T_2} - 2\omega_2 \frac{\partial^2 \xi_1}{\partial \eta_2 \, \partial T_2} - \omega_1' \frac{\partial \xi_1}{\partial \eta_1} - \omega_2' \frac{\partial \xi_1}{\partial \eta_2}$$

$$+ i \frac{p_0 I_x}{I} \frac{\partial \xi_1}{\partial T_2} + \gamma \, |\xi_1|^2 \, \xi_1 + (\mu_1 + \mu_2 \, |\xi_1|^2)\left(\omega_1 \frac{\partial \xi_1}{\partial \eta_1} + \omega_2 \frac{\partial \xi_1}{\partial \eta_2}\right)$$

$$+ i(\chi_1 + \chi_2 \, |\xi_1|^2)\xi_1 \tag{6.4.116}$$

$$\omega_1 \frac{\partial p_2}{\partial \eta_1} + \omega_2 \frac{\partial p_2}{\partial \eta_2} + p_0 \frac{\partial p_2}{\partial \phi} = -\frac{dp_0}{dT_2} + \nu_0 + \nu_2 p_0 + \nu_1 \text{ Imaginary } (\xi_1 e^{-i\phi}) \tag{6.4.117}$$

where

$$L = \left(\omega_1 \frac{\partial}{\partial \eta_1} + \omega_2 \frac{\partial}{\partial \eta_2} + p_0 \frac{\partial}{\partial \phi}\right)^2$$

$$- i \frac{p_0 I_x}{I}\left(\omega_1 \frac{\partial}{\partial \eta_1} + \omega_2 \frac{\partial}{\partial \eta_2} + p_0 \frac{\partial}{\partial \phi}\right) + \omega_0^2 \tag{6.4.118}$$

The solution of (6.4.115) is

$$\xi_1 = A_1(T_2)e^{i\eta_1} + A_2(T_2)e^{i\eta_2} \tag{6.4.119}$$

Then (6.4.117) becomes

$$\omega_1 \frac{\partial p_2}{\partial \eta_1} + \omega_2 \frac{\partial p_2}{\partial \eta_2} + p_0 \frac{\partial p_2}{\partial \phi}$$

$$= -\frac{dp_0}{dT_2} + \nu_0 + \nu_2 p_0$$

$$+ \nu_1[a_1 \sin(\eta_1 - \phi + \theta_1) + a_2 \sin(\eta_2 - \phi + \theta_2)] \tag{6.4.120}$$

where $A_n = a_n \exp(i\theta_n)$ with real a_n and θ_n. Since $p_0 \approx \omega_1$, $\eta_1 - \phi$ is a slowly varying function of time, and we consider it a function of T_2. Now the solution of (6.4.120) contains terms that tend to infinity as η_1, η_2, or $\phi \to \infty$ (i.e., $t \to \infty$), thereby invalidating our expansion for long times unless

$$\frac{dp_0}{dT_2} = \nu_0 + \nu_2 p_0 + \nu_1 a_1 \sin(\eta_1 - \phi + \theta_1) \tag{6.4.121}$$

Then p_2 becomes

$$p_2 = \frac{a_2 \nu}{p_0 - \omega_2} \cos(\eta_2 + \theta_2 - \phi) \tag{6.4.122}$$

With ξ_1 known (6.4.116) becomes

$$L(\xi_3) = Q_1 e^{i\eta_1} + Q_2 e^{i\eta_2} + (i\omega_1\mu_2 + i\chi_2 + \gamma)A_1^2\bar{A}_2 e^{i(2\eta_1-\eta_2)}$$
$$+ (i\omega_2\mu_2 + i\chi_2 + \gamma)\bar{A}_1 A_2^2 e^{i(2\eta_2-\eta_1)} \tag{6.4.123}$$

where

$$Q_1 = -i(\omega_1 - \omega_2)\frac{dA_1}{dT_2} - i\omega_1' A_1$$
$$+ i\{(\omega_1\mu_1 + \chi_1) + (-i\gamma + \chi_2 + \omega_1\mu_2)a_1^2$$
$$+ [-2i\gamma + 2\chi_2 + (\omega_1 + \omega_2)\mu_2]a_2^2\}A_1 + ke^{i(\phi+\phi_0-\eta_1)} \tag{6.4.124}$$

$$Q_2 = i(\omega_1 - \omega_2)\frac{dA_2}{dT_2} - i\omega_2' A_2$$
$$+ i\{(\omega_2\mu_1 + \chi_1) + [-2i\gamma + 2\chi_2 + (\omega_1 + \omega_2)\mu_2]a_1^2$$
$$+ (-i\gamma + \chi_2 + \omega_2\mu_2)a_2^2\}A_2 \tag{6.4.125}$$

Secular terms in (6.4.123) will be eliminated if $Q_1 = Q_2 = 0$. Letting $A_n = a_n \exp(i\theta_n)$ with real a_n and θ_n in (6.4.124) and (6.4.125) with $Q_1 = Q_2 = 0$ and separating real and imaginary parts, we obtain

$$\frac{da_1}{dT_2} = \lambda_{11}a_1 + \lambda_{12}a_1^3 + \lambda_{13}a_1 a_2^2 + \frac{k}{\omega_1 - \omega_2}\sin\Gamma \tag{6.4.126}$$

$$\frac{da_2}{dT_2} = \lambda_{21}a_2 + \lambda_{22}a_2^3 + \lambda_{23}a_1^2 a_2 \tag{6.4.127}$$

$$\frac{d\theta_1}{dT_2} = -\frac{\gamma}{\omega_1 - \omega_2}(a_1^2 + 2a_2^2) - \frac{k}{(\omega_1 - \omega_2)a_1}\cos\Gamma \tag{6.4.128}$$

$$\frac{d\theta_2}{dT_2} = \frac{\gamma}{\omega_1 - \omega_2}(2a_1^2 + a_2^2) \tag{6.4.129}$$

where

$$\Gamma = \phi - \eta_1 - \theta_1 + \phi_0 \tag{6.4.130}$$

$$[\lambda_{11}, \lambda_{12}, \lambda_{13}] = \frac{1}{\omega_1 - \omega_2}[\omega_1\mu_1 + \chi_1 - \omega_1', \omega_1\mu_2 + \chi_2, (\omega_1 + \omega_2)\mu_2 + 2\chi_2]$$
$$\tag{6.4.131}$$

$$[\lambda_{21}, \lambda_{22}, \lambda_{23}] = -\frac{1}{\omega_1 - \omega_2}[\omega_2\mu_1 + \chi_1 - \omega_2', \omega_2\mu_2 + \chi_2, (\omega_1 + \omega_2)\mu_2 + 2\chi_2]$$
$$\tag{6.4.132}$$

Combining (6.4.128) and (6.4.130) and introducing the detuning parameter σ defined by

$$p_0 = \omega_1 + \epsilon^2 \sigma$$

we have

$$\frac{d\Gamma}{dT_2} = \sigma + \frac{\gamma}{\omega_1 - \omega_2}(a_1{}^2 + 2a_2{}^2) + \frac{k}{(\omega_1 - \omega_2)a_1}\cos\Gamma \qquad (6.4.133)$$

6.4.7. THE EARTH-MOON-SPACESHIP PROBLEM

The next example is the one-dimensional earth-moon-spaceship problem discussed in Sections 2.4.2, 3.2.2, and 4.1.7 and given by

$$\frac{1}{2}\left(\frac{dx}{dt}\right)^2 = \frac{1-\mu}{x} + \frac{\mu}{1-x}, \qquad t(0) = 0 \qquad (6.4.134)$$

The straightforward expansion for t, for small μ, in terms of x is singular at $x = 1$, and the region of nonuniformity is $1 - x = O(\mu)$. Thus to determine an expansion valid for all x using the method of multiple scales, we introduce the two variables (Nayfeh, 1964, 1965a)

$$\xi = x, \qquad \eta = \frac{1-x}{\mu}$$

With these variables (6.4.134) becomes

$$\sqrt{2}\left(\frac{\partial t}{\partial \xi} - \mu^{-1}\frac{\partial t}{\partial \eta}\right) = \left(\frac{1-\mu}{\xi} + \frac{1}{\eta}\right)^{-1/2} \qquad (6.4.135)$$

where all functions of x are expressed in terms of ξ except the term $1 - x$, the source of the nonuniformity, which is expressed in terms of η as $\mu\eta$. Now we assume that t possesses the following uniformly valid expansion

$$t = t_0(\xi, \eta) + \mu t_1(\xi, \eta) + \mu^2 t_2(\xi, \eta) + \cdots \qquad (6.4.136)$$

Substituting (6.4.136) into (6.4.135) and equating coefficients of like powers of μ, we obtain

$$\frac{\partial t_0}{\partial \eta} = 0 \qquad (6.4.137)$$

$$\sqrt{2}\left(\frac{\partial t_0}{\partial \xi} - \frac{\partial t_1}{\partial \eta}\right) = \sqrt{\frac{\xi\eta}{\eta + \xi}} \qquad (6.4.138)$$

$$\sqrt{2}\left(\frac{\partial t_1}{\partial \xi} - \frac{\partial t_2}{\partial \eta}\right) = \tfrac{1}{2}\sqrt{\xi}\left(\frac{\eta}{\eta + \xi}\right)^{3/2} \qquad (6.4.139)$$

The general solution of (6.4.137) is

$$\sqrt{2}\,t_0 = A(\xi) \qquad (6.4.140)$$

where A is determined from the condition that t_1/t_0 be bounded for all η. The solution of (6.4.138) becomes

$$-\sqrt{2}\, t_1 = -A'(\xi)\eta + \sqrt{\xi\eta(\eta + \xi)} - \xi^{3/2}\sinh^{-1}\sqrt{\frac{\eta}{\xi}} + B(\xi) \quad (6.4.141)$$

As $\eta \to \infty$, (6.4.141) becomes

$$-\sqrt{2}t_1 = [\sqrt{\xi} - A'(\xi)]\eta + \tfrac{1}{2}\xi\sqrt{\xi} - \xi^{3/2}\ln 2\sqrt{\frac{\eta}{\xi}} + B(\xi) + O(\eta^{-1})$$

$$(6.4.142)$$

Thus t_1 contains two terms which make t_1 singular as $\eta \to \infty$; a term proportional to η and a term proportional to $\ln(\eta)$. The first term can be eliminated if

$$A'(\xi) = \sqrt{\xi} \quad \text{or} \quad A = \tfrac{2}{3}\xi^{3/2} + a \quad (6.4.143)$$

where a is an arbitrary constant. As for the second term, $\ln(\eta)$ is slowly varying with x and μ although η is fast varying. Thus it should be expressed in terms of ξ; that is

$$\xi^{3/2}\ln 2\sqrt{\frac{\eta}{\xi}} = \xi^{3/2}\ln\sqrt{\frac{4(1 - \xi)}{\mu\xi}}$$

Then t_1 is singular as $\xi \to 1$, and it will be bounded as $\xi \to 1$ if

$$B(\xi) = \tfrac{1}{2}\xi^{3/2}\ln(1 - \xi) + C(\xi) \quad (6.4.144)$$

The function $C(\xi)$ is determined by requiring that t_2/t_1 be bounded as $\eta \to \infty$.

Substituting the above solutions into (6.4.139), we obtain

$$\sqrt{2}\frac{\partial t_2}{\partial \eta} = \frac{1}{2}\frac{\eta}{\sqrt{\xi}} - \frac{1}{2}\sqrt{\frac{\eta(\eta + \xi)}{\xi}} - \sqrt{\frac{\eta\xi}{\eta + \xi}} - \tfrac{1}{2}\sqrt{\xi}\left(\frac{\eta}{\eta + \xi}\right)^{3/2}$$

$$+ \tfrac{3}{2}\sqrt{\xi}\sinh^{-1}\sqrt{\frac{\eta}{\xi}} - \tfrac{3}{4}\sqrt{\xi}\ln(1 - \xi) + \frac{1}{2}\frac{\xi^{3/2}}{1 - \xi} - C'(\xi) \quad (6.4.145)$$

As $\eta \to \infty$, (6.4.145) becomes

$$\sqrt{2}\frac{\partial t_2}{\partial \eta} = -\tfrac{7}{4}\sqrt{\xi} + \tfrac{3}{2}\sqrt{\xi}\ln 2\sqrt{\frac{\eta}{\xi}} - \tfrac{3}{4}\sqrt{\xi}\ln(1 - \xi)$$

$$+ \frac{1}{2}\frac{\xi^{3/2}}{1 - \xi} - C'(\xi) + O(\eta^{-1}) \quad \text{as} \quad \eta \to \infty \quad (6.4.146)$$

Here again $\ln \sqrt{\eta}$ is expressed in terms of ξ. Consequently, (6.4.146) becomes

$$\sqrt{2} \frac{\partial t_2}{\partial \eta} = (-\tfrac{7}{4} + \tfrac{3}{2}\ln 2 - \tfrac{3}{4}\ln \mu\xi)\sqrt{\xi} + \frac{1}{2}\frac{\xi^{3/2}}{1 - \xi} - C'(\xi) + O(\eta^{-1})$$

as $\eta \to \infty$

In order that t_2/t_1 be bounded as $\eta \to \infty$

$$C'(\xi) = (-\tfrac{7}{4} + \tfrac{3}{2}\ln 2 - \tfrac{3}{4}\ln \mu\xi)\sqrt{\xi} + \frac{1}{2}\frac{\xi^{3/2}}{1 - \xi}$$

or

$$C = -\tfrac{7}{6}\xi^{3/2} - \sqrt{\xi} - \tfrac{1}{2}\xi^{3/2}\ln \frac{\mu\xi}{4} + \frac{1}{2}\ln \frac{1 + \sqrt{\xi}}{1 - \sqrt{\xi}} + c \quad (6.4.147)$$

where c is a constant of integration.

Expressing t_0 and t_1 in terms of x and using the initial condition $t(x = 0) = 0$, we obtain $a = c = 0$. Hence

$$\sqrt{2}\, t = (1 - \tfrac{1}{3}x)\sqrt{x} - \sqrt{x(1 - x)(1 - x + \mu x)}$$
$$+ \mu \left[x^{3/2} \sinh^{-1} \sqrt{\frac{1 - x}{\mu x}} - \tfrac{1}{2}x^{3/2} \ln \frac{4(1 - x)}{\mu x} \right.$$
$$\left. + \tfrac{7}{6}x^{3/2} + \sqrt{x} - \tfrac{1}{2}\ln \frac{1 + \sqrt{x}}{1 - \sqrt{x}} \right] + O(\mu^2) \quad (6.4.148)$$

We next discuss an alternative method (Nayfeh, 1965a) of determining $A(\xi)$ and $B(\xi)$. Since

$$\sqrt{2}\, t = A(\xi) + \mu \left[A'(\xi)\eta - \sqrt{\xi\eta(\eta + \xi)} + \xi^{3/2} \sinh^{-1} \sqrt{\frac{\eta}{\xi}} - B(\xi) \right] + O(\mu^2)$$

$$(6.4.149)$$

is assumed to be uniformly valid for all x, it must reduce to the straightforward expansion (see Exercise 2.12)

$$\sqrt{2}\, t = \tfrac{2}{3}x^{3/2} + \mu\left(\tfrac{2}{3}x^{3/2} + \sqrt{x} - \tfrac{1}{2}\ln \frac{1 + \sqrt{x}}{1 - \sqrt{x}} \right) + O(\mu^2) \quad (6.4.150)$$

away from $x = 1$. We can use this condition to determine the functions $A(\xi)$ and $B(\xi)$ rather than use the condition that t_n/t_{n-1} be bounded for all ξ and

η. Expressing (6.4.149) in terms of x, and expanding for small μ, we obtain

$$\sqrt{2}\, t = A(x) + \mu\left[A'(x)\,\frac{1-x}{\mu} - \frac{1-x}{\mu}\sqrt{x} - \tfrac{1}{2}x^{3/2} \right.$$

$$\left. + \tfrac{1}{2}x^{3/2}\ln\frac{4(1-x)}{\mu x} - B(x) \right] + O(\mu^2) \quad (6.4.151)$$

In order that the first terms in (6.4.150) and (6.4.151) be the same

$$A(x) = \tfrac{2}{3}x^{3/2} \qquad (6.4.152)$$

Then the second terms are the same if

$$B(x) = -\tfrac{7}{6}x^{3/2} - \sqrt{x} + \tfrac{1}{2}x^{3/2}\ln\frac{4(1-x)}{\mu x} + \tfrac{1}{2}\ln\frac{1+\sqrt{x}}{1-\sqrt{x}} \quad (6.4.153)$$

Substituting these expressions for A and B into (6.4.149) and expressing the result in terms of x, we obtain (6.4.148) exactly.

6.4.8.   A MODEL FOR DISPERSIVE WAVES

We again consider Bretherton's (1964) model equation

$$\phi_{tt} + \phi_{xxxx} + \phi_{xx} + \phi = \epsilon\phi^3 \qquad (6.4.154)$$

The linearized equation admits the uniform wave train solution

$$\phi = a\cos\theta$$
$$\theta = kx - \omega t, \qquad \omega^2 = k^4 - k^2 + 1 \qquad (6.4.155)$$

To determine a wave train slowly varying with position and time, we follow Nayfeh and Hassan (1971) by assuming that

$$\phi = \phi_0(\theta, X_1, T_1) + \epsilon\phi_1(\theta, X_1, T_1) + \cdots \qquad (6.4.156)$$

where

$$\theta = \epsilon^{-1}\zeta(X_1, T_1), \qquad X_1 = \epsilon x, \qquad T_1 = \epsilon t$$
$$k = \theta_x = \frac{\partial\zeta}{\partial X_1}, \qquad \omega = -\theta_t = -\frac{\partial\zeta}{\partial T_1} \qquad (6.4.157)$$

In terms of the new variables θ, X_1, and T_1, the time and space derivatives become

$$\frac{\partial^2}{\partial t^2} = \omega^2\frac{\partial^2}{\partial\theta^2} - 2\epsilon\omega\frac{\partial^2}{\partial\theta\,\partial T_1} - \epsilon\frac{\partial\omega}{\partial T_1}\frac{\partial}{\partial\theta} + \epsilon^2\frac{\partial^2}{\partial T_1^{\,2}}$$

$$\frac{\partial^2}{\partial x^2} = k^2\frac{\partial^2}{\partial\theta^2} + 2\epsilon k\frac{\partial^2}{\partial\theta\,\partial X_1} + \epsilon\frac{\partial k}{\partial X_1}\frac{\partial}{\partial\theta} + \epsilon^2\frac{\partial^2}{\partial X_1^{\,2}}$$

$$\frac{\partial^4}{\partial x^4} = k^4\frac{\partial^4}{\partial\theta^4} + 4\epsilon k^3\frac{\partial^4}{\partial\theta^3\,\partial X_1} + 6\epsilon k^2\frac{\partial k}{\partial X_1}\frac{\partial^3}{\partial\theta^3} + \cdots$$

Substituting (6.4.156) into (6.4.154) and equating coefficients of like powers of ϵ, we have

$$L(\phi_0) \equiv (\omega^2 + k^2) \frac{\partial^2 \phi_0}{\partial \theta^2} + k^4 \frac{\partial^4 \phi_0}{\partial \theta^4} + \phi_0 = 0 \qquad (6.4.158)$$

$$L(\phi_1) = {\phi_0}^3 + 2\omega \frac{\partial^2 \phi_0}{\partial \theta \, \partial T_1} - 2k \frac{\partial^2 \phi_0}{\partial \theta \, \partial X_1} - 4k^3 \frac{\partial^4 \phi_0}{\partial \theta^3 \, \partial X_1}$$

$$+ \left(\frac{\partial \omega}{\partial T_1} - \frac{\partial k}{\partial X_1} \right) \frac{\partial \phi_0}{\partial \theta} - 6k^2 \frac{\partial k}{\partial X_1} \frac{\partial^3 \phi_0}{\partial \theta^3} \qquad (6.4.159)$$

The solution of (6.4.158) is taken to be

$$\phi_0 = A(X_1, T_1)e^{i\theta} + \bar{A}(X_1, T_1)e^{-i\theta} \qquad (6.4.160)$$

where

$$\omega^2 = k^4 - k^2 + 1$$

Substituting for ϕ_0 into (6.4.159), we have

$$L(\phi_1) = Q(X_1, T_1)e^{i\theta} + A^3 e^{3i\theta} + CC \qquad (6.4.161)$$

where

$$Q = 2i\omega \frac{\partial A}{\partial T_1} + 2ik(2k^2 - 1) \frac{\partial A}{\partial X_1} + i \frac{\partial \omega}{\partial T_1} A + i(6k^2 - 1) \frac{\partial k}{\partial X_1} A + 3A^2\bar{A} \qquad (6.4.162)$$

The condition that must be satisfied for there to be no secular terms is $Q = 0$. To simplify this condition, we note that

$$\omega\omega' = 2k^3 - k \qquad (6.4.163)$$

where $\omega' = d\omega/dk$, the group velocity. Differentiating (6.4.163) with respect to X_1, we have

$$\omega\omega'' \frac{\partial k}{\partial X_1} + \omega'^2 \frac{\partial k}{\partial X_1} = (6k^2 - 1) \frac{\partial k}{\partial X_1} \qquad (6.4.164)$$

If ζ is twice continuously differentiable, ω and k satisfy the compatability relationship

$$\frac{\partial k}{\partial T_1} + \frac{\partial \omega}{\partial X_1} = 0$$

or

$$\frac{\partial k}{\partial T_1} + \omega' \frac{\partial k}{\partial X_1} = 0 \qquad (6.4.165)$$

Hence

$$\frac{\partial \omega}{\partial T_1} = \omega' \frac{\partial k}{\partial T_1} = -\omega'^2 \frac{\partial k}{\partial X_1}$$

so that (6.4.164) can be rewritten as

$$\frac{\partial \omega}{\partial T_1} + (6k^2 - 1)\frac{\partial k}{\partial X_1} = \omega\omega''\frac{\partial k}{\partial X_1} \qquad (6.4.166)$$

With (6.4.163) and (6.4.166) the condition $Q = 0$ can be simplified to

$$2\frac{\partial A}{\partial T_1} + 2\omega'\frac{\partial A}{\partial X_1} + \omega''\frac{\partial k}{\partial X_1} A = \frac{3i}{\omega} A^2\bar{A} \qquad (6.4.167)$$

Letting $A = (1/2)a \exp(i\beta)$ in (6.4.167) and separating real and imaginary parts, we have

$$\frac{\partial a^2}{\partial T_1} + \frac{\partial}{\partial X_1}(\omega'a^2) = 0 \qquad (6.4.168)$$

$$\frac{\partial \beta}{\partial T_1} + \omega'\frac{\partial \beta}{\partial X_1} = \frac{3a^2}{8\omega} \qquad (6.4.169)$$

The solution obtained in this section using the method of multiple scales is a different representation of the solution obtained in Section 5.8.1 by averaging the Lagrangian. In fact, the equations governing the amplitude and the wave number have exactly the same form. However, in Section 5.8.1 there is no phase, but the dispersion relationship (5.8.9) is amplitude-dependent; in this section the dispersion relationship is independent of amplitude but the solution describes the phase variation. To show the equivalence of these representations, we expand θ of Section 5.8.1 in the form

$$\theta = \theta_0 - \epsilon\tilde{\beta}$$

so that

$$k = \frac{\partial \theta_0}{\partial x} - \epsilon\frac{\partial \tilde{\beta}}{\partial x} = k_0 - \epsilon\frac{\partial \tilde{\beta}}{\partial x}$$

$$\omega = \omega_0 + \epsilon\frac{\partial \tilde{\beta}}{\partial t} \qquad (6.4.170)$$

Substituting (6.4.170) into (5.8.9) and equating coefficients of like powers of ϵ, we obtain

$$\omega_0^2 = k_0^4 - k_0^2 + 1$$

$$\frac{\partial \tilde{\beta}}{\partial t} + \omega_0'\frac{\partial \tilde{\beta}}{\partial x} = \frac{3a^2}{8\omega_0}$$

The last equation is the same as (6.4.169)

6.4.9. THE NONLINEAR KLEIN–GORDON EQUATION

The last example considered in this chapter is the equation

$$u_{tt} - u_{xx} + V'(u) = 0 \qquad (6.4.171)$$

which was treated in Section 5.8.3 using Whitham's method of averaging the Lagrangian. Our analysis follows that of Luke (1966).

We assume that u possesses a uniformly valid expansion of the form

$$u(x, t) = u_0(\theta, X_1, T_1) + \epsilon u_1(\theta, X_1, T_1) + \cdots \qquad (6.5.172)$$

where θ, X_1, and T_1 are defined in (6.4.157). Substituting (6.4.172) into (6.4.171), using the expressions for the derivatives from the previous section, and equating coefficients of like powers of ϵ, we obtain

$$(\omega^2 - k^2)\frac{\partial^2 u_0}{\partial \theta^2} + V'(u_0) = 0 \qquad (6.4.173)$$

$$(\omega^2 - k^2)\frac{\partial^2 u_1}{\partial \theta^2} + V''(u_0)u_1$$

$$= 2k\frac{\partial^2 u_0}{\partial \theta \, \partial X_1} + 2\omega\frac{\partial^2 u_0}{\partial \theta \, \partial T_1} + \frac{\partial k}{\partial X_1}\frac{\partial u_0}{\partial \theta} + \frac{\partial \omega}{\partial T_1}\frac{\partial u_0}{\partial \theta} \qquad (6.4.174)$$

Equation (6.4.173) can be integrated once to give

$$\tfrac{1}{2}(\omega^2 - k^2)\left(\frac{\partial u_0}{\partial \theta}\right)^2 + V(u_0) = E(X_1, T_1) \qquad (6.4.175)$$

whose solution is

$$\theta = \sqrt{\omega^2 - k^2}\int^{u_0}\frac{d\xi}{\{2[E - V(\xi)]\}^{1/2}} - \eta(X_1, T_1) \qquad (6.4.176)$$

where $E(X_1, T_1)$ and $\eta(X_1, T_1)$ are unknown functions to be determined by examining (6.4.174). Inverting (6.4.176), we find that

$$u_0(\theta, X_1, T_1) = f(\theta + \eta, E, \omega^2 - k^2) \qquad (6.4.177)$$

We assume that f is periodic with a constant period which can be normalized to unity so that

$$\sqrt{\omega^2 - k^2}\oint\frac{d\xi}{\{2[E - V(\xi)]\}^{1/2}} = 1 \qquad (6.4.178)$$

This provides one relationship among ω, k, and E which is a dispersion relationship.

Now the particular solution of (6.4.174) contains terms that make u_1/u_0 unbounded as $\theta \to \infty$ unless the right-hand side of (6.4.174) is orthogonal to the solution of the adjoint homogeneous equation. This condition is sometimes referred to as the solvability condition, and it is a generalization of the condition of elimination of secular terms that has been used extensively in this book. Since (6.4.174) is self-adjoint, the solvability condition demands that its right-hand side be orthogonal to the solution of the homogeneous equation which can easily be shown to be $u_1 = \partial u_0 / \partial \theta$. Thus the solvability condition requires that

$$\oint \left(2k \frac{\partial^2 u_0}{\partial \theta \, \partial X_1} + 2\omega \frac{\partial^2 u_0}{\partial \theta \, \partial T_1} + \frac{\partial k}{\partial X_1} \frac{\partial u_0}{\partial \theta} + \frac{\partial \omega}{\partial T_1} \frac{\partial u_0}{\partial \theta} \right) \frac{\partial u_0}{\partial \theta} \, d\theta = 0$$

which can be rewritten as

$$\frac{\partial}{\partial T_1} \left[\omega \oint \left(\frac{\partial u_0}{\partial \theta} \right)^2 d\theta \right] + \frac{\partial}{\partial X_1} \left[k \int \left(\frac{\partial u_0}{\partial \theta} \right)^2 d\theta \right] = 0 \qquad (6.4.179)$$

Changing the variable of integration from θ to u_0 and substituting for $\partial u_0 / \partial \theta$ from (6.4.175), we can rewrite this condition as

$$\frac{\partial}{\partial T_1} \left\{ \frac{\omega}{\sqrt{\omega^2 - k^2}} \oint \sqrt{2[E - V(u_0)]} \, du_0 \right\}$$

$$+ \frac{\partial}{\partial X_1} \left\{ \frac{k}{\sqrt{\omega^2 - k^2}} \oint \sqrt{2[E - V(u_0)]} \, du_0 \right\} = 0 \qquad (6.4.180)$$

This provides a second relationship among ω, k, and E. The third relationship is the compatability equation (6.4.165).

The results of this section are in agreement with those obtained in Section 5.8.3 using the variational approach.

6.4.10.   ADVANTAGES AND LIMITATIONS OF THE GENERALIZED METHOD

This method can certainly be applied to all problems that can be treated by either the derivative-expansion method or the two-variable expansion procedure. Moreover, it can also be applied to cases in which both of these two methods fail, such as problems requiring nonlinear scales (e.g., an oscillator with slowly varying coefficients) or problems having sharp changes (e.g., the earth-moon-spaceship problem). However, the algebra is more involved, and the derivative-expansion method and the two-variable expansion procedure are preferable for nonlinear oscillation problems with constant coefficients.

The method of multiple scales can be used to obtain uniformly valid expansions for problems that can be treated using the method of strained coordinates. Moreover, the method of multiple scales can be applied to cases in which the method of strained coordinates cannot be applied, such as problems involving damping or sharp changes. In cases in which the method of strained coordinates applies, it may have an advantage because of the implicitness of the solution. For nondispersive hyperbolic equations expansions in terms of exact characteristics are desirable. However, the method of multiple scales can be viewed as a generalization of the method of strained coordinates if the scales are given implicitly rather than explicitly in terms of the original variables.

The examples considered in this chapter demonstrate that the method of multiple scales can be applied to problems that can be treated by the method of matched asymptotic expansions, such as the earth-moon-spaceship problem, as well as to problems that cannot be treated by the latter method, such as nonlinear oscillations. The method of multiple scales yields a single uniformly valid expansion, in contrast with the method of matched asymptotic expansions which yields two expansions that must be matched. Although an ordinary differential equation is transformed into a partial differential equation by the method of multiple scales, the first approximation is not more difficult to solve than the first inner equation. However, the equations for determination of the different scales may be difficult to solve (Mahony, 1962). Moreover, this method has not been applied yet to partial differential equations in which the first term in the expansion is nonlinear, such as viscous flow past a body, or elliptic partial differential equations with inhomogeneous boundary perturbations such as flow past a thin airfoil.

The method of multiple scales is applicable to problems that can be treated by the method of averaging, the method of Krylov–Bogoliubov–Mitropolski, and the Lie transforms as well as to cases that cannot be treated by these methods. If the system is represented by a Hamiltonian, the Lie transforms have an advantage because the higher approximations can be obtained recursively. However, the method of multiple scales can be applied in conjunction with the Lie transforms directly on the Hamiltonian.

Exercises

6.1. Determine a first-order uniform expansion for

$$\ddot{u} + \omega_0^2 u = \epsilon f(u, \dot{u})$$

then specialize your results to the cases: $f = -\dot{u} + \beta u^3$, $\beta u^3 + (1 - u^2)\dot{u}$, and $-|\dot{u}|\dot{u}$.

6.2. Determine second-order uniform expansions for

$$\ddot{u} + (\delta + \epsilon \cos 2t)u = 0$$

for δ near 0 and 4.

6.3. Determine second-order uniform expansions for

$$\ddot{u} + (\delta + \epsilon \cos^3 t)u = 0$$

6.4. Determine first-order uniform expansions for

$$\ddot{u} + (\delta + \epsilon \cos 2t)u = \epsilon f(u, \dot{u})$$

and specialize your results to $f = \beta u^3$, $-|\dot{u}| \dot{u}$, and $(1 - u^2)\dot{u}$.

6.5. Consider the problem

$$\ddot{u} + \omega_0^2 u = \epsilon[u^3 + (1 - u^2)\dot{u}] + K \cos \omega t$$

Determine first-order uniform expansions for

(a) $K = O(1)$ and ω away from ω_0, $3\omega_0$, and $\omega_0/3$

(b) $K = O(1)$ and $\omega \approx 3\omega_0$

(c) $K = O(1)$ and $\omega \approx \omega_0/3$

(d) $K = O(\epsilon)$ and $\omega \approx \omega_0$.

6.6. Consider the problem

$$\ddot{u} + \omega_0^2 u = 2\epsilon \frac{d}{dt} [(1 - z)u] + 2K \cos \omega t$$

$$\tau \dot{z} + z = u^2$$

Determine first-order uniform expansions for the cases enumerated in Exercise 6.5.

6.7. The problem of takeoff of a satellite from a circular orbit with a small thrust can be reduced to

$$u'' + u - v^2 = -\frac{\epsilon v^2}{u^3} (su' + cu)$$

$$v' = -\frac{\epsilon s v^3}{u^3}$$

$$u(0) = 1, \quad u'(0) = 0, \quad \text{and} \quad v(0) = 1$$

where primes denote differentiation with respect to θ, and ϵ, s, and c are constants. For small ϵ show that (Nayfeh, 1966)

$$v = f + 3\epsilon c f^{-3} \ln f + O(\epsilon^2)$$

$$u = f^2 + \epsilon \left\{ f^{7/2} \left[c \cos \left(\theta + \frac{c}{s} \ln f \right) + 2s \sin \left(\theta + \frac{c}{s} \ln f \right) \right] \right.$$

$$\left. + cf^{-2}(6 \ln f - 1) \right\} + O(\epsilon^2)$$

where $f = (1 - 4\epsilon s \theta)^{1/4}$. Is this expansion valid for all θ?

6.8. Consider the problem defined by (6.4.105) and (6.4.106) with constant ω_0, p, γ, μ_i and χ_i.

(a) Determine a first-order uniform expansion for the case $K = O(1)$ and p away from ω_1. (b) Show that this expansion is not valid when $p \approx 0$ or $2\omega_1 - \omega_2$ and determine first-order uniform expansions for these cases (Nayfeh and Saric,

1971b). (c) Determine a first-order uniform expansion for $K = O(\epsilon^2)$ and $p \approx \omega_1$ using the methods of multiple scales (Nayfeh and Saric, 1971b) and the method of averaging (Clare, 1971); then compare both results.

6.9. Use the method of multiple scales (MMS) to determine second-order uniform expansions for the problems

(a) $\epsilon y'' \mp y' + y = 0$

(b) $\epsilon y'' \mp y' = 2x$

(c) $\epsilon y'' \pm (2x + 1)y' = 1$

subject to the boundary conditions

$$y(0) = \alpha, \qquad y(1) = \beta$$

6.10. Determine first-order uniform expansions for the problems

(a) $\epsilon y'' - a(x)y' + b(x)y = 0, \qquad a(x) > 0$

(b) $\epsilon y'' \mp y' + y^2 = 0$

(c) $\epsilon y'' \mp yy' - y = 0$

(d) $\epsilon y'' \mp (2x + 1)y' + y^2 = 0$

(e) $\epsilon y'' \mp y' + y^n = 0$; n is a positive integer

subject to the boundary conditions

$$y(0) = \alpha, \qquad y(1) = \beta$$

6.11. Use the MMS to determine a first-order uniform expansion for the problem

$$\epsilon y'' + a(x)y' = 1$$

$$y(0) = \alpha, \qquad y(1) = \beta$$

if $a(x)$ has a simple zero at μ in $[0, 1]$.

6.12. Use the MMS to determine a first-order uniform expansion for

$$y'' + \lambda^2 (1 - x)^n f(x)y = 0$$

where n is a positive integer, $f(x) > 0$ and $\lambda \gg 1$.

6.13. Determine first-order uniform expansions for (6.4.84) of the form

(a) $u = A(\xi)cn[\eta, \nu(\xi)]$ and (b) $u = A(\xi)dn[\eta, \nu(\xi)]$.

6.14. Consider the problem

$$\ddot{u} + \omega_0^2(\epsilon t)u = \epsilon u^3 + K \cos \phi$$

where $\phi = \omega(\epsilon t)$. Determine first-order uniform expansions for the cases

(a) $K = O(1)$ and ω away from ω_0, $3\omega_0$, and $\omega_0/3$

(b) $K = O(1)$ and $\omega \approx 3\omega_0$

(c) $K = O(1)$ and $\omega \approx \omega_0/3$

(d) $K = O(\epsilon)$ and $\omega \approx \omega_0$

6.15. Consider the problem defined by (6.4.105) through (6.4.107) with variable coefficients. Determine a first-order uniform expansion when $K = O(1)$ and p is away from ω_1. Show that this expansion is not valid when $p \approx 0$ or $2\omega_1 - \omega_2$, and determine uniform expansions for these cases (Nayfeh and Saric, 1972a).

6.16. Solve Exercise 5.14 using the MMS.

6.17. Determine a first-order uniform expansion for small amplitudes for

$$\ddot{x} - \dot{y} + 2x + 3x^2 + 2y^2 = 0$$

$$\ddot{y} + \dot{x} + 2\delta y + 4xy = 0$$

when $\delta \approx 1$.

6.18. Consider the problem

$$u_{tt} - c^2 u_{xx} - \gamma^2 u = \epsilon u^3$$

$$u(x, 0) = a \cos kx, \qquad u_t(x, 0) = 0$$

Determine a first-order uniform expansion when

$$c^2 k^2 \approx \gamma^2$$

6.19. Consider the equation

$$u_{tt} - c^2 u_{xx} + \gamma^2 u = \epsilon u^3$$

Use the MMS to determine first-order expansions for traveling waves if (a) the amplitude and phase vary slowly with position and time and (b) the wave number, frequency, amplitude, and phase vary slowly with position and time.

6.20. Nonlinear transverse oscillations in a homogeneous free-free beam with a nonlinear moment-curvature relationship are given by

$$w_{tt} + c^2 w_{xxxx} = -\epsilon(w_{xx}^3)_{xx}$$

where c and ϵ are constants. Determine first-order uniform expansions for small ϵ using the MMS for the cases enumerated in Exercise 6.19.

6.21. Use the MMS to determine second-order uniform traveling wave solutions for the problem defined in Exercise 5.15 and the cases of Exercise 6.19.

6.22. Consider again Bretherton's model equation

$$\phi_{tt} + \phi_{xxxx} + \phi_{xx} + \phi = \epsilon f(\phi, \phi_t, \phi_x)$$

but with a general nonlinear function f. Use the MMS to determine first-order uniformly valid expansions for small ϵ for the case of nth-harmonic resonance. Specialize your results for (a) second-harmonic resonance and (b) third-harmonic resonance.

6.23. Use the MMS to determine the problems defining the first-order uniform solutions for

    (a) $\epsilon u_{xx} + u_{yy} + u_x = 0$

    (b) $\epsilon u_{xx} + u_{yy} + a(x)u_x = 0$

    (c) $\epsilon u_{xx} + u_{yy} + a(x, y)u_x = 0$

    (d) $\epsilon(u_{xx} + u_{yy}) + a(x, y)u_x + b(x, y)u = 0$

    (e) $\epsilon^2 \nabla^4 u + a(x, y)u_{xx} + b(x, y)u_x + c(x, y)u_y + d(x, y)u = 0$

subject to the conditions

$$u(x, 0) = F_1(x), \qquad u(x, 1) = F_2(x)$$
$$u(0, y) = G_1(y), \qquad u(1, y) = G_2(y)$$

CHAPTER 7

Asymptotic Solutions of Linear Equations

In this chapter we describe new techniques and use some of the techniques described in the previous chapters to obtain asymptotic solutions of linear ordinary and partial differential equations. Interest here is in differential equations with variable coefficients. The approach is to utilize a small or large parameter and develop parameter perturbations or to utilize a small or large coordinate and develop coordinate perturbations.

If the point at infinity is an ordinary point or a regular singular point of an ordinary system of differential equations, convergent solutions can be obtained in inverse powers of the coordinate, except in some special cases of the regular singular problem when one of the solutions may involve the logarithm of the coordinate. Our interest in this chapter is in the irregular singular case, when the solution must be represented by an asymptotic expansion. In the parameter perturbation case, the parameter can be small or large; the first case includes the case of slowly varying coefficients. The expansions in these cases are obtained using the Liouville–Green (WKB) transformation and its generalizations. The resulting expansions are valid everywhere except at certain points called turning or transition points. Expansions valid everywhere including the turning points are obtained using Langer's transformation and its generalizations.

The discussion of partial differential equations is limited to the case of the reduced wave equation with variable index of refraction. The expansion is first developed for the case in which the index of refraction deviates slightly from a constant, using the Born–Neumann procedure, and the solution is then represented by Feynman diagrams. The resulting expansion is valid only for short distances, and its region of uniformity is extended using the renormalization technique. Then the geometrical optics approach and the method of smoothing are described.

Since these problems are linear, a vast amount of literature exists on their asymptotic solutions as well as their mathematical justification. We describe

308

in this chapter the techniques of developing formal asymptotic expansions to the solution of the equations without any mathematical justification. Moreover, a selective number of articles is cited. For further references and mathematical rigor, the reader is referred to Erdélyi (1956), Jeffreys (1962), Cesari (1971), Bellman (1964), Wilcox (1964), Wasow (1965), Feshchenko, Shkil', and Nikolenko (1967), Wasow (1968), and Frisch (1968).

Second-order ordinary differential equations are treated first in Section 7.1, while systems of first-order ordinary differential equations are discussed in Section 7.2. Turning point problems are then taken up in Section 7.3, while the reduced wave equation is treated in Section 7.4.

7.1. Second-Order Differential Equations

In this section we are concerned with the asymptotic development of the solutions of the equation

$$\frac{d^2y}{dx^2} + p(x; \epsilon)\frac{dy}{dx} + q(x; \epsilon)y = r(x; \epsilon) \tag{7.1.1}$$

where ϵ is a parameter which can be either small or large. We assume that p and q do not simultaneously vanish in the interval of interest. We discuss first the asymptotic solutions of this equation near an irregular singular point. Then we describe techniques for determining the asymptotic solutions of (7.1.1) when it contains a large parameter. Next we consider singular perturbation problems in which a small parameter multiplies the highest derivative. Finally, we describe techniques for obtaining asymptotic expansions when p, q, and r are slowly varying functions of x.

7.1.1. EXPANSIONS NEAR AN IRREGULAR SINGULARITY

Let us investigate the asymptotic development of the solutions of

$$\frac{d^2y}{dx^2} + p(x)\frac{dy}{dx} + q(x)y = 0 \tag{7.1.2}$$

as $x \to \infty$, if the point at infinity is an irregular singular point. Before doing so let us define a regular singular and an irregular singular point. Assume that $p(x)$ and $q(x)$ can be developed in ascending powers of $(x - x_0)$, $x_0 < \infty$, as

$$p(x) = p_0(x - x_0)^\alpha[1 + p_1(x - x_0) + \cdots], \quad p_0 \neq 0$$
$$q(x) = q_0(x - x_0)^\beta[1 + q_1(x - x_0) + \cdots], \quad q_0 \neq 0 \tag{7.1.3}$$

The point x_0 is called an *ordinary point* if $\alpha \geq 0$ and $\beta \geq 0$; otherwise it is called a *singular point*. A singular point is called a *regular singular point* if $\alpha \geq -1$ and $\beta \geq -2$; otherwise it is called an *irregular singular point*.

The above definitions indicate that the nature of a finite point such as x_0 can be decided almost at a glance. The nature of the point at infinity can be determined by transforming it first into the origin. Thus we let $x = z^{-1}$ in (7.1.2) and obtain

$$\frac{d^2y}{dz^2} + \left[\frac{2}{z} - \frac{p(z^{-1})}{z^2}\right]\frac{dy}{dz} + \frac{q(z^{-1})}{z^4}y = 0 \qquad (7.1.4)$$

The point at infinity is an ordinary point of the original equation if the origin is an ordinary point of the transformed equation; that is

$$\frac{2}{z} - \frac{p(z^{-1})}{z^2} = O(1)$$

$$\text{as} \quad z \to 0$$

$$\frac{q(z^{-1})}{z^4} = O(1)$$

which correspond to

$$p(x) = 2x^{-1} + O(x^{-2})$$

$$\text{as} \quad x \to \infty \qquad (7.1.5)$$

$$q(x) = O(x^{-4})$$

in the original equation. In order that infinity be a regular singular point of (7.1.2), the origin must be a regular singular point of the transformed equation; that is

$$\frac{2}{z} - \frac{p(z^{-1})}{z^2} = O\left(\frac{1}{z}\right)$$

$$\text{as} \quad z \to 0$$

$$\frac{q(z^{-1})}{z^4} = O\left(\frac{1}{z^2}\right)$$

which correspond to

$$p(x) = O(x^{-1})$$

$$\text{as} \quad x \to \infty \qquad (7.1.6)$$

$$q(x) = O(x^{-2})$$

Thus if $p(x)$ and $q(x)$ can be developed in descending power series of x as

$$p(x) = p_0 x^\alpha + \cdots, \qquad q(x) = q_0 x^\beta + \cdots, \qquad p_0 \text{ and } q_0 \neq 0$$

then the point at infinity is an ordinary point if $\beta \leq -4$, and either $\alpha = -1$ with $p_0 = 2$ or $\alpha \leq -2$. In this case (7.1.2) has two convergent solutions in powers of x^{-1}. If these conditions are not satisfied, and $\alpha \leq -1$ and $\beta \leq -2$, the point at infinity is a regular singular point. In this case (7.1.2) has two convergent solutions in powers of x^{-1} of the form (Frobenius, 1875)

$$y = x^\sigma(1 + a_1 x^{-1} + a_2 x^{-2} + \cdots) \qquad (7.1.7)$$

where σ satisfies what is called the indicial equation

$$\sigma^2 + (p_0 - 1)\sigma + q_0 = 0 \quad \text{if} \quad \alpha = -1 \quad \text{and} \quad \beta = -2$$

except in the special cases in which the roots of this equation are equal or differ by an integer, when one solution may involve $\log x$.

If one or both of the inequalities

$$\alpha > -1, \qquad \beta > -2$$

is satisfied, the point at infinity is an irregular singular point. This is the case of concern to us in this section. In this case (7.1.2) can be satisfied by solutions of the form

$$y(x) = e^{\Lambda(x)} x^\sigma u(x) \tag{7.1.8a}$$

where $u(x) = O(1)$ as $x \to \infty$, which need not converge, and $\Lambda(x)$ is a polynomial in $x^{m/n}$. Letting λx^ν be the leading term in $\Lambda(x)$, substituting the above solution into (7.1.2), and extracting the dominant part of each term, we obtain

$$\lambda^2 \nu^2 x^{2\nu-2} + p_0 \lambda \nu x^{\nu+\alpha-1} + q_0 x^\beta = 0 \tag{7.1.8b}$$

Therefore

$$\nu = \alpha + 1 \quad \text{or} \quad 2\nu = \beta + 2 \tag{7.1.8c}$$

whichever furnishes the greater value of ν. If ν is an integer, the above solution is called a *normal solution* (Thomé, 1883), and it has the form

$$y = \exp(\lambda_\nu x^\nu + \lambda_{\nu-1} x^{\nu-1} + \cdots + \lambda_1 x) x^\sigma (1 + a_1 x^{-1} + \cdots) \tag{7.1.8d}$$

If ν is not an integer, the above solution is called a *subnormal solution*, and Λ is a polynomial in $x^{1/2}$ while u is an ascending series in $x^{-1/2}$. So far, we have assumed that p and q are represented by power series of x so that α and β are integers. If α and β are not integers, ν is a rational function which can be expressed in its lowest terms as $\nu = n/k$. Then the subnormal solutions have the general form

$$y = \exp(\lambda_n x^{n\tau} + \cdots + \lambda_2 x^{2\tau} + \lambda_1 x^\tau) x^\sigma (1 + a_1 x^{-\tau} + a_2 x^{-2\tau} + \cdots)$$

$$\text{with} \quad \tau = k^{-1} \tag{7.1.8e}$$

To determine a normal or subnormal solution, we substitute the corresponding form of the solution into (7.1.2), equate the coefficients of like powers of x, and obtain equations which, in turn, can be solved for λ_m, σ, and a_m in succession.

Let us consider as an example the special case

$$p(x) = \sum_{n=0}^{\infty} p_n x^{-n}, \qquad q(x) = \sum_{n=0}^{\infty} q_n x^{-n} \quad \text{as} \quad x \to \infty \tag{7.1.9}$$

where p_n and q_n are independent of x. In this case, $\nu = 1$ according to (7.1.8c) and (7.1.2) and (7.1.9) have a formal asymptotic solution of the form

$$y = e^{\lambda x} x^\sigma \sum_{n=0}^{\infty} c_n x^{-n} \tag{7.1.10a}$$

according to (7.1.8d), where λ is a root of

$$\lambda^2 + p_0 \lambda + q_0 = 0 \tag{7.1.10b}$$

on account of (7.1.8b). By substituting (7.1.10a) into (7.1.2), using (7.1.9), and equating coefficients of like powers of x, we find that

$$\sigma = -\frac{\lambda p_1 + q_1}{2\lambda + p_0} \tag{7.1.10c}$$

and we arrive at a recurrence relation for c_n.

Jacobi (1849) developed normal solutions for Bessel functions of large argument and first order, while Stokes (1857) developed them for Airy's equation. Horn (1903) gave justification for the asymptotic solutions in the form of products of exponentials and descending series in x.

7.1.2. AN EXPANSION OF THE ZEROTH-ORDER BESSEL FUNCTION FOR LARGE ARGUMENT

The zeroth-order Bessel function is given by

$$\frac{d^2 y}{dx^2} + \frac{1}{x}\frac{dy}{dx} + y = 0 \tag{7.1.11}$$

Here

$$p_1 = 1 \quad \text{and} \quad p_m = 0 \quad \text{for} \quad m \neq 1$$
$$q_0 = 1 \quad \text{and} \quad q_m = 0 \quad \text{for} \quad m \neq 0$$

Hence (7.1.10) indicates that

$$y = e^{ix} x^{-1/2} \sum_{m=0}^{\infty} c_m x^{-m} \quad \text{as} \quad x \to \infty \tag{7.1.12}$$

Substituting this expansion into (7.1.11) and equating like powers of x, we obtain the following recurrence relationship

$$c_{m+1} = -i\frac{(m + \frac{1}{2})^2}{2(m + 1)} c_m \tag{7.1.13}$$

Hence if we take $c_0 = 1$

$$y = e^{ix} x^{-1/2}\left[1 - \frac{1}{4 \cdot 2x} i - \frac{1 \cdot 3^2}{4^2 \cdot 2^2 \cdot 2! \cdot x^2} + \frac{1 \cdot 3^2 \cdot 5^2}{4^3 \cdot 2^3 \cdot 3! \cdot x^3} i \right.$$
$$\left. + \frac{1 \cdot 3^2 \cdot 5^2 \cdot 7^2}{4^4 \cdot 2^4 \cdot 4! \cdot x^4} + \cdots \right] \quad \text{as} \quad x \to \infty \tag{7.1.14}$$

Since the ratio of two successive terms

$$-\frac{i(2m+1)^2}{8(m+1)x} \to \infty \quad \text{as} \quad m \to \infty$$

the right-hand side of (7.1.14) diverges for all values of x. However, for large x, it is an asymptotic expansion because the leading terms diminish very rapidly as m increases.

Another linear independent expansion $\tilde{y}$ can be obtained by replacing i by $-i$ in (7.1.14); that is

$$\tilde{y} = e^{-ix}x^{-1/2}\left(1 + \frac{1}{4\cdot 2x}i - \frac{1\cdot 3^2}{4^2\cdot 2^2\cdot 2!\cdot x^2} - \frac{1\cdot 3^2\cdot 5^2}{4^3\cdot 2^3\cdot 3!\cdot x^3}i\right.$$
$$\left. + \frac{1\cdot 3^2\cdot 5^2\cdot 7^2}{4^4\cdot 2^4\cdot 4!\cdot x^4} + \cdots\right) \quad (7.1.15)$$

Real solutions can be obtained by linearly combining (7.1.14) and (7.1.15) according to

$$y_1 \sim \frac{y + \tilde{y}}{2} = x^{-1/2}(u\cos x + v\sin x)$$
$$y_2 \sim \frac{y - \tilde{y}}{2i} = x^{-1/2}(u\sin x - v\cos x) \quad (7.1.16)$$

where

$$u(x) = 1 - \frac{1\cdot 3^2}{4^2\cdot 2^2\cdot 2!\cdot x^2} + \frac{1\cdot 3^2\cdot 5^2\cdot 7^2}{4^4\cdot 2^4\cdot 4!\cdot x^4} + \cdots \quad (7.1.17a)$$

$$v(x) = \frac{1}{4\cdot 2x} - \frac{1\cdot 3^2\cdot 5^2}{4^3\cdot 2^3\cdot 3!\cdot x^3} + \cdots \quad (7.1.17b)$$

Therefore the asymptotic form of the Bessel function J_0 is given by

$$J_0 \sim Ay_1 + By_2 \text{ as } x \to \infty \quad (7.1.18)$$

where A and B are constants. From (7.1.16) through (7.1.18)

$$\lim_{x\to\infty} x^{1/2}J_0(x) = A\cos x + B\sin x \quad (7.1.19)$$

$$\lim_{x\to\infty} x^{1/2}J_0'(x) = -A\sin x + B\cos x \quad (7.1.20)$$

Hence

$$A = \lim_{x\to\infty} x^{1/2}[J_0(x)\cos x - J_0'(x)\sin x] \quad (7.1.21)$$

$$B = \lim_{x\to\infty} x^{1/2}[J_0(x)\sin x + J_0'(x)\cos x] \quad (7.1.22)$$

However, J_0 has the integral representation (e.g., Ince, 1926, Section 8.22)

$$J_0 = \frac{1}{\pi} \int_0^\pi \cos (x \cos \theta) \, d\theta \qquad (7.1.23)$$

Substituting this expression into (7.1.21), we obtain

$$A = \lim_{x \to \infty} \frac{x^{1/2}}{\pi} \int_0^\pi [\cos x \cos (x \cos \theta) + \sin x \cos \theta \sin (x \cos \theta)] \, d\theta$$

$$= \lim_{x \to \infty} \frac{x^{1/2}}{\pi} \int_0^\pi \cos \left(2x \sin^2 \frac{\theta}{2}\right) \cos^2 \frac{\theta}{2} \, d\theta$$

$$+ \lim_{x \to \infty} \frac{x^{1/2}}{\pi} \int_0^\pi \cos \left(2x \cos^2 \frac{\theta}{2}\right) \sin^2 \frac{\theta}{2} \, d\theta$$

If we let $\sqrt{2x} \sin \theta/2 = \phi$ in the first integral, and $\sqrt{2x} \cos \theta/2 = \alpha$ in the second integral, we obtain

$$A = \frac{\sqrt{2}}{\pi} \left[\int_0^\infty \cos \phi^2 \, d\phi + \int_0^\infty \cos \alpha^2 \, d\alpha \right] = \frac{1}{\sqrt{\pi}}$$

Similarly, we find that $B = 1/\sqrt{\pi}$. Therefore combining (7.1.16) and (7.1.18), we obtain

$$J_0 = \sqrt{\frac{2}{\pi x}} [u \cos (x - \tfrac{1}{4}\pi) + v \sin (x - \tfrac{1}{4}\pi)] \quad \text{as} \quad x \to \infty \quad (7.1.24)$$

7.1.3.   LIOUVILLE'S PROBLEM

Liouville (1837) and Green (1837) simultaneously considered the behavior of the solutions of

$$\frac{d^2y}{dx^2} + [\lambda^2 q_1(x) + q_2(x)]y = 0 \qquad (7.1.25)$$

for large λ, where q_1 is a positive and twice continuously differentiable function and $q_2(x)$ is continuous in the interval $[a, b]$ of interest. Using the transformation

$$z = \phi(x), \qquad v = \psi(x)y(x) \qquad (7.1.26)$$

we change (7.1.25) into

$$\frac{d^2v}{dz^2} + \frac{1}{\phi'^2}\left(\phi'' - \frac{2\phi'\psi'}{\psi}\right)\frac{dv}{dz} + \frac{1}{\phi'^2}\left[\lambda^2 q_1(x) + q_2(x) - \psi\left(\frac{\psi'}{\psi^2}\right)'\right]v = 0 \quad (7.1.27)$$

Choosing ϕ and ψ such that

$$\phi'' - \frac{2\phi'\psi'}{\psi} = 0 \quad \text{and} \quad q_1 = \phi'^2 \qquad (7.1.28)$$

or

$$\phi = \int [q_1(x)]^{1/2}\, dx, \qquad \psi = [q_1(x)]^{1/4} \qquad (7.1.29)$$

we reduce (7.1.27) to

$$\frac{d^2v}{dz^2} + \lambda^2 v = \delta v \qquad (7.1.30)$$

where

$$\delta = \frac{1}{4}\frac{q_1''}{q_1{}^2} - \frac{5}{16}\frac{q_1'^2}{q_1{}^3} - \frac{q_2}{q_1} \qquad (7.1.31)$$

Since q_1 is twice continuously differentiable and q_2 is continuous in $[a, b]$, δ is small compared to λ^2. Hence, to first approximation, v is the solution of (7.1.30) with $\delta = 0$; that is

$$v = a \cos \lambda z + b \sin \lambda z \qquad (7.1.32)$$

where a and b are constants. Therefore, to first approximation

$$y = \frac{a \cos \left[\lambda \int \sqrt{q_1(x)}\, dx \right] + b \sin \left[\lambda \int \sqrt{q_1(x)}\, dx \right]}{\sqrt[4]{q_1(x)}} \qquad (7.1.33)$$

If $q_1(x)$ is negative rather than positive, (7.1.33) is replaced by

$$y = \frac{a \exp \left[\lambda \int \sqrt{-q_1(x)}\, dx \right] + b \exp \left[-\lambda \int \sqrt{-q_1(x)}\, dx \right]}{\sqrt[4]{-q_1(x)}} \qquad (7.1.34)$$

These expansions are in full agreement with those obtained in Section 6.4.3 using the method of multiple scales. It should be mentioned that these expansions break down at or near the zeros of $q_1(x)$. These zeros are called turning or transition points. Turning point problems are discussed in Section 7.3.

The transformation (7.1.26) and (7.1.29) is called the Liouville–Green transformation by mathematicians, while the solutions (7.1.33) and (7.1.34) are called the WKB approximation by physicists after Wentzel (1926), Kramers (1926), and Brillouin (1926). However, an approximate solution of this kind was obtained by Carlini (1817) for Bessel functions when the order and argument are both large.

7.1.4. HIGHER APPROXIMATIONS FOR EQUATIONS CONTAINING A LARGE PARAMETER

We consider the asymptotic development of the solutions of

$$\frac{d^2y}{dx^2} + q(x, \lambda)y = 0 \qquad (7.1.35)$$

for large λ when

$$q(x, \lambda) = \lambda^{2k} \sum_{n=0}^{\infty} \lambda^{-n} q_n(x) \quad \text{as} \quad \lambda \to \infty \tag{7.1.36}$$

with $q_0 \neq 0$ in the interval of interest and k a positive integer. The asymptotic solution of this problem can be sought in either of the following two formal expansions

$$y = \left[\sum_{n=0}^{\infty} \lambda^{-n} a_n(x) \right] \exp \left[\lambda^k \sum_{n=0}^{k-1} \lambda^{-n} g_n(x) \right] \tag{7.1.37}$$

$$y = \exp \left[\lambda^k \sum_{n=0}^{\infty} \lambda^{-n} g_n(x) \right] \tag{7.1.38}$$

The justification of these expansions was given by Horn (1899). Substituting either of these formal expansions into (7.1.35) and (7.1.36) and equating coefficients of like powers of λ, we obtain equations for the successive determination of a_n and g_n.

In the case $k = 1$ and the formal expansion (7.1.37), these equations are

$$g_0'^2 + q_0 = 0 \tag{7.1.39}$$

$$2g_0'a_0' + (q_1 + g_0'')a_0 = 0 \tag{7.1.40}$$

and

$$2g_0'a_m' + (q_1 + g_0'')a_m + \sum_{s=1}^{m} a_{m-s}q_{s+1} + a_{m-1}'' = 0 \quad \text{for} \quad m \geq 1 \tag{7.1.41}$$

The solution of (7.1.39) is

$$g_0 = \pm i \int [q_0(x)]^{1/2} \, dx \tag{7.1.42}$$

Then the solution of (7.1.40) is

$$a_0 \propto [g_0']^{-1/2} \exp \left[-\int \frac{q_1(x)}{2g_0'(x)} \, dx \right]$$

which can be rewritten as

$$a_0 = c[q_0(x)]^{-1/4} \exp \pm \left[\frac{i}{2} \int \frac{q_1(x)}{[q_0(x)]^{1/2}} \, dx \right] \tag{7.1.43}$$

where c is a constant. Hence, to first approximation

$$y = \frac{c_1 \cos \beta(x) + c_2 \sin \beta(x)}{[q_0(x)]^{1/4}} [1 + O(\lambda^{-1})] \tag{7.1.44}$$

where c_1 and c_2 are constants and

$$\beta(x) = \lambda \int^x [q_0(x)]^{1/2} \left[1 + \frac{q_1(x)}{2\lambda q_0(x)} \right] dx \tag{7.1.45}$$

Higher approximations can be obtained by solving successively for a_m from (7.1.41).

Had we used the second formal expansion (7.1.38) instead of (7.1.37) for the same case $k = 1$, we would have obtained the following equations for the determination of g_m

$$g_0'^2 + q_0 = 0 \tag{7.1.46}$$

$$2g_0'g_m' + q_m + \sum_{s=1}^{m-1} g_s'g_{m-s}' + g_{m-1}'' = 0 \quad \text{for} \quad m \geq 1 \tag{7.1.47}$$

The solution of (7.1.46) is given by (7.1.42), and for $m = 1$ (7.1.47) becomes

$$2g_0'g_1' + q_1 + g_0'' = 0$$

whose solution is

$$g_1 = - \int \frac{q_1(x)}{2g_0'(x)} \, dx - \tfrac{1}{2} \ln g_0' \tag{7.1.48}$$

Substituting for g_0 and g_1 into (7.1.38), we obtain exactly (7.1.44) and (7.1.45).

7.1.5.   A SMALL PARAMETER MULTIPLYING THE HIGHEST DERIVATIVE

In this section we consider the equation

$$\epsilon \frac{d^2y}{dx^2} + p(x) \frac{dy}{dx} + q(x)y = r(x) \tag{7.1.49}$$

for $\epsilon \to 0$. Asymptotic expansions of the solutions of this equation have been obtained in Section 4.1.3 by using the method of matched asymptotic expansions, in Section 4.2.2 by using the method of composite expansions, and in Section 6.4.2 by using the method of multiple scales.

Using the transformation (Goldstein, 1969)

$$y = v(x) \exp\left[- \frac{M(x)}{2\epsilon} \right] \quad \text{with} \quad M = \int_{x_0}^{x} p(x) \, dx \tag{7.1.50}$$

we transform (7.1.49) into the normal form

$$\frac{d^2v}{dx^2} - \left(\frac{p^2}{4\epsilon^2} + \frac{p' - 2q}{2\epsilon} \right) v = \frac{r}{\epsilon} e^{M/2\epsilon} \tag{7.1.51}$$

For $r \equiv 0$, (7.1.51) has the same form as (7.1.35) and (7.1.36), with $k = 1$, $\lambda = 1/2\epsilon$, $q_0 = -p^2$ and $q_1 = 2q - p'$. Hence the complementary solution of (7.1.51) is

$$v_c = \frac{c_1 e^{\beta(x)} + c_2 e^{-\beta(x)}}{\sqrt{p(x)}} [1 + O(\epsilon)] \tag{7.1.52}$$

where

$$\beta(x) = \frac{1}{2\epsilon} \int_{x_0}^{x} p(x) \left[1 + \epsilon \frac{p' - 2q}{p^2} \right] dx = \frac{M}{2\epsilon} + \tfrac{1}{2} \ln p - \int_{x_0}^{x} \frac{q(x)}{p(x)} dx \quad (7.1.53)$$

An approximate particular integral to (7.1.49) can by obtained by putting $\epsilon = 0$; that is

$$y_p = \frac{1}{E} \int_{x_0}^{x} \frac{Er}{p} dx \quad \text{with} \quad E = \exp \int_{x_0}^{x} \frac{q}{p} dx \quad (7.1.54)$$

Hence to first approximation

$$y = \frac{c_1}{E} + \frac{c_2 E}{p} e^{-M/\epsilon} + \frac{1}{E} \int_{x_0}^{x} \frac{Er}{p} dx \quad (7.1.55)$$

Higher approximations can be obtained by assuming an expansion of the form

$$y = \sum_{n=0}^{\infty} \epsilon^n A_n(x) e^{-M/\epsilon} + \sum_{n=0}^{\infty} \epsilon^n B_n(x) \quad (7.1.56)$$

where

$$A_0 = \frac{c_2 E}{p}, \qquad B_0 = \frac{c_1}{E} + \frac{1}{E} \int_{x_0}^{x} \frac{Er}{p} dx \quad (7.1.57)$$

This form is the same as that assumed when we employed the method of composite expansions in Section 4.2.2 where we determined M, as well as A_n and B_n, by substituting this expansion into the original equation and equating the coefficients of ϵ^n and $\epsilon^n \exp(-M/\epsilon)$ to zero.

For the case $p = p(x, \epsilon)$ and $q = q(x, \epsilon)$, Wasow (1965, Chapter 7) assumed an asymptotic expansion of the form

$$y = \sum_{n=0}^{\infty} A_n(x) \exp \left[\int \lambda_1(x, \epsilon) \, dx \right] + \sum_{n=0}^{\infty} B_n(x) \exp \left[\int \lambda_2(x, \epsilon) \, dx \right] + \sum_{n=0}^{\infty} \epsilon^n C_n(x)$$

$$(7.1.58)$$

where λ_1 and λ_2 are the roots of

$$\epsilon \lambda^2 + p(x, \epsilon)\lambda + q(x, \epsilon) = 0 \quad \text{as} \quad \epsilon \to 0 \quad (7.1.59)$$

Thus if p and q are independent of ϵ, $\lambda_1 = -p/\epsilon$ and $\lambda_2 = 0$, and (7.1.58) takes the form (7.1.56).

7.1.6. HOMOGENEOUS PROBLEMS WITH SLOWLY VARYING COEFFICIENTS

In this section we consider the equation

$$\frac{d^2 y}{dx^2} + p(\epsilon x; \epsilon) \frac{dy}{dx} + q(\epsilon x; \epsilon) y = 0 \quad (7.1.60)$$

where ϵ is a small parameter, and

$$p = \sum_{n=0}^{\infty} \epsilon^n p_n(\xi), \qquad q = \sum_{n=0}^{\infty} \epsilon^n q_n(\xi) \quad \text{with} \quad \xi = \epsilon x \qquad (7.1.61)$$

An asymptotic expansion of the general solution of (7.1.60) has the form (see Feshchenko, Shkil', and Nikolenko, 1967, for history and references)

$$y = \sum_{n=0}^{\infty} \epsilon^n A_n(\xi)e^{\theta_1} + \sum_{n=0}^{\infty} \epsilon^n B_n(\xi)e^{\theta_2} \qquad (7.1.62)$$

where

$$\frac{d\theta_i}{dx} = \lambda_i(\xi) \qquad (7.1.63)$$

and λ_1 and λ_2 are the roots of

$$\lambda^2 + p_0(\xi)\lambda + q_0(\xi) = 0 \qquad (7.1.64)$$

We have assumed that λ_1 and λ_2 are different in the interval of interest. In (7.1.62), θ_i and ξ are assumed to be independent. This is equivalent to the method of multiple scales described in the previous chapter. The derivatives are transformed according to

$$\frac{d}{dx} = \lambda_1 \frac{\partial}{\partial \theta_1} + \lambda_2 \frac{\partial}{\partial \theta_2} + \epsilon \frac{\partial}{\partial \xi}$$

$$\frac{d^2}{dx^2} = \lambda_1^2 \frac{\partial^2}{\partial \theta_1^2} + 2\lambda_1\lambda_2 \frac{\partial^2}{\partial \theta_1 \partial \theta_2} + \lambda_2^2 \frac{\partial^2}{\partial \theta_2^2}$$

$$+ 2\epsilon\lambda_1 \frac{\partial^2}{\partial \theta_1 \partial \xi} + 2\epsilon\lambda_2 \frac{\partial^2}{\partial \theta_2 \partial \xi} + \epsilon\lambda_1' \frac{\partial}{\partial \theta_1} + \epsilon\lambda_2' \frac{\partial}{\partial \theta_2} + \epsilon^2 \frac{\partial^2}{\partial \xi^2}$$

where $\lambda_i' = d\lambda_i/d\xi$.

Let A and B denote the coefficients of $\exp(\theta_1)$ and $\exp(\theta_2)$. Substituting (7.1.62) into (7.1.60) and equating the coefficients of $\exp(\theta_i)$ to zero, we obtain

$$(\lambda_1^2 + \lambda_1 p + q)A + \epsilon(2\lambda_1 + p)A' + \epsilon\lambda_1' A + \epsilon^2 A'' = 0 \qquad (7.1.65)$$

$$(\lambda_2^2 + \lambda_2 p + q)B + \epsilon(2\lambda_2 + p)B' + \epsilon\lambda_2' B + \epsilon^2 B'' = 0 \qquad (7.1.66)$$

Letting again

$$A = \sum_{n=0}^{\infty} \epsilon^n A_n \quad \text{and} \quad B = \sum_{n=0}^{\infty} \epsilon^n B_n \qquad (7.1.67)$$

in (7.1.65) and (7.1.66) and equating coefficients of like powers of ϵ, we obtain equations to determine successively A_n and B_n. The first terms A_0 and B_0 are given by

$$(2\lambda_1 + p_0)A_0' + (\lambda_1' + \lambda_1 p_1 + q_1)A_0 = 0 \qquad (7.1.68)$$

$$(2\lambda_2 + p_0)B_0' + (\lambda_2' + \lambda_2 p_1 + q_1)B_0 = 0 \qquad (7.1.69)$$

Their solutions are

$$A_0, B_0 \propto \exp - \int^\xi \frac{\lambda'_i + \lambda_i p_1 + q_1}{2\lambda_i + p_0} \, d\xi \qquad (7.1.70)$$

In the case of $p \equiv 0$ and $q_n = 0$ for $n \geq 1$

$$\lambda_1, \lambda_2 = \pm i [q_0(\xi)]^{1/2}$$

and

$$A_0 = \frac{a}{\sqrt{\lambda_1}}, \qquad B_0 = \frac{b}{\sqrt{\lambda_1}}$$

where a and b are constants. Therefore, to first approximation

$$y = \frac{c_1 \cos \int [q_0(\xi)]^{1/2} \, dx + c_2 \sin \int [q_0(\xi)]^{1/2} \, dx}{[q_0(\xi)]^{1/4}} \qquad (7.1.71)$$

This is the Liouville–Green or WKB approximation to the solution of

$$\frac{d^2 y}{dx^2} + q_0(\epsilon x) y = 0 \qquad (7.1.72)$$

7.1.7.   REENTRY MISSILE DYNAMICS

For a symmetric missile the complex angle of attack is given by (Nayfeh, 1969a)

$$\ddot{\delta} + \left[ip(1 + \gamma) + \frac{\dot{u}}{u} + \frac{\epsilon Q}{u} - \epsilon D \right] \dot{\delta}$$

$$+ \left[\gamma(p_c^2 - p^2) + ip \left(\frac{\gamma \dot{u}}{u} - \epsilon D - \epsilon M + \frac{\epsilon \gamma Q}{u} \right) \right] \delta = 0 \quad (7.1.73)$$

In this equation u, Q, D, and M are slowly varying functions of time, while p and γ are constants. Hence it has the same form as (7.1.60), with

$$p_0 = ip(1 + \gamma), \qquad p_1 = \frac{u'}{u} + \frac{Q}{u} - D$$

$$q_0 = \gamma(p_c^2 - p^2), \qquad q_1 = ip \left(\frac{\gamma u'}{u} - D - M + \frac{\gamma Q}{u} \right) \qquad (7.1.74)$$

Here primes denote differentiation with respect to the slow time $\xi = \epsilon t$, while overdots denote differentiation with respect to the fast time t. Substituting for p_0 and q_0 into (7.1.64) gives

$$\lambda^2 + ip(1 + \gamma)\lambda + \gamma(p_c^2 - p^2) = 0 \qquad (7.1.75)$$

Hence

$$\lambda = -\tfrac{1}{2}i(1 + \gamma)p \pm i\omega, \qquad \omega = \sqrt{\tfrac{1}{4}(1 - \gamma)^2 p^2 + \gamma p_c^2} \qquad (7.1.76)$$

Then (7.1.70) gives

$$A_0 = \frac{a}{\sqrt{\omega u}}\, e^{-\Lambda + \Delta\Lambda}, \qquad B_0 = \frac{b}{\sqrt{\omega u}}\, e^{-\Lambda - \Delta\Lambda} \qquad (7.1.77)$$

where

$$\Lambda = \frac{1}{2}\int (Q/u - D)\, d\xi, \qquad \Delta\Lambda = \frac{p}{4}\int \frac{\left(\dfrac{u'}{u} + Q - \gamma D\right)(1 - \gamma) + 2M}{\omega}\, d\xi \qquad (7.1.78)$$

Therefore, to first approximation

$$\delta = \frac{1}{\sqrt{\omega u}}\{a \exp\left[-\Lambda + \Delta\Lambda - \tfrac{1}{2}i(1 + \gamma)p + i\omega\right]$$

$$+ b \exp\left[-\Lambda - \Delta\Lambda - \tfrac{1}{2}i(1 + \gamma)p - i\omega\right]\} \qquad (7.1.79)$$

Analysis of missile dynamics was performed by Fowler et al. (1920), Fowler and Lock (1921), Green and Weaver (1961), Murphy (1963), and Coakley (1968) among others.

7.1.8. INHOMOGENEOUS PROBLEMS WITH SLOWLY VARYING COEFFICIENTS

We consider in this section the asymptotic development of the general solution of

$$\frac{d^2 y}{dx^2} + p(\xi, \epsilon)\frac{dy}{dx} + q(\xi, \epsilon)y = r(\xi, \epsilon)e^{i\phi(x,\epsilon)} \qquad (7.1.80)$$

where

$$\frac{d\phi}{dx} = \omega(\xi) \quad \text{with} \quad \xi = \epsilon x \qquad (7.1.81)$$

and

$$p = \sum_{n=0}^{\infty} \epsilon^n p_n(\xi), \qquad q = \sum_{n=0}^{\infty} \epsilon^n q_n(\xi), \qquad r = \sum_{n=0}^{\infty} \epsilon^n r_n(\xi) \qquad (7.1.82)$$

Two cases arise depending on the values of $i\omega$ and the roots λ_1 and λ_2 of

$$\lambda^2 + p_0\lambda + q_0 = 0 \qquad (7.1.83)$$

If $i\omega = \lambda_1$ or λ_2 at one or more points in the interval of interest, we have a resonant case; otherwise we have a nonresonant case. We treat the latter case first.

The Nonresonant Case. In this case we assume that

$$y = A(\xi, \epsilon)e^{\theta_1} + B(\xi, \epsilon)e^{\theta_2} + C(\xi, \epsilon)e^{i\phi} \qquad (7.1.84)$$

where

$$\frac{d\theta_i}{dx} = \lambda_i(\xi) \tag{7.1.85}$$

The equations for A and B are the same as (7.1.65) and (7.1.66). To determine C, we let $y = C \exp(i\phi)$ in (7.1.80), equate the coefficients of $\exp(i\phi)$ on both sides, and obtain

$$(-\omega^2 + i\omega p + q)C + \epsilon(2i\omega + p)C' + i\epsilon\omega'C + \epsilon^2 C'' = r \tag{7.1.86}$$

Letting

$$C = \sum_{n=0}^{\infty} \epsilon^n c_n(\xi) \tag{7.1.87}$$

in (7.1.86) and equating coefficients of like powers of ϵ, we obtain equations to determine c_n successively. The first one is given by

$$c_0 = \frac{r_0}{-\omega^2 + i\omega p_0 + q_0} = \frac{r_0}{(i\omega - \lambda_1)(i\omega - \lambda_2)} \tag{7.1.88}$$

The solution for A and B is the same as in Section 7.1.6. Therefore, to first approximation

$$
\begin{aligned}
y = {} & a \exp\left[\theta_1 - \int \frac{\lambda_1' + \lambda_1 p_1 + q_1}{2\lambda_1 + p_0}\, d\xi\right] \\
& + b \exp\left[\theta_2 - \int \frac{\lambda_2' + \lambda_2 p_1 + q_1}{2\lambda_1 + p_0}\, d\xi\right] \\
& + \frac{r_0}{(i\omega - \lambda_1)(i\omega - \lambda_2)}\, e^{i\phi}
\end{aligned}
\tag{7.1.89}
$$

where a and b are constants.

The Resonant Case. The expansion (7.1.89) breaks down whenever $i\omega$ is equal to either λ_1 or λ_2 at one or more points because the last term is unbounded at such points. We assume that $i\omega$ is equal to λ_1 at one or more points while $i\omega \neq \lambda_2$ in the interval of interest. Asymptotic expansions valid in this case have been developed by Fowler et al. (1920) and Fowler and Lock (1921).

The particular solution in this case has the form

$$y = \eta(x, \epsilon)e^{i\phi} \tag{7.1.90}$$

where

$$\frac{d\eta}{dx} = [G(\xi, \epsilon) - i\omega]\eta + H(\xi, \epsilon) \tag{7.1.91}$$

Substituting (7.1.90) and (7.1.91) into (7.1.80) and equating the coefficients of each of $\eta \exp i\phi$ and $\exp i\phi$ on both sides, we obtain

$$G^2 + pG + q + \epsilon G' = 0 \tag{7.1.92}$$

$$(G + p + i\omega)H + \epsilon H' = r \tag{7.1.93}$$

Letting

$$G = \sum_{n=0}^{\infty} \epsilon^n G_n(\xi) \quad \text{and} \quad H = \sum_{n=0}^{\infty} \epsilon^n H_n(\xi) \tag{7.1.94}$$

in (7.1.92) and (7.1.93) and equating coefficients of like powers of ϵ, we obtain equations for the successive determination of the G_n and H_n. The first two terms are given by

$$G_0{}^2 + p_0 G_0 + q_0 = 0 \tag{7.1.95}$$

$$2G_0 G_1 + p_0 G_1 + p_1 G_0 + q_1 + G_0' = 0 \tag{7.1.96}$$

$$(G_0 + p_0 + i\omega)H_0 = r_0 \tag{7.1.97}$$

$$(G_0 + p_0 + i\omega)H_1 + (G_1 + p_1)H_0 + H_0' = r_1 \tag{7.1.98}$$

Equation (7.1.95) shows that $G_0 = \lambda_1$ or λ_2. We take $G_0 = \lambda_1$ because λ_1 is assumed to be equal to $i\omega$ at one or more points. Equation (7.1.96) gives

$$G_1 = -\frac{p_1 G_0 + G_0' + q_1}{2G_0 + p_0} \tag{7.1.99}$$

The solutions of (7.1.97) and (7.1.98) are

$$H_0 = \frac{r_0}{(G_0 + p_0 + i\omega)}, \quad H_1 = \frac{r_1 - (G_1 + p_1)H_0 - H_0'}{G_0 + p_0 + i\omega} \tag{7.1.100}$$

The general solution of (7.1.81) is then

$$y = A(\xi, \epsilon)e^{\theta_1} + B(\xi, \epsilon)e^{\theta_2} + \eta(x, \epsilon)e^{i\phi} \tag{7.1.101}$$

where A and B are determined as in Section 7.1.6.

The complementary solution of (7.1.80) can be obtained by using the following technique which is an alternative to that used in Section 7.1.6. We assume that

$$y_c = \zeta(x, \epsilon) \quad \text{with} \quad \frac{d\zeta}{dx} = F(\xi, \epsilon)\zeta \tag{7.1.102}$$

Substituting (7.1.102) into the homogeneous part of (7.1.80), we obtain

$$F^2 + pF + q + \epsilon F' = 0 \tag{7.1.103}$$

Letting

$$F = \sum_{n=0}^{\infty} \epsilon^n F_n(\xi) \tag{7.1.104}$$

in (7.1.103) and equating coefficients of like powers of ϵ, we obtain equations to determine F_n. The first two are

$$F_0{}^2 + p_0 F_0 + q_0 = 0 \qquad (7.1.105)$$

$$2F_0 F_1 + p_0 F_1 + p_1 F_0 + q_1 + F_0' = 0 \qquad (7.1.106)$$

Equation (7.1.105) has two roots, λ_1 and λ_2, as (7.1.64). Then (7.1.106) gives

$$F_1 = -\frac{p_1 \lambda_i + q_1 + \lambda_i'}{2\lambda_i + p_0} \qquad (7.1.107)$$

Hence

$$\frac{d\zeta}{dx} = \left[\lambda_i - \epsilon \frac{p_1 \lambda_i + q_1 + \lambda_i'}{2\lambda_i + p_0} + O(\epsilon^2)\right]\zeta \qquad (7.1.108)$$

Integrating (7.1.108), we find that y_c is the same as that obtained in Section 7.1.6. This method of expansion is the same as that represented by (7.1.38).

7.1.9. SUCCESSIVE LIOUVILLE–GREEN (WKB) APPROXIMATIONS

To obtain higher approximations to the solution of

$$\frac{d^2 y}{dx^2} + k^2(\epsilon x) y = 0 \qquad (7.1.109)$$

for small ϵ, Imai (1948) proposed the use of successive Liouville–Green (WKB) transformations. Thus we introduce the transformation

$$dx_1 = k(\epsilon x)\, dx, \qquad y_1 = [k(\epsilon x)]^{1/2} y(x) \qquad (7.1.110)$$

thereby transforming (7.1.109) into

$$\frac{d^2 y_1}{dx_1{}^2} + k_1{}^2 y_1 = 0 \qquad (7.1.111)$$

where

$$k_1{}^2 = 1 - \frac{1}{2k^3}\frac{d^2 k}{dx^2} + \frac{3}{4k^4}\left(\frac{dk}{dx}\right)^2 \qquad (7.1.112)$$

Since k varies slowly with x, $k_1{}^2 \approx 1$, and an approximate solution to (7.1.111) is

$$y_1 = a \cos x_1 + b \sin x_1 \qquad (7.1.113)$$

with constant a and b. Hence a first approximation to the solution of (7.1.109) is

$$y = \frac{a \cos \displaystyle\int k\, dx + b \sin \displaystyle\int k\, dx}{k^{1/2}} \qquad (7.1.114)$$

To determine a second approximation to y, we note that (7.1.111) has the same form as (7.1.109). Thus an improved solution to (7.1.111) can be obtained by introducing the transformation

$$dx_2 = k_1 \, dx_1 \approx \left[1 - \frac{1}{4k^3} \frac{d^2k}{dx^2} + \frac{3}{8k^4}\left(\frac{dk}{dx}\right)^2 \right] k \, dx \qquad (7.1.115)$$

$$y_2 = y_1\sqrt{k_1} \approx y_1\left[1 - \frac{1}{8k^3} \frac{d^2k}{dx^2} + \frac{3}{16k^4}\left(\frac{dk}{dx}\right)^2 \right] \qquad (7.1.116)$$

Then (7.1.111) is transformed into

$$\frac{d^2y_2}{dx_2^{\,2}} + k_2^{\,2} y_2 = 0 \qquad (7.1.117)$$

where

$$k_2^{\,2} = 1 - \frac{1}{2k_1^{\,3}} \frac{d^2k_1}{dx_1^{\,2}} + \frac{3}{4k_1^{\,4}}\left(\frac{dk_1}{dx_1}\right)^2 \qquad (7.1.118)$$

Since $dk_1/dx_1 = k^{-1}(dk_1/dx) = O(\epsilon)$, the last two terms in (7.1.118) are small compared to 1, hence a first approximation to y_2 is

$$y_2 = a \cos x_2 + b \sin x_2 \qquad (7.1.119)$$

Therefore a second approximation to y is given by

$$y = \frac{1 + \frac{1}{8k^3} \frac{d^2k}{dx^2} - \frac{3}{16k^4}\left(\frac{dk}{dx}\right)^2}{k^{1/2}} (a \cos x_2 + b \sin x_2) \qquad (7.1.120)$$

where

$$x_2 = \int \left[1 - \frac{1}{4k^3} \frac{d^2k}{dx^2} + \frac{3}{8k^4}\left(\frac{dk}{dx}\right)^2 \right] k \, dx \qquad (7.1.121)$$

Higher approximations can be obtained in the same manner by introducing the new transformations

$$dx_{n+1} = k_n \, dx_n, \qquad y_{n+1} = \sqrt{k_n}\, y_n \qquad (7.1.122)$$

7.2. Systems of First-Order Ordinary Equations

In this section we also consider first the asymptotic solutions of equations near an irregular singular point which is assumed to be ∞. Then we discuss equations with either a small or a large parameter. Finally, we describe asymptotic expansions for equations with slowly varying coefficients.

7.2.1. EXPANSIONS NEAR AN IRREGULAR SINGULAR POINT

We consider the behavior of the system of n linear equations

$$\frac{d\mathbf{y}}{dx} = x^q A(x)\mathbf{y} \tag{7.2.1}$$

as $x \to \infty$, where q is an integer ≥ -1, and the matrix

$$A(x) = \sum_{m=0}^{\infty} A_m x^{-m} \quad \text{as} \quad x \to \infty \tag{7.2.2}$$

If $q = -1$, $x = \infty$ is a regular singular point of (7.2.1), while if $q > -1$, $x = \infty$ is an irregular singular point of the system. The behavior of the solution near an irregular singularity depends on whether all the eigenvalues of A_0 are distinct or not. We discuss in this section the case of distinct eigenvalues.

If (7.2.1) is a scalar equation, it can be solved explicitly and its solution has the form

$$y(x) = U(x)x^G e^{Q(x)} \tag{7.2.3}$$

where G is generally a complex constant, $Q(x)$ is a polynomial of x having the form

$$Q(x) = \begin{cases} 0 & \text{if} \quad q = -1 \\ \sum_{m=1}^{q+1} Q_m x^m \end{cases} \tag{7.2.4}$$

and

$$U(x) = \sum_{m=0}^{\infty} U_m x^{-m} \tag{7.2.5}$$

In the case of systems of equations, the asymptotic solution still has the form (7.2.3) but G, Q_m, and U_m are constant matrices. This expansion was called a "normal" solution by Thomé (1883).

To calculate the asymptotic expansions for the solutions of (7.2.1) and (7.2.2), we seek formal solutions of the form

$$\mathbf{y} = \mathbf{u}(x)x^\sigma e^{\Lambda(x)} \tag{7.2.6}$$

where σ is a constant,

$$\Lambda(x) = \frac{\lambda_{q+1} x^{q+1}}{q+1} + \frac{\lambda_q x^q}{q} + \cdots + \lambda_1 x \tag{7.2.7}$$

with $\lambda_{-m} = 0$ for $m \geq 0$, and

$$\mathbf{u}(x) = \sum_{m=0}^{\infty} \mathbf{u}_m x^{-m} \quad \text{as} \quad x \to \infty \tag{7.2.8}$$

Here Λ is a scalar quantity, while $\mathbf{y}$ and $\mathbf{u}$ are column vectors. Substituting (7.2.6) through (7.2.8) into (7.2.1) and (7.2.2) and equating the coefficients of equal powers of x, we obtain equations to determine successively λ_n, σ, and $\mathbf{u}_n$. The first equation gives

$$(A_0 - \lambda_{q+1}I)\mathbf{u}_0 = 0, \qquad I\text{: identity matrix} \qquad (7.2.9)$$

For a nontrivial solution the determinant in (7.2.9) must vanish. This condition gives the following nth-order algebraic equation

$$|A_0 - \lambda_{q+1}I| = 0 \qquad (7.2.10)$$

If the eigenvalues of A_0 are distinct, (7.2.10) gives n distinct values for λ_{q+1} which correspond to n linear independent solutions of the form (7.2.6).

7.2.2. ASYMPTOTIC PARTITIONING OF SYSTEMS OF EQUATIONS

Sibuya (1958) developed the following scheme for simplifying the system of equations (7.2.1) by reducing them to some special differential equations whose solutions can be found more readily than those of the original system. To accomplish this we let

$$\mathbf{y} = P(x)\mathbf{v}(x) \qquad (7.2.11)$$

where P is an $n \times n$ nonsingular matrix to be determined and $\mathbf{v}$ is a column vector. Hence (7.2.1) is transformed into

$$\frac{d\mathbf{v}}{dx} = x^q B(x)\mathbf{v} \qquad (7.2.12)$$

where

$$B(x) = [P(x)]^{-1}\left[A(x)P(x) - x^{-q}\frac{dP(x)}{dx}\right]$$

or

$$\frac{dP(x)}{dx} = x^q[A(x)P(x) - P(x)B(x)] \qquad (7.2.13)$$

The essence of the technique is to choose the matrix $P(x)$ such that the matrix $B(x)$ has a canonical Jordan form. To do this we let

$$B = \sum_{m=0}^{\infty} B_m x^{-m} \quad \text{as} \quad x \to \infty$$

$$\qquad \qquad \qquad \qquad \qquad \qquad (7.2.14)$$

$$P = \sum_{m=0}^{\infty} P_m x^{-m} \quad \text{as} \quad x \to \infty$$

where B_m represents a Jordan canonical matrix.

Substituting (7.2.14) into (7.2.13) and equating coefficients of like powers of x, we obtain

$$A_0 P_0 - P_0 B_0 = 0 \qquad (7.2.15)$$

$$A_0 P_m - P_m B_0 = \sum_{s=0}^{m-1}(P_s B_{m-s} - A_{m-s}P_s) - (m - q - 1)P_{m-q-1} \qquad (7.2.16)$$

for $m \geq 1$, with $P_{m-q-1} = 0$ if $m - q - 1 < 0$. If A_0 has distinct eigenvalues, then P_0 can be chosen such that

$$B_0 = P_0^{-1} A_0 P_0 \tag{7.2.17}$$

is diagonal. We multiply (7.2.16) from the left with P_0^{-1} to obtain

$$B_0 W_m - W_m B_0 = B_m + F_m \tag{7.2.18}$$

where

$$W_m = P_0^{-1} P_m \tag{7.2.19}$$

$$F_m = -P_0^{-1} A_m P_0 + P_0^{-1} \sum_{s=1}^{m-1} (P_s B_{m-s} - A_{m-s} P_s) - (m - q - 1) W_{m-q-1} \tag{7.2.20}$$

If we denote the components of F_m by F_m^{ij}, we choose B_m such that

$$B_m^{ii} = -F_m^{ii}, \qquad B_m^{ij} = 0 \quad \text{for} \quad i \neq j \tag{7.2.21}$$

Since B_0 has distinct eigenvalues, (7.2.18) with (7.2.21) can be solved to determine W_m, hence P_m, from (7.2.19).

In the case of multiple eigenvalues, we can partition the system of equations into simpler systems using the same scheme. We assume that A_0 has multiple eigenvalues and that there exists a matrix P_0 such that

$$B_0 = P_0^{-1} A_0 P_0 = \begin{pmatrix} B_0^{11} & 0 \\ 0 & B_0^{22} \end{pmatrix} \tag{7.2.22}$$

where B_0^{11} has the eigenvalues λ_i $(i = 1, 2, \ldots, r)$ while B_0^{22} has the eigenvalues λ_j $(j = r + 1, r + 2, \ldots, n)$ such that $\lambda_i \neq \lambda_j$. Let us partition W_m and F_m according to

$$W_m = \begin{pmatrix} W_m^{11} & W_m^{12} \\ W_m^{21} & W_m^{22} \end{pmatrix}, \qquad F_m = \begin{pmatrix} F_m^{11} & F_m^{12} \\ F_m^{21} & F_m^{22} \end{pmatrix} \tag{7.2.23}$$

where W_m^{11} and F_m^{11} are $r \times r$ matrices while W_m^{22} and F_m^{22} are $(n - r) \times (n - r)$ matrices. We choose

$$W_m^{11} = W_m^{22} = 0 \quad \text{and} \quad B_m^{11} = -F_m^{11}, \qquad B_m^{22} = -F_m^{22} \tag{7.2.24}$$

Then (7.2.18) becomes

$$B_0^{11} W_m^{12} - W_m^{12} B_0^{22} = F_m^{12}$$
$$B_0^{22} W_m^{21} - W_m^{21} B_0^{11} = F_m^{21} \tag{7.2.25}$$

These equations can be solved uniquely for W_m^{12} and W_m^{21} because B_0^{11} and B_0^{22} do not have common eigenvalues.

As an example, we consider Bessel's equation

$$x^2 \frac{d^2y}{dx^2} + x \frac{dy}{dx} + (x^2 - n^2)y = 0 \tag{7.2.26}$$

Letting

$$y = u_1, \qquad \frac{dy}{dx} = u_2 \tag{7.2.27}$$

we transform (7.2.26) into

$$\frac{d\mathbf{u}}{dx} = A(x)\mathbf{u}$$

$$\mathbf{u} = \begin{pmatrix} u_1 \\ u_2 \end{pmatrix} \qquad A = \begin{pmatrix} 0 & 1 \\ -1 + \dfrac{n^2}{x^2} & -\dfrac{1}{x} \end{pmatrix} \tag{7.2.28}$$

Hence $q = 0$, $A_m = 0$ for $m > 2$, and

$$A_0 = \begin{pmatrix} 0 & 1 \\ -1 & 0 \end{pmatrix}, \qquad A_1 = \begin{pmatrix} 0 & 0 \\ 0 & -1 \end{pmatrix}, \qquad A_2 = \begin{pmatrix} 0 & 0 \\ n^2 & 0 \end{pmatrix} \tag{7.2.29}$$

Since the eigenvalues of A_0 are $\pm i$,

$$P_0 = \begin{pmatrix} 1 & 1 \\ i & -i \end{pmatrix} \quad \text{and} \quad P_0^{-1} = \begin{pmatrix} \dfrac{1}{2} & -\dfrac{i}{2} \\ \dfrac{1}{2} & \dfrac{i}{2} \end{pmatrix} \tag{7.2.30}$$

so that

$$B_0 = P_0^{-1} A_0 P_0 = \begin{pmatrix} i & 0 \\ 0 & -i \end{pmatrix} \tag{7.2.31}$$

Hence from (7.2.20)

$$F_1 = -P_0^{-1} A_1 P_0 = -\frac{1}{2} \begin{pmatrix} -1 & 1 \\ 1 & -1 \end{pmatrix} \tag{7.2.32}$$

Therefore from (7.2.21) and (7.2.25) we obtain

$$B_1 = -\frac{1}{2} \begin{pmatrix} 1 & 0 \\ 0 & 1 \end{pmatrix}, \qquad W_1 = \frac{i}{4} \begin{pmatrix} 0 & 1 \\ -1 & 0 \end{pmatrix} \tag{7.2.33}$$

With W_1 and B_1 known

$$F_2 = -\frac{i}{8} \begin{pmatrix} -1 - 4n^2 & 2 - 4n^2 \\ -2 + 4n^2 & 1 + 4n^2 \end{pmatrix} \tag{7.2.34}$$

from (7.2.20). Hence

$$B_2 = -\frac{i(1+4n^2)}{8}\begin{pmatrix}1 & 0\\ 0 & -1\end{pmatrix}, \qquad W_2 = -\frac{1-2n^2}{8}\begin{pmatrix}0 & 1\\ 1 & 0\end{pmatrix} \qquad (7.2.35)$$

Substituting the expressions for B_0, B_1, and B_2 into (7.2.12), we obtain

$$\frac{d\mathbf{v}}{dx} = \left[i\begin{pmatrix}1 & 0\\ 0 & -1\end{pmatrix} - \frac{1}{2x}\begin{pmatrix}1 & 0\\ 0 & 1\end{pmatrix} - \frac{i(1+4n^2)}{8x^2}\begin{pmatrix}1 & 0\\ 0 & -1\end{pmatrix} + O(x^{-3})\right]\mathbf{v}$$

Hence

$$v_1 = \frac{a}{\sqrt{x}}\exp\left[ix + i\frac{1+4n^2}{8x} + O(x^{-2})\right]$$

$$v_2 = \frac{b}{\sqrt{x}}\exp\left[-ix - i\frac{1+4n^2}{8x} + O(x^{-2})\right]$$

$$(7.2.36)$$

where a and b are constants. Since

$$P = P_0 + \frac{1}{x}P_1 + O(x^{-2}) = P_0\left(I + \frac{1}{x}W_1\right) + O(x^{-2})$$

$$P = \begin{pmatrix}1 & 1\\ i & -i\end{pmatrix}\begin{pmatrix}1 & \dfrac{i}{4x}\\ -\dfrac{i}{4x} & 1\end{pmatrix} + O(x^{-2})$$

$$= \begin{pmatrix}1 - \dfrac{i}{4x} & 1 + \dfrac{i}{4x}\\ i - \dfrac{1}{4x} & -i - \dfrac{1}{4x}\end{pmatrix} + O(x^{-2}) \qquad (7.2.37)$$

Therefore

$$\mathbf{u} = P\mathbf{v} = \frac{a}{\sqrt{x}}\begin{pmatrix}1 - \dfrac{i}{4x}\\ i - \dfrac{1}{4x}\end{pmatrix}\exp\left[ix + i\frac{1+4n^2}{8x}\right]$$

$$+ \frac{b}{\sqrt{x}}\begin{pmatrix}1 + \dfrac{i}{4x}\\ -i - \dfrac{1}{4x}\end{pmatrix}\exp\left[-ix - i\frac{1+4n^2}{8x}\right] + O(x^{-5/2}) \qquad (7.2.38)$$

Note that we did not use P_2 in (7.2.37) and (7.2.38) because the error in **v** is $O(x^{-5/2})$. To compare these results with those obtained in Section 7.1.2, we expand $\exp[\pm i(1 + 4n^2)/8x]$ in powers of x^{-1}. Since

$$y = u_1 = \frac{a}{\sqrt{x}}\left(1 - \frac{i}{4x}\right)\exp\left(ix + i\frac{1+4n^2}{8x}\right)$$

$$+ \frac{b}{\sqrt{x}}\left(1 + \frac{i}{4x}\right)\exp\left(-ix - i\frac{1+4n^2}{8x}\right) + O(x^{-5/2})$$

$$y = \frac{a}{\sqrt{x}}\left(1 - i\frac{1-4n^2}{8x}\right)e^{ix} + \frac{b}{\sqrt{x}}\left(1 + i\frac{1-4n^2}{8x}\right)e^{-ix} + O(x^{-5/2}) \qquad (7.2.39)$$

in agreement with (7.1.14) when $n = 0$. Higher approximations can be obtained by calculating the higher values of B_m and W_m in a straightforward, though tedious, way. The technique employed in Section 7.1.2 is very much easier to implement than the technique described in this section.

7.2.3. SUBNORMAL SOLUTIONS

If A_0 of (7.2.1) has multiple eigenvalues, we cannot decouple all these equations by choosing B to be a diagonal matrix. Instead we partition this system of equations to obtain simpler systems of the form

$$\frac{d\mathbf{v}^i}{dx} = x^q B^i(x)\mathbf{v}^i, \qquad B^i = \sum_{m=0}^{\infty} B_m{}^i \, x^{-m} \qquad (7.2.40)$$

where the eigenvalues λ_i of $B_0{}^i$ are different from λ_j of $B_0{}^j$ for $i \neq j$. Thus corresponding to each single eigenvalue λ_m, B^m is a scalar, and (7.2.40) can be solved. If $q = 0$

$$v^m = a x^{B_1{}^m} e^{B_0{}^m x}\left[1 + \sum_{r=2}^{\infty}\frac{c_r}{r - 1} \, x^{-r+1}\right] \qquad (7.2.41)$$

where a is a constant and c_r are known in terms of $B_r{}^m$ for $r \geq 2$. Hence a normal solution exists corresponding to this eigenvalue of the original system (7.2.1). For a multiple eigenvalue λ_s of multiplicity m_s, B^s has a rank of m_s. It turns out that this reduced m_s system of equations, hence the original system, may not have a normal solution corresponding to this eigenvalue. However, it may have what is called a subnormal solution of the form (7.2.3) through (7.2.5), but Q and U are expanded in powers of $x^{1/r}$ with integer r.

As an example, the equation

$$x\frac{d^2y}{dx^2} + 2\frac{dy}{dx} - \left(\frac{1}{4} + \frac{5}{16x}\right)y = 0 \qquad (7.2.42)$$

has the general solution

$$y = ae^{\sqrt{x}}(x^{-3/4} - x^{-5/4}) + be^{-\sqrt{x}}(x^{-3/4} + x^{-5/4}) \qquad (7.2.43)$$

which consists of two subnormal solutions. Equation (7.2.42) is equivalent to the system

$$\frac{d\mathbf{u}}{dx} = A\mathbf{u}, \qquad \mathbf{u} = \begin{pmatrix} y \\ \dfrac{dy}{dx} \end{pmatrix}$$

$$A = \begin{pmatrix} 0 & 1 \\ \dfrac{1}{4x} + \dfrac{5}{16x^2} & -\dfrac{2}{x} \end{pmatrix} \qquad (7.2.44)$$

In this case

$$A_0 = \begin{pmatrix} 0 & 1 \\ 0 & 0 \end{pmatrix}$$

has the eigenvalue $\lambda = 0$ with a multiplicity of two, which precludes the existence of normal solutions.

7.2.4. SYSTEMS CONTAINING A PARAMETER

Let us consider the system of n linear equations

$$\epsilon^h \frac{d\mathbf{y}}{dx} = A(x, \epsilon)\mathbf{y} \qquad (7.2.45)$$

where ϵ is a small positive number, h is an integer, and $A(x, \epsilon)$ is an $n \times n$ matrix which possesses the asymptotic expansion

$$A(x, \epsilon) = \sum_{m=0}^{\infty} \epsilon^m A_m(x) \quad \text{as} \quad \epsilon \to 0 \qquad (7.2.46)$$

If h is zero or negative, $\mathbf{y}$ possesses asymptotic expansions of the form

$$\mathbf{y}(x, \epsilon) = \sum_{m=0}^{\infty} \epsilon^m \mathbf{y}_m(x) \qquad (7.2.47)$$

If $h > 0$ the asymptotic expansions of the solutions of (7.2.45) depend on whether the eigenvalues of $A_0(x)$ are distinct in the whole interval of interest or not. A point at which $A_0(x)$ has multiple eigenvalues is called a turning or transition point; turning point problems are discussed in Section 7.3.

If the eigenvalues of $A_0(x)$ are distinct, the asymptotic representations of n linearly independent solutions of (7.2.45) have the form

$$\mathbf{y} = \mathbf{u}(x, \epsilon) \exp\left[\int^x \lambda(x, \epsilon)\, dx\right] \qquad (7.2.48)$$

where

$$\lambda(x, \epsilon) = \sum_{r=1}^{h} \epsilon^{-r}\lambda_r(x) \tag{7.2.49}$$

$$\mathbf{u}(x, \epsilon) = \sum_{r=0}^{\infty} \epsilon^r \mathbf{u}_r(x) \tag{7.2.50}$$

Substituting (7.2.48) through (7.2.50) into (7.2.45) and (7.2.46) and equating coefficients of like powers of ϵ, we obtain equations that determine successively λ_r and $\mathbf{u}_r$. There are n linear independent solutions of the form (7.2.48) through (7.2.50) corresponding to the n eigenvalues of $A_0(x)$; that is, the solutions of

$$|A_0(x) - \lambda_0(x)I| = 0 \tag{7.2.51}$$

7.2.5. HOMOGENEOUS SYSTEMS WITH SLOWLY VARYING COEFFICIENTS

In this section we consider the asymptotic solutions of

$$\frac{d\mathbf{y}}{dx} = A(\xi, \epsilon)\mathbf{y}, \qquad \xi = \epsilon x \tag{7.2.52}$$

where

$$A(\xi, \epsilon) = \sum_{m=0}^{\infty} \epsilon^m A_m(\xi) \quad \text{as} \quad \epsilon \to 0 \tag{7.2.53}$$

In this problem x is a fast variable while ξ is a slow variable.

As in Section 7.2.2, we assume the existence of a nonsingular matrix $P_0(\xi)$ such that

$$B_0(\xi) = P_0^{-1}(\xi)A_0(\xi)P_0(\xi) = \begin{pmatrix} B_0^{11}(\xi) & 0 \\ 0 & B_0^{22}(\xi) \end{pmatrix} \tag{7.2.54}$$

where B_0^{11} has the eigenvalues λ_i ($i = 1, 2, \ldots, r$) and B_0^{22} has the eigenvalues λ_j ($j = r + 1, r + 2, \ldots, n$) such that $\lambda_i \neq \lambda_j$. In this case we can reduce the original coupled system of equations (7.2.52) into two decoupled systems of orders r and $n - r$. To do this we let

$$\mathbf{y}(x, \epsilon) = P(\xi, \epsilon)\mathbf{v}(x, \epsilon) \tag{7.2.55}$$

which transforms (7.2.52) into

$$\frac{d\mathbf{v}}{dx} = B(\xi, \epsilon)\mathbf{v} \tag{7.2.56}$$

where

$$\frac{dP}{dx} = AP - PB$$

or

$$\epsilon \frac{dP}{d\xi} = A(\xi, \epsilon)P(\xi, \epsilon) - P(\xi, \epsilon)B(\xi, \epsilon) \qquad (7.2.57)$$

We seek asymptotic representations of P and B of the form

$$P = \sum_{m=0}^{\infty} \epsilon^m P_m(\xi), \qquad B = \sum_{m=0}^{\infty} \epsilon^m B_m(\xi) \qquad (7.2.58)$$

where B_m represents a block diagonal matrix. Substituting (7.2.58) into (7.2.57) and equating coefficients of like powers of ϵ, we obtain

$$\begin{aligned} A_0 P_0 - P_0 B_0 &= 0 \\ A_0 P_m - P_m B_0 &= P_0 B_m + \tilde{F}_m \end{aligned} \qquad (7.2.59)$$

where

$$\tilde{F}_m = \sum_{s=1}^{m-1} (P_s B_{m-s} - A_{m-s} P_s) - A_m P_0 + \frac{dP_{m-1}}{d\xi} \qquad (7.2.60)$$

As in Section 7.2.2, we choose B_0 and P_0 in accordance with (7.2.54), multiply the second of equations (7.2.59) from the left by P_0^{-1}, and use (7.2.19) to obtain

$$B_0 W_m - W_m B_0 = B_m + F_m \qquad (7.2.61)$$

where

$$W_m = P_0^{-1} P_m, \qquad F_m = P_0^{-1} \tilde{F}_m \qquad (7.2.62)$$

To solve (7.2.61) we partition F_m and W_m according to

$$F_m = \begin{pmatrix} F_m^{11} & F_m^{12} \\ F_m^{21} & F_m^{22} \end{pmatrix}, \qquad W_m = \begin{pmatrix} W_m^{11} & W_m^{12} \\ W_m^{21} & W_m^{22} \end{pmatrix} \qquad (7.2.63)$$

where F_m^{11} and W_m^{11} are $r \times r$ matrices. If we choose

$$W_m^{11} = W_m^{22} = 0 \quad \text{and} \quad B_m^{11} = -F_m^{11}, \quad B_m^{22} = -F_m^{22} \qquad (7.2.64)$$

(7.2.61) becomes

$$B_0^{11} W_m^{12} - W_m^{12} B_0^{22} = F_m^{12} \qquad (7.2.65)$$

$$B_0^{22} W_m^{21} - W_m^{21} B_0^{11} = F_m^{21} \qquad (7.2.66)$$

These equations can be solved uniquely for W_m^{12} and W_m^{21} since B_0^{11} and B_0^{22} do not have common eigenvalues.

If A_0 has distinct eigenvalues, one can use the above scheme to reduce the original system to an uncoupled system of n equations having the form (7.2.56) with B a diagonal matrix. The details are the same as in Section 7.2.2.

An easier technique can be used to determine the asymptotic solutions of the system (7.2.52) if A_0 has distinct eigenvalues. The asymptotic representation has the form

$$\mathbf{y} = \mathbf{u}(\xi, \epsilon)e^{\theta(x,\epsilon)} \tag{7.2.67}$$

where

$$\mathbf{u}(\xi, \epsilon) = \sum_{s=0}^{\infty} \epsilon^s \mathbf{u}_s(\xi) \tag{7.2.68}$$

$$\frac{d\theta}{dx} = \lambda(\xi) \tag{7.2.69}$$

where $\lambda(\xi)$ is an eigenvalue of $A_0(\xi)$. There are n linearly independent solutions of the form (7.2.67) corresponding to the n eigenvalues of A_0. Substituting (7.2.67) through (7.2.69) into (7.2.52) and (7.2.53) gives equations that determine $\mathbf{u}_m$ successively.

7.3. Turning Point Problems

We found in Section 7.1.3 that the Liouville–Green or WKB approximation to the solutions of

$$\frac{d^2y}{dx^2} + [\lambda^2 q_1(x) + q_2(x)]y = 0 \tag{7.3.1}$$

for large λ is

$$y = \frac{a_1 \cos\left[\lambda \int \sqrt{q_1(x)}\, dx\right] + b_1 \sin\left[\lambda \int \sqrt{q_1(x)}\, dx\right]}{\sqrt[4]{q_1(x)}} \tag{7.3.2}$$

for positive $q_1(x)$ and

$$y = \frac{a_2 \exp\left[\lambda \int \sqrt{-q_1(x)}\, dx\right] + b_2 \exp\left[-\lambda \int \sqrt{-q_1(x)}\, dx\right]}{\sqrt[4]{-q_1(x)}} \tag{7.3.3}$$

for negative $q_1(x)$. As remarked in Section 7.1.3, these approximations are valid as long as x is away from the zeros of $q_1(x)$. Equations (7.3.2) and (7.3.3) show that y is oscillatory on one side of a zero of $q_1(x)$ while it is exponential on the other side, hence such a zero is called a *transition point*. It is also called a *turning point* because in classic mechanics it is the point at which the kinetic energy of an incident particle is equal to its potential energy and the particle therefore reverses direction. The point $x = \mu$ is called a turning point or a transition point of order α where α is the order of the zero of $q_1(x)$ at $x = \mu$. If $q_2(x)$ is singular at a turning point, the turning point is called a *singular turning point*; otherwise it is regular. In this section we describe techniques for determining the asymptotic solutions of turning point problems starting with second-order equations such as (7.3.1).

7.3.1. THE METHOD OF MATCHED ASYMPTOTIC EXPANSIONS

Let us assume that $q_2(x)$ is a regular function while

$$q_1(x) = (x - \mu)f(x) \quad \text{with} \quad \text{positive } f(x) \tag{7.3.4}$$

Hence the approximate solution to (7.3.1) is given by (7.3.2) for $x > \mu$ and by (7.3.3) for $x < \mu$. These expansions are called outer expansions and break down near $x = \mu$. To determine the region of nonuniformity of these expansions, we let $\xi = (x - \mu)\lambda^{\nu}$ with positive ν in (7.3.1) and obtain

$$\frac{d^2y}{d\xi^2} + \{\lambda^{2-3\nu}\xi f[\mu + \xi\lambda^{-\nu}] + \lambda^{-2\nu}q_2[\mu + \xi\lambda^{-\nu}]\}y = 0 \tag{7.3.5}$$

As $\lambda \to \infty$ the third term in (7.3.5) tends to zero for all positive ν; however, the resulting equation depends on the value of ν. As $\lambda \to \infty$ (7.3.5) tends to

$$y = 0 \quad \text{if} \quad \nu < \tfrac{2}{3}$$

$$\frac{d^2y}{d\xi^2} = 0 \quad \text{if} \quad \nu > \tfrac{2}{3} \tag{7.3.6}$$

$$\frac{d^2y}{d\xi^2} + \xi f(\mu)y = 0 \quad \text{if} \quad \nu = \tfrac{2}{3}$$

It is obvious that the first two limits are not acceptable because their solutions do not match the outer expansions (7.3.2) and (7.3.3). Therefore the acceptable limit is the distinguished limit with $\nu = 2/3$ yielding the third equation in (7.3.6). If we let

$$z = -\xi\sqrt[3]{f(\mu)} \tag{7.3.7}$$

the first-order inner solution is governed by

$$\frac{d^2y}{dz^2} - zy = 0 \tag{7.3.8}$$

Its general solution is

$$y = a_3 Ai(z) + b_3 Bi(z) \tag{7.3.9}$$

where $Ai(z)$ and $Bi(z)$ are the Airy functions of the first and second kind, respectively.

Let us now digress to give some properties of the Airy functions which we will need in the ensuing discussion. These functions have the following integral representations see (e.g., Erdélyi, 1956, Section 4.6)

$$Ai(z) = \frac{1}{\pi} \int_0^\infty \cos{(\tfrac{1}{3}t^3 + zt)} \, dt \tag{7.3.10}$$

$$Bi(z) = \frac{1}{\pi} \int_0^\infty [\exp{(-\tfrac{1}{3}t^3 + zt)} + \sin{(\tfrac{1}{3}t^3 + zt)}] \, dt \tag{7.3.11}$$

These functions can also be related to Bessel functions of order $1/3$ according to

$$Ai(z) = \tfrac{1}{3}\sqrt{z}[I_{-1/3}(\zeta) - I_{1/3}(\zeta)] = \frac{1}{\pi}\sqrt{\frac{z}{3}}\, K_{1/3}(\zeta) \tag{7.3.12}$$

$$Bi(z) = \sqrt{\frac{z}{3}}\, [I_{-1/3}(\zeta) + I_{1/3}(\zeta)] \tag{7.3.13}$$

$$Ai(-z) = \tfrac{1}{3}\sqrt{z}[J_{-1/3}(\zeta) + J_{1/3}(\zeta)] \tag{7.3.14}$$

$$Bi(-z) = \sqrt{\frac{z}{3}}\, [J_{-1/3}(\zeta) - J_{1/3}(\zeta)] \tag{7.3.15}$$

where $\zeta = (2/3)z^{3/2}$. For large positive z these functions have the following asymptotic expansions

$$Ai(z) = \frac{1}{2\sqrt{\pi}} z^{-1/4} e^{-\zeta} \tag{7.3.16}$$

$$Ai(-z) = \frac{1}{\sqrt{\pi}} z^{-1/4} \sin{\left(\zeta + \frac{\pi}{4}\right)} \tag{7.3.17}$$

$$Bi(z) = \frac{1}{\sqrt{\pi}} z^{-1/4} e^{\zeta} \tag{7.3.18}$$

$$Bi(-z) = \frac{1}{\sqrt{\pi}} z^{-1/4} \cos{\left(\zeta + \frac{\pi}{4}\right)} \tag{7.3.19}$$

To match the inner solution (7.3.9) with the outer solution (7.3.2), we express the latter in terms of $\xi = (x - \mu)\lambda^{2/3}$ and determine its limit as $\lambda \to \infty$. In this case $x > \mu$ and

$$\lambda \int_\mu^x \sqrt{q_1(\tau)} \, d\tau = \lambda \int_\mu^x \sqrt{\tau - \mu}\sqrt{f(\tau)} \, d\tau = \tfrac{2}{3}\sqrt{f(\mu)}\, \xi^{3/2} + O(\lambda^{-2/3})$$

Hence

$$y = \frac{\lambda^{1/6}}{\sqrt[4]{\xi f(\mu)}} [a_1 \cos{(\tfrac{2}{3}\sqrt{f(\mu)}\, \xi^{3/2})} + b_1 \sin{(\tfrac{2}{3}\sqrt{f(\mu)}\, \xi^{3/2})}] + \cdots \tag{7.3.20}$$

Expressing (7.3.9) in terms of ξ, we obtain

$$y = a_3 Ai\left(-\xi\sqrt[3]{f(\mu)}\right) + b_3 Bi\left(-\xi\sqrt[3]{f(\mu)}\right)$$

Its expansion for large ξ, obtained by using (7.3.17) and (7.3.19), is

$$y = \frac{\xi^{-1/4}f^{-1/12}}{\sqrt{\pi}}\left[a_3 \sin\left(\tfrac{2}{3}\sqrt{f(\mu)}\,\xi^{3/2} + \frac{\pi}{4}\right)\right.$$
$$\left. + b_3 \cos\left(\tfrac{2}{3}\sqrt{f(\mu)}\,\xi^{3/2} + \frac{\pi}{4}\right)\right] + \cdots \quad (7.3.21)$$

Since the matching principle demands that (7.3.20) and (7.3.21) be equal, we obtain

$$a_1 = \frac{\lambda^{-1/6}f^{1/6}}{\sqrt{\pi}}\left[a_3 \sin\frac{\pi}{4} + b_3 \cos\frac{\pi}{4}\right]$$
$$b_1 = \frac{\lambda^{-1/6}f^{1/6}}{\sqrt{\pi}}\left[a_3 \cos\frac{\pi}{4} - b_3 \sin\frac{\pi}{4}\right] \quad (7.3.22)$$

Hence (7.3.2) becomes

$$y = \frac{1}{\sqrt[4]{q_1(x)}}\left\{\tilde{b}_3 \cos\left[\lambda\int_\mu^x \sqrt{q_1(\tau)}\,d\tau + \frac{\pi}{4}\right] + \tilde{a}_3 \sin\left[\lambda\int_\mu^x \sqrt{q_1(\tau)}\,d\tau + \frac{\pi}{4}\right]\right.$$
$$(7.3.23)$$

where

$$(\tilde{a}_3, \tilde{b}_3) = \frac{\lambda^{-1/6}f^{1/6}}{\sqrt{\pi}}(a_3, b_3)$$

To match (7.3.9) with the outer solution (7.3.3), we note that

$$\lambda\int_x^\mu \sqrt{-q_1(\tau)}\,d\tau = \lambda\int_x^\mu \sqrt{(\mu-\tau)f(\tau)}\,d\tau = \tfrac{2}{3}\sqrt{f(\mu)}(-\xi)^{3/2} + \cdots$$

Hence (7.3.3) becomes

$$y = \frac{\lambda^{1/6}}{\sqrt[4]{-f(\mu)\xi}}\{a_2 \exp[\tfrac{2}{3}\sqrt{f(\mu)}(-\xi)^{3/2}] + b_2 \exp[-\tfrac{2}{3}\sqrt{f(\mu)}(-\xi)^{3/2}]\} + \cdots$$
$$(7.3.24)$$

Since ξ is negative in this case, $z = -\xi\sqrt[3]{f(\mu)}$ is positive, and the asymptotic behavior of the inner solution (7.3.9) for large z, obtained from (7.3.16) and (7.3.18), is

$$y = \frac{(-\xi)^{-1/4}f^{-1/12}}{\sqrt{\pi}}\left[\tfrac{1}{2}a_3 \exp\left(-\tfrac{2}{3}\sqrt{f(\mu)}(-\xi)^{3/2}\right)\right.$$
$$\left. + b_3 \exp\left(\tfrac{2}{3}\sqrt{f(\mu)}(-\xi)^{3/2}\right)\right] + \cdots \quad (7.3.25)$$

Equating (7.3.24) and (7.3.25), we obtain

$$(a_2, b_2) = (\tilde{b}_3, \tfrac{1}{2}\tilde{a}_3) \qquad (7.3.26)$$

Hence the outer solution (7.3.3) for negative $q_1(x)$ is

$$y = \frac{1}{\sqrt[4]{-q_1(x)}}\left[\tilde{b}_3 \exp\left(\lambda \int_x^\mu \sqrt{-q_1(\tau)}\, d\tau\right) + \tfrac{1}{2}\tilde{a}_3 \exp\left(-\lambda \int_x^\mu \sqrt{-q_1(\tau)}\, d\tau\right)\right]$$

$$(7.3.27)$$

Therefore an approximate solution to (7.3.1) with a turning point at $x = \mu$ is given by the three separate expansions: (7.3.9) near the turning point, (7.3.23) for $x > \mu$, and (7.3.27) for $x < \mu$. The matching provided the connection between the constants a_1 and b_1, and a_2 and b_2. This connection was first given by Rayleigh (1912) in his investigation of the total reflection of sound waves from a transition layer, and he presented an explicit solution for the exponentially decreasing solution only. Gans (1915) gave connection formulas for both solutions, and Jeffreys (1924) rediscovered them in an application to Mathieu's function. Wentzel (1926), Kramers (1926), and Brillouin (1926) rediscovered them about the same time in their investigations of the Schrödinger equation. Thus in the physics literature this method is usually named with some permutation of the letters W, K, and B, and recently the letter J has been added to these letters for the contribution of Jeffreys.

Zwaan (1929) established the connection formulas by integration along a path in the complex plane that avoids the turning point. This technique was developed further by Kemble (1935).

A disadvantage of this technique is the fact that the solution is given by three different expansions. A single expansion uniformly valid for all x was obtained in Section 6.4.4 by using the method of multiple scales. In the next section we discuss a powerful technique of treating turning point problems originated by Langer (1931, 1934) and developed by Langer and several researchers as indicated in Sections 7.3.2 through 7.3.10.

7.3.2. THE LANGER TRANSFORMATION

The gist of Langer's approach is that approximately identical equations have approximately identical solutions. He realized that any attempt to express the asymptotic expansions of the solutions of turning point problems in terms of elementary functions must fail in regions containing the turning points. A uniformly valid expansion for all x must be expressed in terms of nonelementary functions which have the same qualitative features as the solutions of the equation.

The decisive step in Langer's approach is to introduce a transformation of the dependent and independent variables of the form

$$z = \phi(x), \qquad v = \psi(x)y(x) \tag{7.3.28}$$

which transforms

$$\frac{d^2y}{dx^2} + [\lambda^2 q_1(x) + q_2(x)]y = 0 \tag{7.3.29}$$

into

$$\frac{d^2v}{dz^2} + \frac{1}{\phi'^2}\left(\phi'' - \frac{2\phi'\psi'}{\psi}\right)\frac{dv}{dz} + \frac{1}{\phi'^2}\left[\lambda^2 q_1(x) + q_2(x) - \psi\left(\frac{\psi'}{\psi^2}\right)'\right]v = 0 \tag{7.3.30}$$

The middle term vanishes if

$$\psi = \sqrt{\phi'} \tag{7.3.31}$$

leaving

$$\frac{d^2v}{dz^2} + \left[\lambda^2 \frac{q_1}{\phi'^2} + \frac{q_2}{\phi'^2} + \frac{3}{4}\frac{\phi''^2}{\phi'^4} - \frac{1}{2}\frac{\phi'''}{\phi'^3}\right]v = 0 \tag{7.3.32}$$

If we choose

$$\frac{q_1}{\phi'^2} = 1 \quad \text{or} \quad \phi = \int \sqrt{q_1(\tau)}\, d\tau \tag{7.3.33}$$

we recover the Liouville–Green transformation. The resulting solution is expressed in terms of the circular functions as in Section 7.1.3, and it is singular at the turning points (zeros of q_1). Since $\psi = \sqrt{\phi'}$ and $\phi' = \sqrt{q_1}$, $\psi = \sqrt[4]{q_1}$ and the transformation (7.3.28) is singular at the zeros of $q_1(x)$.

To obtain a uniformly valid expansion for a problem with a turning point at $x = \mu$ such that

$$q_1(x) = (x - \mu)f(x) \quad \text{with} \quad f(x) > 0 \tag{7.3.34}$$

we follow Langer (1931, 1934) by choosing

$$\frac{q_1}{\phi'^2} = \phi \tag{7.3.35}$$

so that (7.3.32) becomes

$$\frac{d^2v}{dz^2} + \lambda^2 zv = \delta v \tag{7.3.36}$$

where

$$\delta = -\frac{q_2}{\phi'^2} - \frac{3}{4}\frac{\phi''^2}{\phi'^4} + \frac{1}{2}\frac{\phi'''}{\phi'^3} \tag{7.3.37}$$

The solution of (7.3.35) is

$$\tfrac{2}{3}\phi^{3/2} = \int_\mu^x \sqrt{(\tau - \mu)f(\tau)}\, d\tau \quad \text{for} \quad x \geq \mu$$

$$\tfrac{2}{3}(-\phi)^{3/2} = \int_x^\mu \sqrt{(\mu - \tau)f(\tau)}\, d\tau \quad \text{for} \quad x \leq \mu \tag{7.3.38}$$

As $x \to \mu$, $\phi \to \sqrt[3]{f(\mu)}(x - \mu)$ and $\psi \to \sqrt[6]{f(\mu)}$, hence $\delta = O(1)$ if q_2 is continuous. Moreover, the transformation (7.3.28) is regular everywhere including the turning point $x = \mu$. Since $\delta = O(1)$ and λ is large, v is given approximately by what Langer calls the related equation

$$\frac{d^2v}{dz^2} + \lambda^2 zv = 0 \qquad (7.3.39)$$

whose solution is

$$v = c_1 Ai(-\lambda^{2/3}z) + c_2 Bi(-\lambda^{2/3}z) \qquad (7.3.40)$$

where c_1 and c_2 are constants of integration. Hence, to first approximation

$$y = \frac{1}{\sqrt{\phi'(x)}} \{ c_1 Ai[-\lambda^{2/3}\phi(x)] + c_2 Bi[-\lambda^{2/3}\phi(x)] \} \qquad (7.3.41)$$

where ϕ is defined in (7.3.38).

This single expansion is uniformly valid for all x including the neighborhood of the turning point $x = \mu$. Using the asymptotic expansions (7.3.16) through (7.3.19) for large argument of the Airy functions Ai and Bi, we obtain

$$y = \frac{\lambda^{-1/6}}{\sqrt{\pi}\sqrt[4]{q_1(x)}} \left\{ c_1 \sin\left[\lambda \int_\mu^x \sqrt{q_1(\tau)}\, d\tau + \frac{\pi}{4}\right] \right.$$

$$\left. + c_2 \cos\left[\lambda \int_\mu^x \sqrt{q_1(\tau)}\, d\tau + \frac{\pi}{4}\right] \right\} \quad \text{for} \quad x > \mu \quad (7.3.42)$$

and

$$y = \frac{\lambda^{-1/6}}{\sqrt{\pi}\sqrt[4]{-q_1(x)}} \left\{ \frac{c_1}{2} \exp\left[-\lambda \int_x^\mu \sqrt{-q_1(\tau)}\, d\tau\right] + c_2 \exp\left[\lambda \int_x^\mu \sqrt{-q_1(\tau)}\, d\tau\right] \right\}$$

$$\text{for} \quad x < \mu \quad (7.3.43)$$

in agreement with the results of the previous section.

Olver (1954) generalized Langer's transformation to the form

$$y = \chi^{-1/4}v$$

$$\zeta = \zeta(z) = \int^x \sqrt{q_1(\tau)}\, d\tau \qquad (7.3.44)$$

$$\chi = \frac{q_1(x)}{\zeta'^2}$$

where the independent variable z is still any undefined function of x. With this transformation (7.3.29) becomes

$$\frac{d^2v}{dz^2} + \lambda^2\zeta'^2v = \delta v \qquad (7.3.45)$$

where

$$\delta = -\frac{q_2}{\chi} - \chi^{-3/4}\frac{d^2(\chi^{-1/4})}{dx^2} \qquad (7.3.46)$$

If we can choose $\zeta(z)$ such that $\delta = O(1)$, then the related equation

$$\frac{d^2v}{dz^2} + \lambda^2\zeta'^2v = 0 \qquad (7.3.47)$$

has solutions which are asymptotically equivalent to the solutions of (7.3.29) for large λ. In order for δ to be $O(1)$, χ must be regular and not vanish in the interval of interest. Consequently, ζ must be chosen so that ζ'^2 has zeros and singularities of the same order and at the same places as those of $q_1(x)$ so that ζ'^2 and $q_1(x)$ are positive or negative together. Moreover, it is desirable to choose ζ'^2 so that the related equation (7.3.47) is solvable in terms of known functions. This general transformation was rediscovered by Moriguchi (1959). If

$$q_1(x) = (x - \mu)f(x) \quad \text{with} \quad f(x) > 0, \qquad \zeta'^2 = \pm z \qquad (7.3.48)$$

which yields Langer's transformation.

7.3.3. PROBLEMS WITH TWO TURNING POINTS
We consider the case in which

$$q_1(x) = (x - \mu_1)(\mu_2 - x)f(x), \qquad \mu_2 > \mu_1 \quad \text{and} \quad f(x) > 0 \qquad (7.3.49)$$

so that (7.3.29) has two simple turning points at $x = \mu_1$ and μ_2. Such problems with two turning points arise, for example, in the solutions of the Schrödinger equation (e.g., Jeffreys, 1962; Pike, 1964) for tunneling or classic oscillator problems and in the determination of heat transfer in a duct (e.g., Jakob, 1949, pp. 451–480).

Applying the results of the previous section to the turning point $x = \mu_1$, we obtain

$$y = \frac{1}{\sqrt{\phi_1'(x)}}\{a_1Ai[-\lambda^{2/3}\phi_1(x)] + b_1Bi[-\lambda^{2/3}\phi_1(x)]\} \qquad (7.3.50)$$

where

$$\frac{2}{3}\phi_1^{3/2} = \int_{\mu_1}^{x}\sqrt{(\tau - \mu_1)(\mu_2 - \tau)f(\tau)}\,d\tau \quad \text{for} \quad x > \mu_1$$

$$\frac{2}{3}(-\phi_1)^{3/2} = \int_{x}^{\mu_1}\sqrt{(\mu_1 - \tau)(\mu_2 - \tau)f(\tau)}\,d\tau \quad \text{for} \quad x < \mu_1$$

$$(7.3.51)$$

However, as $x \to \mu_2$, $\phi_1' = O[(x - \mu_2)^{1/2}]$. Therefore (7.3.50) breaks down in the neighborhood of $x = \mu_2$, and it is valid for $\mu_2 - x > \delta_2$ where δ_2 is a positive small quantity.

Applying the results of the previous section to the turning point $x = \mu_2$, we obtain

$$y = \frac{1}{\sqrt{\phi_2'(x)}} \{a_2 Ai[-\lambda^{2/3}\phi_2(x)] + b_2 Bi[-\lambda^{2/3}\phi_2(x)]\} \quad \text{for} \quad x - \mu_1 > \delta_1$$

(7.3.52)

where δ_1 is a small positive number, and

$$\tfrac{2}{3}\phi_2^{3/2} = \int_x^{\mu_2} \sqrt{(\tau - \mu_1)(\mu_2 - \tau)f(\tau)} \, d\tau \quad \text{for} \quad x < \mu_2$$

$$\tfrac{2}{3}(-\phi_2)^{3/2} = \int_{\mu_2}^x \sqrt{(\tau - \mu_1)(\tau - \mu_2)f(\tau)} \, d\tau \quad \text{for} \quad x > \mu_2$$

(7.3.53)

Since both (7.3.50) and (7.3.52) are valid in the interval $\mu_1 + \delta_1 < x < \mu_2 - \delta_2$, we can connect these expansions by matching. Expanding (7.3.50) for large argument and for $x > \mu_1$, using (7.3.17) and (7.3.19), we obtain

$$y = \frac{\lambda^{-1/6}}{\sqrt{\pi}[(x - \mu_1)(\mu_2 - x)f(x)]^{1/4}} \left[a_1 \sin\left(\tfrac{2}{3}\lambda\phi_1^{3/2} + \frac{\pi}{4}\right) \right.$$
$$\left. + b_1 \cos\left(\tfrac{2}{3}\lambda\phi_1^{3/2} + \frac{\pi}{4}\right) \right] \quad (7.3.54)$$

Similarly, expanding (7.3.52) for large argument for $x < \mu_2$, we obtain

$$y = \frac{\lambda^{-1/6}}{\sqrt{\pi}[(x - \mu_1)(\mu_2 - x)f(x)]^{1/4}} \left[a_2 \sin\left(\tfrac{2}{3}\lambda\phi_2^{3/2} + \frac{\pi}{4}\right) \right.$$
$$\left. + b_2 \cos\left(\tfrac{2}{3}\lambda\phi_2^{3/2} + \frac{\pi}{4}\right) \right] \quad (7.3.55)$$

Equating (7.3.54) and (7.3.55) gives

$$a_1 \sin\left(\tfrac{2}{3}\lambda\phi_1^{3/2} + \frac{\pi}{4}\right) + b_1 \cos\left(\tfrac{2}{3}\lambda\phi_1^{3/2} + \frac{\pi}{4}\right)$$
$$= a_2 \sin\left(\tfrac{2}{3}\lambda\phi_2^{3/2} + \frac{\pi}{4}\right) + b_2 \cos\left(\tfrac{2}{3}\lambda\phi_2^{3/2} + \frac{\pi}{4}\right) \quad (7.3.56)$$

If we let

$$\Delta = \tfrac{2}{3}\lambda(\phi_1^{3/2} + \phi_2^{3/2}) + \frac{\pi}{2} = \lambda \int_{\mu_1}^{\mu_2} \sqrt{(\tau - \mu_1)(\mu_2 - \tau)f(\tau)} \, d\tau + \frac{\pi}{2} \quad (7.3.57)$$

then

$$\tfrac{2}{3}\lambda\phi_2^{3/2} + \frac{\pi}{4} = \Delta - \left(\tfrac{2}{3}\lambda\phi_1^{3/2} + \frac{\pi}{4}\right)$$

hence from (7.3.56)

$$a_1 = b_2 \sin \Delta - a_2 \cos \Delta$$
$$b_1 = a_2 \sin \Delta + b_2 \cos \Delta \qquad (7.3.58)$$

Now if y is a bounded function of x, as is the case in solutions of the Schrödinger equation, b_1 and $b_2 = 0$ because $Bi(z) \to \pi^{-1/2}z^{-1/4} \exp [(2/3)z^{3/2}]$ as $z \to \infty$. Hence from the second equation of (7.3.58)

$$\sin \Delta = 0 \quad \text{or} \quad \Delta = n\pi \quad \text{with } n \text{ an integer} \qquad (7.3.59)$$

Therefore from (7.3.57)

$$\lambda = \frac{(n - \tfrac{1}{2})\pi}{\displaystyle\int_{\mu_1}^{\mu_2} [(\tau - \mu_1)(\mu_2 - \tau)f(\tau)]^{1/2} \, d\tau} \qquad (7.3.60)$$

Rather than representing the solution by two expansions, Miller and Good (1953), Kazarinoff (1958), and Langer (1959b) suggested expressing the solution by a single uniformly valid expansion using parabolic cylinder functions. Using the transformation (7.3.44), we choose ζ'^2 so that it has two simple zeros. We take those to be $z = \pm 1$ with $z = -1$ corresponding to $x = \mu_1$, and we let (Pike, 1964)

$$\zeta'^2 = 4a^2(1 - z^2) \qquad (7.3.61)$$

We choose a so that $z = 1$ corresponds to $x = \mu_2$. Thus from (7.3.44) we obtain

$$\zeta = 2a \int_{-1}^{z} \sqrt{1 - \tau^2} \, d\tau = \int_{\mu_1}^{x} \sqrt{q_1(\tau)} \, d\tau \qquad (7.3.62)$$

where the branches of the square roots should be chosen so that z is a regular function of x and the regions where $q_1(x) > 0$ and $q_1(x) < 0$ transform into $z^2 < 1$ and $z^2 > 1$, respectively. Choosing $z = 1$ to correspond to $x = \mu_2$, we obtain the following equation for a

$$2a \int_{-1}^{1} \sqrt{1 - \tau^2} \, d\tau = \int_{\mu_1}^{\mu_2} \sqrt{q_1(\tau)} \, d\tau$$

Hence

$$a = \frac{1}{\pi} \int_{\mu_1}^{\mu_2} \sqrt{q_1(\tau)} \, d\tau \qquad (7.3.63)$$

With (7.3.61) the related equation is

$$\frac{d^2 v}{dz^2} + 4a^2\lambda^2(1 - z^2)v = 0 \qquad (7.3.64)$$

whose solutions are given by

$$v = W_v(2\sqrt{a\lambda z}), \qquad v + \tfrac{1}{2} = a\lambda \qquad (7.3.65)$$

where W_v is Weber's function of order v. If y is bounded at infinity, v must be bounded, hence $v = n$ where n is an integer. Therefore

$$\lambda = \frac{\pi(n + \tfrac{1}{2})}{\int_{\mu_1}^{\mu_2} \sqrt{q_1(\tau)}\, d\tau} \qquad (7.3.66)$$

in agreement with (7.3.60).

Problems with two turning points have also been analyzed by Olver (1959), and Moriguchi (1959). Several turning point problems were treated by Evgrafov and Fedoryuk (1966), Hsieh and Sibuya (1966), Sibuya (1967), and Lynn and Keller (1970) among others.

7.3.4. HIGHER-ORDER TURNING POINT PROBLEMS

In this section we let

$$q_1 = (x - \mu)^\alpha f(x), \qquad f(x) > 0 \quad \text{and} \quad \alpha \text{ is a positive real number} \qquad (7.3.67)$$

To determine a single uniformly valid expansion, we let $\zeta'^2 = z^\alpha$ so that ζ'^2 has the same number of zeros as q_1. Hence

$$\zeta = \frac{2}{2 + \alpha} z^{(\alpha+2)/2} = \int_\mu^x \sqrt{q_1(\tau)}\, d\tau \qquad (7.3.68)$$

where the branches of the square root are chosen so that the regions where $q_1(x) > 0$ and $q_1(x) < 0$ correspond to $z^\alpha > 0$ and $z^\alpha < 0$, respectively. This transformation leads to the related equation (Langer, 1931)

$$\frac{d^2 v}{dz^2} + \lambda^2 z^\alpha v = 0 \qquad (7.3.69)$$

The solution of (7.3.69) is

$$v = z^{1/2}\left\{ c_1 {}^*J_v\left[\frac{2\lambda}{2 + \alpha} z^{(\alpha+2)/2}\right] + c_2 {}^*J_{-v}\left[\frac{2\lambda}{2 + \alpha} z^{(\alpha+2)/2}\right]\right\} \qquad (7.3.70)$$

where $v = (2 + \alpha)^{-1}$. Hence, to first approximation

$$y = \frac{\left[\int_\mu^x [(\tau - \mu)^\alpha f(\tau)]^{1/2}\, d\tau\right]^{1/2}}{[(x - \mu)^\alpha f(x)]^{1/4}} \left\{ c_1 J_v\left[\lambda \int_\mu^x [(\tau - \mu)^\alpha f(\tau)]^{1/2}\, d\tau\right] \right.$$

$$\left. + c_2 J_{-v}\left[\lambda \int_\mu^x [(\tau - \mu)^\alpha f(\tau)]^{1/2}\, d\tau\right]\right\} \qquad (7.3.71)$$

McKelvey (1955) expressed the asymptotic solutions of a second-order turning point problem (i.e., $\alpha = 2$) in terms of Whittaker's functions. A second-order turning point problem arises in the diffraction by elliptic cylinders whose eccentricities are almost unity (Goodrich and Kazarinoff, 1963), and in the solution of the Schrödinger equation (Voss, 1933). The first analysis of a second-order turning point problem was given by Goldstein (1931) using the method of matched asymptotic expansions as described in Section 7.3.1.

7.3.5. HIGHER APPROXIMATIONS

So far in our presentation, only the first term in the asymptotic expansion has been obtained. There are four different approaches for the determination of the higher-order terms.

Langer's Approach. The gist of this approach is always to relate the solution of the equation to be solved to that of some simpler but structurally similar problem that can be solved explicitly in terms of transcendental functions (Langer, 1949). The drawback to this approach is that it is unsuitable for numerical calculations because the coefficients of the asymptotic expansions are functions of the independent variable as well as the perturbation parameter. Moreover, the expansions are established using several transformations. Equivalent expansions can be obtained in an easier way by using Olver's approach as indicated later.

Cherry's Approach. In 1949 and 1950, Cherry developed a technique for obtaining the higher-order terms of a simple turning point problem, which has been transformed using the Langer transformation (Section 7.3.2) into

$$\frac{d^2v}{dz^2} + [-\lambda^2 z + \lambda g(z, \lambda)]v = 0 \qquad (7.3.72)$$

where

$$g(z, \lambda) = \sum_{n=0}^{\infty} \lambda^{-n} g_n(z) \qquad (7.3.73)$$

In Cherry's analysis all g_n with even n are missing. We assume a formal expansion of the form

$$v = A(z; \lambda)\zeta_i[\lambda^{2/3}\phi(z; \lambda)], \qquad i = 1, 2$$

where ζ_1 and ζ_2 are the Airy functions of the first and second kind, respectively. Since

$$\frac{d^2v}{dz^2} = A''\zeta_i + (2A'\phi' + A\phi'')\frac{d\zeta_i}{d\phi} + A\phi'^2\frac{d^2\zeta_i}{d\phi^2} \qquad (7.3.74)$$

and

$$\frac{d^2\zeta_i}{d\phi^2} = \lambda^2 \phi \zeta_i$$

(7.3.72) becomes

$$(A'' + \lambda^2 \phi {\phi'}^2 A - \lambda^2 z A + \lambda g A)\zeta_i + (2A'\phi' + A\phi'')\frac{d\zeta_i}{d\phi} = 0 \quad (7.3.75)$$

Equating the coefficients of ζ_i and $d\zeta_i/d\phi$ to zero, we obtain

$$2A'\phi' + A\phi'' = 0 \tag{7.3.76}$$

$$\lambda^2(\phi{\phi'}^2 - z) + \lambda g + \frac{A''}{A} = 0 \tag{7.3.77}$$

From (7.3.76)

$$A = \frac{1}{\sqrt{\phi'}} \tag{7.3.78}$$

hence (7.3.77) becomes

$$\lambda^2(\phi{\phi'}^2 - z) + \lambda g + \frac{3}{4}\left(\frac{\phi''}{\phi'}\right)^2 - \frac{\phi'''}{2\phi'} = 0 \tag{7.3.79}$$

To solve this equation we let

$$\phi = z + \lambda^{-1}\phi_1(z) + \lambda^{-2}\phi_2(z) + \cdots \tag{7.3.80}$$

and equate coefficients of like powers of λ to obtain equations for the successive determination of ϕ_n. The first two equations are

$$2z\phi_1' + \phi_1 + g_0 = 0$$
$$2z\phi_2' + \phi_2 = -g_1 - z{\phi_1'}^2 - 2\phi_1\phi_1' \tag{7.3.81}$$

whose solutions are

$$\sqrt{z}\,\phi_1 = -\int_0^z \frac{g_0(\tau)}{2\sqrt{\tau}}\,d\tau$$

$$\sqrt{z}\,\phi_2 = -\int_0^z \frac{g_1(\tau) + \tau{\phi_1'}^2(\tau) + 2\phi_1(\tau)\phi_1'(\tau)}{2\sqrt{\tau}}\,d\tau \tag{7.3.82}$$

The lower limit in these expressions was chosen so that ϕ_1 and ϕ_2 are regular at $z = 0$.

Therefore v is given by the formal expansion

$$v = \frac{\zeta_i(\lambda^{2/3}\phi)}{\sqrt{\phi'}}$$

$$\phi(z; \lambda) = z + \lambda^{-1}\phi_1(z) + \lambda^{-2}\phi_2(z) + \cdots \tag{7.3.83}$$

This formal expansion is a uniform approximation to the solution of the original problem except in small neighborhoods of the zeros of v, where Cherry used the following modified formula

$$v = Ai(\lambda^{2/3}z)\left(1 + \lambda^{-2}a_1 + \lambda^{-4}a_2 + \cdots\right) + \frac{dAi}{dz}(\lambda^{2/3}z)\left(\lambda^{-2}b_1 + \lambda^{-4}b_2 + \cdots\right)$$

(7.3.84)

If

$$\lambda g(z, \lambda) = \sum_{n=0}^{\infty} \lambda^{-2n} g_{2n}(z)$$

a_n and b_n are determined from (7.3.83) by expanding ϕ' and $Ai(\lambda^{2/3}\phi)$ about $\phi = z$ and using the relation $d^2 Ai/dz^2 = \lambda^2 z Ai$.

Olver's Approach.   Olver (1954) proposed to determine a complete asymptotic expansion by assuming that

$$v = A(z; \lambda)\zeta_i(\lambda^{2/3}z) + B(z; \lambda)\zeta_i'(\lambda^{2/3}z)$$ (7.3.85)

This form is the same as the final form of the expansion obtained by Langer (1949) and the modified formula (7.3.84) of Cherry. This expansion can also be regarded as an application of the method of composite expansions described in Section 4.2.

Since

$$\zeta_i'' = \lambda^2 z \zeta_i, \qquad v' = A'\zeta_i + (A + B')\zeta_i' + B\zeta_i''$$
$$= (A' + \lambda^2 zB)\zeta_i + (A + B')\zeta_i'$$

hence

$$v'' = (A'' + \lambda^2 B + \lambda^2 zB')\zeta_i + (2A' + \lambda^2 zB + B'')\zeta_i' + (A + B')\zeta_i''$$
$$= [A'' + \lambda^2 B + \lambda^2 z(A + 2B')]\zeta_i + (2A' + \lambda^2 zB + B'')\zeta_i'$$

Consequently, (7.3.72) becomes

$$(A'' + \lambda^2 B + 2\lambda^2 zB' + \lambda gA)\zeta_i + (2A' + B'' + \lambda gB)\zeta_i' = 0 \quad (7.3.86)$$

Equating the coefficients of ζ_i and ζ_i' to zero, we obtain

$$2A' + B'' + \lambda gB = 0$$
$$A'' + \lambda^2 B + 2\lambda^2 zB' + \lambda gA = 0$$ (7.3.87)

These equations are satisfied by formal expansions of the form

$$A = \sum_{n=0}^{\infty} \lambda^{-n} A_n(z)$$

$$B = \sum_{n=1}^{\infty} \lambda^{-n} B_n(z)$$ (7.3.88)

where

$$2A_0' + g_0 B_1 = 0$$
$$2zB_1' + B_1 + g_0 A_0 = 0 \tag{7.3.89}$$

$$2A_n' + g_0 B_{n+1} = -\sum_{m=1}^{n} g_m B_{n-m+1} = \alpha_n, \qquad n \geq 1$$

$$2zB_{n+1}' + B_{n+1} + g_0 A_n = -A_{n-1}'' - \sum_{m=1}^{n} g_m A_{n-m} = \beta_n, \qquad n \geq 1 \tag{7.3.90}$$

The solution of (7.3.89) is

$$A_0 = \cosh \int_0^z \frac{g_0(\tau)}{2\sqrt{\tau}} \, d\tau$$

$$B_1 = -\frac{\sinh \int_0^z \frac{g_0(\tau)}{2\sqrt{\tau}} \, d\tau}{\sqrt{z}} \tag{7.3.91}$$

hence the solution of (7.3.90) is

$$A_n = a_n(z) \cosh \int_0^z \frac{g_0(\tau)}{2\sqrt{\tau}} \, d\tau + b_n(z) \sinh \int_0^z \frac{g_0(\tau)}{2\sqrt{\tau}} \, d\tau$$

$$\sqrt{z} \, B_{n+1} = -a_n(z) \sinh \int_0^z \frac{g_0(\tau)}{2\sqrt{\tau}} \, d\tau - b_n(z) \cosh \int_0^z \frac{g_0(\tau)}{2\sqrt{\tau}} \, d\tau \tag{7.3.92}$$

where

$$a_n = \tfrac{1}{2} \int_0^z [\alpha_n(\tau) A_0(\tau) - \beta_n(\tau) B_1(\tau)] \, d\tau$$

$$b_n = \tfrac{1}{2} \int_0^z \left[\sqrt{\tau} \, \alpha_n(\tau) B_1(\tau) - \frac{\beta_n(\tau) A_0(\tau)}{\sqrt{\tau}} \right] d\tau \tag{7.3.93}$$

In the case of

$$\lambda g = \sum_{n=0}^{\infty} \lambda^{-2n} g_n(z)$$

(7.3.87) is satisfied by formal expansions of the form

$$A = A_0 + \sum_{n=1}^{\infty} \lambda^{-2n} A_n(z), \qquad A_0 = 1$$

$$B = \sum_{n=1}^{\infty} \lambda^{-2n} B_n(z) \tag{7.3.94}$$

Substituting this expansion into (7.3.87) and equating coefficients of like powers of λ, we obtain

$$2A_n' = -B_n'' - g_0 B_n - \sum_{m=1}^{n-1} g_m B_{n-m} = \alpha_n$$

$$2zB_n' + B_n = -A_{n-1}'' - g_0 A_{n-1} - \sum_{m=1}^{n-1} g_m A_{n-m-1} = \beta_n \tag{7.3.95}$$

Their solutions are

$$\sqrt{z}\, B_n = \int_0^z \frac{\beta_n(\tau)}{2\sqrt{\tau}}\, d\tau$$

$$A_n = \tfrac{1}{2} \int_0^z \alpha_n(\tau)\, d\tau \tag{7.3.96}$$

Successive Langer Transformations.   To determine higher approximations to the solutions of

$$\frac{d^2y}{dx^2} + \lambda^2 q(x)y = 0 \tag{7.3.97}$$

where $q(x)$ vanishes in the interval of interest, Imai (1948, 1950) proposed repeated application of the Langer transformation. This technique has been applied and extended considerably by Moriguchi (1959). For a simple turning point at $x = \mu$, we first introduce the Langer transformation

$$\tfrac{2}{3}z^{3/2} = \int_\mu^x \sqrt{q(\tau)}\, d\tau, \qquad y = \chi^{-1/4}v, \qquad \chi = \frac{q}{z} \tag{7.3.98}$$

in (7.3.97) to obtain

$$\frac{d^2v}{dz^2} + [\lambda^2 z - \delta(z)]v = 0 \tag{7.3.99}$$

where

$$\delta = \chi^{-1/4}\frac{d^2(\chi^{1/4})}{dz^2} = -\chi^{-3/4}\frac{d^2(\chi^{-1/4})}{dx^2}$$

Since $\delta = O(1)$ and λ is large, v is given approximately by

$$\frac{d^2v}{dz^2} + \lambda^2 z = 0 \tag{7.3.100}$$

that is

$$v = \zeta_i(\lambda^{2/3}z) \tag{7.3.101}$$

where ζ_1 and ζ_2 are the Airy functions of the first and second kind.

To improve (7.3.101) we rewrite (7.3.99) in the same form as the original equation (7.3.97) by changing the independent variable from z to x_1 according to

$$x_1 = z - \mu_1 \tag{7.3.102}$$

where μ_1 is the root of $\lambda^2 z - \delta(z) = 0$; that is

$$\lambda^2 \mu_1 - \delta(\mu_1) = 0 \tag{7.3.103}$$

Then (7.3.99) can be written as

$$\frac{d^2v}{dx_1^2} + \lambda^2[a_1 x_1 + R_1(x_1)]v = 0 \tag{7.3.104}$$

where

$$a_1 = 1 - \lambda^{-2}\delta'(\mu_1), \qquad R_1 = -\lambda^{-2}\left[\frac{1}{2}\delta''(\mu_1)x_1{}^2 + \frac{1}{3!}\delta'''(\mu_1)x_1{}^3 + \cdots\right]$$

(7.3.105)

This equation has the same form as (7.3.97), hence an approximate solution can be obtained using the transformation

$$\tfrac{2}{3}z_1^{3/2} = \int_0^{x_1}\sqrt{q_1(\tau)}\,d\tau, \qquad v = \chi_1^{-1/4}v_1$$

(7.3.106)

$$\chi_1 = \frac{q_1}{z_1}, \qquad q_1 = a_1 x_1 + R_1(x_1)$$

Then (7.3.104) becomes

$$\frac{d^2v_1}{dz_1{}^2} + [\lambda^2 z_1 - \delta_1(z_1)]v_1 = 0$$

(7.3.107)

where

$$\delta_1 = -\chi_1^{-3/4}\frac{d^2(\chi_1^{-1/4})}{dx_1{}^2}$$

(7.3.108)

A first approximation to v_1 is the solution of

$$\frac{d^2v_1}{dz_1{}^2} + \lambda^2 z_1 v_1 = 0$$

(7.3.109)

Hence

$$v_1 = \zeta_i(\lambda^{2/3}z_1)$$

(7.3.110)

Now transforming back to x and y yields an improved approximation to the solution of the original equation. From (7.3.106)

$$\tfrac{2}{3}z_1^{3/2} = \int_0^{x_1}\sqrt{a_1\tau}\sqrt{1 + \frac{R_1}{a_1\tau}}\,d\tau = \tfrac{2}{3}\sqrt{a_1}\,x_1^{3/2} + O(\lambda^{-2})$$

Hence

$$z_1 = \sqrt[3]{a_1}\,x_1 + O(\lambda^{-2})$$

$$\chi_1 = a_1^{2/3} + O(\lambda^{-2})$$

(7.3.111)

Since

$$y = \frac{v}{\sqrt[4]{\chi}} = \frac{v_1}{\sqrt[4]{\chi\chi_1}} \approx \frac{z^{1/4}v_1}{q^{1/4}a_1^{1/6}}$$

two independent solutions of (7.3.97) are approximately given by

$$y = \sqrt{\frac{z(x)}{q(x)}}\,\zeta_i\big[\lambda^{2/3}a_1^{1/3}\big(z(x) - \mu_1\big)\big]$$

(7.3.112)

where $z(x)$ is defined in (7.3.98).

Higher approximation can be obtained by means of repeated application of the above procedure.

7.3.6. AN INHOMOGENEOUS PROBLEM WITH A SIMPLE TURNING POINT—FIRST APPROXIMATION

In this section we determine a first approximation to a particular solution of the equation

$$\frac{d^2y}{dx^2} + [\lambda^2 q_1(x) + q_2(x)]y = \lambda^2 G(x) \tag{7.3.113}$$

for large λ when $q_1(x)$ has a simple zero at $x = \mu$. If we divide (7.3.113) by λ^2 and let $\lambda \to \infty$, we obtain

$$y = \frac{G(x)}{q_1(x)} \tag{7.3.114}$$

as an approximate particular solution. This solution is singular at $x = \mu$ unless $G(x)$ has a simple or a higher-order zero at $x = \mu$.

To determine a first approximation to a particular solution in the case of $G(\mu) \neq 0$, we first employ the transformation

$$z = \phi(x), \qquad \tfrac{2}{3}z^{3/2} = \int_\mu^x \sqrt{q_1(\tau)}\, d\tau, \qquad y = \frac{v}{\sqrt{\phi'}} \tag{7.3.115}$$

to transform (7.3.114) into

$$\frac{d^2v}{dz^2} + (\lambda^2 z - \delta)v = \lambda^2 g(z) \tag{7.3.116}$$

where δ is defined by (7.3.37), and

$$g(z) = \{\phi'[x(z)]\}^{-3/2}\, G[x(z)] \tag{7.3.117}$$

Since $\delta = O(1)$ and λ is large, a first approximation to (7.3.116) is

$$\frac{d^2v}{dz^2} + \lambda^2 z v = \lambda^2 g(z) \tag{7.3.118}$$

To determine a particular solution, we write $g(z)$ as the sum of two terms according to

$$g(z) = g(0) + [g(z) - g(0)] \tag{7.3.119}$$

and determine particular solutions corresponding to each term. A particular solution corresponding to the second term is given approximately by

$$v_1 = \frac{g(z) - g(0)}{z} \tag{7.3.120}$$

uniformly for all z if $g(z)$ is differentiable at $z = 0$. To find a particular solution corresponding to the first term, we let $\xi = \lambda^{2/3}z$ so that (7.3.118) becomes

$$\frac{d^2v}{d\xi^2} + \xi v = \lambda^{2/3}g(0) \qquad (7.3.121)$$

with a particular solution

$$v_2 = \lambda^{2/3}g(0)T(\xi) \qquad (7.3.122)$$

where

$$\frac{d^2T}{d\xi^2} + \xi T = 1, \qquad T(\xi) = \frac{1}{\xi} \quad \text{as} \quad |\xi| \to \infty \qquad (7.3.123)$$

Thus $T(\xi)$ may be represented in terms of Lommel functions (e.g., Watson, 1944, pp. 345–351) according to

$$T(\xi) = \tfrac{2}{3}\xi^{1/2}S_{0,1/3}(\tfrac{2}{3}\xi^{3/2}) = \int_0^\infty e^{-\xi t}e^{-t^3/3}\,dt, \qquad |\xi| < \infty$$

$$= 3^{-2/3}\sum_{n=0}^\infty 3^{n/3}\Gamma\left(\frac{n+1}{3}\right)\frac{(-\xi)^n}{n!}, \qquad |\xi| < \infty$$

$$\sim \sum_{n=0}^\infty \frac{(-1)^n(3n)!}{3^n n!}\xi^{-3n-1}, \qquad |\arg \xi| < \frac{2\pi}{3} \qquad (7.3.124)$$

where $S_{0,1/3}$ denotes a specific Lommel function.

Therefore a particular solution to (7.3.118) is approximately given by

$$v = \lambda^{2/3}g(0)T(\lambda^{2/3}z) + \frac{g(z) - g(0)}{z} \qquad (7.3.125)$$

Transforming back to x and y, we obtain

$$y = \lambda^{2/3}\frac{G(\mu)}{[\phi'(x)f(\mu)]^{1/2}}T(\lambda^{2/3}z) + \frac{1}{q_1(x)}\left[G(x) - \frac{G(\mu)\phi'^{3/2}}{\sqrt{f(\mu)}}\right] \qquad (7.3.126)$$

where

$$f(x) = \frac{q_1(x)}{(x - \mu)}$$

Inhomogeneous problems with turning points arise in the analysis of boundary layer stability (Holstein, 1950) and thin elastic toroidal shells and bending of curved tubes (e.g., Clark, 1964). The technique presented above was developed by Holstein (1950), Clark (1958, 1963) and Tumarkin (1959). Steele (1965) obtained a single particular solution for an inhomogeneous second-order equation in terms of general Lommel functions $S_{\mu,\nu}$.

7.3.7. AN INHOMOGENEOUS PROBLEM WITH A SIMPLE TURNING POINT—HIGHER APPROXIMATIONS

To determine a complete asymptotic representation of

$$\frac{d^2y}{dx^2} + [\lambda^2 f(x) + \lambda\chi(x, \lambda)]y = \lambda^2 G(x, \lambda) \tag{7.3.127}$$

where $f(x)$ has a simple zero at $x = \mu$ and

$$\chi(x, \lambda) = \sum_{n=0}^{\infty} \lambda^{-n}\chi_n(x), \qquad G(x, \lambda) = \sum_{n=0}^{\infty} \lambda^{-n}G_n(x) \quad \text{as} \quad \lambda \to \infty$$

we first employ the transformation

$$z = \phi(x), \qquad \tfrac{2}{3}z^{3/2} = \int_{\mu}^{x} \sqrt{f(\tau)}\, d\tau, \qquad y = v/\sqrt{\phi'} \tag{7.3.128}$$

to transform (7.3.127) into

$$\frac{d^2v}{dz^2} + [\lambda^2 z + \lambda q(z, \lambda)]v = \lambda^2 g(z, \lambda) \tag{7.3.129}$$

where

$$q\big(z(x), \lambda\big) = P^4(x)\chi(x, \lambda) + \lambda^{-1}P^3(x)P''(x)$$

$$g\big(z(x), \lambda\big) = P^3(x)G(x, \lambda), \qquad P(x) = \frac{1}{\sqrt{\phi'(x)}} \tag{7.3.130}$$

We assume that

$$g(z, \lambda) = \sum_{n=0}^{\infty} \lambda^{-n}g_n(z), \qquad q(z, \lambda) = \sum_{n=0}^{\infty} \lambda^{-n}q_n(z) \quad \text{as} \quad \lambda \to \infty \tag{7.3.131}$$

and restrict our attention to (7.3.129).

We assume a complete asymptotic expansion of a particular solution of (7.3.129) and (7.3.131) to have the form

$$v = C(z, \lambda) + \lambda^{2/3}A(z, \lambda)T(\xi) + \lambda^{1/3}B(z, \lambda)T'(\xi), \qquad \xi = \lambda^{2/3}z \tag{7.3.132}$$

where $T(\xi)$ is defined in (7.3.124) as the solution of

$$T'' + \xi T = 1, \qquad T = \frac{1}{\xi} \quad \text{as} \quad |\xi| \to \infty \tag{7.3.133}$$

Since

$$\frac{dv}{dz} = C' + \lambda^{2/3}A'T + \lambda^{1/3}(\lambda A + B')T' + \lambda BT''$$

$$= C' + \lambda B + \lambda^{2/3}(A' - \lambda zB)T + \lambda^{1/3}(\lambda A + B')T'$$

$$\frac{d^2v}{dz^2} = C'' + 2\lambda B' + \lambda^2 A + \lambda^{2/3}(A'' - \lambda^2 Az - \lambda B - 2\lambda zB')T$$

$$\qquad + \lambda^{1/3}(2\lambda A' - \lambda^2 zB + B'')T'$$

(7.3.129) becomes

$$[(\lambda^2 z + \lambda q)C + C'' + 2\lambda B' + \lambda^2 A - \lambda^2 g] + \lambda^{2/3}[A'' - \lambda B - 2\lambda z B' + \lambda q A]T$$
$$+ \lambda^{1/3}[2\lambda A' + B'' + \lambda q B]T' = 0 \quad (7.3.134)$$

In order for (7.3.134) to be an identity, each of the coefficients in square brackets must vanish; that is

$$(\lambda^2 z + \lambda q)C + \lambda^2 A - \lambda^2 g + 2\lambda B' + C'' = 0 \quad (7.3.135)$$
$$2\lambda z B' + \lambda B - A'' - \lambda q A = 0 \quad (7.3.136)$$
$$2\lambda A' + \lambda q B + B'' = 0 \quad (7.3.137)$$

To solve (7.3.135) through (7.3.137), we assume formal expansions of the form

$$A = \sum_{n=0} \lambda^{-n} A_n(z), \qquad B = \sum_{n=0} \lambda^{-n} B_n(z), \qquad C = \sum_{n=0}^{\infty} \lambda^{-n} C_n \quad (7.3.138)$$

and equate coefficients of like powers of λ to obtain

$$z C_n = g_n - A_n - 2B'_{n-1} - C''_{n-2} - \sum_{k=0}^{n-1} q_{n-k-1} C_k \quad (7.3.139)$$

$$2z B'_n + B_n - q_0 A_n = A''_{n-1} + \sum_{k=0}^{n-1} q_{n-k} A_k \quad (7.3.140)$$

$$2A'_n + q_0 B_n = -B''_{n-1} - \sum_{k=0}^{n-1} q_{n-k} B_k \quad (7.3.141)$$

where all coefficients with negative subscripts are defined to be zero. Letting

$$M_z A_n = A''_n(z) + \sum_{k=0}^{n} q_{n-k+1}(z) A_k(z)$$
$$u_n(z) = A_n(z) + i\sqrt{z}\, B_n(z) \quad (7.3.142)$$

we combine (7.3.140) and (7.3.141) into

$$2\sqrt{z}\, u'_n - i q_0 u_n = i M_z A_{n-1} - \sqrt{z}\, M_z B_{n-1} \quad (7.3.143)$$

whose general solution is

$$u_n = (\alpha_n + i\beta_n)e^{i\theta(z)} + \frac{i}{2}\int_0^z e^{i\theta(z)-i\theta(\tau)}[\tau^{-1/2} M_\tau A_{n-1} + i M_\tau B_{n-1}]\, d\tau,$$
$$n = 0, 1, 2 \ldots \quad (7.3.144)$$

where α_n and β_n are arbitrary constants and

$$\theta(z) = \int_0^z \frac{q_0(\tau)}{2\sqrt{\tau}}\, d\tau \quad (7.3.145)$$

Since coefficients with negative subscripts are zero, (7.3.144) reduces for $n = 0$ to

$$u_0 = (\alpha_0 + i\beta_0)e^{i\theta} \qquad (7.3.146)$$

Hence

$$A_0 = \alpha_0 \cos \theta - \beta_0 \sin \theta$$
$$\sqrt{z}B_0 = \alpha_0 \sin \theta + \beta_0 \cos \theta \qquad (7.3.147)$$

Since $\theta(0) = 0$, $\beta_0 \equiv 0$ in order that B_0 be bounded as $z \to 0$. Now (7.3.139) reduces for $n = 0$ to

$$zC_0 = g_0 - A_0 \qquad (7.3.148)$$

In order that C_0 be bounded as $z \to 0$

$$A_0(0) = g_0(0) = \alpha_0 \qquad (7.3.149)$$

Therefore the first approximation to v is

$$v = \frac{g_0(z) - g_0(0) \cos \theta}{z} + \lambda^{2/3}g_0(0) \cos \theta \, T(\xi) + \lambda^{1/3}g_0(0)z^{-1/2} \sin \theta \, T'(\xi)$$

$$(7.3.150)$$

which reduces to (7.3.125) when $q_0 = 0$ (i.e., $\theta = 0$).

In general, for B_n to be regular at $z = 0$, we require that $\beta_n = 0$, while for C_n to be regular at $z = 0$, we require that

$$\alpha_n = A_n(0) = g_n(0) - 2B'_{n-1}(0) - C''_{n-2}(0) - \sum_{k=0}^{n-1} q_{n-k-1}(0)C_k(0),$$

$$n = 0, 1, 2, \ldots \quad (7.3.151)$$

Hence

$$A_n(z) = \alpha_n \cos \theta(z) - \tfrac{1}{2} \int_0^z \tau^{-1/2} \sin [\theta(z) - \theta(\tau)]M_r A_{n-1} \, d\tau$$

$$- \tfrac{1}{2} \int_0^z \cos [\theta(z) - \theta(\tau)]M_r B_{n-1} \, d\tau \qquad (7.3.152)$$

$$\sqrt{z} \, B_n(z) = \alpha_n \sin \theta(z) + \tfrac{1}{2} \int_0^z \tau^{-1/2} \cos [\theta(z) - \theta(\tau)]M_r A_{n-1} \, d\tau$$

$$- \tfrac{1}{2} \int_0^z \sin [\theta(z) - \theta(\tau)]M_r B_{n-1} \, d\tau \qquad (7.3.153)$$

This general solution was obtained by Tumarkin (1959) and justified by Clark (1963).

7.3.8.   AN INHOMOGENEOUS PROBLEM WITH A SECOND-ORDER TURNING POINT

In their investigation of toroidal membranes under internal pressure, Sanders and Liepins (1963) encountered an inhomogeneous problem with a turning point of second order of the form

$$\frac{d^2y}{dx^2} + \lambda^2 q(x)y = \lambda^2 G(x) \qquad (7.3.154)$$

where $q(x)$ has a zero of order two at $x = \mu$. To determine a first approximation to a particular solution of this equation, we let

$$z = \phi(x), \qquad \tfrac{1}{2}z^2 = \int_\mu^x \sqrt{q(\tau)}\, d\tau, \qquad y = \frac{v}{\sqrt{\phi'}} \qquad (7.3.155)$$

in (7.3.154) and obtain

$$\frac{d^2v}{dz^2} + (\lambda^2 z^2 - \delta)v = \lambda^2 g(z) \qquad (7.3.156)$$

where δ and g are defined by (7.3.37) and (7.3.117). Since $\delta = O(1)$ and λ is large, a particular solution to (7.3.156) is given approximately by

$$\frac{d^2v}{dz^2} + \lambda^2 z^2 v = \lambda^2 g(z) \qquad (7.3.157)$$

To find an approximate particular solution to (7.3.157), we express $g(z)$ as the sum of three terms according to

$$g(z) = g(0) + g'(0)z + [g(z) - g(0) - g'(0)z] \qquad (7.3.158)$$

and determine particular solutions corresponding to these three terms. A particular solution corresponding to the last term is approximately given by

$$v_1 = \frac{g(z) - g(0) - g'(0)z}{z^2} \qquad (7.3.159)$$

uniformly for all z if $g''(0)$ exists. To find particular solutions corresponding to the other two terms in $g(z)$, we let $\xi = \lambda^{1/2}z$ thereby transforming (7.3.157) into

$$\frac{d^2v}{d\xi^2} + \xi^2 v = \lambda g(0) + \lambda^{1/2}g'(0)\xi \qquad (7.3.160)$$

Sanders and Liepins (1963) defined the two functions $T_1(\xi)$ and $T_2(\xi)$ by

$$\frac{d^2T_1}{d\xi^2} + \xi^2 T_1 = 1, \qquad T_1 = \frac{1}{\xi^2} \quad \text{as} \quad |\xi| \to \infty$$

$$\frac{d^2T_2}{d\xi^2} + \xi^2 T_2 = \xi, \qquad T_2 = \frac{1}{\xi} \quad \text{as} \quad |\xi| \to \infty \qquad (7.3.161)$$

In terms of these functions, a particular solution of (7.3.160) may be written as

$$v_2 = \lambda g(0)T_1(\xi) + \lambda^{1/2}g'(0)T_2(\xi) \qquad (7.3.162)$$

Therefore a particular solution of (7.3.157) is approximately given by

$$v = \frac{g(z) - g(0) - g'(0)z}{z^2} + \lambda g(0)T_1(\xi) + \lambda^{1/2}g'(0)T_2(\xi) \quad (7.3.163)$$

Transforming back to x and y, we can obtain a uniformly valid first approximation to the original equation.

7.3.9.   TURNING POINT PROBLEMS ABOUT SINGULARITIES
We consider the asymptotic expansions of the solutions of

$$\frac{d^2y}{dx^2} + [\lambda^2 q(x) + r(x)]y = 0 \quad (7.3.164)$$

for large λ under the conditions

$$q(x) = q_0(x - \mu)^\alpha[1 + O(x - \mu)] \quad \text{as} \quad x \to \mu \quad (7.3.165)$$
$$r(x) = r_0(x - \mu)^{-2}[1 + O(x - \mu)]$$

The problems considered in the previous sections ($r_0 = 0$ and $\alpha \geq 0$) are special cases of the present problem where $r_0 \neq 0$ and α may be negative. As $x \to \mu$, (7.3.164) tends to

$$\frac{d^2y}{dx^2} + \left[\lambda^2 q_0(x - \mu)^\alpha + \frac{r_0}{(x - \mu)^2}\right]y = 0 \quad (7.3.166)$$

Therefore we choose

$$\frac{d^2v}{dz^2} + \left(\lambda^2 z^\alpha + \frac{r_0}{z^2}\right)v = 0 \quad (7.3.167)$$

as the related equation with solutions expressed in the form

$$v = z^{1/2}\mathscr{C}_\nu(\gamma z^\beta)$$
$$\beta\gamma = \lambda, \qquad \beta = \frac{\alpha + 2}{2}, \qquad \nu = \frac{\sqrt{1 - 4r_0}}{2 + \alpha} \quad (7.3.168)$$

where the cylindrical functions $\mathscr{C}_\nu(t)$ satisfy the differential equation

$$\frac{d^2\mathscr{C}_\nu}{dt^2} + \frac{1}{t}\frac{d\mathscr{C}_\nu}{dt} + \left(1 - \frac{\nu^2}{t^2}\right)\mathscr{C}_\nu = 0 \quad (7.3.169)$$

The Bessel, Neumann, and Hankel functions $(J_\nu(t), Y_\nu(t), H_\nu^1(t), \text{ and } H_\nu^2(t))$ are special cylindrical functions.

Investigation of turning point problems about singularities was started by Langer (1935). The other principal contributors to the investigation of this problem were Cashwell (1951), Olver (1954), Swanson (1956), Kazarinoff and McKelvey (1956), Erdélyi (1960), and Wasow (1965).

To determine the asymptotic expansions of the solutions of (7.3.164), we introduce the transformation

$$z = \phi(x), \qquad v = \psi(x)y(x), \qquad \psi(x) = \sqrt{\phi'} \qquad (7.3.170)$$

thereby transforming (7.3.164) into

$$\frac{d^2v}{dz^2} + \left\{ \frac{\lambda^2 q(x)}{\phi'^2} + \frac{1}{\phi'^2}\left[r(x) + \frac{2\psi'^2}{\psi^2} - \frac{\psi''}{\psi} \right] \right\} v = 0 \qquad (7.3.171)$$

In order that this equation be approximately identical to (7.3.167), we require that

$$\phi^\alpha \phi'^2 = q(x) \qquad (7.3.172)$$

so that

$$\frac{2}{\alpha + 2} \phi^{(\alpha+2)/2} = \int_\mu^x \sqrt{q(\tau)}\, d\tau \qquad (7.3.173)$$

Hence (7.3.171) becomes

$$\frac{d^2v}{dz^2} + \left(\lambda^2 z^\alpha + \frac{r_0}{z^2} \right) v = F(z)v \qquad (7.3.174)$$

where

$$F = \frac{r_0}{\phi^2} - \frac{1}{\phi'^2}\left[r(x(z)) + \frac{2\psi'^2}{\psi^2} - \frac{\psi''}{\psi} \right] \qquad (7.3.175)$$

As $x \to \mu$

$$\frac{2}{\alpha + 2} \phi^{(\alpha+2)/2} \to \frac{2}{\alpha + 2} \sqrt{q_0}(x - \mu)^{(\alpha+2)/2}$$

so that

and

$$\phi \to q_0^{1/(\alpha+2)}(x - \mu), \qquad \phi' \to q_0^{1/(\alpha+2)}, \qquad \psi \to q_0^{1/2(\alpha+2)} \qquad (7.3.176)$$

$$F = O\left(\frac{1}{x - \mu} \right) \qquad (7.3.177)$$

Hence a first approximation to v is given by (7.3.167) whose solution is given by (7.3.168). Therefore y is given approximately by

$$y = \frac{\left[\int_\mu^x \sqrt{q(\tau)}\, d\tau \right]^{1/2}}{\sqrt[4]{q(x)}} \mathscr{C}_v\left[\lambda \int_\mu^x \sqrt{q(\tau)}\, d\tau \right] \qquad (7.3.178)$$

Higher approximations to (7.3.174) can be obtained by using Olver's approach (Section 7.3.5) by assuming that

$$v = A(z, \lambda)\zeta_i(z; \lambda) + B(z, \lambda)\zeta_i'(z; \lambda) \qquad (7.3.179)$$

where ζ_1 and ζ_2 are the independent solutions of (7.3.167).

7.3.10. TURNING POINT PROBLEMS OF HIGHER ORDER

Most of the interest in turning point problems for differential equations of order higher than two arose from the hydrodynamic stability of parallel flows. The linear stability problem for a parallel flow can be reduced to the solution of the so-called Orr–Sommerfeld equation (e.g., Lin, 1955)

$$\phi^{iv} - 2\alpha^2\phi'' + \alpha^4\phi = i\alpha R\{[U(y) - c](\phi'' - \alpha^2\phi) - U''(y)\phi\} \qquad (7.3.180)$$

for the disturbance amplitude $\phi(y)$. In this equation, $U(y)$ is the velocity profile of the undisturbed flow which is a known function. The parameters α and R are positive constants representing, respectively, the disturbance wave number and a flow Reynolds number. The parameter c is a complex constant whose real part c_r determines the wave speed, while its imaginary part c_i determines the damping or growth rate of the disturbance. This equation is supplemented by four homogeneous boundary conditions to form an eigenvalue problem for the eigenfunction ϕ and the eigenvalues c_r and c_i if α and R are known. The system is unstable if $c_i > 0$, stable if $c_i < 0$, and neutrally stable if $c_i = 0$.

For large αR, two independent solutions of this problem can be obtained in the form

$$\phi = \phi_0(y) + (\alpha R)^{-1}\phi_1(y) + \cdots \qquad (7.3.181)$$

while the remaining two solutions can be obtained in the form

$$\phi = e^{\pm\sqrt{\alpha R}\,\zeta}[(U - c)^{-5/4} + (\alpha R)^{-1/2}f_1(y) + \cdots] \qquad (7.3.182)$$

where $\zeta = \int_{y_0}^{y} \sqrt{i(U - c)}\, dy$. The above solution breaks down near the zeros of $U - c$, which are turning points of (7.3.180).

First-order uniformly valid asymptotic solutions were obtained for (7.3.180) by Tollmien (1947) and Wasow (1953), while complete uniformly valid expansions were obtained by Langer (1957, 1959a), Rabenstein (1959), and Lin and Rabenstein (1960, 1969). K. Tam (1968) obtained uniformly valid expansions for (7.3.180) using the method of multiple scales. Turning point problems for equations of the nth order were treated by Sibuya (1963a, b).

7.4. Wave Equations

In this section we describe some of the available techniques for determining approximate solutions of linear wave problems and their related elliptic problems. In our description of these techniques, we use the wave problem

$$c^2(\mathbf{r}) \nabla^2 v - v_{tt} - \omega_0{}^2(\mathbf{r})v = c^2(\mathbf{r})g(\mathbf{r})e^{i\omega t} \qquad (7.4.1)$$

If we let

$$v = u(\mathbf{r})e^{i\omega t} \qquad (7.4.2)$$

we obtain

$$\nabla^2 u + k^2 n^2(\mathbf{r})u = g(\mathbf{r}) \qquad (7.4.3)$$

where k and n are the wave number and refractive index given by

$$k = \frac{\omega}{c_0}, \qquad n^2 = \frac{\omega^2 - \omega_0^2(\mathbf{r})}{\omega^2} \frac{c_0^2}{c^2(\mathbf{r})} \qquad (7.4.4)$$

with c_0 a reference speed. In this problem we assume that g is deterministic, while n can be a random function so that the results may be applied to wave propagation in a random medium.

For a constant n the homogeneous problem admits a plane wave solution of the form

$$u = Ae^{ink\cdot\mathbf{r}} \qquad (7.4.5)$$

with constant A, while the inhomogeneous problem has the integral

$$u = -\int_V g(\boldsymbol{\xi}) \frac{e^{ink|\mathbf{r}-\boldsymbol{\xi}|}}{4\pi |\mathbf{r} - \boldsymbol{\xi}|} d\boldsymbol{\xi} \qquad (7.4.6)$$

where $\boldsymbol{\xi}$ is a variable vector ranging over the scattering volume V. However, if n is not a constant, we seek asymptotic expansions to the solutions of (7.4.3). The choice of an asymptotic technique to obtain an approximate solution depends on the value of k and the spatial variation of n. If n deviates slightly from a constant, we can use the so-called Born expansion developed by physicists and the Neumann expansion developed by mathematicians, renormalization techniques, or the Rytov technique. If k is large or n is a slowly varying function of position, we can use the geometrical optics technique. Although these techniques were developed for deterministic problems, they can also be used for stochastic problems. For the latter problems we also describe the so-called smoothing technique which is the counterpart of the method of averaging discussed in Chapter 5. For an account of the applications of these techniques and for more references, we refer the reader to the books by Chernov (1960), Tatarski (1961), Babich (1970, 1971), and the survey article by Frisch (1968).

7.4.1. THE BORN OR NEUMANN EXPANSION AND THE FEYNMAN DIAGRAMS

This technique is applicable when n deviates slightly from a constant. In this case we assume that k and n have been normalized so that the constant part of n is unity, thereby allowing n^2 to be written as

$$n^2 = 1 + \epsilon\chi(\mathbf{r}) \qquad (7.4.7)$$

with ϵ small and $\chi = O(1)$. For the statistical problem we assume that χ is a centered random function of $\mathbf{r}$ so that its mean, denoted by $\langle\chi\rangle$, is equal to zero. To obtain a Born expansion (Born, 1926), we let

$$u = \sum_{m=0}^{\infty} \epsilon^m u_m \qquad (7.4.8)$$

Substituting this expansion into (7.4.3), using (7.4.7), and equating coefficients of like powers of ϵ, we obtain

$$L(u_0) = g \qquad (7.4.9)$$

$$L(u_m) = -k^2\chi u_{m-1} \quad \text{for} \quad m \geq 1 \qquad (7.4.10)$$

where the operator L is defined by

$$L = \nabla^2 + k^2 \qquad (7.4.11)$$

Equations (7.4.9) and (7.4.10) can be solved successively. Thus for a given m, the right-hand side of (7.4.10) is known from the previous solution. Hence its solution is given by

$$u_m(\mathbf{r}) = -k^2 \int_V \chi(\mathbf{r}_m) u_{m-1}(\mathbf{r}_m) G_0(\mathbf{r};\mathbf{r}_m)\, d\mathbf{r}_m, \qquad m \geq 1 \qquad (7.4.12)$$

where $\mathbf{r}_m$ is a variable vector ranging over the scattering volume V, and G_0 is the free-space Green's function

$$G_0(\mathbf{r}, \boldsymbol{\xi}) = -\frac{e^{ik|\mathbf{r}-\boldsymbol{\xi}|}}{4\pi |\mathbf{r} - \boldsymbol{\xi}|} \qquad (7.4.13)$$

If $g \equiv 0$, (7.4.9) admits the following plane wave solution

$$u_0 = Ae^{i\mathbf{k}\cdot\mathbf{r}} \qquad (7.4.14)$$

where A is a constant. Then from (7.4.12)

$$u_1 = -Ak^2 \int_V \chi(\mathbf{r}_1)e^{i\mathbf{k}\cdot\mathbf{r}_1}G_0(\mathbf{r};\mathbf{r}_1)\, d\mathbf{r}_1 \qquad (7.4.15)$$

$$u_2 = Ak^4 \int_V \chi(\mathbf{r}_1)\chi(\mathbf{r}_2)e^{i\mathbf{k}\cdot\mathbf{r}_1}G_0(\mathbf{r};\mathbf{r}_2)G_0(\mathbf{r}_2;\mathbf{r}_1)\, d\mathbf{r}_1\, d\mathbf{r}_2 \qquad (7.4.16)$$

$$u_m = A(-1)^m k^{2m} \int_V \chi(\mathbf{r}_1)\chi(\mathbf{r}_2)\cdots\chi(\mathbf{r}_m)$$

$$\times\, e^{i\mathbf{k}\cdot\mathbf{r}_1}G_0(\mathbf{r};\mathbf{r}_m)G_0(\mathbf{r}_m;\mathbf{r}_{m-1})\cdots G_0(\mathbf{r}_2;\mathbf{r}_1)\, d\mathbf{r}_1\, d\mathbf{r}_2\cdots d\mathbf{r}_m \qquad (7.4.17)$$

In this expansion, ϵu_1 is called the first Born approximation, while $\epsilon^m u_m$ is called the mth Born approximation.

If χ is a centered random function of $\mathbf{r}$, the mean of u can be obtained by averaging (7.4.8). The result is

$$\langle u \rangle = Ae^{i\mathbf{k}\cdot\mathbf{r}} + A\epsilon^2 k^4 \int_V \langle\chi(\mathbf{r}_1)\chi(\mathbf{r}_2)\rangle e^{i\mathbf{k}\cdot\mathbf{r}_1} G_0(\mathbf{r};\mathbf{r}_2)G_0(\mathbf{r}_2;\mathbf{r}_1)\, d\mathbf{r}_1\, d\mathbf{r}_2 + \cdots$$

$$+ A(-\epsilon)^m k^{2m} \int_V \langle\chi(\mathbf{r}_1)\chi(\mathbf{r}_2)\cdots\chi(\mathbf{r}_m)\rangle$$

$$\times e^{i\mathbf{k}\cdot\mathbf{r}_1} G_0(\mathbf{r};\mathbf{r}_m)G_0(\mathbf{r}_m;\mathbf{r}_{m-1})\cdots G_0(\mathbf{r}_2;\mathbf{r}_1)\, d\mathbf{r}_1\, d\mathbf{r}_2 \cdots d\mathbf{r}_m + \cdots$$

$$(7.4.18)$$

The averaged quantities $\langle\chi(\mathbf{r}_1)\chi(\mathbf{r}_2)\cdots\chi(\mathbf{r}_m)\rangle$ depend on the configuration of the points $\mathbf{r}_1, \mathbf{r}_2, \ldots, \mathbf{r}_m$ because for most random media there exists a correlation length l (i.e., the values of χ at points of separation larger than l are uncorrelated). To express the dependence on the correlation length, we expand these averaged quantities in the following cluster (cumulant) expansions

$$\langle\chi(\mathbf{r}_1)\chi(\mathbf{r}_2)\rangle = R(\mathbf{r}_1, \mathbf{r}_2)$$

$$\langle\chi(\mathbf{r}_1)\chi(\mathbf{r}_2)\chi(\mathbf{r}_3)\rangle = R(\mathbf{r}_1, \mathbf{r}_2, \mathbf{r}_3)$$

$$\langle\chi(\mathbf{r}_1)\chi(\mathbf{r}_2)\chi(\mathbf{r}_3)\chi(\mathbf{r}_4)\rangle = R(\mathbf{r}_1, \mathbf{r}_2)R(\mathbf{r}_3, \mathbf{r}_4) + R(\mathbf{r}_1, \mathbf{r}_3)R(\mathbf{r}_2, \mathbf{r}_4)$$

$$+ R(\mathbf{r}_1, \mathbf{r}_4)R(\mathbf{r}_2, \mathbf{r}_3) + R(\mathbf{r}_1, \mathbf{r}_2, \mathbf{r}_3, \mathbf{r}_4)$$

$$(7.4.19)$$

$$\langle\chi(\mathbf{r}_1)\chi(\mathbf{r}_2)\cdots\chi(\mathbf{r}_m)\rangle = \sum_{k_1+\cdots+k_s=m} R(\boldsymbol{\xi}_1, \ldots, \boldsymbol{\xi}_{k_1})R(\boldsymbol{\zeta}_1, \ldots, \boldsymbol{\zeta}_{k_2})\cdots$$

$$\times R(\boldsymbol{\eta}_1, \ldots, \boldsymbol{\eta}_{k_s})$$

where $k_i \geq 2$. Thus the summation in the last equation is extended over all possible partitions of the set $\mathbf{r}_1, \mathbf{r}_2, \ldots, \mathbf{r}_m$ into clusters of at least two points. If χ is a centered Gaussian random function, all correlation functions vanish except the two-point correlation functions. Using (7.4.19) in (7.4.18), we obtain an expression for $\langle u \rangle$ which exhibits the dependence on the k-point correlation functions.

If $g \not\equiv 0$ the particular solution of (7.4.9) is

$$u_0(\mathbf{r}) = Mg = \int_V g(\mathbf{r}_0)G_0(\mathbf{r};\mathbf{r}_0)\, d\mathbf{r}_0 \qquad (7.4.20)$$

Then from (7.4.12)

$$u_m(\mathbf{r}) = (-1)^m k^{2m} \int_V g(\mathbf{r}_0)\chi(\mathbf{r}_1)\chi(\mathbf{r}_2)\cdots\chi(\mathbf{r}_m)G_0(\mathbf{r};\mathbf{r}_m)$$

$$\times G_0(\mathbf{r}_m;\mathbf{r}_{m-1})\cdots G_0(\mathbf{r}_2;\mathbf{r}_1)G_0(\mathbf{r}_1;\mathbf{r}_0)\, d\mathbf{r}_0\, d\mathbf{r}_1 \cdots d\mathbf{r}_m \qquad (7.4.21)$$

or in operator form

$$u_m = (-k^2 M\chi)^m Mg \qquad (7.4.22)$$

where the operator M is defined in (7.4.20). Therefore

$$u(\mathbf{r}) = Mg + \sum_{m=1}^{\infty} \epsilon^m (-k^2 M\chi)^m Mg \qquad (7.4.23)$$

This series is also called the Neumann series by mathematicians. It can also be obtained by converting (7.4.3) into the integral equation

$$u = Mg - \epsilon k^2 M\chi u \qquad (7.4.24)$$

and solving it by iteration.

From (7.4.23) we can define Green's function G for (7.4.3) as

$$G(\mathbf{r};\mathbf{r}_0) = G_0(\mathbf{r};\mathbf{r}_0) - \epsilon k^2 \int \chi(\mathbf{r}_1) G_0(\mathbf{r};\mathbf{r}_1) G_0(\mathbf{r}_1;\mathbf{r}_0)\, d\mathbf{r}_1$$

$$+ \epsilon^2 k^4 \int \chi(\mathbf{r}_1)\chi(\mathbf{r}_2) G_0(\mathbf{r};\mathbf{r}_2) G_0(\mathbf{r}_2;\mathbf{r}_1) G_0(\mathbf{r}_1;\mathbf{r}_0)\, d\mathbf{r}_1\, d\mathbf{r}_2$$

$$+ \cdots + (-1)^m \epsilon^m k^{2m} \int \chi(\mathbf{r}_1)\chi(\mathbf{r}_2) \cdots \chi(\mathbf{r}_m) G_0(\mathbf{r};\mathbf{r}_m)$$

$$\times\, G_0(\mathbf{r}_m;\mathbf{r}_{m-1}) \cdots G_0(\mathbf{r}_2;\mathbf{r}_1) G_0(\mathbf{r}_1;\mathbf{r}_0)\, d\mathbf{r}_1\, d\mathbf{r}_2 \cdots d\mathbf{r}_m + \cdots$$

$$(7.4.25)$$

or in operator form as

$$G = \sum_{m=0}^{\infty} (G_0 \mathscr{L})^m G_0 \quad \text{with} \quad \mathscr{L}_\phi \to -\epsilon k^2 \chi(\mathbf{r})\phi(\mathbf{r}) \qquad (7.4.26)$$

This series was represented by Frisch (1968) by what he called "bare" diagrams using the following conventions: G_0 is represented by a solid line and $\mathscr{L}$ is represented by a dot. Then G is represented by the diagram series

$$G = \underset{\mathbf{r}\quad\mathbf{r}_0}{\underline{\qquad}} + \underset{\mathbf{r}\quad\mathbf{r}_1\quad\mathbf{r}_0}{\underline{\qquad\bullet\qquad}} + \underset{\mathbf{r}\quad\mathbf{r}_2\quad\mathbf{r}_1\quad\mathbf{r}_0}{\underline{\qquad\bullet\qquad\bullet\qquad}} + \cdots \qquad (7.4.27)$$

This series has a multiple scattering physical interpretation. The mth term corresponds to a wave which propagates freely from $\mathbf{r}_0$ to $\mathbf{r}_1$, is scattered at $\mathbf{r}_1$ by the inhomogeneities, propagates freely to $\mathbf{r}_2$, is scattered at $\mathbf{r}_2$, and so on. Frish (1968) represented the double Green's function (the tensor product of G and its complex conjugate) by

$$G \otimes \bar{G} = G(\mathbf{r};\mathbf{r}_0)\bar{G}(\boldsymbol{\xi};\boldsymbol{\xi}_0)$$

in the following "bare" double-diagram series

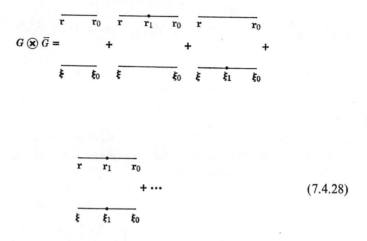

$$(7.4.28)$$

where each double diagram is the tensor product of the operator corresponding to the upper line, with the complex conjugate of the operator corresponding to the lower line.

If χ is a centered random process, $\langle G \rangle$ can be represented by the following "dressed" diagram series (Frisch, 1968).

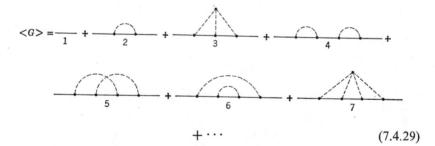

$$(7.4.29)$$

In this diagram series the following conventions were used:

(1) Points belonging to a given cluster were connected by a dotted line.

(2) To every "bare" diagram involving k factors of χ, we associated as many "dressed" diagrams as there are different partitions of $\mathbf{r}_1, \mathbf{r}_2, \ldots, \mathbf{r}_k$ into clusters of at least two points.

(3) To calculate a "dressed" diagram, the solid lines are replaced by G_0, the cluster of dotted lines ending at $\mathbf{r}_1, \mathbf{r}_2, \ldots, \mathbf{r}_s$ by the factors

$$(-\epsilon k^2)^s R(\mathbf{r}_1, \mathbf{r}_2, \ldots, \mathbf{r}_s),$$

and the integration is performed over all intermediate points. Thus

$$\underset{\mathbf{r} \quad \mathbf{r}_2 \quad \mathbf{r}_1 \quad \mathbf{r}_0}{\overparen{}} = \epsilon^2 k^4 \int G_0(\mathbf{r};\mathbf{r}_2)G_0(\mathbf{r}_2;\mathbf{r}_1)G_0(\mathbf{r}_1;\mathbf{r}_0)R(\mathbf{r}_1,\mathbf{r}_2)\,d\mathbf{r}_1\,d\mathbf{r}_2$$

$$(7.4.30)$$

$$\underset{\mathbf{r} \quad \mathbf{r}_4 \quad \mathbf{r}_3 \quad \mathbf{r}_2 \quad \mathbf{r}_1 \quad \mathbf{r}_0}{\overparen{}} = \epsilon^4 k^8 \int G_0(\mathbf{r};\mathbf{r}_4)G_0(\mathbf{r}_4;\mathbf{r}_3)G_0(\mathbf{r}_3;\mathbf{r}_2)G_0(\mathbf{r}_2;\mathbf{r}_1)G_0(\mathbf{r}_1;\mathbf{r}_0)$$

$$\times\; R(\mathbf{r}_4,\mathbf{r}_2)R(\mathbf{r}_3,\mathbf{r}_1)\,d\mathbf{r}_1\,d\mathbf{r}_2\,d\mathbf{r}_3\,d\mathbf{r}_4 \qquad (7.4.31)$$

Similarly, we can express the covariance $\langle G \otimes \bar{G} \rangle$ in the following "dressed" double-diagram series

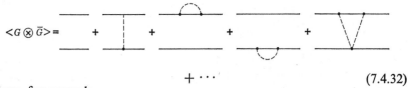

$$+ \cdots \qquad (7.4.32)$$

where, for example

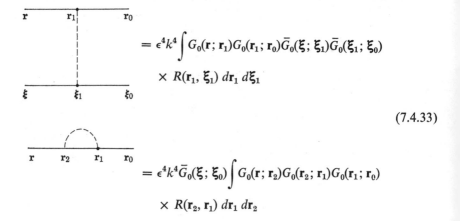

$$= \epsilon^4 k^4 \int G_0(\mathbf{r};\mathbf{r}_1)G_0(\mathbf{r}_1;\mathbf{r}_0)\bar{G}_0(\boldsymbol{\xi};\boldsymbol{\xi}_1)\bar{G}_0(\boldsymbol{\xi}_1;\boldsymbol{\xi}_0)$$

$$\times\; R(\mathbf{r}_1,\boldsymbol{\xi}_1)\,d\mathbf{r}_1\,d\boldsymbol{\xi}_1$$

$$(7.4.33)$$

$$= \epsilon^4 k^4 \bar{G}_0(\boldsymbol{\xi};\boldsymbol{\xi}_0)\int G_0(\mathbf{r};\mathbf{r}_2)G_0(\mathbf{r}_2;\mathbf{r}_1)G_0(\mathbf{r}_1;\mathbf{r}_0)$$

$$\times\; R(\mathbf{r}_2,\mathbf{r}_1)\,d\mathbf{r}_1\,d\mathbf{r}_2$$

For a Gaussian random function, only two-point correlation functions are nonvanishing, hence diagrams such as 3 and 7 in (7.4.29) and 5 in (7.4.32) vanish.

Representation of formal perturbation series by diagrams was first used by Feynman (1948). These diagrams are called Feynman diagrams and are widely used in statistical thermodynamics (see, for example, Prigoginé, 1962), the many-body problem (e.g., Van Hove, Hugenholtz, and Howland, 1961), and quantum electrodynamics (e.g., Balescu, 1963). The first use of Feynman diagrams for solving linear stochastic equations was made by Kraichnan (1961). Feynman diagrams were introduced for the solution of wave propagation in random media by Bourret (1962a, b), Furutsu (1963), and Tatarski (1964) for the case of a Gaussian process, and by Frisch (1965, 1968) for the general case.

The expansions obtained in this section were shown to be divergent by Frisch (1968). Moreover, he also showed that these expansions contain secular terms which restrict the validity of these asymptotic expansions to small arguments. Since $\langle \chi(\mathbf{r}_1)\chi(\mathbf{r}_2) \cdots \chi(\mathbf{r}_{2m}) \rangle$ is the sum of $1.3.5 \cdots (2m - 1)$ two-point correlation functions, this number increases rapidly with increasing m and this is another reason why (7.4.29) and (7.4.32) are divergent. Shkarofsky (1971) modified Born's expansion for backscattering from turbulent plasmas to obtain saturation and cross-polarization. A technique for making such expansions more uniformly valid was developed by Rayleigh (1917) to make a similar expansion which he obtained for first scattering from a thin slab valid for many slabs. This technique is called renormalization, and it has been developed further and extended by several researchers, as discussed in the next section.

7.4.2. RENORMALIZATION TECHNIQUES

To illustrate the nature of the nonuniformity that might arise in the Born (Neumann) expansion and explore methods of remedying it, we discuss the simple example

$$u'' + k^2(\epsilon)u = 0, \qquad u(0) = 1 \qquad (7.4.34)$$

where k is constant and given by

$$k = \sum_{n=0}^{N} \epsilon^n k_n \qquad (7.4.35)$$

Outgoing waves are given exactly by

$$u = e^{ik(\epsilon)x} = \exp\left(ix \sum_{n=0}^{N} \epsilon^n k_n\right) \qquad (7.4.36)$$

However, the Born expansion can be obtained from (7.4.34) to be

$$u = e^{ik_0 x} \sum_{m=0}^{\infty} \frac{1}{m!} \left(ix \sum_{n=1}^{N} \epsilon^n k_n\right)^m$$

$$= e^{ik_0 x}[1 + i\epsilon k_1 x + \epsilon^2(ik_2 x - \tfrac{1}{2}k_1^2 x^2)$$

$$+ \epsilon^3(ik_3 x - k_1 k_2 x^2 - \tfrac{1}{6}ik_1^3 x^3) + \cdots] \qquad (7.4.37)$$

It is clear that this expansion is valid only for short distances, and it breaks down when $k_1 x = O(\epsilon^{-1})$ or larger. The origin of the nonuniformity can be seen by comparing this Born expansion with the exact solution (7.4.36). The Born expansion can be obtained from the exact solution by expanding $\exp (ix \sum_{n=1}^{N} \epsilon^n k_n)$ in Taylor series in terms of $ix \sum_{n=1}^{N} \epsilon^n k_n$, then expanding $(\sum_{n=1}^{N} \epsilon^n k_n)^m$ in powers of ϵ and collecting coefficients of like powers of ϵ. Although the Taylor series expansion of $\exp \phi$ in terms of ϕ converges uniformly and absolutely for all values of ϕ, a finite number of terms cannot be used to represent $\exp \phi$ with a given accuracy for all values of ϕ. Consequently, any uniformly valid expansion of $\exp \phi$ for all ϕ must be $\exp \phi$ itself. Thus to determine an expansion from (7.4.37) valid for x as large as $O(\epsilon^{-1})$, we must sum the sequence

$$1 + i\epsilon k_1 x + \cdots + (i\epsilon k_1 x)^m/m! + \cdots .$$

Its sum is $\exp (ik_1 x)$, and (7.4.37) becomes

$$u = \exp [i(k_0 + \epsilon k_1)x] \sum_{m=0}^{\infty} \frac{1}{m!} \left(ix \sum_{n=2}^{N} \epsilon^n k_n \right)^m$$

$$= \exp [i(k_0 + \epsilon k_1)x](1 + i\epsilon^2 k_2 x + i\epsilon^3 k_3 x - \tfrac{1}{2}\epsilon^4 k_2^2 x^2 + \cdots) \quad (7.4.38)$$

This expansion breaks down when $k_2 x = O(\epsilon^{-2})$. To increase the range of validity of this expansion to values of $x = O(\epsilon^{-2})$, we must sum the sequence $\sum_{m=0}^{\infty} (i\epsilon^2 k_2 x)^m/m!$. An effective technique of summing these sequences without knowing their explicit functional dependence is the method of multiple scales of Chapter 6. Another technique for effecting the summation of the secular terms is the renormalization technique.

The renormalization technique was originally developed by Rayleigh (1917) to generalize his first scattering from a thin slab to scattering from many slabs. He obtained an expansion of the form

$$u = e^{ik_0 x}(1 + i\epsilon \mu x) \quad (7.4.39)$$

for first scattering from one slab. To obtain a solution valid for many slabs, he recast this expansion into an exponential; that is

$$u = \exp [i(k_0 + \epsilon \mu)x] \quad (7.4.40)$$

In this manner he effectively summed the sequence $\sum\limits_{m=1}^{\infty} (i\epsilon\mu x)^m/m!$ of secular terms. The process of summing expansions to make them "more" uniformly valid is called renormalization. This technique was rediscovered by Pritulo (1962) as described in Section 3.4.

To make a more uniformly valid expansion from a two-term Born expansion $u = u_0 + \epsilon u_1$, we recast it into the following exponential

If $u_0 = A_0 \exp iS_0(\mathbf{r})$, then

$$u = u_0 e^{\epsilon u_1/u_0} \tag{7.4.41}$$

$$u = A e^{iS(\mathbf{r})} \tag{7.4.42}$$

where $A = A_0 \exp [\epsilon \text{ Real } (u_1/u_0)]$ and $S = S_0 + \epsilon \text{ Imaginary } (u_1/u_0)$.

This renormalization technique has been extended to obtain the kinetic equations for weakly nonlinear systems (see, for example, Van Hove, 1955, 1957; Prigoginé, 1962; Balescu, 1963, and Al'tshul' and Karpman, 1966). According to this technique, sequences of secular terms are separated and summed with or without the use of Feynman diagrams. The summation of the principal sequences of secular terms leads to quasi-linear equations.

The renormalization technique has also been widely used in the study of wave propagation in random media (e.g., Tatarski, 1961, Chapter 6; Keller, 1962; Karal and Keller, 1964). Thus to determine a more uniformly valid expansion for $\langle G \rangle$ of (7.4.29), we recast it into an exponential

$$\langle G \rangle = G_0 e^{\psi} \tag{7.4.43}$$

Hence the first renormalization gives

$$\psi = \frac{G_2}{G_0} \tag{7.4.44}$$

where

$$G_2 = \epsilon^2 k^4 \int G_0(\mathbf{r}; \mathbf{r}_2) G_0(\mathbf{r}_2; \mathbf{r}_1) G_0(\mathbf{r}_1; \mathbf{r}_0) R(\mathbf{r}_1; \mathbf{r}_2) \, d\mathbf{r}_1 \, d\mathbf{r}_2 \tag{7.4.45}$$

Diagrammatic summation techniques have been used to determine renormalization equations to any order by Bourret (1962a, b), Furutsu (1963), Tatarski (1964), and Frisch (1965). Tatarski (1964) arranged the diagrams for $\langle G \rangle$ and $\langle G \otimes \bar{G} \rangle$ in such a manner that he could recognize that they were the Neumann expansions of two integral equations with two kernels having an infinite number of terms. For a centered Gaussian index of refraction, the diagrammatic expansion of $\langle G \rangle$ is

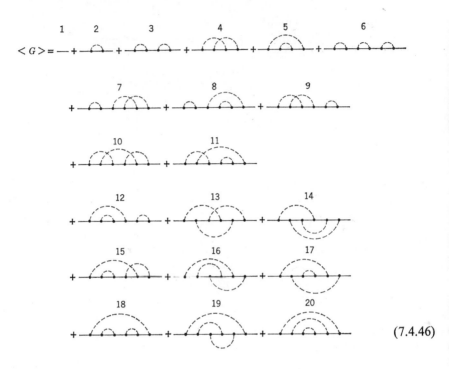

$$(7.4.46)$$

To obtain an integral equation for $\langle G \rangle$ with a kernel consisting of an infinite series, we look at the topology of the diagrams in (7.4.46). Let us make the following definitions.

(1) A diagram without terminals is a diagram stripped of its external solid lines, such as diagram 2 without terminals ⌒ and diagram 4 without terminals ⌒⌒.

(2) A diagram without terminals is connected if it cannot be cut into two or more diagrams without cutting any dotted lines. Diagrams 4, 5, 10, 11, and 13 through 20 are connected, while diagrams 3, 6, 7, 8, 9, and 12 are unconnected. Unconnected diagrams can be factored; for example,

⌒⌒ can be written as the product of the following five diagrams

———— , ⌒ , ———— , ⌒ , ————

(3) The mass operator Q denoted by ● is the sum of all possible connected diagrams contributing to $\langle G \rangle$; that is

$$\bullet = \underbrace{} + \underbrace{} + \underbrace{} + \cdots$$

$$(7.4.47)$$

All unconnected diagrams composed of two connected diagrams occur in the sum of diagrams ●———●, while all unconnected diagrams composed of three connected diagrams occur in the sum of diagrams ●———●———●. From this we note that $\langle G \rangle$, which is denoted by ———, is governed by the following Dyson equation (Tatarski, 1964)

$$\text{———} = \text{———} + \text{———}\bullet\text{———} \qquad (7.4.48)$$

or in analytic form

$$\langle G(\mathbf{r}; \mathbf{r}_0) \rangle = G_0(\mathbf{r}; \mathbf{r}_0) + \int G_0(\mathbf{r}; \mathbf{r}_1) Q(\mathbf{r}_1; \mathbf{r}_2) \langle G(\mathbf{r}_2; \mathbf{r}_0) \rangle \, d\mathbf{r}_1 \, d\mathbf{r}_2 \quad (7.4.49)$$

A similar equation has been used extensively in quantum electrodynamics, quantum field theory, and the many-body problem, and it was first introduced by Dyson (1949). If χ is homogeneous, the mass operator Q is invariant under translations, hence it is a convolution operator whose Fourier transform $Q(\mathbf{k})$ is a multiplication operator of ordinary functions. Hence, taking the Fourier transform of (7.4.49), we obtain

$$\langle G(\boldsymbol{\varkappa}) \rangle = \frac{1}{k^2 - \kappa^2} + \frac{1}{k^2 - \kappa^2} Q(\boldsymbol{\varkappa}) \langle G(\boldsymbol{\varkappa}) \rangle \qquad (7.4.50)$$

because

$$G_0(\boldsymbol{\varkappa}) = \frac{1}{k^2 - \kappa^2} \qquad (7.4.51)$$

Solving (7.4.50) for $\langle G(\boldsymbol{\varkappa}) \rangle$, we obtain

$$\langle G(\boldsymbol{\varkappa}) \rangle = \frac{1}{k^2 - \kappa^2 - Q(\boldsymbol{\varkappa})} \qquad (7.4.52)$$

Thus if Q is known, $\langle G \rangle$ can be obtained by inverting $\langle G(\boldsymbol{\varkappa}) \rangle$. However, the exact expression for Q is as difficult to find as $\langle G \rangle$. Thus one approximates Q. The simplest approximation based on the Dyson equation retains only the first term in the mass operator; that is

$$\text{———} = \text{———} + \underbrace{} \qquad (7.4.53)$$

or in analytic form

$$\langle G(\mathbf{r}; \mathbf{r}_0) \rangle = G_0(\mathbf{r}; \mathbf{r}_0) + \epsilon^2 k^4 \int G_0(\mathbf{r}; \mathbf{r}_1) G_0(\mathbf{r}_1; \mathbf{r}_2) R(\mathbf{r}_1; \mathbf{r}_2) \langle G(\mathbf{r}_2; \mathbf{r}_0) \rangle \, d\mathbf{r}_1 \, d\mathbf{r}_2$$

(7.4.54)

This equation is called the first renormalization equation, and it was introduced in diagrammatic form by Bourret (1962a, b).

It should be mentioned that (7.4.49) and (7.4.54) cannot be solved by iteration, because this would lead to secular terms. An approximate solution to (7.4.54) is given by Varvatsis and Sancer (1971)

$$\langle G(\mathbf{r}; \mathbf{r}_0) \rangle = G_0(\mathbf{r}; \mathbf{r}_0) e^{\epsilon^2 G_2/G_0}$$

(7.4.55)

where

$$G_2 = k^4 \int G_0(\mathbf{r}; \mathbf{r}_1) G_0(\mathbf{r}_1; \mathbf{r}_2) G_0(\mathbf{r}_2; \mathbf{r}_0) R(\mathbf{r}_1; \mathbf{r}_2) \, d\mathbf{r}_1 \, d\mathbf{r}_2$$

(7.4.56)

This solution is valid for homogeneous as well as inhomogeneous random media. It is the same solution that would be obtained if we had recast $\langle G \rangle = G_0 + \epsilon^2 G_2$ into an exponential.

Using diagrammatic techniques, Tatarski (1964) and Frisch (1968) obtained the following Bethe–Salpeter equation for a centered Gaussian and general refractive index

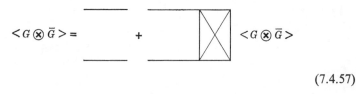

(7.4.57)

where the intensity operator ⬚ consists of all connected diagrams without terminals in the expansion of $\langle G \otimes \bar{G} \rangle$; that is

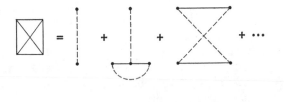

(7.4.58)

This equation was first introduced by Salpeter and Bethe (1951) for relativistic bound-state problems.

7.4.3. RYTOV'S METHOD

To obtain an approximate solution of

$$\nabla^2 u + k^2[1 + \epsilon\chi(\mathbf{r})]u = 0 \tag{7.4.59}$$

we follow Rytov (1937) by assuming that (see also Tatarski, 1961, pp. 121–128; Chernov 1960, pp. 58–67)

$$u = e^\psi \tag{7.4.60}$$

thereby transforming (7.4.59) into

$$\nabla^2\psi + \nabla\psi \cdot \nabla\psi + k^2(1 + \epsilon\chi) = 0 \tag{7.4.61}$$

Now we assume that ψ possesses the following asymptotic expansion

$$\psi = \sum_{n=0}^{\infty} \epsilon^n \psi_n(\mathbf{r}) \tag{7.4.62}$$

Substituting (7.4.62) into (7.4.61) and equating coefficients of like powers of ϵ, we obtain

$$\nabla^2\psi_0 + \nabla\psi_0 \cdot \nabla\psi_0 = -k^2 \tag{7.4.63}$$

$$\nabla^2\psi_1 + 2\nabla\psi_1 \cdot \nabla\psi_0 = -k^2\chi \tag{7.4.64}$$

$$\nabla^2\psi_n + \sum_{m=0}^{n} \nabla\psi_m \cdot \nabla\psi_{n-m} = 0, \qquad n \geq 2 \tag{7.4.65}$$

These equations can be solved successively. The resulting expansion corresponds exactly to the Born expansion $u = \sum_{n=0}^{\infty} \epsilon^n u_n$ of (7.4.59). In fact, the Rytov expansion can be obtained from the Born series by recasting the latter into an exponential. Setting

$$\sum_{n=0}^{\infty} \epsilon^n u_n = \exp\left(\sum_{m=0}^{\infty} \epsilon^m \psi_m\right) \tag{7.4.66}$$

expanding the exponential for small ϵ, and equating coefficients of like powers of ϵ, we obtain for the first terms

$$u_0 = e^{\psi_0}, \qquad u_1 = e^{\psi_0}\psi_1, \qquad u_2 = e^{\psi_0}(\psi_2 + \tfrac{1}{2}\psi_1^2)$$

$$u_3 = e^{\psi_0}(\psi_3 + \psi_1\psi_2 + \tfrac{1}{6}\psi_1^3) \tag{7.4.67}$$

Thus the Rytov expansion is a renormalized Born expansion, hence it should be more uniformly valid than the Born expansion. However, this conclusion is controversial (Hufnagel and Stanley, 1964; deWolf 1965, 1967; Brown, 1966, 1967; Fried 1967; Heidbreder, 1967; Taylor, 1967; Strohbehn, 1968; Sancer and Varvatsis, 1969, 1970).

7.4.4. THE GEOMETRICAL OPTICS APPROXIMATION

The object is to obtain an asymptotic solution for

$$\nabla^2 u + k^2 n^2(\mathbf{r})u = 0 \qquad (7.4.68)$$

for large wave number k (i.e., small wavelength $\lambda = 2\pi/k$). For large k, (7.4.68) has an asymptotic expansion of the form (Keller, 1958)

$$u = e^{ikS(\mathbf{r})} \sum_{m=0}^{\infty} (ik)^{-m} u_m(\mathbf{r}) \qquad (7.4.69)$$

Substituting (7.4.69) into (7.4.68) and equating coefficients of like powers of k, we obtain

$$\nabla S \cdot \nabla S = n^2(\mathbf{r}) \qquad \text{(eiconal equation)} \qquad (7.4.70)$$

$$2\,\nabla S \cdot \nabla u_0 + u_0 \nabla^2 S = 0 \qquad \text{(transport equations)} \qquad (7.4.71)$$

$$2\,\nabla S \cdot \nabla u_m + u_m \nabla^2 S = -\nabla^2 u_{m-1} \quad \text{for} \quad m \geq 1 \qquad (7.4.72)$$

Equation (7.4.70) can be solved using the method of characteristics; that is

$$\frac{d\mathbf{r}}{d\sigma} = 2\lambda\,\nabla S$$

$$\frac{dS}{d\sigma} = 2\lambda n^2 \qquad (7.4.73)$$

$$\frac{d}{d\sigma}(\nabla S) = \lambda\,\nabla(n^2)$$

where σ is a parameter and λ is a proportionality function. Elimination of ∇S from the first and third equations gives

$$\frac{d}{d\sigma}\left(\frac{1}{2\lambda}\frac{d\mathbf{r}}{d\sigma}\right) = 2\lambda n\nabla n \qquad (7.4.74)$$

Then the solution of the second equation in (7.4.73) is

$$S = S_0 + \int_{\sigma_0}^{\sigma} 2\lambda n^2[\mathbf{r}(\tau)]\,d\tau \qquad (7.4.75)$$

where $\mathbf{r}(\sigma)$ is the solution of (7.4.74) subject to the initial conditions $\mathbf{r}(\sigma_0) = \mathbf{r}_0$, $d\mathbf{r}(\sigma_0)/d\sigma = \dot{\mathbf{r}}_0$. Choosing $2\lambda = n^{-1}$ and σ to be the arc length along the rays, we rewrite (7.4.74) and (7.4.75) as

$$\frac{d}{d\sigma}\left[n(\mathbf{r}(\sigma))\frac{d\mathbf{r}(\sigma)}{d\sigma}\right] = \nabla n \qquad (7.4.76)$$

$$S = S_0 + \int_{\sigma_0}^{\sigma} n[\mathbf{r}(\tau)]\,d\tau \qquad (7.4.77)$$

Equation (7.4.71) becomes

$$2n[\mathbf{r}(\sigma)]\frac{du_0}{d\sigma} + u_0\nabla^2 S[\mathbf{r}(\sigma)] = 0$$

along the rays. Its solution is

$$u_0 = u_0[\mathbf{r}(\sigma_0)]\exp\left\{-\int_{\sigma_0}^{\sigma}\frac{\nabla^2 S[\mathbf{r}(\tau)]}{2n[\mathbf{r}(\tau)]}\,d\tau\right\} \qquad (7.4.78)$$

Similarly, the solution of (7.4.72) is

$$u_m = c\,\frac{u_0[\mathbf{r}(\sigma)]}{u_0[\mathbf{r}(\sigma_0)]} - \int_{\sigma_0}^{\sigma}\frac{\nabla^2 u_{m-1}[\mathbf{r}(\tau)]}{2n[\mathbf{r}(\tau)]}\frac{u_0[\mathbf{r}(\sigma)]}{u_0[\mathbf{r}(\tau)]}\,d\tau \qquad (7.4.79)$$

where c is a constant to be determined from the initial conditions.

The expansion obtained in this section is not valid at a caustic (i.e., an envelope of rays), shadow boundaries, foci of the rays, and source points. In such regions neighboring rays intersect and the cross-sectional area of a tube of rays becomes zero. Since the energy is conserved in a tube of rays, the amplitude of the field must be infinite in these regions. The unboundedness of the field at a caustic is shown below, and an expansion valid everywhere including a caustic is obtained in the next section.

To show the breakdown of the expansion of this section at or near a caustic, we specialize it to the case $n(\mathbf{r}) = 1$. In this case the rays are straight lines according to (7.4.76), and $S = S_0 - \sigma_0 + \sigma$ according to (7.4.77). Now we express the solution of (7.4.71) and (7.4.72) in terms of a coordinate system with respect to the caustic which we assume to be smooth and convex. Figure 7-1 shows a point P outside the caustic and two rays passing through it. If we assign a direction to the caustic, this induces a direction for each ray which must be tangent to it at some point. Thus each point P outside the caustic lies on two rays—one has left the caustic and the other is approaching the caustic.

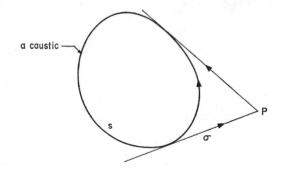

Figure 7-1

In two dimensions we let s measure the arc length along the caustic and σ measure the length from the point of tangency to the point P. Thus the position of P is determined by s_1 and σ_1 or s_2 and σ_2. In terms of these coordinates, the position vector $\mathbf{r}$ of P can be expressed as

$$\mathbf{r} = \mathbf{r}_0(s) + \sigma \mathbf{e}_1 \qquad (7.4.80)$$

where $\mathbf{r} = \mathbf{r}_0(s)$ is the equation of the caustic, and $\mathbf{e}_1 = d\mathbf{r}_0/ds$ is a unit vector tangent to the caustic. Differentiating (7.4.80), we have

$$d\mathbf{r} = \mathbf{e}_1 \, ds + \mathbf{e}_1 \, d\sigma + \frac{\sigma}{\rho} \mathbf{e}_2 \, ds$$

because $d\mathbf{e}_1/ds = \rho^{-1}\mathbf{e}_2$ where ρ^{-1} is the curvature of the caustic. Changing the variables from σ and s to

$$\begin{aligned} \xi &= s \\ \eta &= s + \sigma \end{aligned} \qquad (7.4.81)$$

we obtain

$$d\mathbf{r} = \mathbf{e}_1 \, d\eta + \frac{\eta - \xi}{\rho} \mathbf{e}_2 \, d\xi \qquad (7.4.82)$$

Hence

$$\nabla f = \frac{\partial f}{\partial \eta} \mathbf{e}_1 + \frac{\rho}{\eta - \xi} \frac{\partial f}{\partial \xi} \mathbf{e}_2 \qquad (7.4.83)$$

$$\nabla^2 f = \frac{1}{\eta - \xi} \frac{\partial}{\partial \eta}\left[(\eta - \xi)\frac{\partial f}{\partial \eta} \right] + \frac{\rho}{\eta - \xi} \frac{\partial}{\partial \xi}\left[\frac{\rho}{\eta - \xi} \frac{\partial f}{\partial \xi} \right] \qquad (7.4.84)$$

Since from (7.4.81)

$$\frac{\partial}{\partial \eta} = \frac{\partial}{\partial \sigma} \quad \text{and} \quad \frac{\partial}{\partial \xi} = \frac{\partial}{\partial s} - \frac{\partial}{\partial \sigma}$$

we rewrite (7.4.83) and (7.4.84) as

$$\nabla f = \frac{\partial f}{\partial \sigma} \mathbf{e}_1 + \frac{\rho}{\sigma}\left(\frac{\partial f}{\partial s} - \frac{\partial f}{\partial \sigma} \right) \mathbf{e}_2$$

$$\nabla^2 f = \frac{1}{\sigma}\frac{\partial}{\partial \sigma}\left[\sigma \frac{\partial f}{\partial \sigma} \right] + \frac{\rho}{\sigma}\left(\frac{\partial}{\partial s} - \frac{\partial}{\partial \sigma} \right)\left[\frac{\rho}{\sigma}\left(\frac{\partial f}{\partial s} - \frac{\partial f}{\partial \sigma} \right) \right] \qquad (7.4.85)$$

If we take the special case $S(s, \sigma_0) = s + \sigma_0$, then

$$S(s, \sigma) = s + \sigma \qquad (7.4.86)$$

which is double-valued corresponding to either the ray that has left or the one that is approaching the caustic. Using (7.4.85) and (7.4.86) in (7.4.78) and (7.4.79), we obtain

$$u_0(s, \sigma) = u_0(s, \sigma_0)\sqrt{\frac{\sigma_0}{\sigma}} \tag{7.4.87}$$

$$u_m(s, \sigma) = u_m(s, \sigma_0)\sqrt{\frac{\sigma_0}{\sigma}} - \frac{1}{2}\int_{\sigma_0}^{\sigma}\sqrt{\frac{\tau}{\sigma}}\nabla^2 u_{m-1}(s, \tau)\, d\tau \tag{7.4.88}$$

which are unbounded as $\sigma \to 0$ (i.e., a caustic). A modified expansion valid everywhere including the neighborhood of the caustic is obtained in the next section.

7.4.5. A UNIFORM EXPANSION AT A CAUSTIC

To determine an expansion valid at the caustic, we first must determine the size of the region of nonuniformity and the form of the solution in this region. To do this we let

$$u = \psi(\mathbf{r}, k)e^{ikS(\mathbf{r})} \tag{7.4.89}$$

in (7.4.68) with $n(\mathbf{r}) = 1$ and obtain

$$k^2\psi[1 - (\nabla S)^2] + ik(2\nabla S \cdot \nabla\psi + \psi\nabla^2 S) + \nabla^2\psi = 0 \tag{7.4.90}$$

Assuming that the k^2 term is the leading term, we obtain

$$\nabla S \cdot \nabla S = 1 \qquad \text{(eiconal equation)} \tag{7.4.91}$$

$$ik(2\nabla S \cdot \nabla\psi + \psi\nabla^2 S) + \nabla^2\psi = 0 \tag{7.4.92}$$

We take the solution of (7.4.91) as $S = s + \sigma$. Hence (7.4.92) becomes

$$\left\{ik\left(2\frac{\partial}{\partial\sigma} + \frac{1}{\sigma}\right) + \frac{1}{\sigma}\frac{\partial}{\partial\sigma}\left(\sigma\frac{\partial}{\partial\sigma}\right) + \frac{\rho}{\sigma}\left(\frac{\partial}{\partial s} - \frac{\partial}{\partial\sigma}\right)\left[\frac{\rho}{\sigma}\left(\frac{\partial}{\partial s} - \frac{\partial}{\partial\sigma}\right)\right]\right\}\psi = 0 \tag{7.4.93}$$

To analyze (7.4.93) near the caustic, we introduce the stretching transformation

$$\tau = k^\lambda\sigma, \qquad \lambda > 0$$

and obtain

$$ik^{1+\lambda}\left(2\frac{\partial}{\partial\tau} + \frac{1}{\tau}\right)\psi$$

$$+ \left\{\frac{k^{2\lambda}}{\tau}\frac{\partial}{\partial\tau}\left(\tau\frac{\partial}{\partial\tau}\right) + \frac{\rho k^{2\lambda}}{\tau}\left(\frac{\partial}{\partial s} - k^\lambda\frac{\partial}{\partial\tau}\right)\left[\frac{\rho}{\tau}\left(\frac{\partial}{\partial s} - k^\lambda\frac{\partial}{\partial\tau}\right)\right]\right\}\psi = 0 \tag{7.4.94}$$

The parameter λ is determined by requiring that the highest power of k within the braces (terms neglected in the leading term of the straightforward expansion of the previous section) be equal to the power of k in the first term (i.e., the leading term in the ray expansion). That is, we let

$$1 + \lambda = 4\lambda \quad \text{or} \quad \lambda = \tfrac{1}{3} \tag{7.4.95}$$

Then as $k \to \infty$, (7.4.94) tends to

$$i\left(2\frac{\partial \psi}{\partial \tau} + \frac{\psi}{\tau}\right) + \frac{\rho^2}{\tau}\frac{\partial}{\partial \tau}\left(\frac{1}{\tau}\frac{\partial \psi}{\partial \tau}\right) = 0 \tag{7.4.96}$$

Letting

$$\psi = e^{-i\tau^3/3\rho^2} V(z), \qquad z = \frac{\tau^2}{\rho \sqrt[3]{4\rho}} \tag{7.4.97}$$

we obtain

$$\frac{d^2V}{dz^2} + zV = 0 \tag{7.4.98}$$

whose solutions are the Airy functions $Ai(-z)$ and $Bi(-z)$ (see Section 7.3.1). Had we analyzed the behavior of the solution near a shadow boundary, we would have found that it must be represented by Weber or parabolic cylinder functions. Now we can either match this inner expansion with the outer expansion obtained in the previous section and with another outer expansion inside the caustic (Buchal and Keller, 1960), or form a single uniformly valid expansion imitating the expansions for turning point problems (see Section 7.3).

Following Kravtsov (1964a, b), Ludwig (1966), and Zauderer (1970b), we assume an asymptotic expansion of the form

$$u = e^{ik\theta(\mathbf{r})}\left\{g(\mathbf{r}, k)V[k^{2/3}\phi(\mathbf{r})] + \frac{1}{ik^{1/3}}h(\mathbf{r}, k)V'[k^{2/3}\phi(\mathbf{r})]\right\} \tag{7.4.99}$$

where θ and ϕ are determined from the analysis and $V(z)$ is given by (7.4.98). Substituting (7.4.99) into (7.4.68), using the fact that $V''' + zV' + V = 0$, and equating the coefficients of V and V' to zero, we obtain

$$-k^2g\left[(\nabla\theta)^2 + \phi(\nabla\phi)^2 - 1\right] - 2k^2\phi h\nabla\theta \cdot \nabla\phi$$
$$+ ik[2\nabla\theta \cdot \nabla g + g\nabla^2\theta + 2\phi\nabla\phi \cdot \nabla h + \phi h\nabla^2\phi + h(\nabla\phi)^2] + \nabla^2 g = 0 \tag{7.4.100}$$

$$-k^2h\left[(\nabla\theta)^2 + \phi(\nabla\phi)^2 - 1\right] - k^2g\nabla\theta \cdot \nabla\phi$$
$$+ ik[2\nabla\phi \cdot \nabla g + g\nabla^2\phi + 2\nabla\theta \cdot \nabla h + h\nabla^2\theta] + \nabla^2 h = 0 \tag{7.4.101}$$

The coefficients of k^2 in (7.4.100) and (7.4.101) vanish if

$$(\nabla\theta)^2 + \phi(\nabla\phi)^2 = 1$$
$$\nabla\theta \cdot \nabla\phi = 0 \tag{7.4.102}$$

To solve the resulting equations, we let

$$g(\mathbf{r}, k) = g_0(\mathbf{r}) + k^{-1}g_1(\mathbf{r}) + \cdots$$
$$h(\mathbf{r}, k) = h_0(\mathbf{r}) + k^{-1}h_1(\mathbf{r}) + \cdots \qquad (7.4.103)$$

and equate coefficients of like powers of k^{-1}. The first-order equations are

$$2\nabla\theta \cdot \nabla g_0 + g_0\nabla^2\theta + 2\phi\nabla\phi \cdot \nabla h_0 + \phi h_0\nabla^2\phi + h_0(\nabla\phi)^2 = 0$$
$$2\nabla\phi \cdot \nabla g_0 + g_0\nabla^2\phi + 2\nabla\theta \cdot \nabla h_0 + h_0\nabla^2\theta = 0 \qquad (7.4.104)$$

Equations (7.4.102) are equivalent to the eiconal equation (7.4.70), while (7.4.104) are equivalent to the transport equation (7.4.71). Similar expansions were obtained by Fowkes (1968, Part II) by using the method of multiple scales.

The system of equations (7.4.102) is a nonlinear system of equations which is elliptic where $\phi < 0$ (shadow region), hyperbolic where $\phi > 0$ (illuminated region), and parabolic where $\phi = 0$ (caustic curve or surface). For a convex analytic caustic, the systems (7.4.102) and (7.4.104) can be solved by expanding θ and ϕ in power series in terms of a coordinate system with respect to the caustic.

To see the connection between the system (7.4.102) and the eiconal equation (7.4.70), we multiply the second equation in (7.4.102) by $\pm 2\sqrt{\phi}$ (illuminated region where $\phi > 0$ is considered) and add the result to the first of these equations to obtain

$$(\nabla\theta \pm \sqrt{\phi}\,\nabla\phi)^2 = 1$$

or

$$(\nabla S)^2 = 1 \qquad (7.4.105)$$

where

$$S^{\pm} = \theta \pm \tfrac{2}{3}\phi^{3/2} \qquad (7.4.106)$$

Similarly, we multiply the second equation in (7.4.104) by $\pm\sqrt{\phi}$, add the result to the first of these equations, and use $\nabla\theta \cdot \nabla\phi = 0$ to obtain

$$2\nabla S^{\pm} \cdot \nabla\psi^{\pm} + [\nabla^2 S^{\pm} \mp \tfrac{1}{2}\phi^{-1/2}(\nabla\phi)^2]\psi^{\pm} = 0 \qquad (7.4.107)$$

where

$$\psi^{\pm} = g_0 \pm \sqrt{\phi}\,h_0 \qquad (7.4.108)$$

Equation (7.4.107) differs from the transport equation (7.4.71) only by the term $\pm(1/2)\phi^{-1/2}(\nabla\phi)^2$ which makes the coefficient of $\psi^{\pm}$ bounded near the caustic.

Replacing V and V' by their asymptotic expansions for large argument, we recover the geometrical optics expansion of the previous section ($\phi > 0$, illuminated region). Replacing V and V' by their asymptotic expansions for large argument for the case $\phi < 0$, we obtain an expansion in the shadow region which can be interpreted in terms of complex rays, phases, and transport coefficients.

Uniform asymptotic expansions for wave propagation and diffraction problems were reviewed by Ludwig (1970b) and Babich (1970, 1971). The role of coordinate systems in rendering the expansions uniformly valid was investigated by Zauderer (1970a).

A single uniformly valid expansion for a point source problem was obtained by Babich (1965), while three matched expansioñs for this problem were presented by Avila and Keller (1963).

Problems of diffraction by convex objects were treated by Zauderer (1964b), Buslaev (1964), Grimshaw (1966), Ludwig (1967), and Lewis, Bleistein and Ludwig (1967) among others. Diffraction by a transparent object was analyzed by Rulf (1968).

Uniform expansions for the problem of diffraction near the concave side of an object (whispering galley modes) were obtained by Kravtsov (1964b), Matkowsky (1966), and Ludwig (1970a) among others.

Characteristic transition regions in which two caustics are near each other are analogous to second-order turning points. Their uniform expansions involve Weber or parabolic cylinder functions. Some of these problems were treated by Kravtsov (1965), Babich and Kravtsova (1967), Weinstein (1969), and Zauderer (1970a, b).

Diffraction by a thin screen (Fresnel diffraction) was investigated by Wolfe (1967), Kersten (1967), and Ahluwalia, Lewis, and Boersma (1968).

Multiple transition regions arise from the tangential intersection of two or more caustics and shadow boundaries, such as near terminated caustics (Levey and Felsen, 1967), cusps on caustics (Ludwig, 1966), and points of diffraction for smooth objects (Ludwig, 1967). Problems involving transition regions were treated by Zauderer (1964a, 1970a), Fock (1965), Rulf (1967), and Bleistein (1967) among others.

7.4.6. THE METHOD OF SMOOTHING

To apply this technique to (7.4.3) with $n^2(\mathbf{r})$ a centered random function, we first convert it to the integral equation

$$u = Mg - \epsilon k^2 M\chi u \qquad (7.4.109)$$

where the operator M is defined by

$$Mg = \int_V g(\mathbf{r}_0)G_0(\mathbf{r}; \mathbf{r}_0) \, d\mathbf{r}_0 \qquad (7.4.110)$$

with G_0 the free-space Green's function. Frisch (1968) pointed out that many other linear equations of mathematical physics, such as the Liouville equation for an ensemble of classic interacting particles, the Hopf equation for turbulence, and the Fokker–Planck equation, can be put in the above integral equation form.

In general, we are not interested in the function u but in its projection Pu on a subspace of the original space over which u is defined. For example, in the problem of wave propagation in a random medium, $Pu = \langle u \rangle$, while in the case of N interacting particles, Pu is the N-particle velocity distribution function obtained by integrating u over all position coordinates. Let

$$Pu = u_c, \quad \text{and} \quad u_i = (I - P)u \tag{7.4.111}$$

For wave propagation in a random media, u_c is the coherent part of the field (mean part) while u_i is the incoherent part (fluctuating part). For a deterministic g and a centered random χ

$$Pg = g, \quad PM = MP, \quad P\chi P = 0 \tag{7.4.112}$$

Applying the operator P from the left on (7.4.109), we obtain

$$
\begin{aligned}
u_c &= Mg - \epsilon k^2 MP\chi(Pu + u_i) \\
&= Mg - \epsilon k^2 MP\chi u_i
\end{aligned}
\tag{7.4.113}
$$

Applying the operator $I - P$ from the left on (7.4.109) also, we obtain

$$u_i = -\epsilon k^2 M(I - P)\chi(u_c + u_i) = -\epsilon k^2 M(I - P)\chi u_c - \epsilon k^2 M(I - P)\chi u_i \tag{7.4.114}$$

We solve (7.4.114) by formal iteration for u_i in terms of u_c to obtain

$$u_i = \sum_{n=1}^{\infty} [-\epsilon k^2 M(I - P)\chi]^n u_c \tag{7.4.115}$$

Substituting for u_i into (7.4.113) gives

$$u_c = Mg - \left\{ \epsilon k^2 MP\chi \sum_{n=1}^{\infty} [-\epsilon k^2 M(I - P)\chi]^n \right\} u_c$$

Since $u_c = Pu_c$, this equation can be written as

$$u_c = Mg + MQu_c \tag{7.4.116}$$

where

$$Q = -\sum_{n=1}^{\infty} \epsilon k^2 P\chi[-\epsilon k^2 M(I - P)\chi]^n P \tag{7.4.117}$$

For random operators (7.4.116) is just the Dyson equation (7.4.49) with Q the mass operator.

This technique was called the method of smoothing by Frisch (1968) because it is the counterpart of the method of averaging (Chapter 5) in which the dependent variables have short- and long-period parts. This technique was first introduced by Primas (1961), Ernst and Primas (1963), and Tatarski and Gertsenshtein (1963) for random equations, and by Zwanzig (1964) for the Liouville equation.

The first-order smoothing is given by (7.4.116), with $n = 1$ in (7.4.117); that is

$$Q = \epsilon^2 k^4 P\chi M(I - P)\chi P = \epsilon^2 k^4 P\chi M\chi P$$

Hence

$$u_c = Mg + \epsilon^2 k^4 MP\chi M\chi Pu_c \qquad (7.4.118)$$

which coincides with the Bourett equation (7.4.54) for linear random media, and with the Landau equation for the Liouville problem (see, for example, Prigoginé 1962). For linear random media this first approximation was also obtained by Keller (1962) and Kubo (1963).

Exercises

7.1. Determine asymptotic expansions for large x for the solutions of

(a) $y'' - \left(1 + \dfrac{1}{x} + \dfrac{2}{x^2}\right)y = 0$

(b) $y'' - \left(4x^2 + 1 + \dfrac{1}{x} + \dfrac{2}{x^2}\right)y = 0$

(c) $x^2 y'' + xy' + (x^2 - n^2)y = 0$

(d) $y'' + \left(1 + \dfrac{1}{x} + \dfrac{2}{x^2}\right)y = 0$

(e) $y'' + xy' + y = 0$

7.2. Determine asymptotic expansions for large x for the solutions of

(a) $y'' \pm \left(\dfrac{1}{4x} + \dfrac{2}{x^2}\right)y = 0$

(b) $y'' \pm (\tfrac{9}{4}x + 1)y = 0$

(c) $y'' + xy' + \tfrac{25}{4}x^3 y = 0$

(d) $y'' + x^{-1/3}y' + x^{-2}y = 0$

7.3. Consider the equation

$$y'' - \lambda^2 x^{-2} y = 0$$

(a) Show that the exact solution is

$$y = ax^{m_1} + bx^{m_2}$$

where $m_{1,2} = \tfrac{1}{2} \pm \sqrt{\lambda^2 + \tfrac{1}{4}}$.

(b) Determine the first WKB approximation.

(c) Compare this approximation with the exact solution (Jeffreys, 1962).

7.4. Determine a first approximation to the eigenvalue problem

$$u'' + \lambda^2 f(x)u = 0, \qquad f(x) > 0$$

$$u(0) = u(1) = 0$$

for large λ.

7.5. Consider the problem

$$u'' + \left[f(x) + \sum_{n=1}^{\infty} \epsilon^n g_n(x) \right] u = 0$$

Let

$$u = w(x)e^{\phi(x,\epsilon)}, \qquad \phi = \sum_{n=1}^{\infty} \epsilon^n \phi_n(x)$$

where

$$w'' + f(x)w = 0$$

Determine an equation for ϕ and then determine the ϕ_n (Brull and Soler, 1966).

7.6. Determine a uniform asymptotic expansion for

$$\ddot{u} + \omega^2(\epsilon t)u = f(\epsilon t)$$

where ϵ is a small parameter and f is a bounded nonperiodic function of t.

7.7. Consider the problem

$$\ddot{u} + \omega^2(\epsilon t)u = k \cos \phi$$

where $\dot{\phi} = \lambda(\epsilon t)$. Determine uniform asymptotic expansions for the cases (a) λ is not near ω, and (b) $\lambda = \omega$ for some $t = t_0 > 0$ (Kevorkian, 1971).

7.8. Consider the problem

$$\ddot{u} - i[\omega_1(\epsilon t) + \omega_2(\epsilon t)]\dot{u} - \omega_1(\epsilon t)\omega_2(\epsilon t)u = f(\epsilon t)$$

where overdots denote differentiation with respect to t and ϵ is a small parameter. Determine uniform asymptotic approximations to the solution of this equation if (a) $f(\epsilon t) = 0$, (b) $f(\epsilon t)$ is a bounded nonperiodic function of t, (c) $f(\epsilon t) = k \exp(i\phi)$ with $\dot{\phi} = \lambda(\epsilon t) \neq \omega_i$, and (d) $f(\epsilon t)$ has the same form as in (c) but $\lambda = \omega_1$ at some $t = t_0$.

7.9. Determine a uniform asymptotic expansion for the general solution of

$$\epsilon^2 y'' + \epsilon(2x + 1)y' + 2xy = 0, \qquad 0 \leq x \leq 1$$

7.10. Bessel functions $J_n(nx)$, $Y_n(nx)$, $H_n^{(1,2)}(nx)$ are solutions of the differential equation

$$x^2 y'' + xy' + n^2(x^2 - 1)y = 0$$

Determine a uniform asymptotic expansion for the general solution of this equation for large n.

7.11. Determine uniform asymptotic expansions for small ϵ for the solutions of

$$y'' + \frac{x(1 + x) - \epsilon\alpha}{(1 + x)^2} y = 0, \qquad -1 < x < \infty$$

where α is a constant.

7.12. Consider the Graetz problem for heat transfer in a duct

$$u'' + \lambda^2(1 - x^2)f(x)u = 0$$

$$u(1) = u(-1) = 0$$

where $\lambda \gg 1$ and $f(x) = f(-x) > 0$. Determine a first approximation to λ and the eigenfunctions using (a) the method of matched asymptotic expansions (Sellers, Tribus, and Klein 1956), and (b) the method of multiple scales or Langer's transformation (Nayfeh, 1965b).

7.13. Given that $f(x) > 0$ and $\lambda \gg 1$, determine first approximations to the following eigenvalue problems

(a) $y'' + \lambda^2(1 - x)f(x)y = 0$
$y(0) = 0$ and $y(\infty) < \infty$

(b) $xy'' + y' + \lambda^2 xf(x)y = 0$
$y(1) = 0$ and $y(0) < \infty$

(c) $y'' + \lambda^2(1 - x)^n f(x)y = 0$, $\quad n$ is a positive integer
$y(0) = 0$ and $y(\infty) < \infty$

(d) $xy'' + y' + \lambda^2 x^n f(x)y = 0$, $\quad n$ is a positive integer
$y(1) = 0$ and $y(0) < \infty$

(e) $y'' + \lambda^2(x - 1)^n(2 - x)^m f(x)y = 0$, $\quad m$ and n are positive integers
$y < \infty$ $\quad$ for all x.

7.14. Consider the Graetz problem for heat transfer in a pipe

$$ru'' + u' + \lambda^2 r(1 - r^2)f(r)u = 0$$

$$u(1) = 0, \quad u(0) < \infty$$

where $\lambda \gg 1$ and $f(r) > 0$. (a) Determine an expansion valid away from $r = 0$ and $r = 1$; (b) determine expansions valid near $r = 0$ and $r = 1$; (c) match these three expansions, hence determine λ, and form a uniformly valid composite expansion (Sellers, Tribus, and Klein, 1956).

7.15. Consider Exercise 7.14 again. (a) Determine an expansion valid away from $r = 0$ using Langer's transformation; (b) determine an expansion valid away from $r = 1$ using Olver's transformation; (c) match these two expansions to determine λ, (d) form a composite uniformly valid expansion; and (e) compare these results with those of Exercise 7.14.

7.16. Consider the problem

$$y'' - \lambda^2(x - 1)(2 - x)y = 0$$

Determine an approximate solution to y if $\lambda \gg 1$, and $y < \infty$ for all x.

7.17. Consider the problem

$$u'' + \lambda^2(1 - x)^n(x - \mu)^m f(x)u = 0$$

$$u(1) = u(\mu) = 0$$

where $\lambda \gg 1$, $f(x) > 0$, n and m are positive integers, and $\mu < 1$. Determine expansions valid away from $x = \mu$ and $x = 1$ using Olver's transformation, match them to determine λ, and then form a composite expansion.

7.18. Consider the problem

$$ru'' + \mu u' + \lambda r^n(1 - r)^m f(r)u = 0, \qquad u(1) = 0 \text{ and } u(0) < \infty$$

where $\lambda \gg 1$, $f(r) > 0$, μ is a real number, and n and m are positive integers. Determine expansions valid away from $r = 0$ and 1 using Olver's transformation, match them to determine λ, and then form a composite expansion (Nayfeh, 1967a).

7.19. Given that $f(x) > 0$ and $\lambda \gg 1$, determine first approximations to the eigenvalue problems

(a) $xy'' + y' + \lambda^2 x(1 - x)f(x)y = 0$

(b) $xy'' + y' + \lambda^2 x^n(1 - x)^m f(x) = 0$, m and n are positive integers subject to the condition that $y < \infty$ for $x \geq 0$.

7.20. How would you go about determining a particular solution for

$$u'' + \lambda^2 z^3 u = \lambda^2 g(z)$$

when $\lambda \gg 1$ if (a) $g(0) \neq 0$, (b) $g(0) = 0$ but $g'(0) \neq 0$, (c) $g(0) = g'(0) = 0$ but $g''(0) \neq 0$, and (d) $g(0) = g'(0) = g''(0) = 0$?

References and Author Index

Ablowitz, M. J., and D. J. Benney, (1970). The evolution of multi-phase modes for nonlinear dispersive waves. *Stud. Appl. Math.*, 49, 225-238. [234]

Abraham-Shrauner, B. (1970a). Suppression of runaway of electrons in a Lorentz plasma. I. Harmonically time varying electric field. *J. Plasma Phys.*, 4, 387-402. [235]

Abraham-Shrauner, B. (1970b). Suppression of runaway of electrons in a Lorentz plasma. II. Crossed electric and magnetic fields. *J. Plasma Phy.* 4, 441-450. [235]

Ackerberg, R. C., and R. E. O'Malley, Jr. (1970). Boundary layer problems exhibiting resonance. *Stud. Appl. Math.*, 49, 277-295. [233]

Ahluwalia, D. S., R. M. Lewis, and J. Boersma, (1968). Uniform asymptotic theory of diffraction by a plane screen. *SIAM J. Appl. Math.*, 16, 783-807 [380]

Akins, R. G. See Chang, Akins, and Bankoff.

Akinsete, V. A., and J. H. S. Lee, (1969). Nonsimilar effects in the collapsing of an empty spherical cavity in water. *Phys. Fluids*, 12, *428-434 [78]*

Albright, N. W. (1970). Quasilinear stabilization of the transverse instability. *Phys. Fluids*, 13 1021-1030.[235]

Alfriend, K. T. (1970). The stability of the triangular Lagrangian points for commensurability of order two. *Celestial Mech.*, 1, 351-359. [233]

Alfriend, K. T. (1971a). Stability and motion in two-degree-of-freedom Hamiltonian systems for two-to-one commensurability. *Celestial Mech.*, 3, 247-265. [233]

Alfriend, K. T. (1971b). Stability of and motion about L_4 at three-to-one commensurability. *Celestial Mech.*, 4, 60-77. [233]

Alfriend, K. T. See Davis and Alfriend.

Alfriend, K. T., and R. H. Rand, (1969). Stability of the triangular points in the elliptic restricted problem of three bodies. *AIAA J.*, 7, 1024-1028. [233, 260, 262, 276]

Al'tshul', L. M., and V. I. Karpman, (1966). Theory of nonlinear oscillations in a collisionless plasma.*Soviet Phys. JETP* (English transl.), 22, 361-369. [369]

Alzheimer, W. E., and R. T. Davis, (1968). Unsymmetrical bending of prestressed annular plates. *J. Eng. Mech. Div. Proc. ASCE*, 4, 905-917. [35, 128]

Amazigo, J. C., B. Budiansky, and G. F. Carrier (1970). Asymptotic analyses of the buckling of imperfect columns on nonlinear elastic foundations. *Int. J. Solids Structures*, 6, 1341-1356. [233]

Anderson, W. J. See Spriggs, Messiter, and Anderson.

Asano, N. (1970). Reductive perturbation method for nonlinear wave propagation in inhomogeneous media. III. *J. Phys. Soc. Japan*,29, 220-224. [78]

Asano, N. and T. Taniuti, (1969). Reductive perturbation method for nonlinear wave propagation in inhomogeneous media. I. *J. Phys. Soc. Japan*, 27, 1059-1062.

387

Asano, N., and T. Taniuti (1970).   Reductive perturbation method for nonlinear wave propagation in inhomogeneous media II. *J. Phys. Soc. Japan*, 29, 209-214. [78]

Ashley, H. (1967). Multiple scaling in flight vehicle dynamic analysis – a preliminary look. AIAA Paper No. 67-560. [233]

Avila, G. S. S., and J. B. Keller, (1963). The high frequency asymptotic field of a point source in an inhomogeneous medium. *Comm. Pure Appl. Math.*, 16, 363-381. [380]

Babich, V. M. (1965). A point source in an inhomogeneous medium. *Z. Vycisl. Mat. i Mat. Fiz.*, 5, 949-951; English transl., *USSR Computer Math. Math. Phys.*, 5, 247-251. [380]

Babich, V. M. (1970). *Mathematical Problems in Wave Propagation Theory*, Part I. Plenum, New York. [361, 380]

Babich, V. M. (1971). *Mathematical Problems in Wave Propagation Theory*. Part II. Plenum, New York. [361, 380]

Babich, V. M., and T. S. Kravtsova, (1967). Propagation of wave film type oscillations of quantized thickness. *PMM*, 31, 204-210. [380]

Balescu, R. (1963). *Statistical Mechanics of Charged Particles*. Wiley, New York. [367, 369]

Ball, R. H. (1964). Nonlinear theory of waves in a cold homogeneous plasma without magnetic field. Stanford Microwave Laboratory Rept. No. 1200. [234]

Bankoff, S. G. See Chang, Akins, and Bankoff.

Barakat, R., and A. Houston, (1968). Nonlinear periodic capillary-gravity waves on a fluid of finite depth. *J. Geophys. Res.*, 73 6545-6554. [77]

Barcilon, V. (1970). Some inertial modifications of the linear viscous theory of steady rotating fluid flows. *Phys. Fluids*, 13, 537-544. [235]

Barrar, R. B. (1970). Convergence of the von Zeipel procedure. *Celestial Mech.*, 2, 494-504. [192]

Barua, S. N. (1954). Secondary flow in a rotating straight pipe. *Proc. Roy. Soc. (London)*, A227, 133-139. [79]

Bauer, H. F. (1968). Nonlinear response of elastic plates to pulse excitations. *J. Appl. Mech.*, 35, 47-52. [58]

Bellman, R. (1955). Research problems. *Bull. Am. Math Soc.*, 61, 192. [22]

Bellman, R. (1964). *Perturbation Techniques in Mathematics, Physics, and Engineering*. Holt, New York. [309]

Benney, D. J. (1965). The flow induced by a disk oscillating about a state of steady rotation. *Quart. J. Mech. Appl. Math.*, 18, 333-345. [235]

Benney, D. J. (1966). Long waves on liquid films. *J. Math. and Phys.* 45, 150-155. [41]

Benney, D. J.(1967). The asymptotic behavior of nonlinear dispersive waves. *J. Math. and Phys.*, 46, 115-132. [234]

Benney, D. J. See Ablowitz and Benney.

Benney, D. J., and A. C. Newell, (1967). Sequential time closures for interacting random waves. *J. Math. and Phys.*, 46, 363-393. [234]

Benney, D. J., and A. C. Newell, (1969). Random wave closures. *Stud. Appl. Math.*, 48, 29-53. [234]

Benney, D. J., and G. J. Roskes, (1969). Wave instabilities. *Stud. Appl. Math.*, 48, 377-385. [235]

Benney, D. J. and P. G. Saffman, (1966). Nonlinear interactions of random waves in a dispersive medium. *Proc. Roy. Soc. (London)*, A289, 301-320. [234]

Bethe, H. A. See Salpeter and Bethe.

Bisshopp, F. (1969). A modified stationary principle for nonlinear waves. *J. Diff. Equations,* 5, 592-605. [216]

Bleistein, N. (1967). Uniform asymptotic expansions of integrals with many nearby stationary points and algebraic singularities. *J. Math. Mech.,* 17, 533-559. [380]

Bleistein, N. See Lewis, Bleistein, and Ludwig.

Boersma, J. See Ahluwalia, Lewis, and Boersma.

Bogoliubov, N. N. See Krylov and Bogoliubov.

Bogoliubov, N. N., and Y. A. Mitropolski, (1961). *Asymptotic Methods in the Theory of Nonlinear Oscillations.* Gordon and Breach, New York. [168, 174.]

Bohlin, K. P. (1889). Über eine neue Annaherungsmethode in der Storungstheorie, *Akad Handl. Bihang.,* 14, (Afdl, Stokholm). [56]

Born, M. (1926). Quantum mechanics of impact processes. *Z. Phys.,* 38, 803-827. [362]

Bourret, R. C. (1962a). Propagation of randomly perturbed fields. *Can. J. Phys.,* 40, 782-790. [367, 369, 372]

Bourret, R. C. (1962b) Stochastically perturbed fields, with applications to wave propagation in random media. *Nuovo Cimento,* 26, 1-31. [ ] 367, 369, 372]

Brau, C. A. (1967). On the stochastic theory of dissociation and recombination of diatomic molecules in inert dilutens. *J. Chem. Phys.,* 47, 1153-1163. [235]

Breakwell, J. V., and L. M. Perko, (1966). Matched asymptotic expansions, patched conics, and the computation of interplanetary trajectories. In *Progress in Astronautics and Aeronautics,* Vol. 17, *Method in Astrodynamics and Celestial Mechanics* (R. L. Duncombe and V. G. Szebehely, Eds.), Academic, New York, pp. 159-182. [138]

Breakwell, J. V., and R. Pringle, Jr.(1966). Resonances affecting motion near the earth-moon equilateral libration points. In *Progress in Astronautics and Aeronautics,* Vol. 17, *Methods in Astrodynamics and Celestial Mechanics* (R. L. Duncombe and V. G. Szebehely, Eds.), Academic, New York, pp. 55-74. [199]

Bretherton, F. P. (1962). Slow viscous motion round a cylinder in a simple shear. *J. Fluid Mech.,* 12, 591-613. [110]

Bretherton, F. P. (1964). Resonant interactions between waves. The case of discrete oscillations. *J. Fluid Mech.,* 20, 457-479. [217, 227, 267, 298]

Bretherton, F. P. (1968). Propagation in slowly varying waveguides *Proc. Roy. Soc. (London),* A 302, 555-576. [216]

Bretherton, F. P., and C. J. R. Garrett (1968). Wavetrains in inhomogeneous moving media. *Proc. Roy. Soc. (London),*A302, 529-554. [216]

Brillouin, L. (1926). Remarques sur la mecanique ondulatoire. *J. Phys. Radium,* 7, 353-368. [114, 315, 339]

Brofman, S. See Ting and Brofman.

Brofman, W. (1967). Approximate analytical solution for satellite orbits subjected to small thrust or drag. *AIAA J.,* 5, 1121-1128. [233]

Bromberg, E. (1956). Nonlinear bending of a circular plate under normal pressure. *Comm. Pure Appl. Math.,* 9, 633-659. [144, 146]

Brown, W. P., Jr. (1966). Validity of the Rytov approximation in optical propagation calculations. *J. Opt. Soc. Am.,* 56, 1045-1052. [373]

Brown, W. P., Jr. (1967). Validity of the Rytov approximation. *J. Opt. Soc. Am.* 57, 1539-1543. [373]

Brull M. A., and A. I. Soler, (1966). A new perturbation technique for differential equations with small parameters. *Quart. Appl. Math.*, **24**, 143-151. [383]

Buchal, R. N., and J. B. Keller, (1960). Boundary layer problems in diffraction theory. *Comm. Pure Appl. Math.*, **13**, 85-144. [378]

Budiansky, B. See Amazigo, Budiansky, and Carrier.

Burnside, R. R. (1970). A note on Lighthill's method of strained coordinates. *SIAM J. Appl. Math.*, **18**, 318-321. [79, 107]

Buslaev, N. S. (1964). On formulas of short wave asymptotics in the problem of diffraction by convex bodies. *Akad. Nauk SSSR Trudy Mat. Inst. Steklov*, **73**, 14-117. [380]

Butler, D. S., and R. J. Gribben (1968). Relativistic formulation for non-linear waves in a non-uniform plasma. *J. Plasma Phys.*, **2**, 257-281. [234]

Caldirola, P., O. De Barbieri, and C. Maroli, (1966). Electromagnetic wave propagation in a weakly ionized plasma. I. *Nuovo Cimento*, B42, 266-289. [236]

Campbell, J. A., and W. H. Jefferys, (1970). Equivalence of the perturbation theories of Hori and Deprit. *Celestial Mech.* **2**, 467-473. [202]

Carlini, F. (1817). Ricerche sulla convergenza della serie che serve alla soluzione del problema di Keplero. Memoria di Francesco Carlini, Milano; also *Jacobi's Ges Werke*, **7**, 189-245; also *Astron. Nachrichten*, **30**, 197-254. [315]

Carrier, G. F. (1953). Boundary layer problems in applied mechanics. *Advan. Appl. Mech.*, **3**, 1-19. [119]

Carrier, G. F. (1954). Boundary layer problems in applied mathematics. *Comm. Pure Appl. Math.* , **7**, 11-17 [114]

Carrier, G. F. (1966). Gravity waves on water of variable depth. *J. Fluid Mech.*, **24**, 641-659. [234]

Carrier, G. F. (1970). Singular pertubation theory and geophysics. *SIAM Rev.*, **12**, 175-193. [110]

Carrier, G. F. See Amazigo, Budiansky, and Carrier.

Cashwell, E. D. (1951). The asymptotic solutions of an ordinary differential equation in which the coefficient of the parameter is singular. *Pacific J. Math.*, **1**, 337-353. [358]

Caughey, T. K. See Tso and Caughey.

Caughey, T. K., and H. J. Payne, (1967). On the response of a class of self-excited oscillators to stochastic excitation. *Int. J. Non-Linear Mech.*, **2**, 125-151. [235]

Cesari, L. (1971). *Asymptotic Behavior and Stability Problems in Ordinary Differential Equations.* Springer-Verlag, New York. [309]

Chang, K. S., R. G. Akins, and S. G. Bankoff, (1966). Free convection of a liquid metal from a uniformly heated vertical plate. *Ind. Eng. Chem. Fundam.*, **5**, 26-37. [79]

Chen, C. S. (1971). Parametric excitation in an inhomogeneous plasma. *J. Plasma Phys.*, **5**, 107-113. [235]

Chen, C. S., and G. J. Lewak, (1970). Parametric excitation of a plasma. *J. Plasma Phys.*, **4**, 357-369. [235]

Cheng, H. and T. T. Wu (1970). An aging spring. *Stud. Appl. Math.*, **49**, 183-185. [232]

Cheng, H. K., J. W. Kirsch, and R. S. Lee, (1971). On the reattachment of a shock layer produced by an instantaneous energy release. *J. Fluid Mech.*, **48**, 241-263, [235]

Cheng, S. I. See Goldburg and Cheng.

Chernov, L. A. (1960). *Wave Propagation in a Random Medium.* Dover, New York. [361, 373]

Cherry, T. M. (1927). On the transformation of Hamiltonian systems of linear differential equations with constant or periodic coefficients. *Proc. Lond. Math. Soc.*, **26**, 211-230. [199]

Cherry, T. M. (1949). Uniform asymptotic expansions. *J. London Math. Soc.*, **24**, 121-130. [346]

Cherry, T. M. (1950). Uniform asymptotic formulae for functions with transition points. *Trans. Am. Math. Soc.*, **68**, 224-257. [346]

Chong, T. H., and L. Sirovich, (1971). Non-linear effects in steady supersonic dissipative gas dynamics. Part 1 –Two-dimensional flow. *J. Fluid Mech.*, **50**, 161-176. [235]

Chu, B. T. (1963). Analysis of a self-sustained thermally driven nonlinear vibration. *Phys. Fluids,* **6** 1638-1644. [89]

Chu, B. T., and S. J. Ying, (1963). Thermally driven nonlinear oscillations in a pipe with travelling shock waves. *Phys. Fluid* **6,** 1625-1637. [89]

Chu, V. H., and C. C. Mei (1970). On slowly-varying Stokes waves. *J. Fluid Mech.*, **41**, 873-887. [234]

Chudov, L. A. (1966). Some shortcomings of classical boundary layer theory. NASA (transl.) TT F-360, TT 65-50138. [145]

Clare, T. A. (1971). Resonance instability for finned configurations having nonlinear aerodynamic properties. *J. Spacecraft Rockets*, **8**, 278-283. [305]

Clark, R. A. (1958). Asymptotic solutions of toroidal shell problems. *Quart. Appl. Math.*, **16**, 47-60. [353]

Clark, R. A. (1963). Asymptotic solutions of a nonhomogeneous differential equation with a turning point. *Arch. Rat. Mech. Anal.*, **12**, 34-51. [353, 356]

Clark, R. A. (1964). Asymptotic solutions of elastic shell problems. in *Asymptotic Solutions of Differential Equations and Their Applications* (C. H. Wilcox, Ed.), Wiley, New York, pp. 185-209. [353]

Coakley, T. J. (1968). Dynamic stability of symmetric spinning missiles. *J. Spacecraft Rockets,* **5,** 1231-1232. [321]

Cochran, J. A. (1962). Problems in singular perturbation theory. Ph. D. Thesis, Stanford Universiy. [232, 233, 234, 280, 285]

Coddington, E. A., and N. Levinson, (1955). *Theory of Ordinary Differential Equations.* McGraw-Hill, New York. [60, 62]

Coffey, T. P., Jr. (1969). Invariants to all orders in classical perturbation theory. *J. Math. Phys.*, **10**, 426-438. [199]

Cole, J. D. (1968). *Perturbation Methods in Applied Mathematics.* Blaisdell, Waltham, Mass.

Cole, J. D. See Lagerstrom and Cole.

Cole, J. D., and J. Kevorkian, (1963). Uniformly valid asymptotic approximations for certain nonlinear differential equations. *Nonlinear Differential Equations and Nonlinear Mechanics* (J. P. LaSalle and S. Lefschetz, Eds.), Academic, New York, pp. 113-120. [231, 232, 273]

Comstock, C. (1968). On Lighthill's method of strained coordinates. *SIAM J. Appl. Math.*, **16**, 596-602. [79, 107]

Comstock, C. (1971). Singular perturbations of elliptic equations. I. *SIAM J. Appl. Math.*, **20**, 491-502. [234]

Contopoulos, G. (1963). On the existence of a third integral of motion. *Astron. J.*, **68**, 1-14. [199]

Crane, L. J. (1959). A note on Stewartson's paper "On asymptotic expansions in the theory of boundary layers." *J. Math. and Phys.*, **38**, 172-174. [79]

Crapper, G. D. (1970). Nonlinear capillary waves generated by steep gravity waves. *J. Fluid Mech.*, **40**, 149-159. [217]

Crawford, F. W. See Galloway and Crawford.

Crocco, L. See Sirignano and Crocco.

Das, K. P. (1971). Interaction among three extraordinary waves in a hot electron plasma. *Phys. Fluids*, **14**, 124-128. [234]

Davidson, R. C. (1967). The evolution of wave correlations in uniformly turbulent, weakly nonlinear systems. *J. Plasma Phys.*, **1**, 341-359. [234]

Davidson, R. C. (1968). Nonlinear oscillations in a Vlasov-Maxwell plasma. *Phys. Fluids*, **11**, 194-204. [234]

Davidson, R. C. (1969). General weak turbulence theory of resonant fourwave processes. Phys. Fluids, **12**, 149-161. [234]

Davis, R. T. See Alzheimer and Davis; Wingate and Davis.

Davis, R. T., and K. T. Alfriend, (1967). Solutions to van der Pol's equation using a perturbation method. *Int. J. Non-Linear Mech.*, **2**, 153-162. [232]

Davison, L. (1968). Perturbation theory of nonlinear elastic wave propagation. *Int. J. Solids Structures*, **4**, 301-322. [90, 91, 94]

De Barbieri, O. See Caldirola, De Barbieri, and Maroli.

De Barbieri, O., and C. Maroli, (1967). On the dynamics of weakly ionized gases. *Ann. Phys.*, **42**, 315-333. [236]

De Bruijn, N. G. (1958). *Asymptotic Methods in Analysis.* North-Holland, Amsterdam; Interscience, New York. [18]

Deprit, A. (1969). Canonical transformations depending on a small parameter. *Celestial Mech.*, **1**, 12-30. [199, 201, 202, 205, 212]

Dewar, R. L. (1970). Interaction between hydromagnetic waves and a time-dependent, inhomogeneous medium. *Phys. Fluids.*, **13**, 2710-2720. [217]

DeWolf, D. A. (1965). Wave propagation through quasi-optical irregularities. *J. Opt. Soc. Am.*, **55**, 812-817. [373]

DeWolf, D. A. (1967). Validity of Rytov's approximation. *J. Opt. Soc. Am.*, **57**, 1057-1058. [373]

DiPrima, R. C. (1969). Higher order approximations in the asymptotic solution of the Reynolds equation for slider bearings at high bearing numbers. *J. Lubrication Technology*, **91**, 45-51. [125]

Dirac, P. A. M. (1926). On the theory of quantum mechanics. *Proc. Roy. Soc. (London)*, A112, 661-677. [161]

Dobrowolny, M., and A. Rogister, (1971). Non-linear theory of hydromagnetic waves in a high β plasma. *J. Plasma Phys.*, **6**, 401-412. [235]

Dorodnicyn, A. A. (1947). Asymptotic solution of van der Pol's equation. *J. Appl. Math. Mech.*, **11**, 313-328; *Am. Math. Soc. Transl.*, **88**, 1953. [114]

Dougherty J. P. (1970). Lagrangian methods in plasma dynamics. I. General theory of the method of the averaged Lagrangian. *J. Plasma Phys.*, **4**, 761-785. [217]

Drazin, P. G. (1969). Nonlinear internal gravity waves in a slightly stratified atmosphere. *J. Fluid Mech.*, **36**, 433-446. [216]

Dyson, F. J. (1949). The radiation theories of Tomonaga, Schwinger, and Feynman. *Phys. Rev.*, **75**, 486-502. [371]

Eckhaus, W. (1965). *Studies in Nonlinear Stability Theory*. Springer-Verlag, New York. [162]

Eckstein, M. C. See Shi and Eckstein.

Eckstein, M. C. and Y. Y. Shi (1967). *Low-thrust elliptic spiral trajectories of a satellite of variable mass. AIAA J.*, **5**, 1491-1494. [233]

Eckstein, M. C., Y. Y. Shi, and J. Kevorkian, (1966a). A uniformly valid asymptotic representation of satellite motion around the smaller primary in the restricted three-body problem. In *Progress in Astronautics and Aeronautics*, Vol. 17, *Methods in Astrodynamics and Celestial Mechanics* (R. L. Duncombe and V. G. Szebehely, Eds.), Academic, New York, pp. 183-198. [233, 276]

Eckstein, M. C., Y. Y. Shi, and J. Kevorkian, (1966b). Satellite motion for arbitrary eccentricity and inclination around the smaller primary in the restricted three-body problem. *Astron. J.*, **71**, 248-263. [233]

Eckstein, M. C., Y. Y. Shi, and J. Kevorkian (1966c). Use of the energy integral to evaluate higher-order terms in the time history of satellite motion. *Astron. J.*, **71**, 301-305. [233]

Einaudi, F. (1969). Singular perturbation analysis of acoustic-gravity waves. *Phys. Fluids*, **12**, 752-756. [78]

Einaudi, F. (1970). Shock formation in acoustic gravity waves. *J. Geophys. Res.*, **75**, 193-200. [78]

Emanuel, G. (1966). Unsteady, diffusing, reacting tubular flow with application to the flow in a glow discharge tube. Aerospace Corporation Rept. No. TR-669 (6240-20)-9. [99]

Emery, V. J. (1970). Eikonal approximation for nonlinear equations. *J. Math. Phys.*, **11**, 1893-1900. [234]

Erdélyi, A. (1956). *Asymptotic Expansions*. Dover, New York. (18, 309, 337]

Erdélyi, A. (1960). Asymptotic solutions of differential equations with transition points or singularities. *J. Math. Phys.*, **1**, 16-26. [358]

Erdélyi, A. (1961). An expansion procedure for singular perturbations. *Atti. Accad. Sci. Torino, Cl. Sci. Fis. Mat. Nat.*, **95**, 651-672. [113, 121]

Ernst, R. R., and H. Primas (1963). Nuclear magnetic resonance with stochastic high-frequency fields. *Helv. Phys. Acta.*, **36**, 583-600. [382]

Espedal, M. S. (1971). The effects of ion-ion collision on an ion-acoustic plasma pulse. *J. Plasma Phys.*, **5**, 343-355. [78]

Euler, L. (1754). *Novi Commentarii Acad. Sci. Petropolitanae*, **5**, 205-237; *Opera Omnia*, Ser. I. **14**, 585-617. [11]

Evensen, D. A. (1968). Nonlinear vibrations of beams with various boundary conditions. *AIAA J.*, **6**, 370-372. [106]

Evgrafov, M. A., and M. B. Fedoryuk, (1966). Asymptotic behavior of solutions of the equation $w''(z) - p(z', \lambda)w(z) = 0$ as $\lambda \to \infty$ in the complex z-plane. *Usp. Mat. Nauk*, **21**, 3-50. [345]

Faa De Bruno, F. (1857). Note sur une nouvelle formule de calcul differentiel. *Quart. J. Pure Appl. Math.*, **1**, 359-360. [192]

Fedoryuk, M. B. See Evgrafov and Fedoryuk.

Felsen, L. B. See Levey and Felsen.

Feshchenko, S. F., N. I. Shkil', and L. D. Nikolenko, (1967). *Asymptotic Methods in the Theory of Linear Differential Equations.* American Elsevier, New York. [309, 319]

Feynman, R. P. (1948). Space-time approach to nonrelativistic quantum mechanics. *Rev. Modern Phys.*, 20, 367-387. [366]

Fife, P. See Pierson and Fife.

Fock, V. A. (1965). *Electromagnetic Diffraction and Propagation Problems.* Pergamon, New York. [380]

Fowkes, N. D. (1968). A singular perturbation method. Parts I and II. *Quart. Appl. Math.*, 26, 57-69, 71-85. [232, 234, 379, 285]

Fowler, R. H., and C. N. H. Lock, (1921). Approximate solutions of linear differential equations. *Proc. London Math. Soc.*, 20, 127-147. [321, 322]

Fowler, R. H., E. G. Gallop, C. N. H. Lock, and H. W. Richmond (1920). The aerodynamics of a spinning shell. *Phil. Trans. Roy. Soc. London*, 221, 295-387. [321, 322]

Fox, P. A. (1953). On the use of coordinate perturbations in the solution of physical problems. Project DIC 6915 Technical Rept. No. 1, Massachusetts Institute of Technology, Cambridge. [98]

Fox, P. A. (1955). Perturbation theory of wave propagation based on the method of characteristics. *J. Math. and Phys.*, 34, 133-151. [57, 87, 89]

Fraenkel, L. E. (1969). On the method of matched asymptotic expansions, Parts I-III. *Proc. Cambridge Phil. Soc.*, 65, 209-231, 233-251, 263-284. [119]

Freeman, N. C., and R. S. Johnson, (1970). Shallow water waves on shear flows. *J. Fluid Mech.*, 42, 401-409. [234]

Fried, D. L. (1967). Test of the Rytov approximation. *J. Opt. Soc. Am.*, 57, 268-269. [373]

Friedrichs, K. O. (1942). Theory of viscous fluids. *Fluid Dynamics.* Brown University Press, Providence, R. I., Chapter 4. [119]

Friedrichs, K. O. (1955). Asymptotic phenomena in mathematical physics. *Bull. Am. Math. Soc.*, 61, 485-504. [18]

Frieman, E. A. (1963). On a new method in the theory of irreversible processes. *J. Math. Phys.*, 4, 410-418. [230]

Frieman, E. A., and P. Rutherford, (1964). Kinetic theory of a weakly unstable plasma. *Ann. Phys.*, 28, 134-177. [235]

Frisch, U. (1965). *Wave Propagation in Random Media.* Institut d'Astrophysique, Paris. [367, 369]

Frisch, U. (1968). Wave propagation in random media. *Probability Methods in Applied Mathematics*, Vol. 1 (A. T. Bharucha-Reid, Ed.), Academic, New York, pp. 75-198. [309, 361, 364, 365, 367, 372, 380, 381]

Frobenius, G. (1875). Veber die regulären Integrale der linearen Differentialgleichungen. *J. Reine Angew. Math.*, 80, 317-333. [310]

Furutsu, K. (1963). On the statistical theory of electromagnetic waves in a fluctuating medium. *J. Res. Natl. Bur. Standards, Sec. D*, 67, 303-323. [367, 369]

Gallop, E. G. See Fowler, Gallop, Lock, and Richmond.

Galloway, J. J., and F. W. Crawford, (1970). Lagrangian derivation of wave-wave coupling coefficients. Proc. 4th European Conference on Controlled Fusion and Plasma Physics, Rome (CHEN), 161. [217]

Galloway, J. J. and H. Kim, (1971). Lagrangian approach to non-linear wave interactions in a warm plasma. *J. Plasma Phys.*, 6 53-72. [217]

Gans, R. (1915). Propagation of light through an inhomogeneous media. *Ann. Phys.*, 47, 709-736. [114, 339]

Garrett, C. J. R. (1968). On the interaction between internal gravity waves and a shear flow. *J. Fluid Mech.* 34, 711-720. [216]

Garrett, C. J. R. See Bretherton and Garrett.

Germain, P. (1967). Recent evolution in problems and methods in aerodynamics. *J. Roy. Aeron. Soc.*, 71, 673-691. [110, 235]

Gertsenshtein, M. E. See Tatarski and Gertsenshtein.

Giacaglia, G. E. O. (1964). Notes on von Zeipel's method. NASA X-547-64-161. [191]

Golay, M. J. E. (1964). Normalized equations of the regenerative oscillator-noise, phase-locking, and pulling. *Proc. IEEE*, 52, 1311-1330. [258]

Goldberg, P., and G. Sandri, (1967). Power series of kinetic theory. Parts I and II. *Phys. Rev.*, L154, 188-209. [236]

Goldburg, A., and S. I. Cheng, (1961). The anomaly in the application of Poincaré-Light-hill-Kuo and parabolic coordinates to the trailing edge boundary layer. *J. Math. Mech.*, 10, 529-535. [79]

Goldstein H. (1965). *Classical Mechanics*. Addison-Wesley, Reading, Mass. [181]

Goldstein. S. (1929). The steady flow of viscous fluid past a fixed spherical obstacle at small Reynolds numbers. *Proc. Roy. Soc. (London)*. A-123, 225-235. [141]

Goldstein, S. (1931). A note on certain approximate solutions of linear differential equations of the second order. *Proc. London Math. Soc.*, 33, 246-252. [346]

Goldstein, S. (1969). Applications of singular perturbations and boundary-layer theory to simple ordinary differential equations. In *Contributions to Mechanics* (D. Abir, Ed.), Pergamon, Oxford, pp. 41-67. [317]

Good, R. H. See Miller and Good.

Goodrich, R. F., and N. D. Kazarinoff, (1963). Diffraction by thin elliptic cylinders. *Michigan Math. J.*, 10, 105-127. [346]

Gorelik, G., and A. Witt, (1933). Swing of an elastic pendulum as an example of two parametrically bound linear vibration systems. *J. Tech. Phys. (USSR,)* 3, 244-307. [185]

Green, G. (1837). On the motion of waves in a variable canal of small depth and width. *Trans. Cambridge Phil. Soc.*, 6, 457-462. [314]

Green, G. S., and A. K. Weaver, (1961). The estimation of the three-dimensional gyrations of a ballistic missile descending through the atmosphere. *Royal Aircraft Establishment Tech. Note* G. W. 596 (London). [321]

Gretler, W. (1968). Eine indirekte Methode zur Berechnung der ebenen Unterschallstromung. *J. Mécanique*, 7, 83-96. [89]

Gribben, R. J. See Butler and Gribben.

Grimshaw, R. (1966). High-frequency scattering by finite convex regions. *Comm. Pure Appl. Math.*, 19, 167-198. [380]

Grimshaw, R. (1970). The solitary wave in water of variable depth. *J. Fluid Mech.* 42, 639-656. [216]

Guiraud, J. P. (1965). Acoustique geométrique, bruit balistique des avions supersoniques et focalisation. *J. Mécanique*, 4, 215-267. [89]

Gundersen, R. M. (1967). Self-sustained thermally driven nonlinear oscillations in one-dimensional magnetohydrodynamic flow. *Int. J. Eng. Sci.*, 5, 205-211. [89]

Glydén, H. (1893). Nouvelles recerches sur les séries employées dans les théories des planèts. *Acta Math.* 9, 1-168. [56]

Han, L. S. (1965). On the free vibration of a beam on a nonlinear elastic foundation. *J. Appl. Mech.*, 32, 445-447. [58, 105]

Hanks, T. C. (1971). Model relating heat-flow value near, and vertical velocities of mass transport beneath, ocean rises. *J. Geophys. Res.*, 76, 537-544. [156]

Hasegawa, H. S. See Savage and Hasegawa.

Hassan, S. D. See Nayfeh and Hassan.

Heidbreder, G. R. (1967). Multiple scattering and the method of Rytov. *J. Opt. Soc. Am.*, 57, 1477-1479. [373]

Henrard, J. (1970). On a perturbation theory using Lie transforms. *Celestial Mech.*, 3, 107-120. [202]

Heynatz, J. T. See Zierep and Heynatz.

Hirschfelder, J. O. (1969). Formal Rayleigh-Schrödinger perturbation theory for both degenerate and non-degenerate energy states. *Int. J. Quantum Chem.*, 3, 731-748. [71]

Holstein, V. H. (1950). Uber die aussere and innere Reibungsschicht bei Storungen laminarer Stromungen. *ZAMM*, 30, 25-49. [353]

Holt, M. (1967). The collapse of an imploding spherical cavity. *Rev. Roumaine, Sci. Tech. Ser. Mec. Appl.*, 12, 407-415. [78]

Holt, M., and N. J. Schwartz, (1963). Cavitation bubble collapse in water with finite density behind the interface. *Phys. Fluids.*, 6, 521-525. [78]

Hoogstraten, H. W. (1967). Uniformly valid approximations in two-dimensional subsonic thin airfoil theory. *J. Eng. Math.*, 1, 51-65. [99]

Hoogstraten, H. W. (1968). Dispersion of non-linear shallow water waves. *J. Eng. Math.*, 2, 249-273. [234]

Hori, G. I. (1966). Theory of general perturbations with unspecified canonical variables. *Publ. Astron. Soc. Japan*, 18, 287-296, [199, 201, 202, 212]

Hori, G. I. (1967). Nonlinear coupling of two harmonic oscillations. *Publ. Astron. Soc. Japan*, 19, 229-241. [199, 201, 212]

Hori, G. I. (1970). Comparison of two perturbation theories based on canonical transformations. *Publ. Astron. Soc. Japan*, 22, 191-198. [202]

Horn, J. (1899). Ueber eine lineare Differentialgleichung zweiter Ordnung mit einem willkurlichen Parameter. *Math. Ann.* 52, 271-292. [316]

Horn, J. (1903). Untersuchung der Intergrale einer linearen Differentialgleichung in der Umgebung einer Unbestimmtheitsstelle vermittelst successiver Annaherungen. *Arch. Math. Phys.*, 4, 213-230. [312]

Hoult, D. P. (1968). Euler-Lagrange relationship for random dispersive waves. *Phys. Fluids.*, 11, 2082-2086. [234]

Houston, A. See Barakat and Houston.

Howe, M. S. (1967). Nonlinear theory of open-channel steady flow past a solid surface of finite-wave-group shape. *J. Fluid Mech.* 30, 497-512. [216]

Howland, L. See Van Hove, Hugenholtz, and Howland.

Hsieh, P. F., and Y. Sibuya, (1966). On the asymptotic integration of second order linear ordinary differential equations with polynomial coefficients. *J. Math. Anal. Appl.*, 16, 84-103. [345]

Hufnagel, R. E., and N.R. Stanley (1964). Modulation transfer function associated with image transmission through turbulent media. *J. Opt. Soc. Am.*, 54, 52-61. [373]

Hugenholtz, N. See Van Hove, Hugenholtz, and Howland.

Imai, I. (1948). On a refinement of the W. K. B. method. *Phys. Rev.*, 74, 113. [324, 350]

Imai, I. (1950). Asymptotic solutions of ordinary linear differential equations of the second order. *Phys. Rev.*, 80, 1112. [350]

Ince, E. L. (1926). *Ordinary Differential Equations.* Longmans, Green, London. [5, 314]

Jacobi, C. G. J. (1849). Versuch einer Berechnung der grossen Ungleichheit des Saturns nach einer strengen Entwickelung. *Astron. Nachr.* 28, 65-94. [312]

Jacobs, S. J. (1967). An asymptotic solution of the tidal equations. *J. Fluid Mech.*, 30, 417-438. [234]

Jahsman, W. E. (1968). Collapse of a gas-filled spherical cavity. *J. Appl. Mech.*, 35, 579-587. [78]

Jakob, M. (1949). *Heat Transfer*, Vol. 1. Wiley, New York. [342]

Jefferys, W. H. See Campbell and Jefferys.

Jeffreys, H. (1924a). On certain approximate solutions of linear differential equations of the second order. *Proc. London Math. Soc.*, 23, 428-436. [114, 339]

Jeffreys, H. (1924b). On certain solutions of Mathieu's equation. *Proc. London Math. Soc.* 23, 437-476. [114, 339]

Jeffreys, H. (1962). *Asymptotic Approximations.* Oxford University Press, Oxford. [309, 342, 382]

Johansen, K. F., and T. R. Kane, (1969). A simple description of the motion of a spherical pendulum. *J. Appl. Mech.*, 36, 408-411. [224]

Johnson, R. S. See Freeman and Johnson.

Kain, M. E. See Kane and Kahn.

Kamel, A. A., (1969a). Expansion formulae in canonical transformations depending on a small parameter. *Celestial Mech.*, 1, 190-199. [199, 202, 206, 212]

Kamel, A. A. (1969b). Perturbation theory based on Lie transforms and its application to the stability of motion near sun-perturbed earth-moon triangular libration points. SUDAAR Rept. No. 391, Stanford University. [199, 202, 206]

Kamel, A. A. (1970). Perturbation method in the theory of nonlinear oscillations. *Celestial Mech.*, 3, 90-106. [199, 201, 202, 206]

Kamel, A. A. (1971). Lie transforms and the Hamiltonization of non-Hamiltonian systems. *Celestial Mech.*, 4, 397-405. [202]

Kamel, A. A. See Nayfeh and Kamel.

Kane, T. R. See Johansen and Kane.

Kane, T. R., and M. E. Kahn, (1968). On a class of two-degree-of-freedom oscillations. *J. Appl. Mech.*, 35, 547-552. [185]

Kaplun, S. (1954). The role of coordinate systems in boundary-layer theory. *Z. Angew. Math. Phys.*, 5, 111-135; also Chapter 1 of Kaplun (1967). [50, 114]

Kaplun, S. (1957). Low Reynolds number flow past a circular cylinder. *J. Math. Mech.*, 6, 595-603; also Chapter 3 of Kaplun (1967). [114]

Kaplun, S. (1967). *Fluid Mechanics and Singular Perturbations* (a collection of papers by S. Kaplun, edited by P. A. Lagerstrom, L. N. Howard, and C. S. Liu), Academic, New York. [114, 119]

Kaplun, S., and P. A. Lagerstrom, (1957). Asymptotic expansions of Navier-Stokes solutions for small Reynolds numbers. *J. Math. Mech.*, **6**, 585-593; also Chapter 2 of Kaplun (1967). [114]

Karal, F. C., and J. B. Keller, (1964). Elastic, electromagnetic, and other waves in a random medium. *J. Math. Phys.*, **5**, 537-547. [369]

Karpman, V. I. See Al'tshul' and Karpman.

Karpman, V. I., and E. M. Krushkal' (1969). Modulated waves in nonlinear dispersive media. *Soviet Phys.*, **28**, 277-281. [216]

Kawakami, I. (1970). Perturbation approach to nonlinear Vlasov equation. *J. Phys. Soc. Japan*, **28**, 505-514. [216]

Kawakami, I., and T. Yagishita, (1971). Pertubation approach to nonlinear Vlasov equation. II. Nonlinear plasma oscillation of finite amplitude. *J. Phys. Soc. Japan.* **30**, 244-253. [216]

Kazarinoff, N. D., (1958). Asymptotic theory of second order differential equations with two simple turning points. *Arch. Rat. Mech. Anal.*, **2**, 129-150. [344]

Kazarinoff, N. D. See Goodrich and Kazarinoff.

Kazarinoff, N. D., and R. W. Mc Kelvey, (1956). Asymptotic solution of differential equations in a domain containing a regular singular point. *Can. J. Math.*, **8**, 97-104. [358]

Keller, J. B. (1958). A geometrical theory of diffraction, calculus of variations and its applications. *Proc. Symp. Appl. Math.*, **8**, 27-52. [374]

Keller, J. B. (1962). Wave propagation in random media. *Proc. Symp. Appl. Math.*, **13**, 227-246. [369, 382]

Keller, J. B. (1968). Perturbation Theory. Lecture notes, Mathematics Department, Michigan State University. [58, 150]

Keller, J. B. See Avila and Keller, Buchal and Keller; Karal and Keller; Lynn and Keller; Millman and Keller.

Keller, J. B., and S. Kogelman, (1970). Asymptotic solutions of initial value problems for nonlinear partial differential equations. *SIAM J. Appl. Math.*, **18**, 748-758. [234]

Keller, J. B., and M.H. Millman, (1969). Perturbation theory of nonlinear electromagnetic wave propagation. *Phys. Rev.*, **181**, 1730-1747. [77]

Keller, J. B., and L. Ting, (1966). Periodic vibrations of systems governed by nonlinear partial differential equations. *Comm. Pure. Appl. Math.* **19**, 371-420. [77]

Kelly, R. E. (1965). Stability of a panel in incompressible, unsteady flow. *AIAA J.*, **3**, 1113-1118. [233]

Kelly, R. E. (1967). On the stability of an inviscid shear layer which is periodic in space and time. *J. Fluid Mech.*, **27**, 657-689. [235]

Kelly, R. E. See Maslowe and Kelly.

Kemble, E. C. (1935). A contribution to the theory of the B.W.K. method. *Phys, Rev.*, **48**, 549-561. [339]

Kersten, P. H. M. (1967). Diffraction of an electromagnetic wave by a plane screen. Ph. D. Thesis. Technische Hochschull, Eindhoven. [380]

Kevorkian, J. (1966a). The two variable expansion procedure for the approximate solution of certain nonlinear differential equations. *Space Mathematics.* Part 3. (J.B. Rosser, Ed.) American Mathematical Society, Providence, R.I., pp. 206-275. [231, 232, 233, 273]

Kevorkián, J. (1966b). von Zeipel method and the two-variable expansion procedure. *Astron. J.*, 71, 878-885. [231]

Kevorkian, J. (1971). Passage through resonance for a one-dimensional oscillator with slowly varying frequency. *SIAM J. Appl. Math.*, 20, 364-373. [233, 383]

Kevorkian, J. See Cole and Kevorkian; Eckstein, Shi, and Kevorkian; Lagerstrom and Kevorkian.

Kiang, R. L. (1969). Nonlinear theory of inviscid Taylor instability near the cutoff wave number. *Phys. Fluids*, 12, 1333-1339. [235]

Kim, H. See Galloway and Kim.

Kirchhoff, G. (1877). *Zur Theorie des Condensators*, Berlin, Akad., Monatsber. pp. 144-162. [114]

Kirsch, J. W. See Cheng, Kirsch, and Lee.

Klein, J. S. See Sellers, Tribus, and Klein.

Klimas, A., R. V., Ramnath, and G. Sandri, (1970). On the compatibility problem for the uniformization of asymptotic expansions. *J. Math. Anal. Appl.*, 32, 482-504. [232]

Kogelman, S. See Keller and Kogelman.

Kraichnan, R. H. (1961). Dynamics of nonlinear stochastic systems. *J. Math. Phys.*, 2, 124-148. [367]

Kramers, H. A. (1926). Wellenmechanik und halbzahlige Quantisierung. *Z. Phys.*, 39, 828-840. [114, 315, 339]

Kravtsov, Y. A. (1964a). A modification of the geometrical optics method. *Radiofizika*, 7, 664-673 (in Russian). [378]

Kravtsov, Y. A. (1964b). Asymptotic solutions of Maxwell's equations near a caustic. *Radiofizika*, 7, 1049-1056 (in Russian). [378, 380]

Kravtsov, Y. A. (1965). Modification of the method of geometrical optics for a wave penetrating a caustic. *Radiofizika*, 8, 659-667. [380]

Kravtsova, T. S. See Babich and Kravtsova.

Krushkal', E. M. See Karpman and Krushkal'.

Kruskal, M. (1962). Asymptotic theory of Hamiltonian and other systems with all solutions nearly periodic. *J. Math. Phys.*, 3, 806-828. [168]

Krylov, N., and N. N. Bogoliubov (1947). *Introduction to Nonlinear Mechanics.* Princeton University Press, Princeton, N. J. [165, 174]

Kubo, R. (1963). Stochastic Liouville equation. *J. Math. Phys.*, 4, 174-183. [382]

Kuiken, H. K. (1970). Inviscid film flow over an inclined surface originated by strong fluid injection. *J. Fluid Mech.*, 42, 337-347. [99]

Kuo, Y. H. (1953). On the flow of an incompressible viscous fluid past a flat plate at moderate Reynolds numbers. *J. Math and Phys.*, 32, 83-101. [78]

Kuo, Y. H. (1956). Viscous flow along a flat plate moving at high supersonic speeds. *J. Aeron. Sci.*, 23, 125-136. [78]

Kuzmak, G. E. (1959). Asymptotic solutions of nonlinear second order differential equations with variable coefficients. *J. Appl. Math. Mech.*, 23, 730-744. [232, 287]

Lacina, J. (1969a). New canonical perturbation method for complete set of integrals of motion. *Czech. J. Phys.*, **B19**, 130-133. [199]

Lacina, J. (1969b). New canonical perturbation method for complete set of integrals of motion. *Ann. Phys.*, **51**, 381-391. [199]

Lagerstrom, P. A. See Kaplun and Lagerstrom.

Lagerstrom, P. A., and J. D. Cole, (1955). Examples illustrating expansion procedures for the Navier-Stokes equations. *J. Rat. Mech. Anal.*, **4**, 817-882. [140]

Lagerstrom, P. A., and J. Kevorkian, (1963a). Earth-to-moon trajectories in the restricted three-body problem. *J. Mécanique*, **2**, 189-218. [138]

Lagerstrom, P. A., and J. Kevorkian, (1963b). Matched-conic approximation to the two fixed force-center problem. *Astron. J.* **68**, 84-92. [44, 138]

Landahl, M. T. See Rubbert and Landahl.

Langer, R. E. (1931). On the asymptotic solutions of differential equations, with an application to the Bessel functions of large complex order. *Trans. Am. Math. Soc.*, **33**, 23-64. [339, 340, 345]

Langer, R. E. (1934). The asymptotic solutions of certain linear ordinary differential equations of the second order. *Trans. Am. Math. Soc.*, **36**, 90-106. [339, 340]

Langer, R. E. (1935). On the asymptotic solutions of ordinary differential equations with reference to the Stokes phenomenon about a singular point. *Trans. Am. Math. Soc.*, **37**, 397-416. [358]

Langer, R. E. (1949). The asymptotic solutions of ordinary linear differential equations of the second order, with special reference to a turning point. *Trans. Am. Math. Soc.*, **67**, 461-490. [346, 348]

Langer, R. E. (1957). On the asymptotic solutions of a class of ordinary differential equations of the fourth order, with special reference to an equation of hydrodynamics. *Trans. Am. Math. Soc.*, **84**, 144-191. [360]

Langer, R. E. (1959a). Formal solutions and a related equation for a class of fourth order differential equations of a hydrodynamic type. *Trans. Am. Math. Soc.*, **92**, 371-410. [360]

Langer, R. E. (1959b). The asymptotic solutions of a linear differential equation of the second order with two turning points. *Trans. Am. Math. Soc.*, **90**, 113-142. [344]

Laplace, P. S. (1805). On the figure of a large drop of mercury, and the depression of mercury in a glass tube of a great diameter. In *Celestial Mechanics* (transl. by Nathaniel Bowditch, Boston, 1839), Chelsea, New York, 1966. [114]

Latta, G. E. (1951). Singular perturbation problems. Ph.D. Thesis, California Institute of Technology. [114, 145]

Latta, G. E. (1964). *Advanced Ordinary Differential Equations.* Lecture notes, Stanford University. [317]

Ledovskaja, L. B. See Zabreiko and Ledovskaja.

Lee, D. H., and L. M. Sheppard, (1966). An approximate second-order wing theory. *AIAA J.*, **4**, 1828-1830. [78, 89]

Lee, J. H. S. See Akinsete and Lee.

Lee, R. S. See Cheng, Kirsch, and Lee.

Legras, J. (1951). Application de la méthode de Lighthill à un écoulement plan supersonique. *Compt. Rend.*, **233**, 1005-1008. [78, 86]

Legras, J. (1953). Nouvelles applications de la méthode de Lighthill à l'étude des ondes de choc. O.N.E.R.A. Publ. No. 66. [78, 86]

Lesser, M. B. (1970). Uniformly valid perturbation series for wave propagation in an inhomogeneous medium. *J. Acoust. Soc. Am.*, **47**, 1297-1302. [90]

Le Verrier, U. J. J. (1856). Sur la determination des longitudes terrestres. Paris, *Compt. Rend.*, **43**, 249-257. [168]

Levey, H. C. (1959). The thickness of cylindrical shocks and the PLK method. *Quart. Appl. Math.*, **17**, 77-93. [52, 99, 100]

Levey, L., and L. B. Felsen, (1967). On transition functions occurring in the theory of diffraction in inhomogeneous media. *J. Inst. Mat. Appl.*, **3**, 76-97. [380]

Levinson, N. (1969). Asymptotic behavior of solutions of nonlinear differential equations. *Stud. Appl. Math.*, **48**, 285-297. [21, 22]

Levinson, N. See Coddington and Levinson.

Lewak, G. J. (1969). More uniform perturbation theory of the Vlasov equation. *J. Plasma Phys.*, **3**, 243-253. [78]

Lewak, G. J. (1971). Interaction of electrostatic waves in collisionless plasmas. *J. Plasma Phys.*, **5**, 51-63. [235]

Lewak, G. J. See Chen and Lewak; Zawadzki and Lewak.

Lewis, R. M. See Ahluwalia, Lewis, and Boersma.

Lewis, R. M., N. Bleistein, and D. Ludwig, (1967). Uniform asymptotic theory of creeping waves. *Comm. Pure Appl. Math.*, **20**, 295-328. [380]

Lick, W. (1969). Two-variable expansions and singular perturbation problems. *SIAM J. Appl. Math.*, **17**, 815-825. [89]

Lick, W. (1970). Nonlinear wave propagation in fluids. *Annual Review of Fluid Mechanics*, Vol. 2 (M. van Dyke, W. G. Vincinti, and J. V. Wehausen, Eds.), Annual Reviews, Palo Alto, Calif., pp. 113-136. [235]

Liepins, A. A. See Sanders and Liepins.

Lighthill, M. J. (1949a). A technique for rendering approximate solutions to physical problems uniformly valid. *Phil. Mag.*, **40**, 1179-1201. [42, 57, 77, 80, 87, 108]

Lighthill, M. J. (1949b). The shock strength in supersonic "conical fields." *Phil. Mag.*, **40**, 1202-1223. [78]

Lighthill, M. J. (1951). A new approach to thin airfoil theory. *Aeron. Quart.*, **3**, 193-210. [98]

Lighthill, M. J. (1961). A technique for rendering approximate solutions to physical problems uniformly valid. *Z. Flugwiss.*, **9**, 267-275. [57, 77, 99]

Lighthill, M. J. (1965). Contributions to the theory of waves in non-linear dispersive systems. *J. Inst. Math. Appl.*, **1**, 269-306. [216]

Lighthill, M. J. (1967). Some special cases treated by the Whitham theory. *Proc. Roy. Soc., (London)*, A**299**, 28-53. [216]

Lin, C. C. (1954). On a perturbation theory based on the method of characteristics. *J. Math and Phys.*, **33**, 117-134. [57, 87, 89]

Lin, C. C. (1955). *The Theory of Hydrodynamic Stability*. Cambridge University Press, Cambridge. [360]

Lin, C. C., and A. L. Rabenstein (1960). On the asymptotic solutions of a class of ordinary differential equations of the fourth order. *Trans. Am. Math. Soc.*, **94**, 24-57. [360]

Lin, C. C., and A. L. Rabenstein (1969). On the asymptotic theory of a class of ordinary differential equations of the fourth order. II. Existence of solutions which are approximated by the formal solutions. *Stud. Appl. Math.*, 48, 311-340. [360]

Lindstedt, A. (1882). Ueber die Integration einer fur die strorungstheorie wichtigen Differentialgleichung. *Astron, Nach.*, 103, Col. 211-220. [56]

Lindzen, R. S. (1971). Equatorial planetary waves in shear: Part I. *J. Atmos. Sci.*, 28, 609-622. [234]

Liouville, J. (1837). Second mémoire sur le développement des fonctions en séries dont divers termes sont assujettis a satisfaire à une même équation différentielle du second ordre contenant un paramètre variable. *J. Math. Pure Appl.*, 2, 16-35. [314]

Lock, C. N. H. See Fowler and Lock; Fowler, Gallop, Lock, and Richmond.

Lowell, S. C.(1970). Wave propagation in monatomic lattices with anharmonic potential. *Proc. Roy. Soc. (London)*, A318, 93-106. [217]

Ludwig, D. (1966). Uniform asymptotic expansions at a caustic. *Comm. Pure Appl. Math.*, 19, 215-250. [378, 380]

Ludwig, D. (1967). Uniform asymptotic expansion of the field scattered by a convex object at high frequencies. *Comm. Pure Appl. Math.*, 20, 103-138. [380]

Ludwig, D. (1970a). Diffraction by a circular cavity. *J. Math. Phys.*, 11, 1617-1630. [380]

Ludwig, D. (1970b). Uniform asymptotic expansions for wave propagation and diffraction problems. *SIAM Rev.*, 12, 325-331. [380]

Ludwig, D. See Lewis, Bleistein and Ludwig.

Luke, J. C. (1966). A perturbation method for nonlinear dispersive wave problems. *Proc. Roy. Soc. (London)*, A292, 403-412. [234, 301]

Lynn, R. Y. S., and J. B. Keller (1970). Uniform asymptotic solutions of second order linear ordinary differential equations with turning points. *Comm. Pure Appl. Math.*, 23, 379-408. [345]

Lyusternik, L. A. See Višik and Lyusternik.

Mc Goldrick, L. F. (1970). On Wilton's ripples: A special case of resonant interactions. *J. Fluid Mech.*, 42, 193-200. [234]

Mc Intyre, J. E. (1966). Neighboring optimal terminal control with discontinuous forcing functions. *AIAA J.*, 4, 141-148. [79]

Mc Kelvey, R. W. (1955). The solution of second order linear ordinary differential equations about a turning point of order two. *Trans. Am. Math. Soc.*, 79, 103-123. [346]

Mc Kelvey, R. W. See Kazarinoff and Mc Kelvey.

Mc Namara, B., and K. J. Whiteman (1967). Invariants of nearly periodic Hamiltonian systems. *J. Math. Phys.*, 8, 2029-2038. [199]

Mahony, J. J. (1962). An expansion method for singular perturbation problems. *J. Australian Mat. Soc.*, 2, 440-463. [232, 303]

Malkus, W. V. R., and G. Veronis (1958). Finite amplitude cellular convection. *J. Fluid Mech.*, 4, 225-260. [77]

Maroli, C. (1966). Kinetic theory of high-frequency resonance gas discharge breakdown. *Nuovo Cimento*, B41, 208-224. [236]

Maroli, C. See Caldirola, De Barbieri, and Maroli; De Barbieri and Maroli.

Maroli, C., and R. Pozzoli (1969). Penetration of high-frequency electromagnetic waves into a slightly ionized plasma. *Nuovo Cimento*, B61, 277-289. [235]

Maslowe, S. A., and R. E. Kelly (1970). Finite amplitude oscillations in a Kelvin-Helmholtz flow. *Int. J. Non-Linear Mech.*, 5, 427-435. [77]

Matkowsky, B. J. (1966). Asymptotic solution of partial differential equations in thin domains. Ph. D. Thesis, New York University. [380]

Matkowsky, B. J., See Reiss and Matkowsky.

Maxwell, J. C. (1866). On the viscosity or internal friction of air and other gases. *Phil. Trans. Roy. Soc. London*, 156, 249-268. [114]

Mei, C. C. See Chu and Mei.

Meirovitch, L. (1970). *Methods of Analytical Dynamics*. McGraw-Hill, New York. [181]

Melnik, R. E. (1965). Newtonian entropy layer in the vicinity of a conical symmetry plane. *AIAA J.*, 3, 520-522. [79]

Mendelson, K. S. (1970). Perturbation theory for damped nonlinear oscillations. *J. Math. Phys.*, 11, 3413-3415. [224]

Mersman, W. A. (1970). A new algorithm for the Lie transformation. *Celestial Mech.*, 3, 81-89. [202]

Mersman, W. A. (1971). Explicit recursive algorithms for the construction of equivalent canonical transformations. *Celestial Mech.*,, 3, 384-389. [202]

Messiter, A. F. See Spriggs, Messiter, and Anderson.

Mettler, E. (1959). Stabilitätsfragen bei freien Schwingungen mechanischer Systeme. *Ingenieur-Archiv.*, 28, 213-228. [168, 188]

Meyer, J. W. (1971). Rayleigh scattering of a laser beam from a massive relativistic two-level atom. *Phys. Rev.*, A3, 1431-1443. [235]

Miller, S. C., and R. H. Good (1953). A WKB-type approximation to the Schrödinger equation. *Phys. Rev.*, 91, 174-179. [344]

Millman, M. H. See Keller and Millman.

Millman, M. H., and J. B. Keller (1969). Perturbation theory of nonlinear boundary-value problems. *J. Math. Phys.*, 10, 342-361. [77]

Mitchell, C. E. (1971). Analysis of a combustion instability problem using the technique of multiple scales. *AIAA J.*, 9, 532-533. [235]

Mitropolski, Y. A. (1965). *Problems of the Asymptotic Theory of Non-stationary Vibrations*. Daniel Davey, New York. [174]

Mitropolski, Y. A. See Bogoliubov and Mitropolski.

Montgomery, D., and D. A. Tidman (1964). Secular and nonsecular behavior for the cold plasma equations. *Phys. Fluids*, 7, 242-249. [178]

Moriguchi, H. (1959). An improvement of the WKB method in the presence of turning points and the asymptotic solutions of a class of Hill equations. *J. Phys. Soc. Japan*, 14, 1771-1796. [342, 345, 350]

Morino, L. (1969). A perturbation method for treating nonlinear panel flutter problems. *AIAA J.*, 7, 405-411. [234]

Morris, W. D. (1965). Laminar convection in a heated vertical tube rotating about a parallel axis. *J. Fluid Mech.*, 21, 453-464. [79]

Morrison, H. L. See Richmond and Morrison.

Morrison, J. A. (1966a). Comparison of the modified method of averaging and the two variable expansion procedure. *SIAM Rev.*, 8, 66-85. [231]

Morrison, J. A. (1966b). Generalized method of averaging and the von Zeipel method. In *Progress in Astronautics and Aeronautics, Vol. 17, Methods in Astrodynamics and Celestial Mechanics* (R. L. Duncombe and V. G. Szebehely, Eds.), Academic, New York, pp. 117-138. [168, 191]

Mortell, M. P. (1968). Traveling load on a cylindrical shell. *J. Acoust. Soc. Am.,* 44, 1664-1670. [233]

Mortell, M. P. (1969). Waves on a spherical shell. *J. Acoust. Soc. Am.,* 45, 144-149. [233]

Mortell, M. P. (1971). Resonant thermal-acoustic oscillations. *Int. J. Eng. Sci.,* 9, 175-192. [89]

Mortell, M. P., and E. Varley (1971). Finite amplitude waves in bounded media: Nonlinear free vibrations of an elastic panel. *Proc. Roy. Soc. (London),* A318, 169-196. [90]

Morton, B. R. (1959). Laminar convection in uniformly heated horizontal pipes at low Rayleigh numbers. *Quart. J. Mech. Appl. Math.,* 12, 410-426. [79]

Mulholland, R. J. (1971). Nonlinear oscillations of a third-order differential equation. *Int. J. Non-Linear Mech.,* 6, 279-294. [105]

Murphy, C. H. (1963). Free flight motion of symmetric missiles. Ballistic Research Laboratories Rept. No. 1216, Aberdeen Proving Grounds, Md. [321]

Murray, J. D. (1968). On the effect of drainage on free surface oscillations. *Appl. Sci. Res.,* 19, 234-249. [234]

Musa, S. A. (1967). Integral constraints in weakly nonlinear periodic systems. *SIAM J. Appl. Math.,* 15, 1324-1327. [232]

Musen, P. (1965). On the high order effects in the methods of Krylov-Bogoliubov and Poincaré. *J. Astron. Sci.,* 12, 129-134. [168, 192, 199]

Nair, S., and S. Nemat-Nasser (1971). On finite amplitude waves in heterogeneous elastic solids. *Int. J. Eng. Sci.,* 9, 1087-1105. [94]

Nayfeh, Adnan, and S. Nemat-Nasser (1971). Thermoelastic waves in solids with thermal relaxation. *Acta Mechanica,* 12, 53-69. [45, 136]

Nayfeh, A. H. (1964). A generalized method for treating singular perturbation problems. Ph. D. Thesis, Santford University. [232, 276, 280, 285, 295]

Nayfeh, A. H. (1965a). A comparison of three perturbation methods for the earth-moon-spaceship problem. *AIAA J.,* 3, 1682-1687. [82, 103, 233, 295, 297]

Nayfeh, A. H. (1965b). An expansion method for treating singular perturbation problems. *J. Math. Phys.,* 6, 1946-1951. [232, 234, 276, 280, 285, 384]

Nayfeh, A. H. (1965c). A perturbation method for treating nonlinear oscillation problems. *J. Math. and Phys.,* 44, 368-374. [230, 232]

Nayfeh, A. H. (1965d). Nonlinear oscillations in a hot electron plasma. *Phys. Fluids.,* 8, 1896-1898. [230, 234]

Nayfeh, A. H. (1966). Take-off from a circular orbit by a small thrust. In *Progress in Astronautics and Aeronautics,* Vol. 17, *Methods in Astrodynamics and Celestial Mechanics* (R. L. Duncombe and V. G. Szebehely, Eds.), Academic, New York, pp. 139-157. [99, 109, 233, 305]

Nayfeh, A. H. (1967a). Asymptotic solutions of an eigenvalue problem with two turning points—heat transfer in a tube. *J. Math and Phys.,* 46, 349-354. [385]

Nayfeh, A. H. (1967b). The van der Pol oscillator with delayed amplitude limiting. *Proc. IEEE,* 55, 111-112. [231, 232, 258]

Nayfeh, A. H. (1968). Forced oscillations of the van der Pol oscillator with delayed amplitude limiting. *IEEE Trans. Circuit Theory*, 15, 192-200. [230, 232, 258]

Nayfeh, A. H. (1969a). A multiple time scaling analysis of re-entry vehicle roll dynamics. *AIAA J.*, 7, 2155-2157. [233, 320]

Nayfeh, A. H. (1969b). On the nonlinear Lamb-Taylor instability. *J. Fluid Mech.*, 38, 619-631. [235]

Nayfeh, A. H. (1970a). Characteristic exponents for the triangular points in the elliptic restricted problem of three bodies. *AIAA J.*, 8, 1916-1917. [68]

Nayfeh, A. H. (1970b). Finite amplitude surface waves in a liquid layer. *J. Fluid Mech.*, 40, 671-684. [234]

Nayfeh, A. H. (1970c). Nonlinear stability of a liquid jet. *Phys. Fluids*, 13, 841-847. [99, 235]

Nayfeh, A. H. (1970d). Triple- and quintuple-dimpled wave profiles in deep water. *Phys. Fluids*, 13, 545-550. [234]

Nayfeh, A. H. (1971a). Third-harmonic resonance in the interaction of capillary and gravity waves. *J. Fluid Mech.*, 48, 385-395. [234]

Nayfeh, A. H. (1971b). Two-to-one resonances near the equilateral libration points. *AIAA J.*, 9, 23-27. [233]

Nayfeh, A. H., and S. D. Hassan (1971). The method of multiple scales and nonlinear dispersive waves. *J. Fluid Mech.*, 48, 463-475. [234, 298]

Nayfeh, A. H., and A. A. Kamel (1970a). Stability of the triangular points in the elliptic restricted problem of three bodies. *AIAA J.*, 8, 221-223. [64, 66]

Nayfeh, A. H. and A. A. Kamel (1970b). Three-to-one resonances near the equilateral libration points. *AIAA J.*, 8, 2245-2251. [233]

Nayfeh, A. H., and W. S. Saric, (1971a). Nonlinear Kelvin-Helmholtz instability. *J. Fluid Mech.*, 46, 209-231. [235]

Nayfeh, A. H., and W. S. Saric (1971b). Nonlinear resonances in the motion of rolling re-entry bodies. AIAA Paper No. 71-47. [233, 305]

Nayfeh, A. H., and W. S. Saric (1972a). An analysis of asymmetric rolling bodies with nonlinear aerodynamics. *AIAA J.*, 10, 1004-1011. [233, 291, 292, 306]

Nayfeh, A. H., and W. S. Saric (1972b). Nonlinear waves in a Kelvin-Helmholtz flow. *J. Fluid Mech.*, 55, 311-327. [234]

Nemat-Nasser, S. See Adnan Nayfeh and Nemat-Nasser; Nair and Nemat-Nasser.

Neubert, J. A. (1970). Asymptotic solution of the stochastic Helmholtz equation for turbulent water. *J. Acoust. Soc. Am.*, 48, 1203-1211. [234]

Newell, A. C. (1968). The closure problem in a system of random gravity waves. *Rev. Geophys.*, 6, 1-31. [234]

Newell, A. C. (1969). Rossby wave packet interactions. *J. Fluid Mech.*, 35, 255-271. [234]

Newell, A. C. See Benney and Newell.

Newell, A. C., and J. A. Whitehead (1969). Finite bandwidth, finite amplitude convection. *J. Fluid Mech.*, 38, 279-303. [235]

Nienhuis, G. (1970). On the microscopic theory of Brownian motion with a rotational degree of freedom. *Physica*, 49, 26-48. [236]

Nikolenko, L. D. See Feshchenko, Shkil', and Nikolenko.

Noerdlinger, P. D., and V. Petrosian (1971). The effect of cosmological expansion on self-gravitating ensembles of particles. *Astrophys. J.*, 168, 1-9. [232]

Ockendon, J. R. (1966). The separation of Newtonian shock layers. *J. Fluid Mech.*, **26**, 563-572. [79]

Olver, F. W. J. (1954). The asymptotic solution of linear differential equations of the second order for large values of a parameter and the asymptotic expansion of Bessel functions of large order. *Phil. Trans. Roy. Soc. London Ser. A.*, **247**, 307-368. [341, 348, 358]

Olver, F. W. J. (1959). Uniform asymptotic expansions for Weber parabolic cylinder functions of large orders. *J. Res. Natl. Bur. Standards*, **63B**, 131-169. [345]

O'Malley, R. E., Jr. (1968a). A boundary value problem for certain nonlinear second order differential equations with a small parameter. *Arch. Rat. Mech. Anal.*, **29**, 66-74. [233]

O'Malley, R. E., Jr. (1968b). Topics in singular perturbations. *Advan. Math.*, **2**, 365-470. [110, 233]

O'Malley, R. E., Jr. (1971). Boundary layer methods for nonlinear initial value problems. *SIAM Rev.*, **13**, 425-434. [145]

O'Malley, R. E., Jr. See Ackerberg and O'Malley.

Oseen, C. W. (1910). Über die Stokessche Formel and über eine verwandte Aufgabe in der Hydrodynamik. *Ark. Mat. Astron. Fys.*, **6**, No. 29. [140]

Oswatitsch, V. K. (1965). Ausbreitungsprobleme. *ZAMM*, **45**, 485-498. [89]

Pandey, B. C. (1968). Study of cylindrical piston problem in water using PLK method. *ZAMP*, **19**, 962-963. [86]

Parker, D. F. (1969). Nonlinearity, relaxation and diffusion in acoustics and ultrasonics. *J. Fluid Mech.*, **39**, 793-815. [234]

Parker, D. F., and E. Varley (1968). The interaction of finite amplitude deflection and stretching waves in elastic membranes and strings. *Quart. J. Mech. Appl. Math.*, **21**, 329-352. [90]

Payne, H. J. See Caughey and Payne.

Pearson, J. R. A. See Proudman and Pearson.

Pedlowsky, J. (1967). Fluctuating winds and the ocean circulation. *Tellus*, **19**, 250-257. [77]

Perko, L. M. (1969). Higher order averaging and related methods for perturbed periodic and quasi-periodic systems. *SIAM J. Appl. Math.*, **17**, 698-724. [231]

Perko, L. M. See Breakwell and Perko.

Pertosian, V. See Noerdlinger and Petrosian.

Peyret, R. (1966). Écoulement quasi unidimensionnel dans un accélérateur de plasma à ondes progressives. *J. Mécanique*, **5**, 471-515. [234]

Peyret, R. (1970). Étude de l'écoulement d'un fluide conducteur dans un canal par la méthode des échelles multiples. *J. Mécanique*, **9**, 61-97. [235]

Pierson, W. J., and P. Fife (1961). Some nonlinear properties of long-crested periodic waves with lengths near 2.44 centimeters. *J. Geophys. Res.*, **66**, 163-179. [77]

Pike, E. R. (1964). On the related-equation method of asymptotic approximation. *Quart. J. Mech. Appl. Math.*, **17**, 105-124, 369-379. [242, 344]

Poincaré, H. (1892). *New Methods of Celestial Mechanics*, Vol. I-III (English transl.), NASA TTF-450, 1967. [10, 56]

Potter, M. C. See Reynolds and Potter.

Pozzoli, R. See Maroli and Pozzoli.

Prandtl, L. (1905). Über Flüssigkeitsbewegung bei sehr kleiner Reibung. Proceedings Third Internat. Math. Kongr., Heidelberg, pp. 484-491. Motion of fluids with very little viscosity. Tech. Memo N.A.C.A. (English transl.), No. 452, 1928. [113]

Prasad, R. (1971). Effect of ion motion on parametric oscillations of a cold plasma in a magnetic field. J. Plasma Phys., 5, 291-302. [235]

Prigoginé, I. (1962). Nonequilibrium Statistical Mechanics. Wiley, New York. [367, 369, 382]

Primas, H. (1961). Über quantenmechanische Systeme mit einem stochastischen Hamilton-operator. Helv. Phys. Acta, 34, 36-57. [382]

Primas, H. See Ernst and Primas.

Pringle, R. Jr. See Breakwell and Pringle.

Pritulo, M. F. (1962). On the determination of uniformly accurate solutions of differential equations by the method of perturbation of coordinates. J. Appl. Math. Mech., 26, 661-667. [57, 95, 369]

Proudman, I. (1960). An example of steady laminar flow at large Reynolds number. J. Fluid Mech., 9, 593-602. [54, 156]

Proudman, I., and J. R. A. Pearson (1957). Expansions at small Reynolds numbers for the flow past a sphere and a circular cylinder. J. Fluid Mech., 2, 237-262. [114, 141, 144]

Puri, K. K. (1971). Effect of viscosity and membrane on the oscillations of superposed fluids. J. Appl. Phys., 42, 995-1000. [235]

Rabenstein, A. L. (1959). The determination of the inverse matrix for a basic reference equation for the theory of hydrodynamic stability. Arch. Rat. Mech. Anal., 2, 355-366. [360]

Rabenstein, A. L. See Lin and Rabenstein.

Rajappa, N. R. (1970). Nonlinear theory of Taylor instability of superposed fluids. J. Phys. Soc. Japan, 28, 219-224. [77]

Ramanathan, G. V., and G. Sandri (1969). Model for the derivation of kinetic theory. J. Math. Phys., 10, 1763-1773. [236]

Ramnath, R. V. (1970a). A new analytical approximation for the Thomas-Fermi model in atomic physics. J. Math. Anal. Appl., 31, 285-296. [235]

Ramnath, R. V. (1970b). Transition dynamics of VTOL aircraft. AIAA J., 8, 1214-1221. [233]

Ramanth, R. V. (1971). On a class of nonlinear differential equations of astrophysics. J. Math. Anal. Appl., 35, 27-47. [235]

Ramnath, R. V. See Klimas, Ramnath, and Sandri.

Ramnath, R. V., and G. Sandri (1969). A generalized multiple scales approach to a class of linear differential equations. J. Math. Anal. Appl., 28, 339-364. [232]

Rand, R. H. See Alfriend and Rand.

Rand, R. H., and S. F. Tseng (1969). On the stability of a differential equation with application to the vibrations of a particle in the plane. J. Appl. Mech., 36, 311-313. [104]

Rao, P. S. (1956). Supersonic bangs. Aeron. Quart., 7, 135-155. [78]

Rarity, B. S. H. (1969). A theory of the propagation of internal gravity waves of finite amplitude. J. Fluid Mech., 39, 497-509. [216]

Rasmussen, M. L. (1970). Uniformly valid approximations for non-linear oscillations with small damping. *Int. J. Non-Linear Mech.*, 5, 687-696. [232]

Rayleigh, Lord (1912). On the propagation of waves through a stratified medium, with special reference to the question of reflection. *Proc. Roy. Soc. (London)*, A86, 208-226. [114, 339]

Rayleigh, Lord, (1917). On the reflection of light from a regularly stratified medium. *Proc. Roy. Soc. (London)*, A93, 565-577. [95, 367, 368]

Rehm, R. G. (1968). Radiative energy addition behind a shock wave. *Phys. Fluids*, 11, 1872-1883. [89]

Reiss, E. L. (1971). On multivariable asymptotic expansions. *SIAM Rev.*, 13, 189-196. [232]

Reiss, E. L., and B. J. Matkowsky (1971). Nonlinear dynamic buckling of a compressed elastic column. *Quart. Appl. Math.*, 29, 245-260. [233]

Reissner, E., and H. J. Weinitschke (1963). Finite pure bending of circular cylindrical tubes. *Quart. Appl. Math.*, 20, 305-319. [54]

Reynolds, W. C., and M. C. Potter (1967). Finite-amplitude instability of parallel shear flows. *J. Fluid Mech.*, 27, 465-492. [162]

Richmond, H. W. See Fowler, Gallop, Lock, and Richmond.

Richmond, O., and H. L. Morrison (1968). Application of a perturbation technique based on the method of characteristics to axisymmetric plasticity. *J. Appl. Mech.*, 35, 117-122. [90]

Rogister, A. (1971). Parallel propagation of nonlinear low-frequency waves in high-β plasma. *Phys. Fluids* 14, 2733-2739. [235]

Rogister, A. See Dobrowolny and Rogister.

Roskes, G. J. See Benney and Roskes.

Ross, L. W. (1970). Perturbation analysis of diffusion-coupled biochemical reaction kinetics. *SIAM J. Appl. Math.*, 19, 323-329. [79]

Rubbert, P. E., and M. T. Landahl (1967). Solution of the transonic airfoil problem through parametric differentiation. *AIAA J.*, 5, 470-479. [235]

Rulf, B. (1967). Relation between creeping waves and lateral waves on a curved interface. *J. Math. Phys.*, 8, 1785-1793. [380]

Rulf, B. (1968). Uniform asymptotic theory of diffraction at an interface. *Comm. Pure Appl. Math.*, 21, 67-76. [380]

Rutherford, P. See Frieman and Rutherford.

Rytov, S. M. (1937). Diffraction of light by ultrasonic wave. *Izv. Akad. Nuak SSSR Ser. Fiz.* No. 2, 223-259 (in Russian). [373]

Saffman, P. G. See Benney and Saffman.

Sakurai, A. (1965). Blast wave theory. In *Basic Developments in Fluid Dynamics*, Vol. I (M. Holt, Ed.), Academic, New York, pp. 309-375. [78]

Sakurai, T. (1968). Effect of the plasma impedance on the time variation of the inverse pinch. *J. Phys. Soc. Japan*, 25, 1671-1679. [78]

Salpeter, E. E., and H. A. Bethe (1951). A relativistic equation for bound-state problems. *Phys. Rev.*, 84, 1232-1242. [372]

Sancer, M. I. See Varvatsis and Sancer.

Sancer, M. I., and A. D. Varvatsis (1969). An investigation of the renormalization and Rytov methods as applied to propagation in a turbulent medium. Northrop Corporate Laboratories. Rept. No. 69-28R. [373]

Sancer, M. I., and A. D. Varvatsis (1970). A comparison of the Born and Rytov methods. *Proc. IEEE*, **58**, 140-141. [373]

Sanders, J. L., Jr., and A. A. Liepins (1963). Toroidal membrane under internal pressure. *AIAA J.*, **1**, 2105-2110. [356, 357]

Sandri, G. (1965). A new method of expansion in mathematical physics. *Nuovo Cimento*, **B36**, 67-93. [230]

Sandri, G. (1967). Uniformization of asymptotic expansions. In *Nonlinear Partial Differential Equations: A Symposium on Methods of Solutions* (W. F. Ames, Ed.), Academic, New York, pp. 259-277. [230]

Sandri, G. See Goldberg and Sandri; Klimas, Ramnath, and Sandri; Ramanathan and Sandri; Ramnath and Sandri.

Saric, W. S. See Nayfeh and Saric.

Savage, J. C., and H. S. Hasegawa (1967). Evidence for a linear attenuation mechanism. *Geophysics*, **32**, 1003-1014. [78]

Schechter, H. B. (1968). The effect of three-dimensional nonlinear resonances on the motion of a particle near the earth-moon equilateral libration points. *Second Compilation of Papers on Trajectory Analysis and Guidance Theory*. NASA PM-67-21, 229-344. [199]

Schrödinger, E. (1926). Quantisierung als Eigenwertproblem. *Ann. Phys.*, **80**, 437-490. [56, 71]

Schwartz, N. J. See Holt and Schwartz.

Schwertassek, V. R. (1969). Grenzen von Mitnahmeberichen. *ZAMM*, **49**, 409-421. [232]

Scott, A. C. (1970). Propagation of magnetic flux on a long Josephson tunnel junction. *Nuovo Cimento*, **B69**, 241-261. [221]

Scott, P. R. (1966). Equations of the oscillator with delayed amplitude limiting. *Proc. IEEE*, **54**, 898-899. [258]

Searl, J. W. (1971). Expansions for singular perturbations.*J. Inst. Math. Appl.*, **8**, 131-138. [233]

Sellers, J. R., M. Tribus, and J. S. Klein (1956). Heat transfer to laminar flow in a round tube or flat conduit–The Graetz problem extended. *Trans. ASME*, **78**, 441-448. [384]

Sethna, P. R. (1963). Transients in certain autonomous multiple-degree-of-freedom nonlinear vibrating systems. *J. Appl. Mech.*, **30**, 44-50. [168]

Sethna, P. R. (1965). Vibrations of dynamical systems with quadratic nonlinearities. *J. Appl. Mech.*, **32**, 576-582. [188, 225]

Shabbar, M. (1971). Side-band resonance mechanism in the atmosphere supporting Rossby waves. *J. Atmos. Sci.*, **28**, 345-349. [234]

Shen, C. N. (1959). Stability of forced oscillations with nonlinear second-order terms. *J. Appl. Mech.*, **26**, 499-502. [104]

Sheppard, L. M. See Lee and Sheppard.

Shi, Y. Y. See Eckstein and Shi; Eckstein, Shi, and Kevorkian.

Shi, Y. Y., and M. C. Eckstein (1966). Ascent or descent from satellite orbit by low thrust. *AIAA J.* **4**, 2203-2209. [233]

Shi, Y. Y., and M. C. Eckstein (1968). Application of singular perturbation methods to resonance problems. *Astron. J.*, **73**, 275-289. [233, 276]

Shkarofsky, I. P. (1971). Modified Born back scattering from turbulent plasmas: Attenuation leading to saturation and cross-polarization. *Radio Sci.*, **6**, 819-831. [367]

Shkil', N. I. See Feshchenko, Shkil', and Nikolenko.

Shniad, H. (1970). The equivalence of von Zeipel mappings and Lie transforms. *Celestial Mech.*, **2**, 114-120. [202]

Shrestha, G. M. See Terrill and Shrestha.

Sibuya, Y. (1958). Sur réduction analytique d'un système d'équations différentielles ordinaires linéaires contenant un paramètre. *J. Fac. Science, Univ. Tokyo*, **7**, 527-540. [327]

Sibuya, Y. (1963a). Asymptotic solutions of a linear ordinary differential equation of nth order about a simple turning point. In *International Symposium Differential Equations and Nonlinear Mechanics* (J. P. La Salle and S. Lefschetz, Eds.), Academic, New York, pp. 485-488. [360]

Sibuya, Y. (1963b). Simplification of a linear ordinary differential equation of the nth order at a turning point. *Arch. Rat. Mech. Anal.*, **13**, 206-221. [360]

Sibuya, Y. (1967). Subdominant solutions of the differential equation $y'' - \lambda^2 (x -a_1) (x - a_2) ... (x - a_n) y = 0$. *Acta. Math.*, **119**, 235-271. [345]

Sibuya, Y. See Hsieh and Sibuya.

Simmons, W. F. (1969). A variational method for weak resonant wave interactions. *Proc. Roy. Soc. (London)*, A309, 551-575. [216]

Sirignano, W. A., and L. Crocco (1964). A shock wave model of unstable rocket combustors. *AIAA J.*, **2**, 1285-1296. [78]

Sirovich, L. See Chong and Sirovich.

Sivasubramanian, A. See Tang and Sivasubramanian.

Soler, A. I. See Brull and Soler.

Spriggs, J. H. A. F. Messiter, and W. J. Anderson (1969). Membrane flutter paradox—An explanation by singular-perturbation methods. *AIAA J.*, **7**, 1704-1709. [234]

Stanley, N. R. See Hufnagel and Stanley.

Steele, C. R. (1965). On the asymptotic solution of nonhomogeneous ordinary differential equations with a large parameter. *Quart. Appl. Math.*, **23**, 193-201. [353]

Stern, D. P. (1970a). Direct canonical transformations. *J. Math. Phys.*, **11**, 2776-2781. [199]

Stern, D. P. (1970b). Kruskal's perturbation method. *J. Math Phys.*, **11**, 2771-2775. [169]

Stern, D. P. (1971a). A new formulation of canonical perturbation theory. *Celestial Mech.*, **3**, 241-246. [199]

Stern, D. P. (1971b). Classical adiabatic theory. *J. Math. Phys.*, **12**, 2231-2242. [169]

Stern, D. P. (1971c). The canonization of nice variables. *J. Math. Phys.*, **12**, 2226-2231. [91]

Stewartson, K., and J. T. Stuart, (1971). A non-linear instability theory for a wave system in plane Poiseuille flow. *J. Fluid Mech.*, **48**, 529-545. [235]

Stoker, J. J. (1957) *Water Waves*. Wiley, New York. [57, 77]

Stokes, G. G. (1851). On the effect of the internal friction of fluids on the motion of pendulums. *Trans. Cambridge Phil. Soc.,* **9**, 8-106. [30]

Stokes, G. G. (1857). On the discontinuity of arbitrary constants which appear in divergent developments. *Cambridge Phil. Trans.,* **10**, 106-128; *Coll. Papers,* **4**, 77-109. [312]

Stone, P. H. (1969). The meridional structure of baroclinic waves. *J. Atmos. Sci.,* **26**, 376-389. [234]

Strohbehn, J. W. (1968). Comments on Rytov's method. *J. Opt. Soc. Am.,* **58**, 139-140. [373]

Struble, R. A. (1962). *Nonlinear Differential Equations.* McGraw-Hill, New York. [171, 223]

Stuart, J. T. (1958). On the nonlinear mechanics of hydrodynamic stability. *J. Fluid Mech.,* **4**, 1-21. [162]

Stuart, J. T. (1960a). Nonlinear effects in hydrodynamic stability. Proc. Xth Int. Cong. Appl. Mech. Stresa, Italy. [162]

Stuart, J. T. (1960b). On the nonlinear mechanics of wave disturbances in stable and unstable parallel flows. Part 1. The basic behaviour in plane Poiseuille flow. *J. Fluid Mech.,* **9**, 353-370. [162]

Stuart, J. T. (1961). On three-dimensional nonlinear effects in the stability of parallel flows. *Advan. Aeron. Sci.,* **3**, 121-142. Pergamon, Oxford. [162]

Stuart, J. T. See Stewartson and Stuart.

Sturrock, P. A. (1957). Nonlinear effects in electron plasmas. *Proc. Roy. Soc. (London),* A**242**, 277-299. [230]

Sturrock, P. A. (1958). A variational principle and an energy theorem for small amplitude disturbances of electron beams and of electron-ion plasmas. *Ann. Phys.,* **4**, 306-324. [216]

Sturrock, P. A. (1962). *Plasma Hydromagnetic,* Stanford University Press, Stanford, California. [216]

Sturrock, P. A. (1963). Nonlinear theory of electromagnetic waves in plasmas. Stanford University Microwave Laboratory Rept. No. 1004. [230]

Swanson, C. A. (1956). Differential equations with singular points. Tech. Rept. 16, Contract Nonr.-220(11), Department of Mathematics, California Institute of Technology. [358]

Sweet, J. (1971). Impulse of a ring with nonlinear material behavior. *AIAA J.,* **9**, 332-334. [58]

Tam, C. K. W. (1969). Amplitude dispersion and nonlinear instability of whistlers. *Phys. Fluids,* **12**, 1028-1035. [234]

Tam, C. K. W. (1970). Nonlinear dispersion of cold plasma waves. *J. Plasma. Phys.,* **4**, 109-125. [234]

Tam, K. K. (1968). On the asymptotic solution of the Orr-Sommerfeld equation by the method of multiple scales. *J. Fluid Mech.,* **34**, 145-158. [233, 360]

Tang, T. W., and A. Sivasubramanian, (1971). Nonlinear instability of modulated waves in a magnetoplasma. *Phys. Fluids,* **14**, 444-446. [217]

Taniuti, T. See Asano and Taniuti.

Tatarski, V. I. (1961). *Wave Propagation in a Turbulent Medium.* McGraw-Hill, New York. [361, 369, 373]

Tatarski, V. I. (1964). Propagation of electromagnetic waves in a medium with strong dielectric-constant fluctuations. *Soviet Phys. JETP* (English transl.), **19**, 946-953. [367, 369, 371, 372]

Tatarski, V. I., and M. E. Gertsenshtein, (1963). Propagation of waves in a medium with strong fluctuation of the refractive index. *Soviet Phys. JETP* (English transl.), **17**, 458-469. [382]

Taussig, R. T. (1969). Macroscopic quasilinear theory of high-frequency radiation in a cold plasma. *Phys. Fluids*, **12**, 914-922. [234]

Taylor, L. S. (1967). On Rytov's method. *Radio Sci.*, **2**, 437-441. [373]

Temple, G. (1958). Linearization and delinearization. Proceedings of the International Congress of Mathematics, Edinburgh, pp. 233-247. [94]

Terrill, R. M., and G. M. Shrestha, (1965). Laminar flow through a channel with uniformly porous walls of different permeability. *Appl. Sci. Res.*, A15, 440-468. [54, 156]

Thomé, J. (1883). Über Integrale zweiter Gattung. *J. Reine Angew. Math.*, **95**, 241-250. [311, 326]

Tidman, D. A. See Montgomery and Tidman.

Timoshenko, S., and S. Woinowsky-Krieger, (1959). *Theory of Plates and Shells*, 2nd ed., McGraw-Hill, New York. [36]

Ting, L. See Keller and Ting.

Ting, L., and S. Brofman, (1964). On take-off from circular orbit by small thrust. *ZAMM*, **44**, 417-428. [233]

Tollmien, W. (1947). Asymptotische Integration der Störungsdifferentialgleichung ebener laminarer Strömungen bei hohen Reynoldsschen Zahlen. *ZAMM*, **25/27**, 33-50, 70-83. [300]

Tribus, M. See Sellers, Tribus, and Klein.

Tseng, S. F. See Rand and Tseng.

Tsien, H. S. (1956). The Poincaré-Lighthill-Kuo method. *Advan. Appl. Mech.*, **4**, 281-349. [78, 80, 99]

Tso, W. K., and T. K. Caughey, (1965). Parametric excitation of a nonlinear system. *J. Appl. Mech.*, **32**, 899-902. [224]

Tumarkin, S. A. (1959). Asymptotic solution of a linear nonhomogeneous second order differential equation with a transition point and its application to the computations of toroidal shells and propeller blades. *Appl. Math. Mech. (Prikl. Mat. Mech., ASME transl.)* **23**, 1549-1565. [353, 356]

Usher, P. D. (1968). Coordinate stretching and interface location. II. A new PL expansion. *J. Computer Phys.*, **3**, 29-39. [95]

Usher, P. D. (1971). Necessary conditions for applicability of Poincaré-Lighthill perturbation theory. *Quart, Appl. Math.*, **28**, 463-471. [79]

Vaglio-Laurin, R. (1962). On the PLK method and the supersonic blunt-body problem. *J. Aeron. Sci.*, **29**, 185-206. [99]

Van der Corput, J. G. (1956). Asymptotic developments I. Fundamental theorems of asymptotics. *J. Anal. Math.*, **4**, 341-418. [18]

Van der Corput, J. G. (1962). *Asymptotic Expansions*. Lecture notes, Stanford University. [12]

Van der Pol, B. (1922). On a type of oscillation hysteresis in a simple triode generator. *Phil. Mag.*, **43**, 177-193. [3]

REFERENCES AND AUTHOR INDEX 413

Van der Pol, B. (1926). On oscillation hysteresis in a simple triode generator. *Phil. Mag.*, **43**, 700-719. [164]

Van der Pol, B. (1927). Über Relaxations schwingungen. *Jahrb. Drahtl. Telegr. Teleph.*, **28**, 178-184. [34]

Van Dyke, M. D. (1952). A study of second-order supersonic flow theory. N.A.C.A. Rept. No. 1081. [27]

Van Dyke, M. (1964). *Perturbation Methods in Fluid Mechanics.* Academic, New York. [110, 114, 119, 130]

Van Hove, L. (1955). Quantum-mechanical perturbations giving rise to a statistical transport equation. *Physica*, **21**, 517-540. [369]

Van Hove, L. (1957). The approach to equilibrium in quantum statistics. *Physica*, **23**, 441-480. [369]

Van Hove, L., N. Hugenholtz, and L. Howland, (1961). *Quantum Theory of Many Particle Systems.* Benjamin, New York. [367]

Van Wijngaarden, L. (1968). On the oscillations near and at resonance in open pipes. *J. Eng. Math.*, **2**, 225-240. [89]

Van Wijngaarden, L. See Verhagen and Van Wijngaarden.

Varley, E. See Mortell and Varley; Parker and Varley.

Varvatsis, A. D. See Sancer and Varvatsis.

Varvatsis, A. D., and M. I. Sancer, (1971). On the renormalization method in random wave propagation. *Radio Sci.*, **6**, 87-97. [372]

Vasil'eva, A. B. (1959). On repeated differentiation with respect to the parameter of solutions of systems of ordinary differential equations with a small parameter in the derivative. *Mat. Sb.*, **48**, 311-334 (in Russian). [114, 121]

Vasil'eva, A. B. (1963). Asymptotic behavior of solutions of certain problems for ordinary nonlinear differential equations with a small parameter multiplying the highest derivatives. *Usp. Mat. Nauk*, **18**, 15-86 (in Russian); *Russian Math. Surveys*, **18**, (1963) 13-81. [114]

Verhagen, J. H. G., and L. Van Wijngaarden, (1965). Non linear oscillations of fluid in a container. *J. Fluid Mech.*, **22**, 737-751. [89]

Veronis, G. See Malkus and Veronis.

Visik, M. I., and L. A. Lyusternik, (1957). Regular degeneration and boundary layer for linear differential equations with small parameter. *Usp. Mat. Nauk*, **12**, 3-122 (in Russian); Am. Math. Soc. Transl., Serv. 2, **20**, 239-364, 1962. [114, 144]

Volosov, V. M. (1961). Higher approximations in averaging. *Soviet Math. Dokl.*, **2**, 221-224. [168]

Volosov, V. M. (1962). Averaging in systems of ordinary differential equations. *Russian Math. Surveys*, **7**, 1-126. [168]

Von Zeipel, H. (1916). Movements of minor planets. *Ark. Mat. Astron. Fysik, Stockholm*, **11**, No. 1, 1-58, No. 7, 1-62. [189]

Voss, W. (1933). Bedingungen Für das Auftreten des Ramsauereffektes. *Z. Phys.*, **83**, 581-618. [346]

Wasow, W. A. (1953). Asymptotic solution of the differential equation of hydrodynamic stability in a domain containing a transition point. *Ann. Math.*, **58**, 222-252. [360]

Wasow, W. A. (1955). On the convergence of an approximation method of M. J. Lighthill. *J. Rat. Mech. Anal.*, **4**, 751-767. [79]

Wasow, W. A. (1965). *Asymptotic Expansions for Ordinary Differential Equations* Wiley, New York. [110, 309, 318, 358]

Wasow, W. A. (1968). Connection problems for asymptotic series. *Bull. Am. Math. Soc.*, **74**, 831-853. [309]

Watson, G. N. (1944). *A Treatise on the Theory of Bessel Functions.* Macmillan, New York. [353]

Watson, J. (1960). On the nonlinear mechanics of wave disturbances in stable and unstable parallel flows. Part 2. The development of a solution for plane Poiseuille flow and for plane Couette flow. *J. Fluid Mech.*, **9**, 371-389. [162]

Weaver, A. K. See Green and Weaver.

Weinitchke, H. J. See Reissner and Weinitschke.

Weinstein, L. A. (1969). *Open Resonators and Open Waveguides.* Golem Press, Boulder, Colorado. [380]

Wentzel, G. (1926). Eine Verallgemeinerung der Quantenbedingung fur die Zwecke der Wellenmechanik. *Z. Phys.*, **38**, 518-529. [114, 315, 339]

Weyl, H. (1942). On the differential equations of the simplest boundary-layer problems. *Ann. Math.*, **43**, 381-407. [114]

Whitehead, A. N. (1889). Second approximations to viscous fluid motion. *Quart. J. Math.*, **23**, 143-152. [31]

Whitehead, J. A. See Newell and Whitehead.

Whiteman, K. J. See Mc Namara and Whiteman.

Whitham, G. B. (1952). The flow pattern of a supersonic projectile. *Comm. Pure Appl. Math.*, **5**, 301-348. [57, 78, 87]

Whitham, G. B. (1953). The propagation of weak spherical shocks in stars. *Comm. Pure Appl. Math.*, **6**, 397-414. [57, 78, 87]

Whitham, G. B. (1965a). A general approach to linear and nonlinear waves using a Lagrangian. *J. Fluid Mech.*, **22**, 273-283. [216, 221]

Whitham. G. B. (1965b). Nonlinear dispersive waves. *Prov. Roy. Soc. (London)*, **A283**, 238-261. [226]

Whitham, G. B. (1967a). Nonlinear dispersion of water waves. *J. Fluid Mech.*, **27**, 399-412. [216]

Whitham, G. B. (1967b). Variational methods and applications to water waves. *Proc. Roy. Soc. (London)*, **A299**, 6-25. [216]

Whitham, G. B. (1970). Two-timing, variational principles and waves. *J. Fluid Mech.*, **44**, 373-395. [216]

Whittaker, E. T. (1914). On the general solution of Mathieu's equation. *Edinburgh Math. Soc. Proc.*, **32**, 75-80. [62]

Whittaker, E. T. (1916). On the adelphic integral of the differential equations of dynamics. *Proc. Roy. Soc. Edinburgh*, **37**, 95-116. [199]

Whittaker, E. T. (1937). *Analytical Dynamics of Particles and Rigid Bodies*, 4th ed., Cambridge University Press, Cambridge. [199]

Wilcox, C. H. (1964). *Asymptotic Solutions of Differential Equations and Their Applications.* Wiley, New York. [309]

Wilcox, C. H. (1966). *Perturbation Theory and its Applications in Quantum Mechanics.* Wiley, New York. [71]

Wingate, R. T., and R. T. Davis, (1970). Perturbation solution of a hyperbolic equation governing longitudinal wave propagation in certain nonuniform bars. *J. Acoust. Soc. Am.*, 47, 1334-1337. [234]

Witt, A. See Gorelik and Witt.

Woinowsky-Krieger, S. See Timoshenko and Woinowsky-Krieger.

Wolfe, P. (1967). A new approach to edge diffraction. *SIAM J. Appl. Math.*, 15, 1434-1469. [380]

Wu, T. T. See Cheng and Wu.

Wu, Y. T. (1956). Two-dimensional sink flow of a viscous, heat-conducting, compressible fluid; cylindrical shock waves. *Quart. Appl. Math.*, 13, 393-418. [100]

Yagishita, T. See Kawakami and Yagishita.

Ying, S.J. See Chu and Ying.

Yuen, M. C. (1968). Nonlinear capillary instability of a liquid jet. *J. Fluid Mech.*, 33, 151-163. [99]

Zabreiko, P. P., and L. B. Ledovskaja, (1966). Higher approximations of the Bogolinbov-Krylov averaging method. *Dokl. Akad. Nauk. SSSR*, 171, 1453-1456. [168]

Zauderer, E. (1964a). Wave propagation around a convex cylinder. *J. Math. Mech.*, 13, 171-186. [380]

Zauderer, E. (1964b). Wave propagation around a smooth object. *J. Math. Mech.*, 13, 187-199. [380]

Zauderer, E. (1970a). Boundary layer and uniform asymptotic expansions for diffraction problems. *SIAM J. Appl. Math.*, 19, 575-600. [380]

Zauderer, E. (1970b). Uniform asymptotic solutions of the reduced wave equation. *J. Math. Anal. Appl.*, 30, 157-171. [378, 380]

Zawadzki, E. M., and G. J. Lewak, (1971). Penetration to second order of an electrostatic field into a warm plasma. *J. Plasma Phys.*, 5, 73-87. [78]

Zierep, V. J., and J. T. Heynatz, (1965). Ein analytisches Verfahren zur Berechnung der nichtlinearen Wellenausbreitung. *ZAMM*, 45, 37-46. [89]

Zwaan, A. (1929). Intensitaten im Ca-Funkenspektrum. Ph. D. Thesis, Utrecht. [339]

Zwanzig, R. (1964). On the intensity of three generalized master equations. *Physica*, 30, 1109-1123. [382]

Subject Index

Acoustic, 77, 78
Aerodynamic, 110
Airfoil theory, 98, 99, 235, 303; *see* Supersonic airfoil theory
Airy's equation, 312, 336
Airy's functions, 49, 336, 378
Algebraic equation, 2, 57, 74, 95, 327
Algorithm, 168, 171, 199, 209
  for canonical systems, 212-214
  generalized, 202-206
  simplified, 206-208
Anomaly, 79
Aperiodic motion, 189
Astrophysics, 235
Asymptotic expansion, 23, 78
  of Airy's functions, 337
  of Bessel's function, 16
  definition of, 12
  divergent, 16
  elementary operation on, 18-19
  uniform, 17-19
  uniqueness of, 14
  see also Asymptotic series
Asymptotic matching principle; *see* Matching
Asymptotic partitioning, 327-331
Asymptotic sequence, 12, 14, 16, 18, 19
  factorial, 12
  fractional powers in, 136, 137
  logarithms in, 137, 144
Asymptotic series, 10-12
  definition of, 11
  versus convergent series, 15-16
  see Asymptotic expansion
Attenuation, 46, 78
Averaging, method of, 159-227
  generalized method of, 168-171, 191, 211, 223-225, 231
  Krylov-Bogolinbov, method of, 165-168

Krylov-Bogolinbov-Mitropolski, method of, 174-179, 183, 194, 211, 212, 223, 224, 246, 248
  Struble's method of, 171-174, 176, 183, 194, 223
  using, canonical variables, 179-189
  Lagrangian, 216-222
  Lie series and transforms, 200-216
  von Zeipel transformation, 189-200
  van der Pol's method of, 164-165
  see also Smoothing

Beam, vibration of, 105, 106, 155, 226, 306
Bearing, slider, 54, 125-128
Bellman, equation of, 22
Bénard problem, 77
Bending of, shells and tubes, 54, 353; *see also* Unsymmetrical bending of plates
Benney's technique, 38-42
Bernoulli's equation, 83
Bessel functions, 1, 5, 6, 15, 312, 315, 383
  asymptotic expansion of, 16, 312-314, 329-331
  integral representation of, 16, 314
  zeros of, 21
Bethe-Salpeter equation, 372
Blunt body problem, 99
Boltzmann's equation, 236
Born approximation, 362
Born expansion, 308, 361-367, 373
  renormalization of, 367-372
Boundary conditions, loss of, 31, 34, 37, 38, 54, 111, 114, 122
  transfer of, 27
Boundary layer, 18, 23, 79, 111, 112, 147, 233
  location of, 114-116, 122
  Prandtl, 34

problem with two, s, 128-133
  stability of, 353
Branch point, 82
Bretherton's equation, exercises involving,
    227, 306
  treated, by variational approach, 217-221
    by method of multiple scales, 266-269,
      298-300
  wave-wave interaction for, 219-221, 266-
    269
Brownian motion, 236
Buckling, 233

Canonical, averaging, variables, 179-189
  equations, 180, 199
  Jordan, form, 327
  mixed, variables, 199
  system, 190, 201
  transformation, 181, 187, 191
  variables, 181, 183, 184, 195, 202, 216,
    224
Caustic, 234, 375-380
Change, 111, 113
  of characteristics, 89
  see also Sharp change; Type change of
Characteristic, parameters, 89, 94
  wave speeds, 91
Characteristic exponent, 58, 62, 66
Characteristics, expansion in terms of, 57,
    86-94, 303
  method of, 374
Circuit, electronic, see van der Pol oscillator
Cluster expansions, 363
Compatability relationship, 217, 222
Composite expansion, 114, 144, 384, 385
  construction of, 121
  for earth-moon-spaceship problem, 139
  for equation with variable coefficients,
    125
  for simple example, 121-122
  for slider bearing, 128
  for thermoelastic waves, 136
  for unsymmetrical bending of plates, 133
Composite expansions, applied to turning
    point problems, 348-350
  method of, 144-154, 317, 318
Composite solution, 113
Conservation form of equations, 226
Coordinate, optimal, 50-51
  parabolic, 79

perturbations, 1, 4-7, 21, 309-314, 379
  role of, systems, 23, 49-52, 54, 380
  see also Strained coordinates
Correlation, 363
Cosmological expansion, 233
Cycle, limit, 35
Cylinder, elliptic, 346
  a solid, expanding, 83-86
Cylindrical, functions, 358, 378, 380
  jet, 99

Degenerate, 72, 74
Derivative–expansion procedure, 302
  applications of, 243-269
  description of, 230, 236-240
  limitations of, 269-270
  see also Multiple scales, method of
Detuning, 250
Diagram, 366, 367
  bare, 364
  connected, 370, 371, 372
  double, 365
  dressed, 365
Diffraction, 346, 380
Diffusion equation, 38
Dirichlet problem, 38
Discontinuity, see Singularity
Dispersion relationship, 178, 217, 218, 219,
    220, 222, 227, 266, 301
Dispersive waves, 78, 234, 303
  long nonlinear, 38-42
  see also Bretherton's equation; Klein-
    Gordon equation; Thermo-elastic
    waves; Wave-wave interaction
Divisor, small, 196
Domain, effect of, on nonuniformity of
    expansions, 38, 42
  infinite, 24-31
Duffing equation, 50, 51
  exercises involving, 54, 104, 105, 223,
    224, 304
  with slowly vary coefficients, 286
  straightforward expansion for, 24-25
  treated, by averaging using canonical
    variables, 182-183
  by derivative-expansion procedure, 243-
    245
  by generalized method of multiple
    scales, 286-291
  by Krylov-Bogoliubov method, 167

by Krylov-Bogoliubov-Mitropolski method, 175-176
by Lindstedt-Poincaré method, 58-60
by renormalization, 95-96
by Struble's method, 171-174
by two-variable expansion procedure, 271-273
by von Zeipel's procedure, 192-194
Dyson equation, 371, 381

Earth-moon-spaceship problem, 233, 302, 303
exercises involving, 53, 107
illustrating limitations of method of strained coordinates, 102-103
straightforward expansion for, 43-45
treated, by Lighthill's technique, 82-83
by method of composite expansions, 153-154
by method of matched asymptotic expansions, 137-139
by method of multiple scales, 295-298
Eccentricity, 64, 233, 346
Edge layer, 111
Eiconal equation, 374, 377, 379
Eigenvalue, 56
Eigenvalue problem, linear, 68-71
quasilinear, 71-76
Elastic, 46, 58, 90, 353
waves, 89-94
Elliptic equation, 37, 42, 98, 189, 234, 303, 360
Energy level, 56, 58
Entropy layer, 79
Euler-Lagrange equation, 216, 217, 218, 220, 222

Faa de Bruno operators, 192
Feynman diagrams, 308, 361-372
Flight mechanics, 233
Floquet theory, 60, 62, 64, 66
Flow, down an inclined plane, 38-42, 99
hypersonic, 79
jet, 99
past a body, 33-34, 113
through a channel, 54, 56, 216
see also Flow past a sphere; Supersonic airfoil theory
Flow past a sphere, exercise involving, 158
straightforward expansion for, 28-31

treated by method of matched asymptotic expansions, 139-144
Flutter, 234
Foci, 253
Fokker-Planck equation, 235, 381
Fourier, 45, 167, 178
Frequency, 56, 58, 96, 165, 252
Fresnel diffraction, 380
Frobenius, method of, 5, 310

Gauge, function, 7, 8
transformation, 232
Gaussian, 363, 367, 369, 372
Generalized expansion, see Composite expansion
Generalized version of method of multiple scales, applications of, 276-302
description of, 232, 241-243
limitations of, 302-303
Generalized method of averaging, see Averaging, method of
Generalized vector, 179
Generating function, 181, 184, 189, 190, 192, 195, 196, 200, 202, 215
Generating vector, 201
Geometrical optics, 308, 361, 374-377, 379
Geophysics, 110
Graetz problem, 384
Green's function, 362, 364, 380
double, 364
Group velocity, 179, 219, 220, 267, 299

Hamilton-Jacobi equation, 181, 182, 183, 186, 190, 199
Hamiltonian, 223, 224, 225
definition of, 180
for Duffing equation, 182
for Mathieu equation, 184
for swinging spring, 186
transformation of, using von Zeipel procedure, 189-200
using Lie transforms, 202, 212-216
Harmonic balance, 218
Harmonic resonance, 219-221, 234, 262-269
Harmonic wave, 76
Heat, 45, 79, 84, 156, 342, 384
problem for, equation, 150-152
Helmholtz equation, 234
Hill's equation, 60

Hopf equation, 381
Hydraulic jump, 89
Hyperbolic equation, 37, 42, 57, 99, 379.
    See also Dispersive waves; Elastic
    waves; and Supersonic airfoil theory

Inclined plane, *see* Flow, down an inclined
    plane
Indicial equation, 311
Induction, 13
Infinite domain, as a source, of nonuni-
    formity, 23-31, 229
    of uniformity, 38, 42
Inhomogeneous, 361
    problems with turning points, 352-359
Initial, boundary value problem for heat
    equation, 150-152
    layer, 23
Inner and our expansions, *see* Matched
    asymptotic expansions, method of
Inner expansion, 110, 112, 114, 119, 145,
    146, 148
    for bending of plates, 129-132
    definition, 117-118
    for earth-moon-spaceship problem, 139
    for equation with variable coefficients,
        122-124
    for simple example, 117-118
    for slider bearing, 127
    for thermoelastic waves, 134-135
    for turning point problems, 336
Inner limit, 112, 118, 119, 129, 131, 139
Inner region, 113, 122, 146, 153
Inner solution, 112, 113
Inner variables, 119, 120, 121, 144, 149,
    153, 154
    choice of, 114-116, 122-124, 126-127,
        134-135, 137-138
    generalized, 145
Instability, 78, 269. *See also* Model for
    nonlinear instability; Stability
Integral, 10, 11, 12
    differential equation, 18
    equation, 370, 380
    of motion, 24, 199
    relations, 99
Intermediate limit, 119
Intermediate matching, *see* Overlapping

Jacobian matrix, 202

Jacobi elliptic functions, 189
Jerky oscillations, 34
Jordan, form, 327
    matrix, 327
Josephson tunnel, 221

Kamel's algorithm, 171
Kamel's method, 224
Kelvin-Helmholtz, 77, 235
Kernel, 370
Klein-Gordon equation, 175, 234
    exercises involving, 105, 226
    treated, by averaging using Lagrangian,
        221-222
        by Krylov-Bogoliubov-Mitropolski
        method, 178-179
        by method of multiple scales, 301-302
        by method of strained parameters, 76-
        77
Krylov-Bogoliubov technique, 165-168, 223
Krylov-Bogoliubov-Mitropolski technique,
    174-179, 183, 194, 211, 212, 223,
    224, 246, 248, 303
Kruskal's technique, 168, 191

Lagrange equations, 179, 180
Lagrangian, 179
    averaging using, 216-222, 301
    for Bretherton's equations, 217
    for Klein-Gordon equation, 221
    for swinging spring, 186
Lamé coefficients of elasticity, 45, 90
Laminar, 39, 54
Landau, equation, 382
    symbols, 8, 9
Langer's transformation, 308, 346, 384
    for first-order turning point problems,
        339-341
    for generalization of, 341-342
    successive, 350
Latta's technique, 144-154
Layer, 23, 79, 111; *see also* Boundary layer
Libration points, *see* Stability of elliptic tri-
    angular points
Lie series and transforms, 171, 199, 200-
    216, 223, 225, 303
Lie triangle, 205, 206
Lighthill's technique, 57, 77-95
    exercises involving, 107, 108
    limitations of, 79, 98-100, 107, 109

Limit, 7
  cycle, point, or solution, 35, 99, 109
    distinguished, 336
  Oseen's, 140
  Stokes', 140
  see also Inner limit; Intermediate limit;
    and Outer limit
Limitations of, method, of composite ex-
    pansions, 153-154
  of matched asymptotic expansions, 144-
    145, 155, 156, 303, 339
  of multiple scales, 269-270, 275-276,
    302-303
  of strained coordinates, 79, 98-103,
    107, 109, 110, 302, 303
  Struble's method, 174
  von Zeipel's procedure, 199
Lindstedt-Poincaré technique, 56, 57, 58-
    60, 95, 96, 185, 200
Linear damped oscillator, treated by mul-
    tiple scales, 228-243
Linearization, method of, 57, 94-95
Liouville, equation, 236, 380, 382
  problem, 314
Liouville-Green approximations, 49, 320,
    335
  higher, 315-317
  successive, 324-325
Liouville-Green transformation, 308, 315,
    340
Logarithms, 7, 12, 45, 83, 137, 144, 308,
    311
Lommel functions, 353
Long period part, 169, 191, 209, 214
Lunar motion, 60; see also Earth-moon-
    spaceship problem

Mach number, 26, 83, 85
Matched asymptotic expansions, method of,
    37, 48, 51, 78, 110-144, 148, 153,
    154, 235, 317, 346, 378, 384
  applied to turning point problems, 336-
    339
  limitations of, 144, 155, 156, 303
Matching, 51, 110, 114, 115, 116
  asymptotic, principle, see van Dyke's
    principle
  as guide to forms of expansions, 141
  intermediate, 119
  Kaplun's, principle, 119

Prandtl, procedure, 112, 118
  refined, 118-119
Matching of inner and outer expansions,
  bending of plates, 130-132
  caustic problem, 378
  earth-moon-spaceship problem, 139
  equation with variable coefficients, 124
  flow past a sphere, 141-144
  simple example, 120-121
  slider bearing, 127
  term by term, 130
  thermoelastic waves, 135-136
  turning point problems, 337-339, 343-
    344
Mathien equation, 339
  exercises involving, 53, 104, 223, 303,
    304
  treated, by averaging using canonical var-
    iables, 183-185
    by Lindstedt-Poincaré technique, 60-62
    by method of multiple scales, 253-257
    by von Zeipel's procedure, 194-200
    by Whittaker's method, 62-64
Maxwell's heat conduction law, 45
Membrane, 356
Method, of strained coordinates, see
    Strained coordinates, method of
  of strained parameters, see Strained para-
    meters, method of
  of multiple scales, see Multiple scales,
    method of
Missile dynamics, 233
  exercises involving, 304, 305, 306
  linear, 320-321
  nonlinear, 291-295
Model for nonlinear instability, 50
  exercises involving, 55
  illustrating limitations of method of
    strained coordinates, 99-100
  straightforward expansion for, 25-26
  treated, by method of multiple scales,
    264-266
    by method of renormalization, 96-97
Momenta, 179, 182, 190, 199, 224
Moon, see Earth-moon-spaceship problem;
    Lunar motion
Multiple scales, method of, 51, 97, 100,
    153, 221, 228-307, 315, 317, 339,
    368, 384
  applied, to caustic, 379

to equations with slowly varying coefficients, 318-320
to Orr-Sommerfeld equation, 360

Navier-Stokes equations, 28, 33, 39
Neumann, expansion, 361; *see also* Born
expansion
series, 364
Newtonian theory, 79
Nondispersive waves, 78, 269-270, 303. *See
also* Dispersive waves; Elastic; Shock
waves; and Supersonic airfoil theory
Nonuniformity, in airfoil theory, 28, 50
in asymptotic expansions, 16-18
in bending of plates, 37
in Born's expansion, 367
dependence of, on coordinates, 49-51
on size of domain, 24-31, 38, 42
in Duffing's equation, 24-25, 49-50
in earth-moon-spaceship problem, 45
in equation with constant coefficients, 32
in flow past a sphere, 31
in geometrical optics approximation, 376-
377
in interior of domain, 137
in jet instability, 99
in linear oscillator, 229
in model for nonlinear instability, 26, 50
in relaxation oscillations, 35
in slider bearing, 126
in shift in singularity, 43, 79
in thermoelastic waves, 47-48
in turning point problems, 49, 284, 315,
336
variable exhibiting, 57, 285
see also Boundary layer; Region of nonuniformity; Sources of nonuniformity; and Type change of
Node, 252
Normal solution, 58, 60, 66, 311, 326, 331,
332
No–slip condition, 33, 34

Olver's transformation, 341, 384, 385
Operator, 206
adjoint, 164
Faa de Bruno, 192
intensity, 372
mass, 371, 381
self-adjoint, 163, 164

Optics, *see* Geometrical optics
Optimal, control, 79
coordinate, 50, 51
Orbit, 64, 99
Orbital mechanics, 233
Order symbols, 8, 9
Orr-Sommerfeld equation, 233, 360
Oscillations, 25, 34, 89, 226, 232, 303
Oseen's equation, 141
Oseen's expansion, 140-144
Oseen's limit process, 140, 141
Oseen's variable, 143
Outer expansion, 110, 112, 114, 119, 146
for bending of plates, 128-129
definition of, 117
for earth-moon spaceship problem, 45,
138
for equation with variable coefficients,
124
for simple example, 117
for slider bearing, 126
for thermoelastic waves, 47-48
Outer limit, 112, 117, 119, 128, 138, 144
Outer region, 113, 122
Outer solution, 111, 113, 115
Outer variable, 119, 121, 144, 145
Overlapping, 113, 119

Parabolic coordinates, 79
Parabolic cylinder function, 344, 378
Parabolic equation, 37, 38, 99, 379
Paradox, 31
Parameter, 89
large, 315; *see also* Turning point prob-
lems
perturbation, 1-4, 16, 308
see also Small parameter multiplying high-
est derivative; Strained parameters,
method of
Parametric resonance, *see* Mathieu equation;
Stability of elliptic triangular points
Parameters, 201; *see also* Variation of para-
meters
Parametrization, 88, 92
Partitioning, *see* Asymptotic partitioning
Pascal triangle, 205
Pendulum, 103, 224; *see* Swinging spring
Period, 169, 191, 192
Periodic, 58, 60
motion, 189

orbit, 99
solutions, 34, 77
see Mathieu equation; Stability of elliptic
triangular points
Perturbations, coordinate, 4-7, 308-314
parameter, 1-4, 16, 308
Phase, 178, 380
rapidly rotating, 168, 201, 234
speed, 76, 77
Plasma, 58, 78, 216, 217, 226, 234, 235,
367
Plasticity, 90
Plates, see Unsymmetrical bending of plates
Poincaré-Lighthill-Kuo method, see Light-
hill's technique
Poisson ratio, 36, 47
Potential function, 26, 83, 89
Prandtl's boundary layer, 113
Prandtl's technique, 111, 114
Pritulo's technique, see Renormalization,
method of

Quantum, 58, 371; see also Schrödinger
equation

Random, 361, 362, 363, 365, 367, 369,
372, 380, 381, 382
Rankine-Hugonoit relation, 84
Ratio test, 6, 7, 10, 11, 16
Rayleigh-Schrödinger method, 56, 71
applied to eigenvalue problem, 68-71
Rayleigh-Taylor instability, 77, 235
Rayleigh wave speed, 47, 134
Reaction kinetics, 79
Reentry dynamics, see Missile dynamics
Region of nonuniformity, 17, 18, 19, 23,
79
for earth-moon-spaceship problem, 137-
138
for equation with constant coefficients,
116
for flow past a sphere, 140
near a caustic, 377-378
for thermoelastic waves, 134-135
for turning point problems, 284, 336
Related equation, 341
Relaxation oscillations, 34-35
Renormalization, method of, 57, 95-98,
308, 361, 367-372
exercises involving, 103, 106, 107, 108

Resonance, 89, 233, 248
external, 225
internal, 185, 225
in linear systems, 321-324
near, 195
perfect, 189
parametric, 235
passage through, 233
see also Harmonic resonance
Reynold's equation, see Slider bearing
Reynolds number, 40, 360
high, 33, 130
small, 28, 29, 139
Rigidity, flexural, 36
Ritz-Galërkin procedure, 58
Rossby wave, 77, 234
Rytov's method, 361, 373

Saddle point, 252
Satellite, 233, 276
Scales, 51, 110, 230, 231, 232, 233, 242,
303; see also Multiple scales, method
of
Scattering, 57, 95, 235, 361, 364, 367, 368
Schrödinger equation, 56, 71, 160, 339,
342, 344, 346
Secular terms, 23-26
elimination of, 65-68, 261, 288, 302
Self-sustained oscillations, 89; see also
van der Pol oscillator
Series, see Asymptotic series; Lie series and
transforms
Shadow boundary, 378
Sharp change, 103, 110, 303
Shell, 233
Shift, 42, 77, 79, 106
in singularity, 103
exercises involving, 52, 53, 106, 107
straightforward expansion for, 42-43
treated, by Lighthill's technique, 79-82
by renormalization, 98
by Temple's technique, 94-95
Shock waves, 52, 78, 83, 84, 85, 99, 100,
235
Short period part, 169, 191, 214
Singular, 17, 57, 78, 81, 231
Singular perturbation, 17, 99, 340
dependence on region size, 38, 42
see also Nonuniformity
Singular point, definition of, 309, 310

expansion near an irregular, 309-312, 326-327
regular, 5, 308
Singularity, 85, 86
as a jump discontinuity, 32, 35
branch point, 82
essential, 7
growing, 23, 42, 43, 45, 95
logarithmic, 45, 83, 102, 137
turning point with, 358-359
worst, 81, 98
see also Nonuniformity
Skin layer, 111
Slider bearing, 54, 125-128
Small parameter multiplying highest derivative, 23, 31-37, 99, 317-318
in limitations of method of strained coordinates, 99, 100-102
Smoothing, method of, 308, 361, 380-382
Solvability condition, 152, 288, 302; *see* Secular terms
Sonic boom, 78
Sources of nonuniformity, 23-55, 140
Spaceship, *see* Earth-moon-spaceship problem
Speed, 56, 57, 58, 76, 77, 83, 90, 91
Sphere, *see* Flow past a sphere
Spherical, 28, 78, 224
Spring, 232. *See also* Duffing equation; Swinging spring
Stability, 60, 77, 78, 99, 100, 162-164, 235, 252, 353, 360. *See also* Mathieu equation; Model for nonlinear stability; Stability of elliptic triangular points
Stability of elliptic triangular points, treated, by method of multiple scales, 259-262, 275
by method of strained parameters, 64-66
by Whittaker's technique, 66-68
Statistical mechanics, 236
Stochastic, 361, 367
Stokes' expansion, 140
Stokes' limit process, 140
Stokes' variable, 143
Stokes' solution, 30
Strained coordinates, method of, 51, 52, 56-109, 110, 114, 156, 235, 302, 303

Strained parameters, method of, 56, 58-77, 78, 79, 97, 99, 100, 103, 105, 106
Straining, of characteristics, 87, 89
dependent variable, 99, 108
function, 57, 78, 79, 81, 87, 99, 101, 102, 103, 107
Stratification, 216
Stream function, 28, 29, 33, 40, 57, 113, 140
Stretching transformation, for bending of plates, 128
for caustic, 377-378
for dependent and independent variables, 134
for earth-moon-spaceship problem, 137-138
for equation with variable coefficients, 122, 123
for heat equation, 151
for turning point problems, 284, 336
Struble's method, 171-174, 176, 183, 194, 223
Stuart-Watson-Eckhaus technique, 162-164
Subnormal solution, 311, 331-332
Supersonic airfoil theory, 50, 78, 93
straightforward expansion for, 26-28
treated, by Lighthill's technique, 86-89
by renormalization, 97-98
Swinging spring, exercises involving, 105, 225
treated, by averaging Hamiltonian, 185-189
by Lie series and transforms, 214-216
by method of multiple scales, 262-264

Temple's technique, 57, 94-95
Thermoelastic waves, straightforward expansion for, 45-48
treated by method of matched asymptotic expansions, 133-137
Thomas-Fermi model, 235
Three body problem, 199, 233, 276. *See also* Earth-moon-spaceship problem; Stability of elliptic triangular points
Triangle, 205, 206
Triangular points, 233; *see also* Stability of elliptic triangular points
Transfer of boundary condition, 27
Transform, *see* Lie series and transforms
Transformation, canonical, 181, 190,

191, 199
contracting, 140
Deprit, 202
Hori, 202
Lie, 201
Liouville-Green, 49, 315, 340
near identity, 51, 57, 168, 190, 201
stretching, 111, 113, 114, 116, 117, 377
von Zeipel, 171, 191, 199, 202
Transition, 380; see Turning point problems
Transition curves, exercises involving, 104,
    105, 225, 303, 304
for libration points, 64-68, 259-262, 275
for Mathieu's equation, 60-64, 183-185,
    194-200, 253-257
Transport equation, 374, 379
Tunneling effect, 342
Turning point problems, 114, 308, 309,
    335-360
definition of, 49, 122, 284, 335
exercises involving, 53, 305, 383, 384,
    385
near caustic, 377-380
treated by method of multiple scales, 232,
    233, 284-286
Two-body problem, 201
Two-variable expansion procedure, 243,
    302
applications of, 270-275
description of, 231, 240-241
limitations of, 275-276
see also Multiple scales, method of
Type change of, 23, 37-42

Uniformity, see Nonuniformity
Uniformization, 94, 107, 232
Unsymmetrical bending of plates, exercises
    involving, 155
straightforward expansion for, 35-37
treated by method of multiple scales, 128-
    133

van der Pol, method of, 164-165, 166
van der Pol oscillator, exercises involving,
    52, 54, 104, 223, 224, 303, 304
straightforward expansion for, 3-4, 34-35

treated by generalized method of averag-
    ing, 169-171
by Krylov-Bogoliubov technique, 167-
    168
by Krylov-Bogoliubov-Mitropolski tech-
    nique, 176-177
by Lie series and transforms, 209-212
by method of multiple scales, 245-253,
    272-275
van der Pol oscillator with delayed ampli-
    tude limiting, 104, 257-259, 304
van Dyke's matching principle, 114, 119
mechanics of, 120-122
Variables, see Canonical variables; Inner
    variables; and Outer variables
Variation of parameters, 52, 159-222
Variational, approach, 216-222, 227
equations, 172, 183, 184, 190
Vibrations, 114, 232. See also Oscillations;
    Waves
Vlasov's equation, 78, 216
von Zeipel's procedure, 189-200, 202, 231

Water waves, 56, 57, 58, 77, 216, 217, 234
Wave equations, 360-382
Wave number, 58, 77, 99, 360
Waves, 58, 60, 77, 78, 89, 90, 94, 216, 217,
    233, 234, 235. See also
    Bretherton's equation; Dispersive
    waves; Elastic waves; Flow down an
    inclined plane; Klein-Gordon equa-
    tion; Non dispersive waves; Plasma;
    Shock waves; Supersonic airfoil
    theory; and Thermoelastic waves
Weber function, 345, 378, 380
Whitham's method, 216-222
Whittaker's functions, 346
Whittaker's method, 62-64, 66-68, 104,
    105, 185, 200
WKBJ approximation, 49, 308, 315, 320,
    335, 339, 382
successive, 324-325
Wronskian, 160

Young's modulus, 36

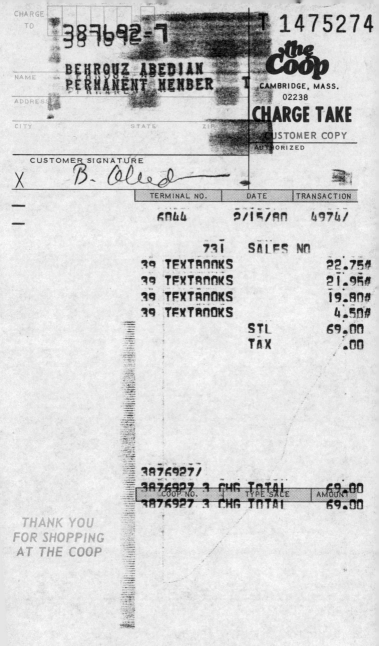

CHARGE TO

387692-7

T 1475274

NAME BEHROUZ ABEDIAN
PERMANENT MEMBER

ADDRESS

CITY   STATE   ZIP

the Coop
CAMBRIDGE, MASS.
02238

CHARGE TAKE

CUSTOMER COPY
AUTHORIZED

CUSTOMER SIGNATURE

X B. Abed

TERMINAL NO.	DATE	TRANSACTION
6044	2/15/80	49747

731 SALES NO

39 TEXTBOOKS	22.75#
39 TEXTBOOKS	21.95#
39 TEXTBOOKS	19.80#
39 TEXTBOOKS	4.50#
STL	69.00
TAX	.00

3876927/

3876927 3 CHG TOTAL	69.00
COOP NO.	TYPE SALE	AMOUNT
3876927 3 CHG TOTAL	69.00

THANK YOU
FOR SHOPPING
AT THE COOP

CHARGE TAKE-WITH

SUB   9/17/99   1:49/

751   SALES NO
99 TEXTBOOKS   00.25H
99 TEXTBOOKS   71.89H
99 TEXTBOOKS   19.80H
99 TEXTBOOKS   1.50H
S.TL   69.00
TAX   .00

9076987/
9076987 3 CHG Total   69.00
9076987 3 CHG Total   69.00

CHARGE TAKE WITH